D1567607

Biochemistry
of Copper

BIOCHEMISTRY OF THE ELEMENTS

Series Editor: Earl Frieden
Florida State University
Tallahassee, Florida

Recent volumes in this series:

Volume 4	BIOCHEMISTRY OF DIOXYGEN Lloyd L. Ingraham and Damon L. Meyer
Volume 5	PHYSICAL METHODS FOR INORGANIC BIOCHEMISTRY John R. Wright, Wayne A. Hendrickson, Shigemasa Osaki, and Gordon T. James
Volume 6	BIOCHEMISTRY OF SULFUR Ryan J. Huxtable
Volume 7	GENERAL PRINCIPLES OF BIOCHEMISTRY OF THE ELEMENTS Ei-Ichiro Ochiai
Volume 8	BIOCHEMISTRY OF THE LANTHANIDES C. H. Evans
Volume 9A	BIOCHEMISTRY OF THE ELEMENTAL HALOGENS AND INORGANIC HALIDES Kenneth L. Kirk
Volume 9B	BIOCHEMISTRY OF HALOGENATED ORGANIC COMPOUNDS Kenneth L. Kirk
Volume 10	BIOCHEMISTRY OF COPPER Maria C. Linder

A Continuation Order Plan is available for this series. A continuation order will bring delivery of each new volume immediately upon publication. Volumes are billed only upon actual shipment. For further information please contact the publisher.

Biochemistry of Copper

Maria C. Linder
Department of Chemistry and Biochemistry
California State University
Fullerton, California

with a contribution by
Christina A. Goode
Department of Chemistry and Biochemistry
California State University
Fullerton, California

PLENUM PRESS • NEW YORK AND LONDON

Library of Congress Cataloging-in-Publication Data

Linder, Maria C.
  Biochemistry of copper / Maria C. Linder.
      p.   cm. -- (Biochemistry of the elements ; v. 10)
  Includes bibliographical references and index.
  ISBN 0-306-43658-2
  1. Copper--Metabolism.   I. Title. II. Series.
QP535.C9L56 1991
596'.019214--dc20   91-28381
                                                                  CIP

ISBN 0-306-43658-2

© 1991 Plenum Press, New York
A Division of Plenum Publishing Corporation
233 Spring Street, New York, N.Y. 10013

All rights reserved

No part of this book may be reproduced, stored in a retrieval system, or transmitted in any form or by any means, electronic, mechanical, photocopying, microfilming, recording, or otherwise, without written permission from the Publisher

Printed in the United States of America

To my dear husband, Gordon Nielson, without whose unflagging support it would not have been possible to carry this book through

Preface

Copper has long been known as essential to living systems, in part through its fundamental role in electron transport and respiration. Over the years into the present, its involvement in an ever increasing number of processes in all kinds of organisms has become apparent, and new and exciting vistas of its roles in such areas as the central nervous system, and in humoral functions, are appearing on the horizon. Although the biochemistry of this element has not been studied nearly as much as that of many others, a formidable amount of work *has* been carried out. It has thus been a challenge to produce a summary of what has been found that provides both breadth and depth.

My goal has been to try to be as comprehensive as possible, within some limitations. I have tried to provide basic information and basic data that should continue to be useful for a long time. The goal has also been to interpret where we currently stand in our knowledge of the structure, function, regulation, and metabolism of Cu-dependent processes and substances, especially proteins. Thus, I have tried to make this a source book for historic as well as current information on all aspects of copper biochemistry, and a summary of our current knowledge of copper-dependent proteins and processes.

Most of the research on copper has been carried out on vertebrates, especially mammals. This has played a role in the organization of the book. After an introductory chapter on the distribution of copper in living and nonliving systems (Chapter 1) including mammalian tissues, Chapter 2 addresses how copper enters from the intestine. This is followed by discussions of how the entering copper is distributed to cells, via the blood and extracellular fluids (Chapter 3), and an enumeration and exposition of what is known about the copper-binding substituents (mostly proteins) found in these fluids (Chapter 4). Chapter 5 then deals with how copper is lost from the body.

Chapter 6 is about copper within cells; how it is distributed to intracellular components and compartments; the roster of known intracellular copper-binding proteins; and how a copper deficiency (or its excess) is handled by the cell. Chapter 7 concerns itself with regulation, both of the levels of copper (and copper components) within and outside cells, and how cellular processes may be regulated by copper. This is followed (in Chapter 8) by changes in copper distribution and function that occur in embryogenesis and development of the mammalian fetus, and Chapter 9, on copper diseases and the metabolism of copper in disease states (contributed by my colleague, Christina A. Goode).

Chapter 10 addresses nonvertebrate copper metabolism and proteins, especially those proteins not already covered under previous chapters. [Some of the same proteins (like cytochrome oxidase and metallothionein) are present in most cells and organisms.] This brings in information about unicellular organisms (from bacteria to yeast) and multicellular, nonvertebrate organisms (like plants), where much new and interesting information is beginning to appear.

I am most grateful to Dr. Earl Frieden for having given me the opportunity to tackle this sometimes daunting project, and for his support and encouragement throughout. His own contributions to copper biochemistry have been of great value, and he and his colleagues continue to bring insight to this field.

A number of colleagues in copper biochemistry have been most generous in advising me about different parts of the book. I am particularly grateful to Amu (Bibudhendra) Sarkar, Joseph Prohaska, Ian Bremner, Joan Selverstone Valentine, Dennis Winge, Murray Ettinger, David Danks, Julian Mercer, Harry McArdle, and Ed Harris, all of whom gave me valuable comments and very useful information. I sincerely thank them for their wonderful help. I would also like to thank Carolyn Young and Deborah Hawkins for their valuable assistance with the preparation of the manuscript (especially the references), and Joanne (Jody) Wallick for her loyal and hard work on all the manuscript drafts, from first to last. Without them, this volume would not have been completed.

This book is dedicated to my dear husband, Gordon Nielson, whose unflagging love, support, and concern have seen me through this task.

Maria C. Linder

Acknowledgments

I am most sincerely grateful to Dr. Earl Frieden for his encouragement, patience, and help in preparing this volume; to Dr. Christina Goode for the contribution of a chapter on copper and disease; to Drs. Ian Bremner, Jim Camakaris, David Danks, Murray Ettinger, Joseph Prohaska, Amu Sarkar, Joan Valentine, and Dennis Winge for their helpful comments and suggestions; and to Ms. Joanne Wallick, Ms. Deborah Hawkins, and Ms. Carolyn Young for assistance in preparing the manuscript.

Contents

1. **Introduction and Overview of Copper as an Element Essential for Life** 1
   - 1.1 Abundance 1
   - 1.2 Essential Need for Copper. 3
   - 1.3 Distribution of Copper in Living Organisms 4
   - 1.4 Overview of Copper Metabolism in Mammals. 11

2. **Absorption of Copper from the Digestive Tract** 15
   - 2.1 Overall Copper Absorption. 15
   - 2.2 Exogenous Factors Affecting Copper Absorption 21
   - 2.3 Mechanism of Copper Absorption. 26
   - 2.4 Endogenous Factors Regulating Copper Absorption 34
   - 2.5 Copper-Binding Components and Copper Distribution in the Mucosa and Gastrointestinal Tract. 37
   - 2.6 Summary of Gastrointestinal Copper Absorption 40

3. **Copper Uptake by Nongastrointestinal Vertebrate Cells**. 43
   - 3.1 Cell Uptake of Ionic Copper in the Presence of Albumin and/or Amino Acids. 43
     - 3.1.1 Uptake by Hepatocytes 43
     - 3.1.2 Uptake by Nonhepatic Cells 52
   - 3.2 Cell Uptake of Copper from Other Plasma Sources 54
     - 3.2.1 Glycylhistidyllysine. 54
     - 3.2.2 Histidine-Rich Glycoprotein 55
     - 3.2.3 Transcuprein. 56
     - 3.2.4 Uptake of Copper from Ceruloplasmin 58

4. Extracellular Copper Substituents and Mammalian Copper Transport 73

4.1 Components of the Blood Plasma and Other Extracellular Fluids Associated with Copper.......................... 73
    4.1.1 Ceruloplasmin....................................... 73
    4.1.2 Albumin.. 94
    4.1.3 Low-Molecular-Weight Copper Ligands in Vertebrate Plasma... 98
    4.1.4 Other Copper-Binding Proteins in Blood Plasma/Serum 105
4.2 Amounts of Copper Associated with Various Substituents in Blood Plasma and Extracellular Fluid 116
4.3 Origins of Extracellular Copper Components................ 123
4.4 Copper Transport and Delivery to Organ Systems 124

5. Excretion of Copper in the Mammal............................ 135

5.1 Routes of Excretion 135
5.2 Absorbability of Endogenously Secreted Copper............ 139
5.3 Nature of Copper Components in Urine, Bile, and Other Alimentary Secretions 142
5.4 Sources of Copper for Excretion from the Body............. 149
5.5 Overall Turnover of Copper in the Body 154
5.6 Regulation of Copper Excretion............................ 158
5.7 Summary.. 161

6. Copper within Vertebrate Cells................................. 163

6.1 Distribution of Copper among Cell Organelles 163
6.2 Copper Components Associated with Different Fractions of the Cell 169
    6.2.1 The Nuclear Fraction................................ 169
    6.2.2 The Mitochondrial Fraction 170
    6.2.3 The Lysosomal and Microsomal Fractions............ 173
    6.2.4 The Cytosol Fraction................................ 175
6.3 Specific Copper-Binding Components in Most Vertebrate Cells... 182
    6.3.1 Metallothioneins.................................... 182
    6.3.2 Superoxide Dismutase 194
    6.3.3 Cytochrome c Oxidase 204
6.4 Specific Copper Components in Special Tissues............. 212
    6.4.1 Connective Tissue and Lysyl Oxidase 212
    6.4.2 The Brain and Dopamine ß-Hydroxylase 220

6.5	Other Intracellular Enzymes Requiring Copper	231
	6.5.1 Tyrosinase and Pigmentation	231
	6.5.2 Phenylalanine Hydroxylase	235
	6.5.3 Tryptophan Oxygenase	237
	6.5.4 Diamine Oxidases	238
6.6	Summary	239

7. Copper and Metabolic Regulation ... 241

7.1	Regulation of Copper Metabolism by Hormones	241
	7.1.1 Regulation of Plasma Cu and Ceruloplasmin	242
	7.1.2 Regulation of Other Aspects of Copper Metabolism by Hormones	263
7.2	Copper Regulation of Metabolism and Hormones	266
	7.2.1 Effects of Copper Deficiency	266
	7.2.2 Other Potential Actions of Copper	296

8. Copper in Growth and Development ... 301

8.1	Gestation	301
	8.1.1 Maternal Responses	301
	8.1.2 The Fetus	305
	8.1.3 Fetal Liver	309
8.2	After Birth	313
	8.2.1 Serum Copper and Ceruloplasmin	313
	8.2.2 Lactation	315
	8.2.3 Copper in Milk	316
	8.2.4 Copper Intake in Infancy	318
	8.2.5 Copper Absorption	319
8.3	Changes in Tissue Copper during Growth	320
	8.3.1 Liver and Blood	320
	8.3.2 Intestine	324
	8.3.3 Brain and Kidney	326
	8.3.4 Bone and Skin	329

9. Copper and Disease (*by Christina A. Goode*) ... 331

9.1	Inherited Diseases of Copper Metabolism	331
	9.1.1 Wilson's Disease	331
	9.1.2 Menkes' Disease	342

	9.2	Other Disorders of Copper Metabolism	350
		9.2.1 Copper and Inflammation	350
		9.2.2 Alterations of Copper Metabolism in Cancer	358
		9.2.3 Coronary Artery Disease	364

10. Copper in Nonvertebrate Organisms 367

	10.1	Copper in Bacteria	367
		10.1.1 Copper Transport and Resistance	367
		10.1.2 Bacterial Copper-Binding Proteins	369
	10.2	Copper and Fungi	376
	10.3	Copper in Yeast	379
	10.4	Copper in Plants (and Algae)	383
		10.4.1 Copper Absorption and Distribution	383
		10.4.2 Copper and Roots	383
		10.4.3 Copper Transport in the Plant	384
		10.4.4 Copper in Other Plant Parts	385
	10.5	Plant Copper Proteins and Their Functions	386
		10.5.1 Plastocyanins (and Cytochrome Oxidase)	386
		10.5.2 Other Small, Blue Copper Proteins in Plants	389
		10.5.3 Laccase	392
		10.5.4 Ascorbate Oxidase	398
		10.5.5 Tyrosinase (Phenolase, Catechol Oxidase)	400
		10.5.6 Low-Molecular-Weight Copper Ligands	402
	10.6	Copper in Mollusks and Arthropods	405

Appendix A: Copper Contents of Foods 409

Appendix B: Copper Content of Human and Animal Tissues 413

References 415

Index 505

Introduction and Overview of Copper as an Element Essential for Life

Copper, in the form of bracelets and ointments, has been a part of folk medicine in the treatment of painful muscles and joints, probably for a long time. However, only in the very recent past has the need for copper in the normal growth, development, and function of living organisms become apparent. Indeed, it is now clear that traces of this versatile transition metal are necessary for all life forms. Copper functions as a cofactor for cytochrome c oxidase, the terminal enzyme of the respiratory electron transport chain in most cells; as part of plastocyanin in the photosynthetic apparatus of plants and algae; as part of the oxygen-carrying hemocyanin of arthropod and gastropod blood; as a cofactor for numerous other (sometimes less fundamental) enzymes; and most probably also as part of smaller peptide and nonpeptide factors that may be involved in regulatory processes (such as cell proliferation and angiogenesis) and in some aspects of body defense in man and animals.

1.1 Abundance

Copper (atomic number 29) is not among the most abundant elements in the periodic table. Nevertheless, Nriager (1980) has estimated that the Earth's crust may contain a total of 15×10^{17} kg, of which only small proportions are present in tillable soils and water (6.7×10^{12} and 5×10^{12} kg, respectively), and even more minute proportions in living organisms (2.9×10^{10} kg in plants and 2.4×10^5 kg in animals). A small but significant portion is in aerosols (Cattell, 1978), particularly above the oceans (2.6×10^6 kg). Surface seawater contains about 1 μg of Cu/liter (range 0.9–1.6) (Owen, 1982c), with increasing concentrations as one probes deeper and deeper (about 100 μg/liter more per 1000 m). Values for freshwater generally vary from 0.1 to 1.0 μg/liter with larger concentrations where there are effluents form various industries. Based on the observation

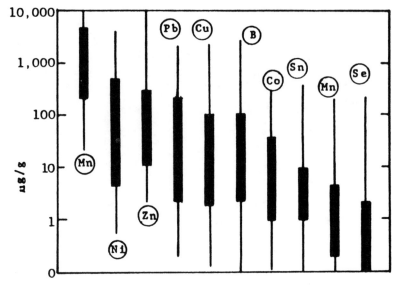

Figure 1-1. The content of copper and other trace elements in soils. The solid bars reflect the "usual range" found for an element. Reproduced, with permission, from Stevenson (1986).

that Arctic snow has about the same copper content today as in 1914, it would appear that the environment has not in a general way become "polluted" with this metal.

The content of copper in soils is not tied to that of other trace elements and varies as widely (Fig. 1-1) as it does in different types of rocks

Table 1-1. Copper and Other Trace Elements in Igneous and Sedimentary Rock[a]

		Concentration (μg/g)				
Element	Principal forms	Granite	Basalt	Limestone	Sandstone	Shale
Fe	Ferromagnesian minerals	27,000	86,000	3,800	9,800	47,000
Mn	Substituted ion in ferromagnesian minerals	400	1500	1100	<100	850
Zn	Sulfides, substituted ion in common minerals	40	100	20	16	95
Cu	Sulfides, substituted ion in common minerals	10	100	4	30	45
B	Tourmaline, soluble borates	15	5	20	35	100
Mo	Sulfides, substituted ion in common minerals	2	1	0.4	0.2	2.6

[a] Rearranged, from Sauchelli (1969).

(Table 1-1). This has implications for the health of animals, plants, and humans dependent upon soil for copper. Indeed, gross deficiencies of copper in the muck soils of Florida (McHargue, 1925–1927; Felix, 1927; Neal et al., 1931) and Holland (Sjollema, 1933, 1938), as well as in mixed soils of Western Australia (with a yellow clay subsoil) (Bennetts and Chapman, 1937), were instrumental in leading to the discoveries of an essential need for copper by animals and by plants. The heavy organic matter and humate content of muck soils is thought to bind the Cu and render it less available to plants (Stevenson, 1986). Thus, inorganic soils are considered deficient if they contain less than 4 ppm of Cu ($\mu g/g$), and organic soils if they contain less than 20–30 ppm.

1.2 Essential Need for Copper

Hints that copper plays a vital role in living systems appeared already in the first half of the last century (Owen, 1982c), when it was discovered, first, that the ash of plants contained the element (Meissner, 1816, 1817) and then that bovine blood ash did as well (Sarzeau, 1830). This was followed by the identification of copper in a number of foods [and later in the blood of snails (Harless, 1847), leading eventually to the discovery of the oxygen carrier protein hemocyanin (Underwood, 1971)]. By the mid to late 1800s, scientists had convinced themselves that the results were not due to contamination (Chevreul, 1868; Owen, 1982c), but the possibility that copper was essential for life had to await further work.

It was not until the 1920s that evidence for an essential need was obtained by J. S. McHargue (1925–1927) in Kentucky, who found that plants and animals needed copper for optimal growth and health. McHargue (1925, 1926) first demonstrated that various plants grew less well in copper-deficient soils and responded positively to copper administration. This led him to propose the element as a factor vital for life. In the same period, he provided similar suggestive evidence for the need for copper in mollusks and rats (McHargue, 1927a and b). The need for copper by plants was soon confirmed by two other groups (Allison et al., 1927; Felix., 1927), working with Florida muck soils, and this was followed by the elegant work of Sommer (1930) on sunflower and tomato plants and finally that of Lipman and McKinney (1931) on barley plants in culture.

In animals, the need for copper, separate from iron, in hematopoiesis was established by Hart et al. in 1928, and this was soon followed by the discovery that copper-cured "salt-sickness" (Neal et al., 1931), "Lechsucht" (Sjollema, 1933, 1938), and enzoic neonatal ataxia (Bennetts and Chapman, 1937), in cattle and sheep raised on copper-deficient soils around the world. Reviews of the data by Hoagland in 1932, and later by Arnon and Stout (1939), led to the acceptance of copper as a nutrient essential for plant and animal life and probably for most (or all) living systems (Owen, 1982c).

Later work on the ataxia of cattle, sheep, pigs, and other experimental animals led to a recognition of the fundamental need for copper in energy metabolism (via cytochrome c oxidase) and its functions in connective tissue, particularly the cross-linking of elastin (O'Dell *et al.*, 1961b) and collagen by extracellular lysyl oxidase. Many other copper-containing enzymes were discovered, including tyrosinase, needed for melanin pigment formation (in sheep's wool, but also elsewhere in animals and humans); dopamine β-hydroxylase, necessary for catecholamine production and therefore nerve and metabolic function; superoxide dismutase, aiding in the disposal of potentially damaging radicals produced in normal metabolic reactions; the plastocyanin of plant chloroplasts, involved in electron transport for photosynthesis; ceruloplasmin, a potential extracellular free-radical scavenger as well as a ferroxidase, in blood plasma and other extracellular fluids; and other, more specialized enzymes such as tryptophan oxygenase, ascorbate oxidase, and laccase, in plants and animals. Copper-containing, nonenzymatic proteins were also identified, and the association of copper with amino acids, small peptides, and perhaps other metabolites has been reported. Reviews of the known and potential functions of these components are a major focus of this book. However, many copper-binding components in living systems remain to be identified, and it is possible that we currently understand only a small but essential part of the function of this element in living systems. In this regard, the possibility that copper (or lack of it) is a factor in the development of major human diseases, from atherosclerosis and inflammation to cancer, has also been suggested. The status of work to identify disease connections at the molecular level, though not very far advanced at this time, will also be reviewed.

1.3 Distribution of Copper in Living Organisms

The *overall* concentration of copper in living organisms is somewhat uniform, in the range of 2 ppm (μg/g), with a few exceptions, such as in shellfish, ducks, and some microorganisms (Table 1-2). In general, the concentrations are less than those of tillable soils and rocks, but much greater than those of the water in lakes, rivers, and oceans.

Within multicellullar organisms, the distribution of copper varies from one part to another. Thus, in the edible plant, the highest concentrations

Table 1-2. Concentrations of Copper in Living Organisms

Organism	Copper concentration (ppm)[a]	Reference(s)
Human, adult	1.7	Underwood (1971)
Animals, adult	1.5–2.5	Underwood (1971)
Fowl	0.5–3	Souci et al. (1981); USDA
(ducks)	(2–5)	Souci et al.)1981); USDA
Fish (sea)	2–3	Souci et al. (1981)
Freshwater; unpolluted	0.4–3[b]	Harrison (1987)
Shellfish		
Oysters (sea)	12–37	Souci et al. (1981)
Various (freshwater)	3–22[b]	Harrison (1987)
Insects	3–31[b]	Harrison (1987)
Arthropods	4–30	Harrison (1987)
Plants (parts used for food)		
Roots	0.5–1.5	Souci et al. (1981); USDA
Stems	0.5–1.5	Souci et al. (1981); USDA
Leaves	0.3–1.7[c]	Souci et al. (1981); Jones (1973)
Flowers	0.3–3	Souci et al. (1981); USDA
Fruits	0.4–1.5 (2.1–2.6)[d]	Souci et al. (1981); USDA
Nuts	6–37[e]	Souci et al. (1981); Schroeder (1973)
Seeds	3–8	Souci et al. (1981); USDA
Legumes	3–7	Souci et al. (1981)
Microorganisms		
Algae	0.05–0.29[b]	Harrison (1987)
Yeast	8	Souci et al. (1981)
Soils	3–100 (av. 55)	Owen (1982b)

[a] $\mu g/g$ wet weight (raw/fresh).
[b] Based on dry weight data $\times 0.25$.
[c] Based on dry weight $= 8\%$ of wet weight.
[d] Values for avocados in parentheses.
[e] The highest values are for cashews and coconuts.

by far are in the seeds, as exemplified by the cereal grains, various nuts, and the beans of the legume family (Table 1-2). Here, copper (and many other trace elements) are associated especially with that portion of the seed constituting the embryonic plant [or the germ and the bran (Table 1-3)]. These parts of the seed are largely removed from the food supply of much of the world, through milling, polishing, and processing, resulting in the removal also of a substantial portion of the copper that would otherwise appear in the diet. Other effects of refining and processing are also apparent from the data. A fuller listing of the copper contents of food plants is given in Appendix A, based on data collected by the U.S. Depart-

Table 1-3. Effect of Processing on the Copper Contents of Foods

Food	Cu concentration[a] (μg/g)	Reference(s)
Wheat		
Whole	4.6 ± 0.9 (26)	USDA; Schroeder (1973)
Flour, standard	0.62, 0.63 (2)	Schroeder (1973)
Germ	6.1 ± 3.3 (9)	USDA; Schroeder (1973)
Bran	15.2 ± 3.1 (19)	USDA
Rice		
Whole	4.1, 2.4 (2)	USDA; Schroeder (1973)
Polished	1.7 ± 0.13 (10)	USDA
Sugar		
Cane (wet)	1.00 (1)	Schroeder (1973)
Brown sugar	1.34, 3.4 (2)	Schroeder (1973)
White sugar	0.57 (1)	Schroeder (1973)
Molasses	2.2, 6.8 (2)	Schroeder (1973)
Carrots		
Raw	0.47 ± 0.19 (90)	USDA
Boiled, drained	0.13 ± 0.04 (12)	USDA
Peas, green		
Raw	1.76 ± 0.34 (7)	USDA
Boiled, drained	1.73 ± 0.30 (14)	USDA
Spinach		
Raw	1.30 ± 0.19 (7)	USDA
Boiled, drained	1.74 ± 0.48 (16)	USDA
Potatoes		
Raw, flesh	2.59 ± 0.94 (73)	USDA
Raw, skin	4.23 ± 0.51 (6)	USDA
Apples		
Raw, with skin	0.41 ± 0.21 (119)	USDA
Raw, without skin	0.31 ± 0.13 (164)	USDA
Peaches		
Raw	0.68 ± 0.27 (84)	USDA
Canned, juice pack	0.50 ± 0.11 (14)	USDA

[a] Mean ± SD (N) (recalculated from SE).

ment of Agriculture (USDA), Schroeder (1973), and Pennington and Calloway (1973) for the United States and by Souci *et al.* (1981) at least partly for Europe. It is of potential interest that the copper contents of many European vegetables and fruits appear to be higher than those for the United States (Appendix A, starred values). (This does not appear to be so for the grains.)

Table 1-4. Copper Contents of the Major Parts of the Human Body[a]

Tissue/organ	Copper concentration[b] (μg/g)	Total for average 70-kg person[c] (mg)
Kidney	12 ± 7 (19)	3.2
Liver	6.2 ± 0.8 (9)	9.9
Brain	5.2 ± 1.1 (10)	8.8
Heart	4.8 ± 1.9 (14)	1.6
Skeleton[d]	4.1 ± 1.3 (8)	45.5
GI tract[e]	1.9 ± 0.9 (12)	2.8
Spleen	1.5 ± 0.4 (14)	0.2
Lungs	1.3 ± 0.4 (11)	1.3
Blood	1.1 ± 0.1 (5)	6.2
Muscle	0.9 ± 0.3 (7)	26.2
Skin	0.8 ± 0.4 (9)	3.8
Adipose	0.2, 0.3 (2)	3.0
Hair	2 ± 6 (21)	—
Nails	20 ± 17 (10)	—

[a] Based primarily on data collected from the published literature over many years by Owen (1982a). Mean values shown were calculated from all mean recorded values for a given parameter, excluding only values that fell above or below 2 SDs of means calculated from assays carried out in the 1970s and 1980s.
[b] Mean ± SD of reported mean values for number of reports in parentheses.
[c] Based on data for average tissue weight of 70-kg man (Magnus-Levy, 1910).
[d] Extrapolated from values for whole bone.
[e] Esophagus, stomach, and intestines.

In animals and fowl, the concentration of copper is also higher in the "seeds" (embryos and newborn forms) than in the adults: 2.4–3.5 μg/g in egg yolk (Souci *et al.*, 1981) and 2.9–6.9 μg/g in newborn animals and human infants (Widdowson, 1960). (See Table 1-2 for adult values).

With regard to vertebrates, the distribution of copper among the parts of the human body serves as a good example of the general organ variability observed. Table 1-4 shows the concentration and total content of copper in the major tissues of the average 70-kg (154-lb) adult human. Organs/tissues are listed in the order of decreasing copper concentrations. The values shown have been calculated from the mean values in large numbers of reports, primarily from the 1970s, but also earlier. The data from these reports (comprising almost all the scientific literature on copper through 1980) were gathered by C. A. Owen, Jr., of the Mayo Clinic in Rochester, Minnesota, and published in a series of volumes on copper in

1982 (Owen, 1982a–e). The gathered values shown have been, where necessary, recalculated to a wet weight base, using conversion values derived from data of H. A. Schroeder (also given in the Owen books). Unduly high or low values were excluded, as were values from the 1920s through 1940s unless they fell within two standard deviations of the mean of recent values. Each value thus represents a mean of means for the number of reports given in parentheses. A full listing of human and animal tissue copper concentrations from this data base is given in Appendix B.

Table 1-4 shows that, apart from hair and nails, the highest concentrations of copper are in kidney, liver, brain, heart, and the skeleton. Concentrations are much lower in the other major organs, skin, and blood

Table 1-5. Copper Contents of Human Body Fluids[a]

Fluid	Copper concentration	
	$\mu g/ml$[b]	μM
Blood		
Whole	1.11 + 0.13 (5)	17 ± 2
Plasma	1.13 ± 0.15 (70)	18 ± 2
Lymph	1.17 (1)	18
Sweat[c]		
Men	0.6 (1)	9
Women	1.5 (1)	24
Saliva	0.22 ± 0.08 (4)	3 ± 1
Fallopian secretions	1.1 (1)	17
Seminal fluid	0.5, 1.5 (2)	8, 24
Cerebrospinal fluid	5 ± 2 (4)	78 ± 31
Pleural fluid	0.60 (1)	9
Synovial fluid	0.2, 0.5 (2)	4, 8
Aqueous humor	0.14 (1)	2
Gastric juice	0.4 (1)	6
Pancreatic juice	0.26 (1)[d]	4
Bile	4.0 ± 1.9 (5)	63 ± 30
	1.2 ± 4 (5)[e]	
Gallbladder	(1.5–7.5)	(23–117)
Common bile duct	(0.3–10.5)	(5–160)
Urine	0.05–0.4[f] (1)[g]	1–6

[a] Owen (1982a)
[b] Mean ± SD of reported mean values for number of reports in parentheses; for plasma, reports are from 1950 to 1980.
[c] Sweat induced by heat and collected in arm bag.
[d] Ishihara et al. (1987).
[e] Iyengar et al. (1988).
[f] $\mu g/mg$ creatinine.
[g] Ishihara et al. (1986).

Table 1-6. Copper Concentrations in Tissues of Adult Humans and Animals: Major Organs[a]

Tissue/organ	Copper concentration[b] ($\mu g/g$)				
	Human	Rat	Pig	Mouse	Other
Kidney	12 ± 7 (19)	7.9 ± 5.5 (14)	7.3 ± 4.5 (4)	4.4 ± 1.1 (3)	5.8, 7.9 (2) (chick)
					6.9, 10 (2) (dog)
Liver	6.2 ± 0.8 (9)	4.6 ± 1.1 (23)	5.2 ± 0.7 (5)	4.1[c], 4.7[d] (2)	3.0, 2.9 (2) (chicken)
					67 ± 23 (5) (dog)[e]
Brain	5.2 ± 1.1 (10)	3.1 ± 1.2 (10)	3.9 ± 1.5 (4)	4.0 ± 2.1 (4)	
Heart	4.8 ± 1.9 (14)	4.8[f], 6.2 (2)	4.6 (1)		4.6 (1) (cow)
Bone	4.1 ± 1.3 (8)	2.5 ± 0.6 (3)	1.4, 2.4 (2)		4.4 (1) (sheep)
Stomach	2.2 ± 0.7 (7)		1.6 (1)		
Intestine	1.0, 3.0 (2)	1.7, 2.1 (2)		1.7 (1)	
Spleen	1.5 ± 0.4 (14)	2.3 ± 2.2 (8)	1.4 ± 0.3 (4)	1.2[c], 4.2 (2)	2.3, 2.4 (2) (dog)
Lung	1.3 ± 0.4 (11)	1.8 ± 0.6 (5)	1.2, 1.4 (2)	3.9 (1)	2.6 (1) (dog)
Blood	1.11 ± 0.13 (5)				
Plasma	1.13 ± 0.15 (70)	1.28 ± 0.26 (12)	1.75 ± 0.43 (11)	0.38[c] (1)	0.42 ± 0.20 (9) (chicken)
Muscle	0.9 ± 0.3 (7)	1.0 ± 0.4 (5)	1.0 ± 0.6 (3)		3.7 (1) (cow)
					0.5–0.9 (3) (beef)[g]
Skin	0.8 ± 0.4 (9)	1.7 ± 0.8 (4)	1.0, 1.5 (2)	2.4 (1)	0.4 (1) (cow)
Adipose	0.2, 0.3 (2)	0.35 (1)	0.8, 0.7 (2)		1.2 (1) (whale)

[a] Owen (1982a).
[b] Mean ± SD of reported mean values for number of reports in parentheses.
[c] Keen and Hurley (1980).
[d] M. C. Linder, L. Rigby, and C. Murillo, unpublished results.
[e] Based on recent data, Owen believes that the high values reflect the high Cu content of dog food, which contains a great deal of liver (Owen, 1982a). However, other factors, such as albumin, may be involved. See Chapter 4, Section 4.1.2.
[f] Linder and Munro (1973).
[g] Souci et al. (1981).

and are lowest in the triglyceride-full adipocyte. Based on organ weights for the average 70-kg man (with a lean body mass of 57.4 kg; 18% fat), the skeleton (including the marrow) and muscle stand out as major repositories of the metal. The quantitative importance of skeletal copper has not been generally noted, and little is known about its localization. The liver and brain continue to rank highly in terms of total copper content; also they are noteworthy for the apparent constancy of their copper concentrations: Coefficients of variation (CV) for liver and brain are 13 and 21%, respectively, as compared with values of 31% or more for most other organs. Blood also stands out as an important tissue, with regard to the total quantity of copper it contains and with regard to the constancy of its concentration at about 1 μg/ml (CV of 12% for these data). The same concentration is present in the blood plasma (Table 1-5), and the concentration in serum may be a trace higher [(Rosenthal and Blackburn (1974): 1.22 ± 0.27 versus 1.12 ± 0.18 mg/ml (mean ± SD, $N = 40$)].

Concentrations of copper in body fluids tend to be lower than they are in cells (Table 1-5), with the exception of bile (the major route for copper excretion) and cerebrospinal fluid (for reasons not readily apparent; see Chapter 6). The data on organ Cu contents are generally consistent among vertebrates (Table 1-6). Some species-specific exceptions include the dog and sheep, where liver concentrations are higher. Although Owen suggests that the high dog values are due to a high-copper liver diet, the dog is also

Table 1-7. Plasma Copper Concentrations and Ceruloplasmin Oxidase Activities in Different Species[a]

Species (N)	Plasma copper (μg/ml)		Ceruloplasmin oxidase activity (% of human value)[b]
	Mean	Range	
Pig (5)	2.2	1.5–2.7	140
Human (16)	1.3	0.95–1.7	100
Rat (20)	1.2	1.0–1.4	85
Sheep (16)	1.2	0.7–2.0	43
Cattle (16)	0.9	0.3–1.5	19
Dog (10)	0.67	0.51–1.1	12
Chicken (9)[c]	0.42		
Mouse (5)[d]	0.38		60
Peacock (6)	0.23	0.02–0.5	0
Turkey (16)	0.05	0.0–0.2	0

[a] Modified from Evans and Wiederanders (1967c).
[b] Percent of human activity/ml, measured with p-phenylenediamine.
[c] Owen (1982a).
[d] C. Murillo, L. Rigby, and M. C. Linder, unpublished results.

peculiar in terms of its albumin (which normally is involved in delivery of copper to the liver) (see Chapters 3 and 4). The high liver values of dogs and sheep are the result of a reduced capacity for biliary copper excretion (see Chapter 5). Relative to other species, the human is particularly rich in copper in the kidney and brain. There is considerably more interspecies variation in plasma copper (Table 1-7), which must have implications for copper transport and function within these animals and fowl. Here, the rat is again most like the human, as it tends to be also in terms of the copper in other tissues.

The data in these various tables provide a context within which the functions and metabolism of copper in living organisms and specific organs or tissues may be discussed. The ultimate aim would, of course, be to account for the function of every picogram of the element, in every cell and in every part of the extracellular fluids and solids present in plants, animals, and humans as well as in other living creatures. At this juncture, we can only give a broad outline of the picture and sketch in the details for a significant (but hardly encompassing) portion of it.

1.4 Overview of Copper Metabolism in Mammals

Figure 1-2 presents an overview of human copper metabolism, the details for which are provided in most of the rest of the book. The data on quantities of copper are taken from, or extrapolated to, the situation for the average human adult and refer to concentrations or total quantities in different compartments, as well as daily amounts taken in or excreted/secreted by the body.

Briefly, less than 1 mg of Cu is available from the typical American diet. Of this, more than half is absorbed in conjunction with other copper secreted into the GI tract in various fluids. [Most of the secreted copper is reabsorbed except that in the bile, which is largely excreted (along with dietary copper that is not absorbed).] Absorbed copper first enters the cells of the intestinal mucosa and then the blood. Some regulation of entry occurs at the serosal surface. In the blood plasma, copper immediately binds to albumin and to transcuprein, a newly discovered Cu transport protein. These proteins deliver Cu to the liver (and kidney), where dietary (or otherwise administered) copper is first deposited. From there, some of the copper enters the bile; some is used for production of internally required liver proteins; and much of it is incorporated into ceruloplasmin for resecretion into the blood. Ceruloplasmin is probably the main source of copper for the other tissues, delivering it via specific receptors. It also serves as an extracellular scavenger for oxygen radicals to protect the outer

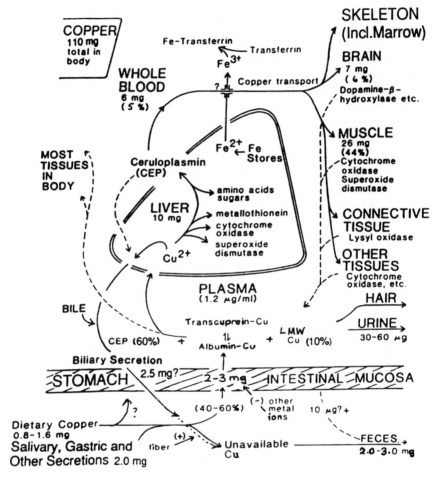

Figure 1-2. Overview of copper intake, absorption, distribution, content, function, and excretion in the average, normal human adult. Quantities are copper concentrations or total contents in tissues or average daily amounts eaten, secreted, absorbed, and excreted. (−), Inhibition of process; (+), promotion of process indicated. Modified from Linder et al. (1989).

parts of cell membranes, and it may facilitate Fe transport, as well. Copper is utilized by almost all cells to form the intracelluar enzymes cytochrome c oxidase and superoxide dismutase. It also attaches to metallothioneins (displacing other metal ions, notably zinc). It is essential for several other enzymes, including extracellular lysyl oxidase (which cross-links elastic fibers and collagen in connective tissues) and dopamine β-hydroxylase (involved in catecholamine formation by the central nervous system and the adrenal medulla). It also has other less well understood vital functions in many tissues, including the brain, and the immune system. Small quantities of copper are lost from the body by various routes (including the urine). A larger amount is lost in the bile, and it is in this form that most of the body copper is excreted.

Previous reviews of aspects of copper biochemistry, metabolism, function, and nutrition are those by Evans (1973), Mason (1979), Bremner (1980), Frieden (1970, 1979, 1980, 1981), Owen (1982a–c), Cousins (1985), and Camakaris (1987), as well as volumes on copper proteins, such as *Copper Proteins*, edited by T. G. Spiro (1981), Vol. 13 of the series *Metal Ions in Biological Systems*, edited by H. Sigel (1981), and volumes on copper metabolism, *Biological Roles of Copper* (Ciba Foundation Symposium 79) and *Copper in Animals and Man*, edited by D. M. Howell and J. M. Gawthorne (1987).

Absorption of Copper from the Digestive Tract

2.1 Overall Copper Absorption

Copper may be one of those rare nutrients for which at least some minimal absorption occurs already in the stomach. In 1965, Van Campen and Mitchell reported that in rats a substantial portion of a dose of ^{64}Cu(II) was absorbed over 2 h, when placed in the stomach ligated at the pyloric valve (Table 2-1). Indeed, a greater percentage was absorbed there than in 7-cm segments of various parts of the small intestine. This contrasted sharply with findings for ^{65}Zn and ^{59}Fe salts, which showed almost no absorption when administered to the gastric fluid (in the tied-off stomach). However, Fields et al. (1986c) recently failed to confirm this effect, reporting less than 0.1% of the dose in the blood and nongastrointestinal tissues between 0.5 and 4 h after gastric or oral intubation of ^{64}Cu(NO$_3$)$_2$ (with the stomach ligated at the pylorus). (Three to 7% of the dose appeared in the washed stomach wall.) Van Barnefeld and Van den Hamer (1984), on the other hand, found 22% of an oral dose in the (washed) stomach wall of mice after 2 h, of which none remained at 6 h. The earlier data of Van Campen and Mitchell (1965) for rats also suggested that there is a lower rate of absorption of copper, the farther down the small intestine one goes. This disagreed with studies of Crampton et al. (1965) (also in rats), in which there was a trend in the opposite direction, with maximum absorption in the third and fourth segment before the cecum (see later). In the chick, Starcher (1969) showed much greater intestinal absorption by the duodenum than lower down, and Wapnir and Stiel (1987) have reported similar rates in the jejunum and ileum of the rat. It is thus not clear whether there are significant and consistent differences in the capacity for copper absorption along the length of the small intestine, within or across species. It is unlikely (though this is not entirely clear) that gastric absorption occurs at all, and, if it does, it is also not clear how important this is in relation to absorption by the small intestine, especially under

Table 2-1. Uptake of Cu(II) and Other Metal Ions by Different Portions of the Gastrointestinal Tract of the Rat[a]

Metal	Dose (μg)	Total uptake as percent of dose in tissues sampled[b]			
		Stomach	Duodenum	Midsection	Ileum
^{64}Cu(II)	0.6–3.8	11.4	7.9	6.8	2.3
^{65}Zn(II)	2.9	0.5	51.8	7.3	17.5
^{59}Fe(II) (+ ascorbate)	0.6	1.1	9.1	4.0	5.3

[a] Data from Van Campen and Mitchell (1965) for radioisotopes placed in the stomach (ligated at the pylorus) or into 7-cm ligated segments at the duodenal, jejunal, and ileal ends of the small intestine, for 2 h.
[b] Tissues sampled for uptake were blood, heart, liver, and kidneys.

"meal conditions": The stomach is the first to receive dietary Cu, but the small intestine has a greater length and surface area that makes it likely to play the major role in dietary uptake. Justifiably or not, the general assumption has been that virtually all absorption occurs in the small intestine.

Table 2-2. Values Reported for Copper Absorption in Man[a]

Isotope and method[b]	Dose (mg)	No. of subjects	Absorption[c] (% of dose)	Reference
^{64}Cu balance	1	4 (healthy young men)	28 ± 7	Bush et al. (1955)
^{64}Cu whole-body counting	0.1–1.5	12 (aged 19–66, male and female)	60 ± 8	Weber et al. (1969)
^{64}Cu balance, corrected	?	7 (normal and hepatic cirrhosis)	54 ± 9	Strickland et al. (1972)
^{65}Cu balance NAA	3.2 (with meal)	22	57 ± 4	King et al. (1978)
^{65}Cu balance, chelate MS	2 (in Trutol or orange juice)	3	75 ± 4	Johnson (1982)
^{65}Cu balance, TIMS	2.9 (with meal)	10 (elderly men)	26 ± 1	Turnland et al. (1982)
^{65}Cu balance, NAA	2 (NPO or with meal)	7	61 ± 3	Ting et al. (1984)

[a] Rearranged from Ting et al. (1984).
[b] NAA, Neutron activation analysis; MS, mass spectrometry; TIMS, thermal ionization MS.
[c] Mean ± SE.

The overall capacity of the digestive tract to absorb various dietary forms of copper and doses of radioactive tracer copper has been studied extensively, especially in laboratory animals, but also to some extent in man. By absorption, we mean the overall percentage of an administered dose of copper entering the body *per se*, after crossing the cells lining the digestive tract. Table 2-2 summarizes the values reported for copper absorption in humans, as measured by various procedures. In interpreting the data, one should bear in mind that the average American consumes in the range of 0.5–1.5 mg of Cu per day (Linder *et al.*, 1985b). Also, it should be noted that long-term balance studies do not necessarily differentiate between absorption and retention (of copper, in this case), as retention results from a combination of absorption and excretion (of newly absorbed material). With two exceptions (notably the first study ever done and one with elderly men), the results are quite similar, and it appears that the metal is quite readily available; a large percentage of the Cu present or administered is absorbed. In more recent work, Turnland *et al.* (1989) found that apparent absorption of ^{65}Cu by young men consuming 1.7 mg of Cu per day was only 36% and was 56% and 12% for the same individuals consuming 0.8 and 7.5 mg of Cu per day, respectively. Similarly, David *et al.* (1989) reported that a low copper diet somewhat enhanced the apparent absorption of copper in both sexes (and in the elderly). The salivary and urinary concentrations of copper in the former subjects did not appear to vary with the diet, at least over the short term (24 days) (Turnlund *et al.*, 1990). This implies that man is capable of considerable adaptation to variations in copper intake, enhancing absorption when copper is less available, and vice versa. [Copper absorption has recently been reviewed by Johnson (1989), and a review by Turnlund is being prepared (J. Turnlund, personal communication).]

Overall, the percentage of oral Cu absorbed by the rat appears to be somewhat lower, although few studies lasting more than 2–3 h have been carried out. Figure 2-1A shows the cumulative absorption, over time, of different doses of Cu given as ^{67}CuCl$_2$ by intragastric intubation. At low and very low doses, about 30% had been absorbed by 8 h. The percentages were much lower at higher doses. (A 300-g adult laboratory rat, eating 15 g of rodent chow a day, may consume about 80 μg of Cu.)

The studies of Marceau *et al.* (1970) (Fig. 2-1A) showed that there is a constant relationship between the amount administered and amount absorbed, for doses up to 12 mg. Above such doses, a different (still linear) relationship pertains: The total amount absorbed goes up, but at a lower rate relative to the amount administered. Similar results were obtained by Farrer and Mistilis (1968) with ^{64}Cu(II) acetate over 24 h: about 40% of intrapyloric doses of 0.5–10 μg was absorbed, and lower per-

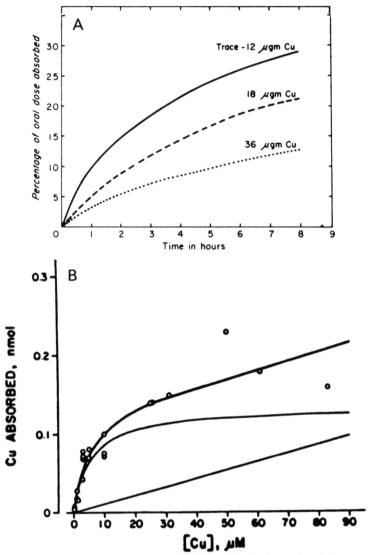

Figure 2-1. Time course and apparent kinetics of intestinal copper absorption at various (low) doses of copper, in the rat. (A) Time course at different copper doses: percent dose of radioactivity in plasma and plasma ceruloplasmin, at various times after intragastric intubation of trace (0–12 μg), 18 μg, and 36 μg doses of ^{67}Cu-labeled $CuCl_2$. Reprinted, with permission, from Marceau et al. (1970). (B) Rates of copper uptake by rat jejunum at different copper concentrations, determined in the presence of 0.14 N NaCl. An apparent diffusion rate is extrapolated from the top of the curve and has been subtracted to give the middle (saturation) curve. (The apparent diffusion rate curve is also indicated, at the bottom.) Reprinted, with permission, from Bronner and Yost (1985).

centages were absorbed at higher doses. A summary of the data for both studies is given in Fig. 2-2A and shows an average of 40–50% absorption at 0.07–12 µg, with descending percentages to 200 µg, and a constant 10% thereafter. At low doses, the absolute relationship is linear (Fig. 2-2B). At higher doses, this is also the case (but at a lower percentage) (Fig. 2-2C).

This differential dose response implies that there may be at least two mechanisms by which copper enters the body from the intestinal lumen (Bremner, 1980). Indeed, Bronner and Yost (1985) and Wapnir and Stiel (1987) have interpreted similar data for mouse duodenum and rat jejunum, respectively (Fig. 2-1B), as arising from a combination of transport via a saturable carrier plus (nonsaturable) diffusion. Their analysis of the saturation curves gives apparent half-maximal rates at Cu concentrations of 4 and 21 μM, respectively, for the two species. Bronner and Yost (1985) found that the slope of the nonsaturable absorption curve was very low (about 1%) and similar to that which could be calculated for the rat from earlier work of Cohen *et al.* (1979). Thus, at low copper concentrations, almost all the Cu is absorbed via the saturable carrier, while at higher concentrations increasing amounts (but a smaller percentage of the doses) would appear to be absorbed by diffusion. Nevertheless, the overall kinetic results are difficult to interpret without further knowledge about the individual steps required for the absorptive process.

Here one must consider that the metal first passes across the mucosal border of cells lining the gastrointestinal tract, thus entering the mucosal cells themselves. (In the case of the intestine, this involves crossing the brush border.) The second step is transfer from the mucosal cells into the interstitial fluid and blood, across the serosal (or basolateral) surface. In the case of iron, steps 1 and 2 are regulated independently: At the brush border, specific receptors may be involved (Kimber *et al.*, 1973; Greenberger *et al.*, 1969), but the most stringent control is on transfer of iron across the serosal surface (Linder and Munro, 1977). Thus, metal ions may enter the mucosal cells and not be transferred any further, eventually being lost with sloughed-off cells carried through the gut. In this connection, Fischer and L'Abbé (1985) studied the uptake of ^{64}Cu(II) by brush-border membrane vesicles and concluded that uptake across the mucosal surface was probably by simple diffusion and controlled by mass action. In conjunction with the other studies, this implies either that the real control of copper absorption, and the saturable carrier, is on the basolateral (serosal) surface of the mucosal cell or that conditions (such as, perhaps, too high doses of copper) were not right for identifying a saturable carrier in the brush border.

Although studies on the mechanism of copper absorption involving

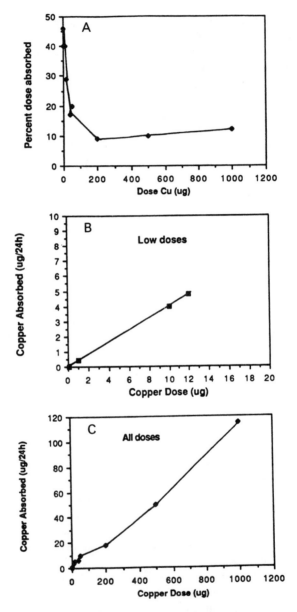

Figure 2-2. Effect of copper dose on rate of intestinal absorption by rats. Mean values for the combined data of Farrer and Mistilis (1968) and Marceau et al. (1970) for intragastric intubation of ^{64}Cu acetate or ^{64}CuCl$_2$ into adult male rats. (A) Percentage absorbed as a function of dose. (B) Amounts absorbed as a function of dose at low doses. (C) Amounts absorbed as a function of dose at all doses.

these steps have not sufficiently advanced for us to speak about the molecular aspects involved, considerable work has been done to study some aspects of overall intestinal transport (especially for rat duodenum), in ligated *in vivo* segments or in everted gut sacs incubated *in vitro*. For obvious reasons, similar studies have not been done with humans or larger vertebrates. Previous reviews of intestinal copper absorption include those by Bremner (1980), Cousins (1985), and Camakaris (1987).

2.2 Exogenous Factors Affecting Copper Absorption

Table 2-3 summarizes what is known about luminal factors that may affect copper absorption, their effects, the species in which the studies were carried out, and the methods used. As shown by a variety of means, several dietary substituents may enhance the overall absorption of copper from the gastrointestinal tract. These include protein and some amino acids (presumable acting through the same mechanism), but apparently not histidine (Wapnir and Balkman, 1990) and methionine (Strain and Lynch, 1990), an excess of which may reduce absorption or body copper content, respectively. Na^+, citrate, phosphate, and other small molecules and ions, including notably oxalate (which inhibits iron and calcium absorption), also stimulate copper absorption, as do chelating agents such as EDTA and nitrilotriacetate. Other dietary factors may have inhibiting effects, including phytate and fiber. Here it should be noted that high-fiber foods in general contain much more copper than normal foods, and the overall absorption from high-fiber-containing foods (whole grains or nuts, for example) is higher than that from non-high-fiber-containing foods (Kelsay *et al.*, 1979). This is in contrast to the results of animal or human studies in which purified fiber has been deliberately added. Apart from these factors, there may be inhibition mainly by other metal ions and by some gastrointestinal secretions, especially those from the pancreas and gallbladder.

If one peruses the methods used to study these questions, one realizes that intestinal absorption *per se* was again not always studied but that overall retention of copper was measured instead. Here, again, it is far from clear whether alterations in absorption or alterations in excretion, or both, are responsible for the retentions observed. (Excretion is covered in Chapter 5.) Studies in which some aspect of absorption was measured, such as removal of copper from the intestinal lumen or blood and tissue radioactivity in the short term, have been noted in Table 2-3 to emphasize their potentially greater validity. With regard to competing metals, there appears to be antagonism especially between copper and zinc, but also

Table 2-3. Luminal Factors That May Affect Overall Intestinal (or Total) Gastrointestinal Copper Absorption[a]

Factor	Apparent effect	Species	Reference	Method
Amino acids (L > D)	Enhancement	Rat	Kirchgessner and Grassman (1970)	Liver Cu after feeding Cu–amino acid mixtures
Tryptophan	Enhancement	Rat	Wapnir and Balkman (1990)[b]	Intestinal perfusion
Histidine	None	Human	Gollan and Deller (1973)[b]	^{64}Cu into intestine (vs. ^{64}Cu acetate)
	Inhibition	Rat	Marceau et al. (1970)[b]	^{67}Cu into intestine
	Inhibition	Rat	Wapnir and Balkman (1990)[b]	Intestinal perfusion
Methionine (excess)	Inhibition	Rat	Strain and Lynch (1990)	Tissue Cu, long-term
High dietary protein	Enhancement	Chick	Davis et al. (1962)	Balance studies
		Human	Engel et al. (1967)	Balance studies
		Human	Greger and Snedeker (1980)	^{64}Cu by stomach tube
		Rat	McCall and Davis (1961)[b]	
	Slight inhibition	Men (elderly)	Turnland et al. (1982)	Stable isotope absorption
Homocysteine	Inhibition	Rat	Brown and Strain (1990)	Tissue Cu
Citrate, phosphate, carbonate, gluconate	Enhancement		Schultze et al. (1936)	Absorption of these Cu salts vs. CuSO$_4$
Na$^+$ (isotonic)	Enhancement	Rat	Wapnir and Stiel (1987)[b]	Loss in jejunal perfusion
Oxalate, EDTA	Enhancement	Cattle	Chapman and Bell (1963)[b]	^{64}Cu absorption
		Rat	Kirchgessner and Grassman (1970)	
Nitrilotriacetate	Enhancement	Mouse	Keen et al. (1980a)	Tissue Cu after feeding
Thiomolybdate	Inhibition	Various	Suttle (review, 1975a, 1980)	Decreased ceruloplasmin
		Pory	Cymbaluk et al. (1981)[b]	^{64}Cu absorption
		Rat	Mills et al. (1981)[b]	Oral ^{64}Cu absorption
		Rat	Wang and Mason (1988)[b]	^{64}Cu, intestinal loops

Absorption of Copper from the Digestive Tract

Factor	Effect	Species	Reference	Comments
Fiber	Inhibition	Human	Drews et al. (1979)	Serum, urinary, and fecal copper
		Human	Kelsay et al. (1979)	Balance studies
Gluten	Slight inhibition	Rat	Brzozowska et al. (1989)	Tissue Cu, long-term feeding studies
Bile	Inhibition	Rat	Owen (1964)	Intestinal ^{64}Cu: fecal excretion
		Rat	Gollan (1975)	Intestinal ^{64}Cu: carcass retention, 76 h
Ascorbic acid (high doses)	Inhibition	Chick	Hill and Starcher (1965)	Oral ^{64}Cu: liver radioactivity; elastin
		Chick	Carlton and Henderson (1965)	Growth inhibition, anemia, and aortic rupture
		Rat	Van Campen and Gross (1968)[b]	^{64}Cu given into ligated duodenal segments (no effect of 1% dietary ascorbate on liver Cu)
		Rabbit	Hunt et al. (1970)	Whole-body retention of ^{64}Cu
		Rat	Van den Berg et al. (1990)	Tissue Cu reduced 20%, more oral ^{64}Cu to feces
	None	Human	Jacob et al. (1987)	Ceruloplasmin concentrations
Fructose	None	Rat	Fields et al. (1986b)	^{67}Cu diet, intubation 8–96 h
		Rat	Fields et al. (1986a)[b]	^{64}Cu in ligated duodenal loops (with NaCl)
		Pig	Schoenemann et al. (1990)	Tissue Cu, feeding trials
High zinc (100–30:1; µg:µg)	Inhibition	Rat	Oestreicher and Cousins (1985)[b]	Perfused intestine
		Rat	Fischer et al. (1981, 1983)[b]	Everted gut sac in vitro
		Rat	Van Campen (1969)[b]	Everted gut sac in vitro
High cadmium	Inhibition	Rat	Davies and Campbell (1977)[b]	^{64}Cu uptake on high-Cd diets
Pancreatic juice protein	Inhibition	Rat	Jamison et al. (1981)[b]	^{64}Cu in duodenal loops, in vivo
Saliva, gastric juice	None, vs. Cu acetate	Rat	Gollan (1975)[b]	^{64}Cu into intestine; carcass radioactivity
Iron	No effect	Rat	Forth and Rummel (1971)[b]	^{64}Cu in everted gut sacs of normal and Fe-deficient rats
		Rat	Bremner and Price (1985)[b]	^{64}Cu in carcass, short term
Cu(I), Cu$_2$O	Less available	Cockerel	McNaughton et al. (1974)	Dietary status vs. CuSO$_4$ (liver Cu, Hb)

[a] Modified from R. J. Cousins, Physiol. Rev. **65**, 239 (1985).
[b] Direct studies of absorption.

between copper and cadmium. (The interrelations with zinc are explored more extensively below.) In contrast, iron did not appear to antagonize copper absorption, although potential effects of simultaneous Fe(II) or Fe(III) addition were not examined. Some kind of antagonism within the body itself is apparent from several studies in which copper and iron status has been varied by dietary means. Hedges and Kornegay (1973) found that a high copper intake by pigs reduced blood hemoglobin concentrations except when extra iron was added to the diet (Table 2-4), but that high iron had no apparent effects on Cu status. Similarly, Bremner et al. (1987) found that in weaned calves iron supplementation enhanced the rates of loss of liver copper (liver copper normally decreases at this early age; see Chapter 8) and somewhat depressed plasma Cu concentrations as well. However, erythrocyte Cu and superoxide dismutase activities (as well as general health) were unaffected. Conversely, there have been several reports that copper stores in internal organs are enhanced by iron deficiency (and vice versa) (Sourkes et al., 1968; Sherman and Tschiember, 1979; Bremner, 1980). The possible mechanism(s) of these effects might include changes in copper compartmentalization (as, for example, between liver and bone) or changes in biliary excretion but is presently unclear (Bremner et al., 1987). (The potential interactions between iron and copper in hematopoiesis are covered in Chapter 7.)

Of some long-standing interest has been the possibility that ascorbic acid inhibits absorption of copper from the diet. This was based partly on studies in which copper status was assessed in the long term, using diets with and without large doses of ascorbic acid, effects being seen especially in chicks (Table 2-3) (Hunt et al.), in guinea pigs (Smith and Bidlack, 1980), and most recently in rats (Van den Berg et al., 1990), in which ascorbate (as 1% of the diet for 6 weeks) reduced the total Cu content of the body by 20%. However, a reduced copper status (as suggested by lower serum ceruloplasmin oxidase activity or deficient elastin production) could be the result of either decreased absorption or increased excretion of copper from the body. (Van den Berg et al., 1990, report no enhancement of ^{64}Cu excretion, implying that reduced absorption might be involved.) In some cases, ceruloplasmin oxidase activity was taken as the measure of copper status. In subjects or animals on a high-ascorbate diet, this is not reliable, as increased levels of circulating ascorbate will reduce ceruloplasmin enzyme activity without altering actual ceruloplasmin concentrations (Jacob et al., 1987). The closest to direct studies of the absorption mechanism involving ascorbate are the studies of Van Campen and Gross (1968) with rats, in which only a small decrease ($\sim 37\%$) in Cu uptake over 3 h was observed upon administration of 2.5 mg of the vitamin with 1 mg of Cu, as ^{64}Cu, to ligated duodenal segments. Preliminary

Table 2-4. Blood Hemoglobin Values in Pigs as a Function of Dietary Copper and Iron[a]

Week	Dietary Fe (ppm)	Hemoglobin (g/dl blood)			
		7 ppm dietary Cu	25 ppm dietary Cu	257 ppm dietary Cu	Mean
1	101	11.3	11.1	11.1	11.2
	312	10.7	11.5	11.4	11.2
3	101	12.3	12.2	11.2[b]	11.9
	312	12.8	12.8	12.7	12.7[c]
6	101	12.7	14.0	11.4[b]	12.3
	312	14.0	12.4	13.9	13.8[c]
9	101	12.7	12.7	10.7[b]	12.1
	312	13.3	12.5	12.8	12.8[c]

[a] Modified from Hedges and Kornegay (1973).
[b] Significant effect of excess Cu ($p < 0.05$–< 0.01) (horizontal).
[c] Significant Fe effect ($p < 0.01$) (vertical).

studies of Allerton and Linder (1985) supported the concept that there is little direct effect of ascorbate on copper absorption at the level of the small intestine, at least as detectable over a short time period. In these studies, ligated intestinal segments of rat duodenum, *in vivo*, were administered 1 μg of copper [as ^{67}Cu(II)-NTA] with or without 1 mg of ascorbate. Uptake (monitored over 1 h) was unchanged. Most recently, Jacob *et al.* (1987) studied retention of stable ^{65}Cu in 11 human subjects given diets containing from 5 to 605 mg of ascorbic acid per day. Although serum ceruloplasmin oxidase activities were depressed significantly (27%) in subjects on the high-ascorbate regimen, there were no differences in serum ceruloplasmin measured immunologically, again implying no effect of the excess vitamin on copper status (at least over a three-week period). There were also no effects on total serum copper. It thus seems unlikely that ascorbate has any marked effects on absorption or retention of copper.

Also of recent interest are the studies of Fields *et al.* (1984, 1986b,c, 1989), in which fructose- (or sucrose-) containing diets were shown to enhance development of copper deficiency more than starch-containing diets. Again, no clear-cut direct effect of fructose on intestinal copper uptake was observed, as monitored acutely or over the long term. Uptake of ^{67}Cu (intubated into the stomach in conjunction with the diet) was no different for the various diets from 2 or 8 to 96 h (5 rats per dietary group). Uptake of ^{64}Cu by *in situ* ligated duodenal loops was also unaffected when NaCl was present. It is noteworthy that Na$^+$ itself may be a variable in some of these studies (Table 2-3) and appears to be necessary for, or to enhance, Cu(II) absorption (Wapnir and Stiel, 1987). [It also abolished apparent differences in Cu absorption aberved for fructose versus starch (Fields et al., 1986c).] The effect of Na^+ is not a function of osmolarity or induced changes in water flux across the mucosal surface (Wapnir and Stiel, 1987), although ionic strength may be a factor. Since most secretions into the gastrointestinal tract contain isotonic saline, it behooves researchers to include NaCl in their studies (as many already do). Clearly, more work is needed to understand exactly what is the effect of fructose on copper metabolism. The most obvious alternative, namely, that fructose enhances Cu excretion, should be checked (after injection of radioisotope). A confounding factor in these studies now also appears to be sorbitol (Fields et al., 1989). In the copper studies of Fields (1986d, 1987), this sugar alcohol was produced in large (and probably damaging) quantities by male (but not female) rats on 60% fructose diets, and copper deficiency was also enhanced by fructose only in the males. This implies that sorbitol itself may be damaging tissue parts vital to copper absorption, distribution, or excretion.

2.3 Mechanism of Copper Absorption

Studies on the mechanism of intestinal absorption of copper, with radioisotopes, were initiated by Crampton et al. in 1965, using everted gut sacs, with and without inserted cannulas. Incubating the sacs in oxygenated Krebs–Ringer bicarbonate buffer (containing glucose), they found that the rate of copper transfer across the mucosal plus serosal surfaces (comparable to transit from the intestinal lumen to the blood) was clearly more rapid than was the flow of copper in the other direction (back into the intestinal lumen). Nevertheless, the flow of copper was bidirectional. Uptake and retention of labeled Cu by the intestinal wall itself was also greater when copper was administered from the luminal side. All of these findings were consistent with a net flow of copper from the intestinal lumen into the blood. The finding of a tendency for net unidirectional flow implies a need for energy mobilization in support of transport.

Direct evidence for an energy requirement was provided by additional studies reported in the same paper (Crampton et al., 1965). As shown in Table 2-5, incubation of gut sacs in a medium where nitrogen had replaced the oxygen almost completely prevented overall copper transport. Similar inhibitions were observed with inhibitors and uncouplers of energy metabolism. Almost complete inhibition was apparent at two (low) copper doses. Thus, virtually all of the copper transport at low (physiological)

Table 2-5. Effects of Anoxia and Effectors of
Energy Metabolism on Cu Transport by
Everted Intestinal Sacs of Hamsters[a]

Additive	Luminal Cu concentration (μg/ml)	^{64}Cu transported (% of control)
N_2	0.25	8
	1.0	4
Dinitrophenol (10^{-4} M)	0.25	8
	1.0	17
Iodoacetate (10^{-3} M)	0.25	10
	1.0	7
F^- (10^{-2} M)	1.0	16

[a] Data of Crampton et al. (1965). Overall transfer of ^{64}Cu(II) from lumen to serosal fluid was measured *in vitro*, after 1-h intubation. Addition of 10^{-3} M CN^- and a dramatic enhancing effect on Cu transport, probably reflecting formation of a highly permeable cyanide–Cu complex.

doses may be energy dependent. We do not know whether that is the case for the stomach (if it transports copper). We also do not know with certainty whether energy is required only for transfer of copper across the basolateral membrane of the intestinal mucosal cells, as implied by the studies of Fischer and L'Abbé (1985), and not for its transport across the mucosal brush border. Also still somewhat unclear is whether the additional, nonsaturable(?) mechanism for absorption (that comes into play with high copper doses) is energy dependent, although the kinetics suggest that this is unlikely.

Studies in which attempts have been made to modify copper absorption by endogenous and exogenous means, for example, by taking advantage of the antagonism between zinc and copper ions, have been useful in further elucidating the pathway copper takes across the intestinal mucosal cell. Animals on high-zinc diets (rather than given excess zinc at the same time as copper) will have a lower copper status, or absorb less of the test dose, than animals on moderate-zinc diets. Table 2-6 shows that rats on diets with a high (240 ppm), versus normal or low, Zn content have reduced levels of serum Cu and liver Cu, the former being one of the better hallmarks of Cu deficiency. Similarly, intestinal transport of Cu across the serosal surface is reduced, and this is accompanied by an increased retention of newly absorbed Cu by mucosal cells. A reciprocity between increased mucosal retention and decreased serosal transfer (the latter equivalent to overall Cu absorption), in animals on very high zinc diets,

Table 2-6. Evidence for Cu Deficiency and Decreased Intestinal Absorption in Rats on High-Zinc Diets[a]

	Concentration of dietary zinc		
	Low (7 ppm)	Medium (30 ppm)	High (240 ppm)
Serum Cu (μg/ml)	114 ± 9 (12)	109 ± 2 (11)	45 ± 10 (12)[b]
Liver Cu (μg/g dry wt.)	13.0 ± 0.6 (9)	12.5 ± 0.6 (9)	10.8 ± 0.6 (8)[c]
Intestinal Cu Transport[d] (ng/mg protein in 30 min)	2.3 ± 0.2	1.8 ± 0.3	1.3 ± 0.1[c]
Retention (ng/mg protein in 30 min)	221 ± 19 (15)	266 ± 18 (16)	305 ± 26 (16)[c]

[a] Data from Fischer et al. (1981); means ± SE (N).
[b] $p < 0.01$ for difference from 7 and 30 ppm Zn.
[c] $p < 0.01$ for difference from 7 ppm Zn only.
[d] Measured with everted duodenal segments incubated in Waymouth medium containing 6 μg of Cu/ml (mean of 9–12 determinations). Transport is overall release across the intestinal segment at the end of incubation.

was first noted by Hall et al. (1979). Their data are shown in Fig. 2-3. Within the mucosal cell, most of the newly absorbed Cu is associated with the cytosol and, except at very low zinc intake (see below), with metallothionein. It is now clear that zinc (Richards and Cousins, 1975a, b, 1977), but probably not copper (Bremner, 1980), induces an increased

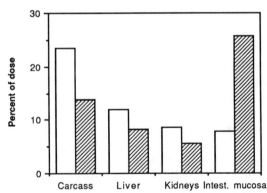

Figure 2-3. Effect of dietary zinc supplementation on intestinal absorption of ^{64}Cu, in rats over 4 h. Open bars, control; hatched bars, excess zinc. Data of Hall et al. (1979), redrawn, with permission, from Bremner (1980; Biological Roles of Copper, Ciba Foundation Symposium, 1979, pp. 23–48, Excerpta Medica, Amsterdam).

synthesis of metallothionein in the intestinal mucosa and in other tissues of the mammal, via increased transcription of mRNA (Menard et al., 1981). Data supporting this concept are given in Table 2-7. There is a 9-fold increase in intestinal metallothionein concentration [measured indirectly by a Cd(II)-binding assay] and a 7-fold increase in metallothionein mRNA with a 36-fold increase in dietary zinc. A similar increase in dietary Cu has no such effect, although it may have produced a slight increase in metallothionein. (There was no change in mRNA). It is well established that the affinity of this small divalent-metal-binding protein for copper is much greater than it is for zinc. Therefore, a plausible explanation for the inhibiting effect of dietary zinc on intestinal copper absorption (and transfer to the blood) is that the high-zinc diet induces intestinal metallothionein production and that, on entering, copper displaces the zinc on the abundant metallothionein in intestinal cells and is bound so firmly that little remains to be transferred across the serosal surface to the blood. Mucosal metallothionein might thus be a "stumbling block" to overall absorption of copper in the digestive tract, and copper retained by the mucosa might eventually be lost from the intestine through the sloughing-off of cells, although it is not known whether this actually occurs. (Moreover, would Cu released as Cu-thionein not be reabsorbed?)

There are other instances in which a reciprocal relationship between intestinal retention (on metallothionein) and transfer to the blood has been noted. This has been so for patients with Menkes' disease (Danks et al., 1972a), for brindled, blotchy, and mottled mutant mice (Evans and Reis, 1978; Mann et al., 1979), considered models for Menkes' disease, and for tumor-bearing and estrogen-treated rats (Fig. 2-4). Whether the

Table 2-7. Apparent Amounts of Metallothionein and Its mRNA in Intestines of Rats on Zinc and Copper Diets

Dietary metal concentration (μg/g)		Intestinal metallothionein	
Cu	Zn	Concentration[a,b] (pg/g)	mRNA[c] (molecules/cell)
1	5	400 ± 31	3 ± 1
1	180	3540 ± 170^d	20 ± 8
36	6	954 ± 31^e	4 ± 2
36	180	4320 ± 520	4 ± 3

[a] Data from Oestreicher and Cousins (1985) for determinations on 4 rats (mean $\pm$ SE).
[b] Determined by a Cd-binding assay (Eaton and Toal, 1982).
[c] Blalock et al. (1988); mean $\pm$ SE for 4 rats.
[d] $p < 0.001$ for difference from 5 ppm Zn.
[e] $p < 0.001$ for difference from 1 ppm Cu.

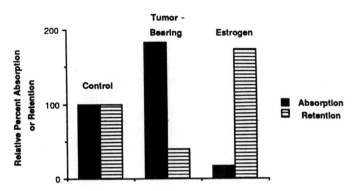

Figure 2-4. Reciprocity between copper absorption and intestinal retention in tumor-bearing and estrogen-treated rats. Modified from Linder (1979).

mechanisms regulating transfer across the serosal surface are identical under these various conditions and whether all involve, at least indirectly, increases or decreases in intestinal metallothionein is not known with certainty, although, in general, the newly absorbed Cu is found with fractions eluting with metallothionein in gel permeation columns. Certainly, one way of inhibiting some overall copper absorption appears to be through manipulation of intestinal metallothionein concentrations. At the same time, we do know that the primary genetic defect in Menkes' disease, for example, has nothing directly to do with the metallothionein gene, located on a different chromosome (Schmidt et al., 1984). (See also Chapter 9.)

At very low dietary levels of zinc, a considerable portion of newly absorbed copper in the intestinal mucosa may associate with larger components (possibly proteins) in the cell cytosol (Table 2-8; Fig. 2-5D). Although these studies may have been flawed by the use of large amounts of copper in the *in vitro* incubations, it appeared that the copper binding to large components became proportionately less as metallothionein levels increased (Table 2-8). It is possible that the larger component(s), also seen by other researchers (see Chapter 6), is more directly involved in copper transport across the basolateral membrane.

The data cited on reciprocity between retention and transport imply that Cu absorption can be controlled to some extent by the mucosal cytosol, and thus prior to transfer across the serosal (or basolateral) barrier. These data come from nonacute comparisons of animals in different dietary or physiological states, some of them rather extreme. Acute studies on Zn/Cu interactions suggest an additional site where regulation of Cu absorption may occur, involving the basolateral (serosal) barrier *per se*. When low concentrations of copper were perfused through the intestine

Table 2-8. Binding of Newly Absorbed Copper to High- and Low-Molecular-Weight Components in the Cytosol of Intestinal Mucosa: Rats on Diets with Three Levels of Zn[a]

Diet		^{67}Cu content (ng/mg protein)[b]		
Cu (ppm)	Zn (ppm)	HMW component	LMW component	Total
6	7.5	6.5 ± 1	6.0 ± 1	12.5
	30	3.0 ± 1	10.0 ± 1.5	13.0
	120	2.5 ± 0.5	10.5 ± 1.5	13.0
12	7.5	15.0 ± 0.5	11.5 ± 0.5	16.5
	20	9.0 ± 0.5	16.0 ± 0.5	15.0
	120	4.0 ± 0.5	19.0 ± 0.5	23.0

[a] Data from Fig. 6 of Fischer *et al.* (1983) for noneverted duodenal sacs incubated with ^{67}CuCl$_2$ (1.5–24 µg Cu/ml) for 45 min. Mucosal cell cytosol was fractionated on Sephacryl S-200. Means ± SE(7) were read to the nearest 0.5 ng.
[b] Copper bound (ng) was determined by dividing the ^{67}Cu cpm by the original specific activity of the isotope. This does not give an accurate value of the actual chemical Cu present, although doses were high enough to almost "swamp out" endogeneous Cu.

of rats on normal diets, *in vivo*, a 30–180-fold greater simultaneous infusion of zinc acutely inhibited overall copper absorption and transfer by about 50% (Oestreicher and Cousins, 1985). At higher amounts of infused copper, a roughly five fold greater amount of zinc also inhibited absorption, but to a much lesser extent. The reverse was also true; namely, increasing amounts of copper in the lumen inhibited zinc absorption, even when the zinc concentration was still about four fold greater than that of copper. As these acute studies eliminate changes in metallothionein as a factor, some sort of direct competitive antagonism between zinc and copper must also be occurring.

The data of Flanagan *et al.* (1984) (Table 2-9) for mice also point to the acute competition being at the basolateral membrane (serosal surface) of the mucosal cell. Here we see that uptake of zinc by intestinal mucosal cells was not inhibited (but enhanced) by the addition of copper but that transfer of zinc from the mucosal cells across the basolateral surface was significantly inhibited when a 10-fold excess of Cu(II) was given simultaneously. It is also of interest that there was some inhibition (at the mucosal as well as the serosal surface) of zinc absorption by two other divalent metal ions (cobalt and iron) (Table 2-9). Preliminary data of Allerton and Linder (unpublished) confirm the concept that the antagonism between copper and zinc is at the serosal rather than the mucosal surface of intestinal cells. Using *in vivo* ligated duodenal segments

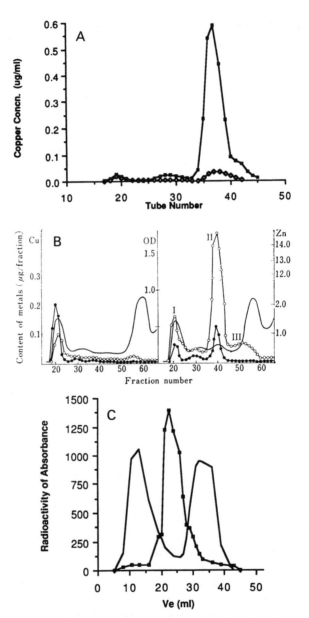

Figure 2-5. Elution of copper from intestinal cell supernatants fractionated on gel permeation columns. (A) Control (◇) and brindled (■) mice; Cu (μg/ml) eluted from Sephadex G-75 [replotted, with permission, from Bremner (1980)]. (B) Control (left) and excess-zinc-fed (right) rats; Cu (●) and Zn (○) (μg/fraction) eluted from Sephadex G-75 (Ogiso *et al.*, 1979). (C) Radioactivity (■) and absorbance at A_{280} nm (———) for normal rats; ^{64}Cu eluted

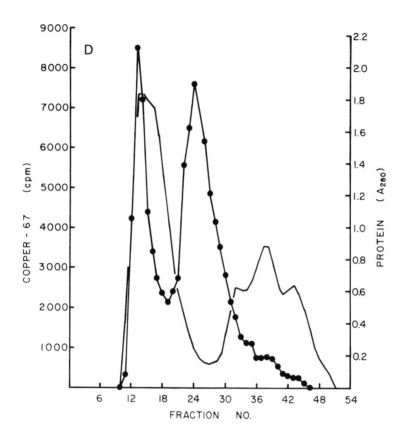

Figure 2-5. *(Continued)* from Sephadex G-75 1 h after administration of intragastric radiocopper [replotted, with permission, from Sonsma *et al.* (1981)]. (D) Duodenal sacs from untreated rats, incubated with large amounts of ^{64}Cu(II) (6 μg/ml); ^{67}Cu (●) and A_{280} (——) elutions (reprinted by permission from Fischer *et al.*, 1983).

in rats, it was found that simultaneous addition of copper and a 10-fold excess of zinc had no effect on overall uptake of copper from the lumen (amount in the mucosa plus amount appearing in the carcass) but that there was a larger percentage of copper retained in the intestinal mucosa when zinc was given simultaneously. The studies of Fischer and L'Abbé (1985) with intestinal brush-border preparations also show that mucosal absorption of copper is not inhibited but is even enhanced for membrane vesicles of mice on a high-zinc diet. This suggests that metallothionein induction may actually enhance mucosal copper uptake, while slowing the transit of copper across the serosal surface.

Table 2-9. Effect of Copper, Iron, and Cobalt on Duodenal Uptake and Transport of ^{65}Zn in Mice[a]

	Condition			
	^{65}Zn(II) alone	+Cu(II)	+Fe(II)	+Co(II)
Duodenal Zn uptake [nmol/(30 min)]	12 ± 2	20 ± 2[b] (+67%)	8 ± 1[b] (−33%)	6 ± 1[b] (−50%)
Serosal Zn transfer [nmol/(30 min)]	3.2 ± 0.7	1.5 ± 0.2[b] (−53%)	1.4 ± 0.2[b] (−56%)	1.0 ± 0.1[b] (−69%)

[a] Data from Fig. 1 in Flanagan et al. (1984) for iron-deficient, otherwise normal mice perfused with ^{65}ZnCl$_2$ (0.2 mM) in situ, with or without 2 mM competing metal ion. Duodenal uptake reflects serosal transfer + intestinal retention of radioisotope. Mean ± SE for 7–9 mice.
[b] $p < 0.05$ for difference from Zn-only control.

In conclusion, it is clear that there can be antagonism between copper and zinc, during the absorption probably at two levels: (a) via regulation of metallothionein (which has a high affinity for Cu), and (b) perhaps via a metal ion carrier in the basolateral membrane of the mucosal cell. However, it is also noteworthy that, within the normal ranges of dietary zinc and copper intakes, there will be relatively little antagonism between the two metal ions during absorption. Only a large preponderance of the one over the other will have significant effects in the rat (Tables 2-6 and 2-8; Hahn and Evans, 1975), as well as in humans (Valberg et al., 1984). Similarly, Bremner et al. (1979) and Bremner (1980) have demonstrated that, under normal dietary conditions, alterations in absorption may occur without major changes in metallothionein. Nevertheless, studies of these antagonisms have made valuable contributions to our understanding of the mechanisms of intestinal copper absorption.

2.4 Endogenous Factors Regulating Copper Absorption

Some of the potential endogenous factors (or physiological states) regulating overall intestinal copper absorption have already been described or alluded to, including metallothionein, tumors, estrogen, and genetic traits inherent in Menkes' disease patients and mutant mice. A fuller listing is given in Table 2-10. A need for increased copper within the body, either to alleviate a tissue deficiency or to provide the element for tissue growth (in the newborn and young), would appear to be reflected in an increased capacity for overall absorption, as an inherent property of the intestine (as

Table 2-10. Endogenous Factors or States Regulating
Intestinal Copper Absorption

Factor/state	Effect on absorption	Method/species	Reference(s)
Copper deficiency	Enhancement	Everted gut sacs, rats	Schwarz and Kirchgessner (1974)
		^{65}Cu retention, man	Turnland et al. (1989)
Mucosal metallothionein elevation	Inhibition	High-zinc diets, isotope absorption, rats	Hall et al. (1979) Oestreicher and Cousins (1985)
Menkes' disease	Inhibition	Oral ^{64}Cu, blood radioactivity, man	Danks et al. (1972a,b; 1973)
Brindled, mottled, and blotchy traits	Inhibition	Intragastric or duodenal ^{64}Cu, mice	Starcher et al. (1978) Evans and Reis (1978) Mann et al. (1979) Bremner (1980) Crane and Hunt (1983)
Estradiol treatment	Inhibition	Intragastric ^{64}Cu, rats	Cohen et al. (1979)
Oral contraceptives (progestagen-rich)	No effects	^{65}Cu balance, man	King et al. (1978)
Cancer (implanted tumors)	Enhancement	Intragastric ^{64}Cu, rats	Cohen et al. (1979)
Pregnancy	Enhancement	Intragastric ^{64}Cu, rats	Davies and Williams (1976) Schwarz et al. (1981)
Newborn	Enhancement	Intragastric ^{64}Cu, rats	Mearrick and Mistilis (1964)
		Dietary ^{64}Cu, lambs	Suttle (1965b)
		Intraduodenal ^{64}Cu, mice	Camakaris et al. (1979)

demonstrated with everted sacs in *in vitro* incubations), though only a few studies have been carried out. Except for pinocytosis in the newborn (Camakaris *et al.*, 1979), the actual mechanisms and sites involved in the regulation, within the intestinal cells or the gastrointestinal tract, are not clear at this time. (Whether there is increased stomach absorption, for example, in certain states has not been examined.) Unless induced to extreme levels by high dietary zinc (or perhaps by other endogenous factors), metallothionein (in the intestinal mucosa) may not play a major inhibitory role (nor a facilitating role) in absorption. In the only studies on Cu-deficient rats (Schwarz and Kirchgessner, 1973, 1974), there appeared to be no reciprocal changes in basolateral (serosal) copper transfer (enhanced 0.7- to 3-fold by deficiency) and retention by the intestinal sac (not significantly changed by deficiency), in contrast to the results for everted duodenal sacs from normal rats. This leaves (a) changes in higher molecular weight copper-binding components or (b) changes in serosal (basolateral) carrier concentrations, or both, as most likely to be responsible for changes in intestinal absorption. In the case of newborn animals (and perhaps also human infants), an alternative (enhancing) mechanism involving pinocytosis appears to be responsible (Mistilis and Mearrick, 1969; Camakaris *et al.*, 1979). [As in the case of many other nutrients, premature maturation of the intestinal mucosa (induced by cortisone treatment) reduces the efficiency of copper absorption, from 100% to that of adult levels.]

The majority of factors (Table 2-10) known to affect intestinal copper absorption thus do so by a mechanism involving decreased or increased retention during passage through the cells of the mucosa. Whether changes in the *gastric* mucosa might also be contributing to the variations in absorption observed in some of the studies is not known. Inhibition by virtue of genetic traits (expressed as Menkes' disease in humans or various mutations in mice) or by estradiol treatment results in all cases in enhanced retention of incoming copper, absorbed from the intestinal lumen by the mucosal cell, and decreased transfer to the blood. Conversely, implantation and growth of the fast-growing, undifferentiated Dunning mammary tumor DMBA-5A, in rats (Cohen *et al.*, 1979) or of the Morris hepatomas 7800 and 7793 (M. C. Linder and D. Cohen, unpublished results) resulted in enhanced transfer of copper to the blood and less retention by the mucosa (Fig. 2-4). This was demonstrable both by intragastric intubation of $^{64}Cu(II)$ *in vivo* and by incubating everted duodenal sacs with $^{64}CuCl_2$ *in vitro*. The observations suggest that, for inhibition, transfer across the basolateral surface of the mucosal cell was somehow limited or that there was an increased, very tight binding of copper to mucosal cell components. However, in all cases, newly absorbed (radioactive) copper was bound to a component of the apparent size and elution of metallothionein (Fig. 2-5). This was also the case in animals in which less copper was retained in the mucosa, and again almost all of the newly absorbed copper in the cytosol appeared to be on metallothionein (Cohen *et al.*, 1979). This does not exclude the possibility that noncystolic factors such as cell organelles may be involved, although the percentage of newly absorbed copper in the cytosol was similar in normal, tumor-bearing, and estrogen-treated rats (Cohen *et al.*, 1979). Thus, again it would seem that metallothionein is not the major block to intestinal Cu absorption, but rather that the regulation occurs at the basolateral (serosal) surface of the mucosal cell, or through intervention by some other subcellular compartment of these cells.

In the case of the neonate, it had been suggested that the copper-binding protein is not metallothionein but a Cu-chelatin with a lower cysteine content. However, it must be remembered that, especially with Cu-thioneins, oxidation can be a major problem that can artifactually result in this kind of "finding" (see Chapter 6, Section 6.3.1). Certainly, we know that metallothionein itself is synthesized by the adult intestinal mucosa and that it has a high affinity for copper. Although this does not prove that newly absorbed copper is bound to it, it seems fairly certain that this is generally the case. [It can, however, currently not be fully excluded that a low-molecular-weight, nonmetallothionein protein is present along with metallothionein (see later).]

As with the studies on Cu-deficient rats (Table 2-10), the enhanced absorptive capacity displayed by everted intestinal sacs from pregnant rats (early and midterm) (Schwarz *et al.*, 1981), if significant, was not accompanied by decreased retention in the intestinal wall. (No statistical comparisons can be made from the data reported.) If valid, these findings again point to the serosal barrier as the point of absorptive regulation. The studies on zinc-deficient rats already cited (Table 2-8), in which newly absorbed copper is associated more with a larger cytosol component (Fischer *et al.*, 1983) (presumably because there is less metallothionein), are also consistent with less of a role for metallothionein in normal absorption regulation.

2.5 Copper-Binding Components and Copper Distribution in the Mucosa and Gastrointestinal Tract

There are few reports on the total copper content of different portions of the gastrointestinal tract (Appendix B), but values for small intestine are in the range of 1–2 µg/g (wet weight) for humans, rats, and mice. Ogiso *et al.* (1979), using 2–3-month-old male Wistar rats (on normal 7 ppm Cu, 50 ppm Zn diets), reported values for segments of the gastrointestinal tract wall, in µg of Cu/g wet weight (mean ± SE for an unspecified number of animals): 2.1 ± 0.3 for stomach, 1.1 ± 0.1 (for duodenum, 1.0 ± 0.1 for jejunum, and 0.7 ± 0.0 for ileum. These are not dissimilar to values found in our laboratory for the duodenum of female Fischer rats (Table 2-11). Ogiso *et al.* found that a high-zinc diet (1% by weight for 8 days) significantly depressed all of the whole-wall values. What is not known is whether this decrease was confined to the nonmucosal cell portion of the intestinal (or gastric) wall or was also in the mucosal cells. Table 2-11 shows that the nonmucosal portion of the intestine, which must receive copper after its transfer from mucosal cells into the blood and interstitial fluid, can indeed have a quite different copper content from that in the

Table 2-11. Copper Content of the Duodenum of Female, Adult Fischer Rats, with and without Mammary Tumors, DMBA-5A[a]

	Intestinal copper concentration (μg/g wet wt.)		
	Whole duodenum	Mucosal portion	Nonmucosal portion[b]
Normal rats (8)	0.67 ± 0.46	0.70 ± 0.34	0.64
Rats with tumors			
5–7 g (4)	1.80 ± 0.85^c	1.09 ± 0.64	2.51
42–48 g (4)	4.57 ± 1.07^d	2.82 ± 0.86^d	6.32

[a] Data of K. Joseph and M. C. Linder (unpublished). Values are means ± SD for numbers of rats shown. Tumor-bearing rats were killed at 2 times after implantation.
[b] Calculated from mean values for whole duodenum and mucosa; weight of nonmucosal portion of intestine was 50% that of whole intestine.
[c] $p < 0.05$ for difference from normal rats.
[d] $p < 0.01$ for difference from normal rats.

mucosa, though in the case examined we are dealing with an instance (tumor implantation) in which the rate of copper absorption has been enhanced, and less copper has been immediately retained by the mucosal cells (Fig. 2-4).

As already intimated, a number of investigators have examined the copper content and labeling (with radiocopper) of components within the intestinal mucosal cell cytosol, prepared by scraping the mucosa, homogenizing in isotonic buffers, and centrifuging, usually for 60 min at $100,000 \times g$. In this procedure, about 20% of newly absorbed copper is sedimented, in the case of the normal rat (Cohen et al., 1979; Sonsma et al., 1981). As shown in Fig. 2-5A for normal brindled mice and in Fig. 2-5B for normal and excess-zinc-fed rats, there are three or four copper-containing components, separable on Sephadex G-75, the first in the void volume [$> 50,000$ daltons (Da)]. The next component is in the range of 35,000 Da, the elution position of superoxide dismutase, although the binding component may be some other protein (see Chapter 6). There is another peak in the range of 10,000 Da (between the main A_{280} peaks), traditionally ascribable to metallothionein (actual molecular weight about 6000). It is clear that not much copper is normally associated with the metallothionein. However, when an absorption inhibition is present (as in the brindled mouse) or when metallothionein has been induced by a very high zinc diet, more copper is associated with this component.

In contrast to the picture for copper components determined by chemical analysis is the picture for newly absorbed radioactive copper. At lower doses of ^{64}Cu(II), association is almost exclusively with the

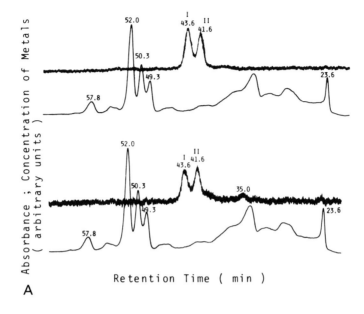

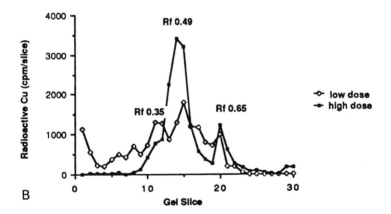

Figure 2-6. Fractionation of mucosal cytosol and metallothionein peaks by HPLC and nondenaturing electrophoresis. (A) HPLC of rat liver supernatant (TSK SW 3000), showing absorbance at 280 nm (——) (lower lines) and Cd (upper half) or Zn (lower half) elution, versus retention time. Metallothioneins I and II are separated. Reprinted, with permission, from Suzuki (1980). (B) Electrophoretic migration of radioactive copper-binding components obtained in the metallothionein peak on Sephadex G-75, derived from duodenal mucosal cytosol of rats given 150 μg of Cu (^{64}Cu) or trace (^{67}Cu) doses of copper by intragastric intubation (T. Sonsma and M. C. Linder, unpublished results). The 0.49 and 0.65 R_f peaks were also labeled by ^{65}Zn(II), ^{109}Cd(II), and [^{35}S] Cys.

metallothionein-like component (Fig. 2-5C), while at higher doses given *in vitro* (see also Table 2-8), binding to the high-molecular-weight component(s) of 80,000 Da or more becomes important (Fig. 2-5D). [Results similar to the latter have been reported by El-Shobaki and Rummel (1979) for (huge) doses of 30 μg of ^{64}Cu placed in ligated duodenal segments.]

Further fractionation of the metallothionein peak has shown that the copper is associated with at least two forms of the protein, separated in high-pressure liquid chromatography (HPLC) (Fig. 2-6A) and in nondenaturing electrophoresis (Fig. 2-6B), and also with a third component (R_f 0.35), only prominent when tracer doses of copper are administered (T. Sonsma and M. Linder, unpublished results). This component does not bind ^{65}Zn(II) or ^{109}Cd(II), nor does it incorporate appreciable amounts of [^{35}S] cysteine (Sonsma et al., 1981). It has a characteristic electrophoretic migration and a fairly high apparent pI, of 6.5. Clearly, much more needs to be done to elucidate the copper-binding components of the intestinal (and gastric) mucosae and to follow, with trace doses, the pathway(s) of newly absorbed copper within the intestinal mucosal cell.

2.6 Summary of Gastrointestinal Copper Absorption

Figure 2-7 is an attempt to summarize our current picture of the mechanisms and regulation of copper absorption, based on the findings cited. Traces of dietary Cu may be absorbed through the wall of the stomach and the rest in the small intestine, across the mucosal cells lining the villi and microvilli. In the intestine, Cu (most probably as the divalent ion) crosses the brush border, probably by (non-energy dependent) diffusion. In certain circumstances, chelates of Cu may also diffuse across (as, for example, copper cyanide). Once in the cell cytosol, some Cu becomes bound to metallothionein, displacing other metal ions (notably some zinc). Small portions may also bind to a protein of about 35,000 Da and to other larger proteins, and some copper also gradually finds its way into cell organelles.

Some Cu(II) continues on to the basolateral cell membrane, where it crosses to the blood and interstitial fluid by an energy-dependent, carrier-mediated, saturable process. With higher doses entering the cell, excess Cu crosses the basolateral surface by a second (less efficient) mechanism that is not energy dependent and may simply involve diffusion down a chemical gradient. The source of intracellular copper in unclear but may simply be Cu(II) or chelated copper in equilibrium with that bound to proteins. The specific carriers in the serosal membrane that are involved in the energy-dependent step would include some that are specific for Cu(II) but perhaps

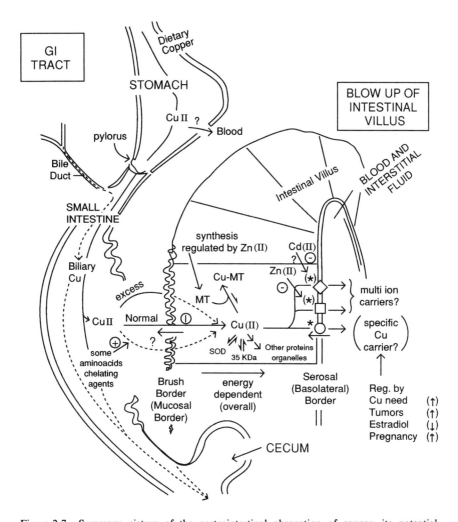

Figure 2-7. Summary picture of the gastrointestinal absorption of copper, its potential mechanisms and regulation, showing events in gastrointestinal tract, overall, and specific events thought to be occurring in absorptive cells of the intestinal mucosa. Multiple carriers in the brush border and serosal (basolateral) border of the cell allow transfer of Cu(II) in the first (1) and second (serosal) steps, by which the element is transferred to the blood. (See text for summary explanation.) MT, Metallothionein; SOD, superoxide dismutase; (*) indicates energy-dependent processes; ⊖ indicates inhibition of a process; ⊕ indicates stimulation of a process.

also some that carry both copper and zinc (and perhaps other divalent metal ions). Multiple carriers here would account for the apparent acute competition between Cu and Zn, Cd, and perhaps other ions that occurs with large excesses of one ion over the other. Alternatively, it may be that there are relatively specific carriers for each ion that are each inhibited by a large excess of another or that each carrier can transport other ions when they are present in very high concentrations. The concentrations of these carriers in the serosal (basolateral) membrane may be regulated, in the case of the Cu carrier, by copper need (as in dietary deficiency or pregnancy) and may be influenced directly or indirectly by the presence of a tumor in the body (perhaps via an interleukin).

A further regulation of copper absorption is via the amounts and rates of synthesis of intestinal metallothionein, known to be influenced by high levels of Zn but perhaps also up- or down-regulated by other factors. Excess metallothionein is capable of holding substantial amounts of copper very tightly (see Chapter 6, Section 6.3.1), more tightly than it can hold most other metal ions. This may divert much of the incoming Cu "flow" from continuing across the serosal surface and slow the absorptive process. (The ions released by Cu "occupation" of metallothionein may secondarily inhibit Cu "flow" by competing for sites on the serosal carriers.)

In Menkes' disease (Chapter 9), and in the mutant mouse models for this syndrome, the concentration of specific Cu carriers for the second step may be greatly reduced, although this may be an indirect effect rather than the primary genetic defect involved, perhaps becoming apparent through a lack of stimulus for carrier production. All this regulation and activity must also be viewed in the context of absorptive cells constantly being generated in the crypts of the villi, migrating up and sloughing off at the tip of the villus, a process spanning only a few days even in the human (2.8 days in the human, 1.5 days in the rat).

Copper Uptake by Nongastrointestinal Vertebrate Cells

3.1 Cell Uptake of Ionic Copper in the Presence of Albumin and/or Amino Acids

3.1.1 Uptake by Hepatocytes

Apart from the intestinal mucosa, most of the work on copper uptake by internal eucaryotic cells has been done with hepatocytes, especially those of the rat. Beginning with studies on liver slices incubated in Krebs–Ringer solution with ^{64}Cu-labeled Cu(II) citrate, it was shown by Saltman *et al.* (1959) that copper uptake obeyed saturation kinetics. Subsequent studies by Harris and Sass-Kortsak (1967), in which 0.1 volume of dialyzed serum was added to the Krebs–Ringer buffer, were consistent with a first-order process but showed that the addition of amino acids in a mixture containing 63 μM histidine (9.7 mg/liter), 71 μM cysteine (8.6), and 370 μM threonine (44), as well as most other amino acids in approximately physiological amounts, stimulated uptake two- to three-fold at Cu concentrations from 6.4 to 200 μM (0.1–5 mg/ml). This was also observed by Neumann and Silverberg (1966), who reported that histidine (4–12 μM) stimulated uptake of ^{64}Cu (10 μM) by rat liver or kidney slices (or by Ehrlich ascites tumor cells) when incubated in a Krebs–Ringer medium containing 4% albumin. Harris and Sass-Kortsak (1967) showed that there was the same degree of uptake (and stimulation of uptake) by amino acids at two different albumin concentrations, corresponding to about 0.3 and 1.5 g/dl. Also, the absence of an oxygen atmosphere had little effect on copper uptake recorded in the absence of amino acids but profoundly inhibited (in fact, more or less abolished) the stimulatory effect of the added amino acids. In contrast, 2,4-dinitrophenol (1×10^{-4}–1.8×10^{-4} M) had no effect on uptake measured in the absence or presence of extra amino acids, implying that the oxygen dependency of the amino acid stimulatory effect (if indeed it is reproducible) is not mediated by intracellular ATP availability.

These studies established, under semiphysiological conditions where albumin and other plasma proteins are present, that hepatocytes probably contain a saturable, specific non-energy-dependent Cu transport system, for which amino acids may be helpful in delivering the copper. These basic findings still hold, although more details have been added, and reinterpretation of all the data may be necessary in the light of other emerging facts.

The most important and extensive information on hepatocyte copper transport comes from the studies of Murray Ettinger and his colleagues, who carefully recorded the kinetics of copper uptake, in the absence or presence of histidine, other amino acids, albumin, and other divalent metal ions. Although the studies were carried out on freshly isolated cells that may have sustained some proteolytic damage, and at 20°C rather than 37°C, cells were viable and took up bile acids by active transport. More persuasively, very similar results have been obtained with cells cultured overnight to allow recovery from proteolytic effects (McArdle et al., 1987; Weiner and Cousins, 1983) and at 37°C (McArdle et al., 1988b). The results of the work of the Ettinger group are summarized by the diagram in Fig. 3-1. In the absence of albumin, other plasma proteins, or amino acids, copper [probably as Cu(II)] is taken up by hepatocytes by a mechanism not dependent upon levels of intracellular ATP (no effects of azide or dinitrophenol or many other uncouplers or inhibitors of electron transport) (Schmitt et al., 1983). Uptake is by a saturable mechanism, with a K_m the range of $10^{-5} M$ and a V_{max} of 1–4 nmol of Cu/min per mg of cellular protein, at 20°C (Fig. 3-2). Efflux appears to be by the same membrane transporter (same K_m, V_{max}) (Darwish et al., 1984a; Ettinger et al., 1986). As the rate of cellular copper accumulation diminished with time, especially at low Cu(II) concentrations, initial rates were used in all kinetic calculations.

The presence of histidine, at molar ratios to Cu of 2 or more, has a slightly inhibitory effect (Table 3-1 and Fig. 3-3) on free Cu(II) uptake. This implies that Cu is not crossing as a histidine complex. Albumin (in the absence of histidine) has a marked inhibitory effect (Fig. 3-4), and, in the presence of albumin, histidine stimulated hepatocyte copper uptake (Table 3-1). It is well known that albumin forms a very tight complex with Cu (K_D $10^{-17} M$), even tighter in the presence of histidine at concentrations equimolar to that of Cu or higher (K_D $10^{-22} M$; Lau and Sarkar, 1971). So this implies that histidine addition aids in the release of Cu from albumin, as suggested by the equilibria drawn in Fig. 3-1. [Recent studies by Gao et al. (1989) indicate that acetate-based ligands remove copper from albumin also, by forming a ternary intermediate.] Other amino acids, including threonine, were ineffective, although cystine may have had a

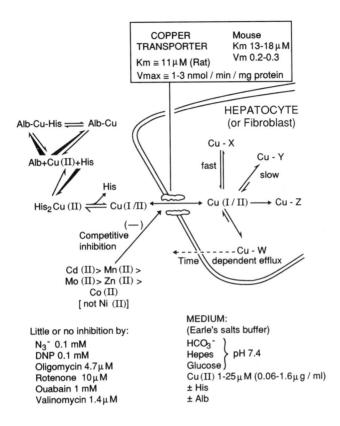

Figure 3-1. Mechanism of uptake of ionic copper by hepatocytes (and fibroblasts). Based on the kinetic studies of Ettinger and colleagues. Modified from Ettinger et al. (1986).

slight effect (Darwish et al., 1984b). Transport of Cu(II) was inhibited by other divalent metal ions, including competitively by Zn(II) (Fig. 3-5). These studies were done under conditions in which the total copper concentration was in the physiological range of that for blood plasma (Fig. 3-1), although of course under physiological conditions most is bound to proteins, and the free concentration is much lower (see below). Similar results have been obtained by McArdle et al. (1988a) for cultures of mouse hepatocytes, although the studies had to be carried out in the presence of albumin and 10% fetal calf serum. As in the early studies of Harris and Sass-Kortsak (1967), mixtures of amino acids (other than histidine, cystine, and threonine) also had a stimulating effect, and there was some enhancement of uptake also by cystine (and perhaps threonine) (statistics not

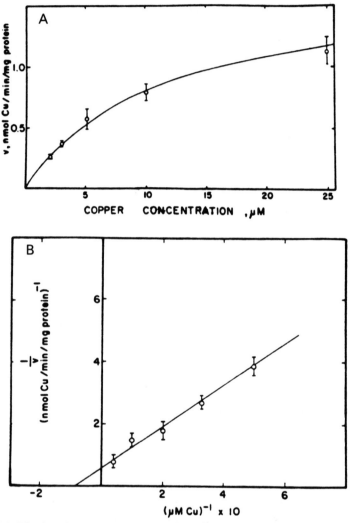

Figure 3-2. Kinetics of copper uptake by rat hepatocytes. Incubations were in HEPES–HCO_3^-, pH 7.4, containing equimolar histidine. (A) Initial rates (v) plotted against Cu concentration in the medium. (B) Lineweaver–Burk treatment of the initial rate data. Reprinted, with permission, from Schmitt *et al.* (1983).

reported). Indeed, the data of McArdle *et al.* (1990a) suggest that the ternary complex of his-copper-albumin may make copper more available than when albumin alone is present, and that the way copper is bound to albumin (to the high affinity site) is also important.

Confirmation that copper is not crossing as a histidine complex was

Table 3-1. Effects of Amino Acids on Copper Uptake in the Presence and Absence of Albumin[a]

	Relative uptake rate[c]	
Amino acid[b]	Cu(II) ± amino acid	Cu(II) + albumin ± amino acid
None	1.00 ± 0.10	0.44 ± 0.02
Histidine	0.64 ± 0.08[d,e]	0.90 ± 0.16[d]
Glutamine	0.69 ± 0.08	0.46 ± 0.04
Threonine	0.73 ± 0.09	0.46 ± 0.04
Cystine		0.63 ± 0.06
Histidine + glutamine	0.87 ± 0.11	0.92 ± 0.07[d]
Histidine + threonine	0.85 ± 0.11	0.94 ± 0.11[d]
Histidine + cystine	1.10 ± 0.08	
Histidine + serine		0.84 ± 0.12[d]

[a] Data from Darwish et al. (1984b).
[b] Amino acids with no effects were alanine, asparagine, aspartate, arginine, glycine, lysine, and serine.
[c] Values are mean relative uptake rates ±SD, for 6 determinations on fresh hepatocytes incubated with 10 μM Cu(II), with or without 20 μM total amino acid (no albumin) or 50 μM each amino acid (+albumin). Values for +albumin were multiplied by 0.44 to reflect the inhibition of Cu uptake effected by a 1:1 molar ratio of this protein to Cu.
[d] $p < 0.01$ for difference from control (value at top of each column).
[e] Values for equimolar histidine were 0.94 ± 0.10.

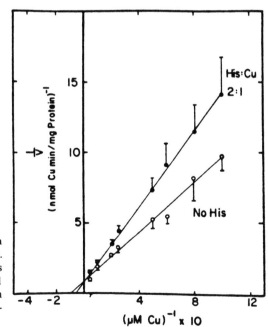

Figure 3-3. Effect of histidine on rates of hepatocyte copper uptake. Double reciprocal plot of results without and with histidine (2:1 molar ratio to Cu); otherwise as in Fig. 3-2. Reprinted, with permission, from Darwish et al. (1984).

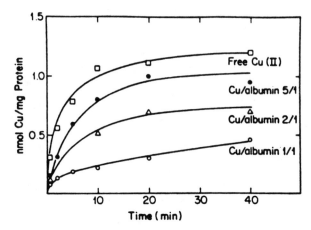

Figure 3-4. Inhibitions of hepatocyte Cu(II) uptake by albumin. Uptake of copper in the presence or absence of albumin at different molar ratios, in the absence of histidine, by rat hepatocytes. Reprinted, with permission, from Darwish et al. (1984b).

first provided by evidence from studies with [^{3}H]-His. The K_m for histidine transport under the same conditions was in the millimolar range versus 11 μM for Cu (Darwish et al., 1984b), and it was shown that half- or equimolar amounts of copper had no effects on histidine transport. Moreover, it is known that transport of histidine is Na$^+$ dependent whereas that of

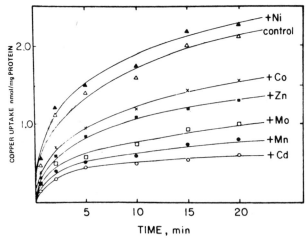

Figure 3-5. Effects of other divalent metal ions on hepatocyte Cu(II) uptake. Incubation was with 10 μM ^{64}Cu and a 50 μM concentration of the competing metal ion, in the presence of 60 μM histidine. Reprinted, with permission, from Schmitt et al. (1983).

Cu(II) is not (M. Ettinger, personal communication). [Intestinal Cu absorption is Na$^+$ dependent (Wapnir and Stiel, 1987).] Also, in the studies of McArdle et al. (1987) with suspensions of mouse hepatocytes, histidine uptake was inhibited by higher concentrations of Cu(II). [If histidine were crossing at least partly as a copper complex, the addition of copper should accelerate the transport rate.] In similar studies, inhibitors of ATP production also greatly reduced histidine but not copper transport (McArdle et al., 1988a).

Once in the cell, there would appear to be at least two compartments from which back flow/release of copper to the medium may be occurring. One is much more readily accessible to outflow (or chelation by external agents) than the other (Ettinger et al., 1986; McArdle et al., 1989) (X, Y, and perhaps W; Fig. 3-1). External albumin and histidine do not, however, stimulate efflux of copper from the cells.

Similar data for uptake have been obtained by the Ettinger group for hepatocytes from normal and brindled mice (Darwish et al., 1983) and more recently for fibroblasts from the same kinds of animals (Waldrop and Ettinger, 1990). The studies of McArdle et al. (1988b), Waldrop and Ettinger (1990), and Waldrop et al. (1988, 1990) with fibroblasts, confirm the inhibitory effects of albumin and indeed indicate a more marked inhibitory effect of albumin and histidine on Cu uptake by fibroblasts (as compared to hepatocytes). With regard to the mouse hepatocytes, although there was no difference in uptake between those derived from normal and brindled mice, there appeared to be a difference in the capacity for intracellular copper accumulation, which was less for the brindled hepatocytes (Fig. 3-6) (Darwish et al., 1983). [In these models for Menkes' disease, in vivo uptake/deposition of copper is decreased in the liver and enhanced in the fibroblasts (see Chapter 9).] Figure 3-6 shows that there is a kind of "overshoot" in cellular accumulation of copper (in the presence of histidine) by normal hepatocytes, after which radiocopper levels decline and attain a steady state, presumably reflecting a delay in efflux of newly absorbed copper. This phenomenon was less prominent in hepatocytes of Cu-deficient mice, suggesting that any copper that entered was needed and retained more tenaciously. In the brindled hepatocytes, on the other hand, the "overshoot" was not as high, and there appeared to be less delay before the efflux. Thus, brindled hepatocytes may have a lower capacity to "hold on to" incoming copper. This may be related to their content of factors (such as X, Y, and W) inside the cell (Fig. 3-1) (see more in Chapter 9). [On the other hand, McArdle et al. (personal communication) have not seen the falloff in rates of uptake seen by Ettinger et al., and this may reflect their use of cells cultured for longer periods.] The V_{max} of copper uptake found by Darwish et al. (1983) for mouse hepatocytes was

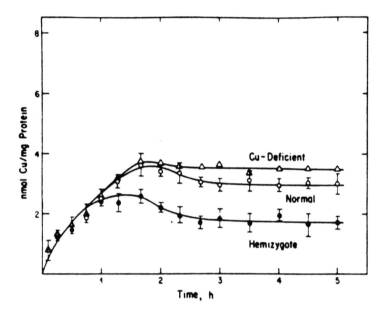

Figure 3-6. Time course of copper uptake by hepatocytes derived from normal, Cu-deficient, and brindled mice. Uptake of copper by isolated hepatocytes from male mice, at 20°C (as in Fig. 3-2). Reprinted, with permission, from Darwish et al. (1983).

about a third or less of that of rat hepatocytes (0.2–0.3 vs. 1 nmol/min per mg of hepatocytic protein), although K_ms were similar. [Mice also have less plasma copper (Tables 1-6 and 1-7) and ceruloplasmin than rats (or humans), as determined by assays of oxidase activity (Evans and Wiederanders, 1967c; Prohaska et al., 1983b; L. Rigby and M. C. Linder, unpublished results).]

As already mentioned, the kinetic data on copper uptake kinetics accumulated by Ettinger's group are (or are based upon) initial rates. The rapid falloff (with time) in apparent rates is illustrated in Fig. 3-4 (uptake of Cu from albumin) and Fig. 3-5 (uptake of Cu in the absence and presence of other metal ions) and also in Fig. 3-6, in which uptake by mouse hepatocytes is followed over even longer times. Part of the explanation for a slowing and eventual stasis of radiocopper uptake lies in the fact that the efflux rate will increase as accumulation progresses. It cannot however be excluded entirely that additional factors are at work. As mentioned by Cousins (1985), there was quite extensive uptake of copper by fresh hepatocytes at 0°C (Schmitt et al., 1983) given a 1:1 molar Cu:His mixture. In fact, uptake at 0°C eventually reached 80% of that obtained at 20°C. McArdle et al. (1988a) have also reported biphasic uptake rates for suspensions but not cultures of mouse hepatocytes. They attributed the very rapid initial rates to binding rather than transport, because of a lack of sensitivity to reduced temperature and a higher sensitivity to partial digestion (of whole cells) by pronase. This suggests that in some cases some of the initial uptake "rate" reflects the rate of binding of Cu to the transporter and also to less specific sites on the plasma membrane. However, Ettinger's group found more than 90% of their copper "uptake" to be intracellular.

The data that Weiner and Cousins (1980) obtained for rat hepatocyte monolayer cultures incubated in a serum-free medium, at 37°C, also suggest that the kinetics of copper uptake might be biphasic, with an initial much faster rate followed by a slower but steady rate of accumulation for up to 12 h, as is the case for suspensions of mouse hepatocytes and rat fibroblasts (Fig. 3-7), respectively, at 37°C in 10% fetal calf serum, as reported by McArdle *et al.* (1988a, b). The results from the Ettinger group did show a slowing down of uptake with time (Fig. 3-6), as already explained, but no continued long-term linear uptake. This may be due to the details of the conditions employed: copper concentration [2 or 5 μM for the study of McArdle *et al.* (1988a, b); an unknown concentration for the study of Weiner and Cousins (1980)], temperature, presence of amino acids and/or serum, and, of course, cell type and condition. Nevertheless, even if the slower (later) rate is used as a basis, the findings of Ettinger *et al.*, McArdle *et al.*, and Weiner and Cousins all lead to a quite similar conclusion about the mechanism of copper transport involving ionic copper, amino acids (histidine), and albumin. The relative ineffectiveness of zinc in inhibiting copper transport [reported by Weiner and Cousins (1980) and earlier by Saltman *et al.* (1959) for liver slices] may be explained on the basis that the zinc concentrations used were too low (only up to 20 μM, with Cu concentrations probably less than 10 μM). [The

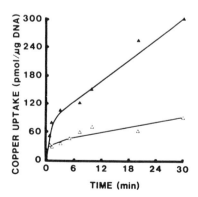

Figure 3-7. Uptake of ionic copper by cultured human fibroblasts in 10% fetal calf serum. Copper concentration was 5 μM, in Eagle medium containing amino acids, including some histidine, and in the presence ($\triangle$) or absence ($\blacktriangle$) of equimolar albumin. Reprinted, with permission, from McArdle *et al.* (1987b).

findings of McArdle et al. (1988) on zinc inhibition (up to 100 μM) at least qualitatively agree with those of Schmitt et al. (1983).]

These various data on hepatocytes, as well as those that follow for other cells, indicate the presence of a rather low affinity but potentially high V_{max} transport system for Cu(II) (and perhaps other divalent metal ions) that comes into play when concentrations of free copper (or its histidine complex) are high. As the K_m is about $10^{-5}\,M$, this transport system may not normally be that active, as concentrations of free Cu(II) and histidine complexes are not estimated to exceed 10^{-11} and $10^{-7}\,M$, respectively (see Chapter 4). However, even if this transport system is functioning at a fractional percent of capacity, and in the presence of albumin, significant amounts of Cu may be absorbed by such a mechanism, although this has not been studied with copper concentrations as low as those encountered physiologically (Ettinger et al., 1986; van den Berg and van den Hamer, 1984; Hartter and Barnea, 1988; McArdle et al., 1987a, b, 1988a, b).

3.1.2 Uptake by Nonhepatic Cells

As already mentioned, the behavior of fibroblasts appears to be almost identical to that of hepatocytes, even as regard the K_m of the ionic copper transport system (Waldrop et al., 1988; McArdle et al., 1987a, b). Similar results have been reported by Herd et al. (1987) for lymphoid cells. The one apparent difference was that histidine (forming a His_2–Cu complex) *markedly* inhibited copper uptake, even in the case of copper added as the Cu–albumin complex, which it does not for hepatocytes (Table 3-1, for example). This suggests that histidine is having an additional (inhibitory) effect on transport in these cells not seen in hepatocytes and related to the equilibrium between albumin–Cu, His_2–Cu, and free Cu(II) (Fig. 3-1). Indeed, amino acids like histidine do *not* stimulate uptake of Cu from Cu–albumin in fibroblasts as they do in hepatocytes. Another finding reported by McArdle et al. (1987b) for fibroblasts was that addition of "cold" Cu(II) to cells incubated with ^{64}Cu–albumin enhanced uptake of radiocopper. This would be expected, as cold Cu(II) should displace ^{64}Cu(II) from albumin, resulting in its increased availability for uptake. (Overall Cu uptake would also be increased, due to mass action and the higher Cu:albumin ratio.) Uptake of excess copper (in the absence of serum) may also involve a second, nonsaturable system (as reported for lymphoid cells by Herd et al., 1987), although this was not detected by Ettinger (personal communication).

Along these lines, Hartter and Barnea (1988) recently studied uptake of ^{67}Cu(II) by rat hypothalamic slices (in a serum-free medium, in the presence of high concentrations of histidine) and reported the presence of two saturable uptake systems not dependent upon metabolic energy. One has an apparent K_m of 6 μM, with a very low capacity (V_{max} 23 pmol/min per mg of protein), the other an apparent K_m of 40 μM, with higher capacity (0.43 nmol/min per mg). It may be that the very high histidine concentrations used resulted in these more complex kinetics (M. Ettinger, personal communication). Very large concentrations of histidine, threonine, and cysteine, as well as histamine and glutathione (molar ratios of 1000:1 or 2000:1 relative to Cu), also resulted in enhanced rates of copper uptake at high and low (20 and 0.2 μM) copper concentrations, and the same was true for alanine, glycine, and lysine at even higher ratios (and 0.1 μM Cu) (Katz and Barnea, 1990). (It is striking how little copper was taken up relative to the amounts added to the incubation medium.)

Finally, uptake and release of copper by nonhepatic (nongastrointestinal) mammalian cells was studied long ago in one other system: the human erythrocyte. Bush *et al.* (1956b) incubated whole blood, diluted with 0.16 volumes of "acid–citrate–dextrose" solution, with about 17 ng of added Cu per ml, as ^{64}Cu. The time course of uptake of radioactivity (Fig. 3-8) was again more rapid initially than later on. Assuming equilibration of radiocopper with endogenous nonceruloplasmin copper in plasma (about 400 ng/ml or 6 μM; Wirth and Linder, 1985), the initial rate of uptake (over 2 h) would appear to be about 0.006 nmol/min per mg of red cell protein. (Other assumptions were a hematocrit of 45% and an erythrocyte protein content of 200 mg/g.) This rate is more than 200-fold slower than that for rat hepatocytes (about 2 nmol/min per mg of protein), suggesting that the "finished" erythrocyte is not normally avid for extra

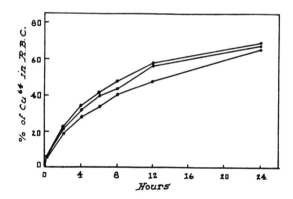

Figure 3-8. Uptake of ^{64}Cu from whole blood by human erythrocytes. About 1 μg of Cu, as ^{64}Cu(II), was added to 50 ml of whole blood mixed with 8 ml of a citrate buffer and incubated at 37°C. Reprinted, with permission, from Bush *et al.* (1956b).

copper. However, since some new copper did slowly enter, and since a slow efflux of copper from prelabeled cells was also measurable (Bush *et al.*, 1956b), some equilibration of erythrocyte and plasma copper pools appears to occur. Amino acid stimulation of uptake into erythrocytes, by a non-energy-dependent mechanism, in the presence (but not absence) of serum, was later confirmed by Smith and Wright (1973), who found the same responses in rat kidney and Ehrlich ascites tumor cells. Unlike fibroblasts, but like the hypothalamus, erythrocytes exhibit similar behavior to that reported for hepatocytes.

3.2 Cell Uptake of Copper from Other Plasma Sources

As detailed in the Chapter 4, albumin and amino acids are not the only substances with which copper is associated in blood plasma and presumably in interstitial fluid. Ceruloplasmin is a major potential source, although it is well accepted that most or all of its copper is "undialyzable" and does not exchange with the copper attached to albumin or amino acids. At least four other copper binding components that are potential sources must also be considered. These are (a) the glycylhistidyllysine tripeptide reported by Pickart *et al.* (1980) to be stimulatory to Cu(II) uptake for hepatoma cells in culture; (b) a histidine-rich glycoprotein (Morgan, 1978, 1981) reported to bind Cu(II) [and Zn(II)] *in vitro* (molecular weight about 67,000); (c) transcuprein (Weiss *et al.*, 1985; Wirth and Linder, 1985), a previously unknown plasma protein reported to bind significant amounts of copper (apparent molecular weight 270,000), and (d) α_2-macroglobulin (molecular weight about 800,000), which may also bind a small portion of the metal (see later). The capacities of these components to deliver copper to various cells have not been studied to nearly the same extent as those of albumin or histidine, and much additional work will be required to establish their roles, if any, as physiologically relevant uptake sources. However, some of the initial work suggests where the functions of these components in transport may lie.

3.2.1 Glycylhistidyllysine

Glycylhistidyllysine (GHL) was first reported present in human plasma by Pickart and Thaler (1973, 1980) and found to be associated with copper and iron (see Chapter 4). Moreover, these authors (1980) reported that it stimulated adhesiveness and growth of hepatoma cells in culture and uptake of Cu(II) when added (separately) to cells cultured with 0.7% fetal calf serum and a 300-fold molar excess of histidine (over Cu). [The recent report of Hartter and Barnea (1988) on hypothalamic brain slices agrees that GHL stimulates copper uptake when present in large amounts (molar ratios of 1000:1 or 2000:1 over Cu), while Antholine et al. (1989) reported poor copper uptake from the GHL complex.] As mentioned by Darwish et al. (1984b), the active copper species in such a mixture is unknown, and some (or even most) of the Cu(II) might be in the form of an albumin–Cu–histidine ternary complex (K_D 10–22 M; Lau and Sarkar, 1971). Thus, the tripeptide could be stimulating copper uptake by a mechanism other than that of delivering copper to a transport system. Indeed, under histidine- and albumin-free conditions, and using hepatocytes rather than hepatoma cells, Darwish et al. (1984b) found that a 1:1 molar ratio of the tripeptide to Cu(II) produced at 50% inhibition of uptake (identical to that produced by equimolar albumin). The inhibitory effect of albumin plus the tripeptide was significantly greater than that of either alone (all concentrations used were 10 mM), suggesting independent, additive inhibitory actions (not involving interactions between albumin and GHL). Histidine addition (10 mM) did not release tripeptide inhibition, indicating that the affinity of the tripeptide for Cu might be greater than that for the histidine monocomplex, as indeed it appears to be (K_D values of 10^{-16} and 10^{-21} M for the mono- and bis-GHL complexes, respectively; Lau and Sarkar, 1981a, b). As the concentration of albumin *in vitro* is about 600 mM, compared with perhaps 0.6 mM for the tripeptide (and 40–90 mM values for histidine), it seems unlikely that the tripeptide participates to any extent in copper binding, under normal conditions. It may, however, function in other ways to regulate copper uptake or have other effects relating to cell growth.

3.2.2 Histidine-Rich Glycoprotein

This 67,000-dalton (3.8 S) histidine-rich glycoprotein (HRG) was isolated from human and rabbit serum (Heimburger et al., 1972; and others) and from platelets (see Chapter 4 for details). In human plasma, the concentration is about 12.5 mg/dl (Morgan et al., 1978), but concentrations are 10-fold higher in rabbit blood (Morgan, 1981). Comprised of a single polypeptide chain of 507 amino acids (Koide et al., 1986), with 14% carbohydrate, it interacts with heparin (Heimburger et al., 1972; Niwa et al., 1985), plasmin (and plasminogen) (Lijnen et al., 1980), and thrombospondin (Leung et al., 1983), suggesting a role in regulation of blood clotting. Characterized by a histidine (and proline) content of more than 12%, especially at the C-terminal end, it has homologies with antithrombin C (at the N-terminus) and with high-molecular-weight kininogen (histidine-rich region) (Koide et al., 1986). HRG has a total of 10 metal binding sites [K_D about 10^{-6} M for Cu(II)], with a preference for Cu(II) and Zn(II), and successfully competes with albumin for the latter (Morgan, 1981; Guthans and Morgan, 1982). Nothing has been done to assess the possibility that this protein can be a source of copper for uptake by mammalian cells, and initial work suggests that it does not serve to transport zinc, at least in man (Failla et al., 1982). It is not known whether significant amounts of copper normally bind to HRG in blood plasma, and, if so, whether that copper equilibrates with albumin or amino acid-copper. However, since HRG elutes in the same position as albumin in gel chromatography, it has not been entirely excluded that some of the copper (radiocopper) in the "albumin peak" is attached to this protein, although the binding constant is much lower than that of albumin. What may be a related histidine-rich protein, which may also bind copper, has recently been isolated from albacore tuna blood (Dyke et al., 1987). It has an apparent molecular weight of 66,000, contains 8.2% histidine by weight, and binds three Zn atoms with a binding constant of $10^{9.4}$.

3.2.3 Transcuprein

Binding of newly absorbed copper to a component of blood plasma eluting in the void volume on columns of Sephadex G-150 (>150,000 Da) was first reported by Campbell et al. (1981) and then investigated further by Weiss et al. (1985), who found that it could be labeled with ^{67}Cu(II) in vitro. Lau and Sarkar (1984) had also reported in vitro labeling of a void volume substituent from Sephadex G-100, one that is especially prominent in the plasma of the human neonate. In earlier studies by Bush et al. (1959a), in vivo in man, a significant portion of radioactivity present at zero time and in the early phase of uptake was associated with the precipitate obtained by addition of ammonium sulfate to 50% saturation. This would be copper associated neither with albumin, amino acids, nor, at such times, ceruloplasmin. Weiss et al. (1985) have confirmed this and found that the void volume component had an apparent molecular weight of 270,000; that it rapidly exchanged copper with albumin (in whole serum); that it was not reactive to antiserum against albumin or ceruloplasmin; and that it appeared to be saturated with copper at much lower concentrations than albumin, implying a higher affinity for the metal. They tentatively named the protein transcuprein. (For further information, see Chapter 4). In preliminary studies, transcuprein has behaved as a good source of copper for Ehrlich ascites tumor cells and BALB/c fibroblasts in culture (Table 3-2) and as a potentially better source (of Cu) for these cells than

Table 3-2. Uptake of ^{67}Cu from Various Serum Proteins by Cultured Cells[a]

	Binding or uptake (pg per 10^6 cells) from:		
	Transcuprein	Ceruloplasmin	Albumin
Ehrlich ascites tumor cells			
0°C (binding)	25	74	94
37°C (uptake)	56	126	141
Net uptake[b]	31	52	47
BALB/c fibroblasts			
0°C (binding)	14	55	65
37°C (uptake)	50	82	103
Net uptake[b]	36	27	38

[a] Data of Weiss et al. (1985) for Ehrlich ascites tumor cells and BALB/c fibroblasts in serum-free medium, cultured at 0° or 37°C for 60 min.
[b] Net uptake (difference between data for 0° and 37°C).

albumin. These studies were done with crude fractions of in vivo ^{67}Cu-labeled material (obtained by separating whole serum on Sephadex G-150). Incubations were in serum-free minimal medium (containing all amino acids, including histidine), and radioactivity associated with washed cells was counted after 60 min at 0° or 37°C. The results are clearly preliminary, but, taken together with repeated findings that transcuprein and albumin copper pools rapidly equilibrate (Weiss et al. 1985; Linder et al., 1987a), they indicate that the possibility that transcuprein is an intermediate in the "flow" of copper from albumin pools to hepatocytes (and other cells) cannot be ignored. Whether this also requires the mediation of amino acids like histidine or whether specific receptors recognizing the transcuprein–copper complex (in the cell membrane) are involved needs to be determined. [The equilibration of copper between transcuprein and albumin does not appear to require free amino acid, as further successful exchange studies were performed with mixtures of the partially purified proteins, in the absence of amino acids (Linder et al., 1987a)]. Indeed, direct interaction of the two proteins appears to facilitate copper exchange (M. T. Tsai and M. C. Linder, unpublished).

Added recent findings (Linder et al., 1987, 1988) are that histidine enhances removal of Cu from transcuprein and that the "off-rate" for Cu (without added histidine) (at 4°C, in 150 mM NaCl–20 mM phosphate, pH 7) is less than 2% per 48 h. Assuming that the "on-rate" is diffusion controlled, this calculates to a dissociation constant for rat transcuprein of 10^{-17} M, very close to that for human albumin. (The K_D for rat albumin may be higher than that for human albumin; certainly, its K_D is higher than that for rat transcuprein.) Transcuprein has now been purified

(Goode et al., 1989) and consists of two subunits (about 10 and 200 kDa) that are currently being sequenced (M. T. Tsai and M. C. Linder, unpublished results). During purification, it was also noted that a small portion of radioactive copper in rat plasma seemed to copurify with α_2-macroglobulin. This very large (756-kDa), four subunit antiproteinase is a major binder of plasma zinc (Adham et al., 1977), with two to five binding sites per subunit (Osterberg and Malmensten, 1984; Pratt and Pizzo, 1984) (K_D about $10^{-7} M$). What function, if any, this additional protein may have in donating copper to cells remains to be examined. (See Chapter 4 for more information on this protein.)

3.2.4 Uptake of Copper from Ceruloplasmin

Ceruloplasmin has long been known as the major copper-containing component of mammalian blood plasma/serum. Comprised of a single polypeptide chain (of 121,000 Da), with several asparagine-linked carbohydrate side chains (about 11,000 Da), this α_2-glycoprotein holds six to seven copper atoms in three- to four-liganded states (see Chapter 4). All but one of the copper atoms that may be present on ceruloplasmin are nondialyzable and do not exchange with other copper in the environment. (Indeed, because they are largely "buried" in the protein structure, attempts to reconstitute the protein after complete copper removal have largely been unsuccessful.) One site with intermediate affinity (K_D $10^{-10} M$) that is potentially accessible for exchange/binding with copper in the fluid environment has been identified (McKee and Frieden, 1971), but it is not known whether this site may play a role in copper transport or delivery of copper to cells. Ceruloplasmin is also an enzyme with a variety of activities including oxidation, ferroxidation, and free-radical interaction. (For further details on all these matters, see Chapter 4.) Along with the fact that ceruloplasmin copper is nondialyzable, this knowledge had been a major reason why ceruloplasmin was discounted as a direct source of copper for cell uptake. Ceruloplasmin also is an "acute phase reactant" (see Chapters 4 and 9), again diverting (for some) any interest in a copper transport and delivery role.

Despite these biases, a variety of investigators over the years have provided evidence supporting ceruloplasmin as a direct copper donor to cells, and cumulative evidence is now persuasive. In 1971, Owen reported that intravenous injection of radiocopper-labeled ceruloplasmin (in whole serum) resulted in uptake of copper by most tissues, 2–72 h after administration. For these studies, "^{64}Cu-ceruloplasmin" was radioactive plasma from donor rats injected intraperitoneally with ^{64}CuCl$_2$, 16–24 h before. [Weiss et al. (1985) have recently confirmed that at such times virtually all of the radioactivity is bound to ceruloplasmin.] On this basis, Owen (1971) proposed that ceruloplasmin is a donor of copper to cells, although he found that ionic copper injected directly as $^{64}CuCl_2$ was also a source. More recently, Campbell et al. (1981) did similar studies, over a shorter time period, during which appreciable degradation of ceruloplasmin (and release of its copper by other mechanisms) would not be expected. Ceruloplasmin copper was taken up by all tissues examined (including the tumors of rats with Dunning mammary carcinomas). In the uptake calculations, the contributions to tissue radioactivity of residual blood were subtracted. Also, plasma with radiolabeled ceruloplasmin was passed through Chelex-100 to remove any free or loosely bound $^{64}Cu(II)$. As in the case of Owen (1971), uptake of radioactive copper from ceruloplasmin was compared with that from nonceruloplasmin sources (similar nonradioactive plasma to which tracer $^{64}CuCl_2$ had been added directly). The results, (together with those of Owen, are summarized in Table 3-3.

With either source of copper, the liver was the main recipient of the radioactivity administered, implying that either form of copper is available to the hepatocyte. Of special interest were findings that the percentages of radioactivity absorbed by the cells of various tissues differed depending upon the source. In particular, the liver and kidney appeared to be more avid for nonceruloplasmin copper, while the heart and brain appeared to have a preference for the copper of ceruloplasmin (Table 3-3). It was also notable that the rate of uptake of intravenously administered radiocopper was rapid, with about 80–85% of both forms absorbed (in normal rats) over 1 h in the studies of Campbell et al. (1981) and probably about 90% of ionic copper, but much less ceruloplasmin copper, in 2 h in the studies of Owen (1971) (assuming that blood plasma is 4% body weight and rats weighed about 500 g). Any differences in uptake rates between these two studies may reflect differences in the age, strain, and sex of the rats (younger Fischer females versus older Sprague–Dawley males), as well as the doses of plasma substituents and radioisotope. [Addition of excess ionic Cu(II) rather than tracer amounts might stimulate removal from the blood, especially by the liver, while excess ceruloplasmin copper might be removed more slowly.]

The data shown are in terms of percent of administered radioactive dose, which does not take into account the relative sizes of the pools labeled by the injected substituents. Indeed, Campbell et al. (1981) calculated that the general contribution of ceruloplasmin copper as a tissue source must be much greater than that of the nonceruloplasmin sources (Table 3-4). (The ceruloplasmin-copper pool within which the radiotracer

Table 3-3. Distribution of Radiocopper from Ceruloplasmin and Nonceruloplasmin (Non-Cp) Sources among Rat Tissues after Intravenous Injection

Tissue	^{67}Cu concentration (% of dose/g)		Percent of total uptake (for given tissues)	
	Ceruloplasmin	Non-Cp	Ceruloplasmin	Non-Cp
I. Campbell et al. (1981); 1 h				
Liver	6.6	9.4^a	74	64
Kidney	4.1	12.2^a	10	27^a
Heart	5.6^a	2.9	6^a	2
Spleen	8.6^a	4.9	9^a	5
Brain	0.5^a	0.3	2^a	0.9
II. Owen (1971)^b; 2 h				
Liver	0.5	2.4	8.5	42
Kidney	0.7	2.9	3.0	12
Heart	0.3	0.2	0.5	0.5
Skeletal muscle	0.05	0.08	7.3	13
Bone	0.2	0.3	16	23
Brain	0.001	0.02	0.007	0.14

^{a} Significantly higher value than for other copper source ($p<0.01$); no statistics available for the Owen data.
^{b} No statistics available for these data. Values for percent of total uptake estimated from values for % dose/g, assuming a body weight of 500 g and organ weights as 3.5, 0.8, 0.4, 30, 15, and 1.3% of body weight, from liver to brain, respectively.

is diluted is much greater.) The original calculations were done on the basis of the long-standing assumption that 90% of plasma copper is with ceruloplasmin. Recent work (Wirth and Linder, 1985; Barrow and Tanner, 1988) shows that only about 60% is actually with ceruloplasmin, so the data have also been calculated on the latter base (Table 3-4). Either way, the importance of ceruloplasmin remains clear.

Also of interest was the finding that pure ^{125}I- or ^{3}H-labeled ceruloplasmin (reflecting ceruloplasmin protein) was taken up by tissues but at a slower rate than ceruloplasmin copper (Campbell et al., 1981), suggesting that some of the copper uptake may involve endocytosis of the whole protein. However, it also implies that a major portion of uptake does *not* involve protein endocytosis. Other evidence supporting some internalization was provided by Linder and Moor (1977), who showed that a ceruloplasmin-like protein with *p*-phenylenediamine oxidase activity could be detected and partially purified from cytosol extract of heart and other tissues and, also, that residual blood plasma in the tissue could not account

Table 3-4. Total Uptake of Copper from Ceruloplasmin and Nonceruloplasmin Sources in the Rat, Based on Estimates of the Sizes of Plasma Copper Pools[a]

	Total copper uptake (μg) from:			
	Ceruloplasmin		Nonceruloplasmin	
	A	B	A	B
Normal rats	8.3	5.5	1.0	3.9
Tumor-bearing rats	22	14	1.7	9.9

[a] Data of Campbell et al. (1981) for 1-h uptake (% dose) per 173–188 g of body weight, using values of 1.2 and 2.2 μg of total Cu/ml plasma in the normal and tumor-bearing rats, respectively; plasma volumes of 10% body weight × (1-hematocrit) (as fraction); hematocrits of 0.50 and 0.35, respectively; and assumptions that ceruloplasmin Cu is 90% (A) or 60% (B) of the total in plasma, the latter based on Wirth and Linder (1985).

for the oxidase activity present. It cannot, of course, be said with full certainty that this ceruloplasmin is inside the cells, rather than, for example, tightly bound (and concentrated) at the cell surface (from which it may be removed during homogenization in isotonic saline), but this seems unlikely. Endogenous production (by these nonliver cells) is also unlikely (Aldred et al., 1987), although there are some preliminary data, using a ceruloplasmin cDNA probe, which indicate production of ceruloplasmin mRNA in heart and lung tissue during inflammation (J. D. Gitlin, personal communication) (see Chapters 4 and 9).

Circumventing to some extent the potential problem of plasma/blood contamination in tissues, Campbell et al. (1981) and, more recently, Weiss et al. (1985), Orena et al (1986), and Percival and Harris (1989) have shown that ceruloplasmin copper is taken up by cultured cells in serum-free medium (Table 3-2). Again, uptake at 37°C was much greater than at 0°C, implying that it involved more than just binding to the cell membrane. Even more persuasively, Harris and Dameron (1987; Dameron and Harris, 1987) have now reported that ceruloplasmin, labeled by in vitro addition of ^{67}Cu in the presence of ascorbate (known to involve the "blue" copper sites), was able to donate copper to the intracellular Cu/Zn enzyme, superoxide dismutase, precipitated with specific antiserum, during 24-h incubation of Cu-deficient chick aorta in serum-free medium. Uptake was temperature dependent and inhibited by nonradioactive ceruloplasmin but not by azide, dinitrophenol, or cyanide. Also, Percival and Harris (1989) showed temperature-dependent uptake by human K562 cells, using the same in vitro labeled ceruloplasmin. Uptake did not involve endocytosis, but transfer at the cell surface (Percival and Harris, 1990). In both studies, uptake was markedly enhanced by 10 μM L-ascorbate. This is not surprising, in that ascorbate is known to release copper from ceruloplasmin, via reduction, but it also suggests that a reduction step is involved in the transfer. These results must still be confirmed whith authentic ceruloplasmin, as Goode *et al.* (1989) have reported the *in vitro* labeled ceruloplasmin to be labile to Cu(II) exchange, in contrast to the *in vivo* labeled protein.

In the intervening years, further indirect evidence for ceruloplasmin as a copper source for cells was provided by Hsieh and Frieden (1975) and by Harris and DiSilvestro (1981), who showed, respectively, that intravenous ceruloplasmin restored liver cytochrome *c* oxidase activity more rapidly in copper-deficient rats than did ionic copper and that, in chicks recovering from copper deficiency, there was a parallelism between restoration of plasma ceruloplasmin levels (which had been markedly depressed) and connective tissue lysyl oxidase activity, as if restoration of copper to this latter enzyme had to await its arrival on ceruloplasmin. Marceau and Aspin (1973) also administered intravenous ^{67}Cu-ceruloplasmin and compared its tissue uptake and intracellular hepatic distribution with that of ionic ^{64}Cu(II) added to whole serum. Although uptake was observed from both sources, and apparent differences in intracellular distribution were observed, the times chosen for examination were such that few, if any, conclusions may be drawn: uptake of "albumin Cu" was examined after 10–45 min, and that of ceruloplasmin Cu after 20–138 h, in the latter case making any direct donor effect uncertain and preventing any comparisons with regard to source. The more recent work of Weiss *et al.* (1985), following the path of injected ^{67}Cu(II), also supports a role of ceruloplasmin, particularly in the distribution to nonhepatic, nonkidney tissues. Little or no uptake of ^{67}Cu was detected in heart, muscle, and brain at times before ^{67}Cu was found in blood ceruloplasmin. Taken all together, these various data make a good case for ceruloplasmin as a direct donor of copper to cells, for a variety of animal tissues.

Additional interest in ceruloplasmin as a copper donor has arisen from recent reports on the presence of specific receptors. Using ^{125}I-labeled ceruloplasmin from the chick, Stevens *et al.* (1984) provided evidence that plasma membrane fractions of chicken aorta and heart bind radioactive ceruloplasmin in a manner that can be competed for by a 500–600-fold excess of nonradioactive protein. Figure 3-9A shows the results for total (●), nonspecific (○), and "specific" binding (inset), the latter being defined as the difference between binding in the presence and absence of added, excess nonradioactive ligand. (This commonly accepted procedure makes a distinction between high-affinity or "specific" sites—which may be

Copper Uptake by Nongastrointestinal Vertebrate Cells

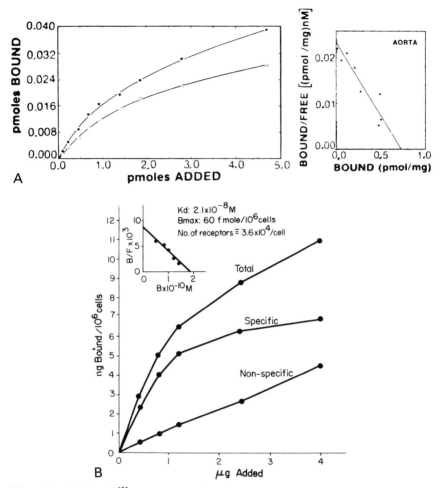

Figure 3-9. Binding of ^{125}I-ceruloplasmin to cell membranes in the presence and absence of excess nonradioactive ligand. (A) Data (reprinted with permission) of Stevens et al. (1984) for plasma membranes from chick aorta, showing amount of ^{125}I bound as a function of radioceruloplasmin concentration, in the presence (○) and absence (●) of a 500-molar excess of nonradioactive ceruloplasmin, and (inset) Scatchard analysis of data for specific binding. (B) Data (reprinted with permission) of Kataoka and Tavassoli (1985) for human blood lymphocytes.

competed for by the nonradioactive ligand,—and low-affinity or "nonspecific" sites—which, because of their abundance, will bind all other ligand, whether radioactive or not.) In the studies of Stevens et al. (1984), as indeed in most subsequent work, one sees that a significant portion of the binding is specific. Also, an excess of other unrelated proteins or asialoceruloplasmin does not appreciably affect radioceruloplasmin binding. On that basis, data for specific binding over a range of ^{125}I-ceruloplasmin concentrations were used to calculate an apparent dissociation constant of about $10^{-8}M$.

The more preliminary work of Barnes and Frieden (1984) with human ^{125}I-ceruloplasmin and human and bovine erythrocytes (again with 500–1000-fold excesses of "cold" ligand) yielded a similar K_D value, as did that of Kataoka and Tavassoli (1985) using various human white cell types (Fig. 3-9B), this time with only a 100-fold excess of cold ligand (the more usual procedure). (More recent studies by Saenko *et al.* (1990; Saenko and Yaropolov, 1990), as well as Stern and Frieden (1990) and Goode (C. A. Goode and A. Fischer, personal communication), confirm the existence of erythrocyte receptors and show a dependence of binding on Ca(II).) The reasons for the rather high nonspecific binding of ceruloplasmin, seen in two of three of these studies, are still unclear, although many possibilities may be advanced. These include the special lability of isolated ceruloplasmin and the likely existence of several forms containing different amounts of sialic acid as well as Cu (see more below).

A summary of all the receptor evidence in relation to cell type is given in Table 3-5 and includes mention of the electron microscope demonstration of gold-labeled ceruloplasmin bound to the surface of liver endothelial cells (Tavassoli, 1985). Together, these reports would seem to establish the presence of cell surface receptors that recognize the ceruloplasmin molecule. They also suggest that ceruloplasmin is, in general, somewhat sticky (lots of nonspecific binding; Fig. 3-9). If one assumes a K_D value of $10^{-8}M$ and an interstitial fluid ceruloplasmin concentration of about $10^{-6}M$ (about 2.2 mM in blood plasma), it would seem that ceruloplasmin receptors on the various cell surfaces should be under more or less constant full occupation. (This is even more likely to be true for diferric transferrin, present in blood at concentrations of about 7 mM, the receptors for which have a K_D of $10^{-9}M$.) Whether this in all cases is related to the delivery of copper (or iron) to those cells is another question, not yet sufficiently studied. For example, in the case of blood cells, especially erythrocytes, binding of ceruloplasmin to the cell surface may be important for protection of the cell membrane (see Chapter 4, Section 4.1.1), whereas, in other tissues, membrane binding might be important both for protection and for Cu transport. Indeed, Lovstad has reported that ceruloplasmin (and albumin) protects erythrocytes against lysis by copper (Lovstad, 1982; Caffrey, *et al.*, 1990) and iron ions (Lovstad, 1981, 1983) and by hemin (Lovstad, 1986). Preliminary information from Vassos and Newman (1988) suggests that the M_r of ceruloplasmin receptor subunits may be 78,000 and 84,000, at least in

Table 3-5. Summary of Evidence Relating to Specific Ceruloplasmin (Cp) Receptors in Various Mammalian Cells

Tissue	System	Specific binding	Apparent dissociation constant (M)	Reference
Aorta (chick)	Plasma membranes, ^{125}I-Cp, ±Cp	Yes	10^{-8}	Stevens et al. (1984)
Heart (chick)	Plasma membranes, ^{125}I-Cp, ±Cp	Yes	10^{-8}	Stevens et al. (1984)
Heart (rat)	Microsomes, ^{67}Cu-Cp, ±Cp	Yes	—	Orena et al. (1986)
	Microsomes, ^{67}Cu-Cp, ±Cu(II)	Yes	10^{-7}	Orena et al. (1986)
Brain (rat)	Microsomes, ^{67}Cu-Cp, ±Cp	Yes	—	Orena et al. (1986)
	Microsomes, ^{67}Cu-Cp, ±Cu(II)	Yes	10^{-7}	Orena et al. (1986)
Liver (rat)	Microsomes, ^{67}Cu-Cp, ±Cp	Yes[a]	—	Orena et al. (1986)
	Microsomes, ^{67}Cu-Cp, ±Cu(II)	Yes[a]	—	Orena et al. (1986)
Liver (rat)	Hepatocytes, ^{125}I-Cp, ±Cp	Very little, if any	—	Tavassoli et al. (1986)
	Hepatocytes, ^{125}I-Cp, ±asialo-Cp	Yes	—	Tavassoli (1985)
	Kupffer cells, ^{125}I-Cp, ±Cp	Yes[b]	—	Tavassoli (1985)
	Endothelial cells, ^{125}I-Cp, ±Cp	Yes	10^{-7}	Tavassoli (1985)
Leukocytes (human)	Granulocytes, ^{125}I-Cp, ±Cp	Yes	10^{-8}	Kataoka and Tavassoli (1985)
	Monocytes, ^{125}I-Cp, ±Cp	Yes	10^{-8}	Kataoka and Tavassoli (1985)
	Lymphocytes, ^{125}I-Cp, ±Cp	Yes	10^{-8}	Kataoka and Tavassoli (1985)
Erythrocytes (human)	Whole cells, ^{125}I-Cp, ±Cp	Yes	10^{-8}	Barnes and Frieden (1984); Saenko et al. (1990); Saenko and Yaropolov (1990)

[a] Less capacity than in heart and brain.
[b] Less capacity than in liver endothelial cells.

hematopoietic cells in culture, and that the receptor contains carbohydrate, while reports from Omoto and Tavassoli (1989, 1990) indicate that the receptor on rat liver endothelial cells (where ceruloplasmin may be desialylated) is comprised of a single subunit type of about 35,000 Da., with a 3% carbohydrate content and a pI of 5.2.

Returning to the matter of cell type, we have already cited *in vivo* evidence for the apparent preference of nonhepatic, nonkidney cells for ceruloplasmin over nonceruloplasmin copper. This is consistent with reports that ceruloplasmin receptors are present on heart, aorta, and various blood cells (Table 3-5). Moreover, it fits with recent histological/histochemical work of Tavassoli (1985; Tavassoli *et al.*, 1986) in which hepatic receptors for nondesialylated ceruloplasmin are reported to be confined to the endothelial cells lining the liver sinusoids, whereas asialoceruloplasmin binds to hepatocytes via the galactose receptor that can be competed for by other asialo proteins. This work also appears to indicate that ceruloplasmin is endocytosed by the endothelial cells and regurgitated (without sialic acid) in the space of Disse, thus providing a means for the return of ceruloplasmin copper to the liver, for excretion in the bile. (Further discussions of these findings will be found in Chapter 5 on copper excretion.) It also confirms previous evidence that some ceruloplasmin, and probably most asialoceruloplasmin (containing some copper), is taken up (internalized) by liver cells (also discussed in Chapter 5).

A further twist to the matter of ceruloplasmin receptors and copper transport is provided by the studies of Orena *et al.* (1986), in which ceruloplasmin labeled with ^{67}Cu (*in vivo*) was used as a probe to study its binding (and that of its copper) to membranes from several rat tissues. This was studied in the presence and absence of a 100–150-fold excess of nonradioactive ceruloplasmin or of a 3300-fold molar excess (versus ceruloplasmin protein) of Cu(II), in the form of a nitrilotriacetate (NTA) complex. As shown in Table 3-6, total binding of ^{67}Cu-ceruloplasmin to microsomal membranes from heart and brain was three- to six-fold greater than binding to microsomal membranes of liver and kidney. This is consistent with the conclusions of other workers that ceruloplasmin is the major copper source for nonliver, nonkidney cells but that nonceruloplasmin copper is the main usual source for liver and kidney (see Chapter 4). In the presence of excess "cold" ceruloplasmin (Table 3-6), ^{67}Cu binding was cut in half in heart and brain and also somewhat reduced in liver, indicating that specific, high-affinity binding of ceruloplasmin, involving the protein and/or carbohydrate moieties, was occurring. An unexpected additional finding was that Cu(II) itself was able to prevent the binding of a substantial portion of the ^{67}Cu from ceruloplasmin. The effect was at least partly additive with that of "cold" excess ceruloplasmin, imply-

Table 3-6. Binding of ^{67}Cu from Ceruloplasmin to
Various Rat Tissue Membranes[a,b]

Membrane source	Total binding (pmol)			Percent binding in the presence of excess nonradioactive		
	Per mg of protein	Per g of tissue	Corrected for 5'-nucleotidase	Ceruloplasmin	Cu-NTA	Both
Heart	3.15 ± 0.60 (17)	76.4 ± 9.3 (5)	50	56 ± 7 (15)	44 ± 11 (17)	20 ± 3 (9)
Brain	7.71 ± 1.50 (6)	65.6 ± 2.9 (3)	118	54 ± 15 (6)	37 ± 9 (6)	31, 32 (32)
Liver	1.16 ± 0.16 (8)	16.7 ± 3.0 (4)	25	70 ± 8 (9)	72 ± 15 (8)	49 ± 4 (4)
Kidney	—	16.5, 13.1 (2)	—	—	116 (1)	—

[a] Reprinted, with permission, from Orena et al. (1986).
[b] Membranes (0.56–2.64 mg of protein) from heart, brain, or liver were incubated with 48–50 pmol of ^{67}Cu-ceruloplasmin alone or with nonradioactive ceruloplasmin (100–150-fold excess) or Cu-NTA (3300-fold excess) or both, for 30–90 min of 25°C. Values are means ± SD (N) for pmoles of ^{67}Cu-ceruloplasmin bound/mg membrane protein or % total binding.

ing two sites of competition for ^{67}Cu binding. Addition of Cu-NTA or NTA alone to ^{67}Cu-ceruloplasmin, in the absence of cell membranes, did not result in exchange of ^{67}Cu(II) between ceruloplasmin and NTA (Orena et al., 1986) (as expected from earlier work on ceruloplasmin). Therefore, the result implies that binding to a protein-recognizing site on the membranes allows release of ^{67}Cu(II) from ceruloplasmin to a site from which it can be displaced (or blocked directly) by ionic Cu(II). The specificity of this effect was further substantiated by showing no specific (Cu-NTA competable) binding to membranes that had been preboiled and no effect of excess Zn(II)-NTA rather than Cu-NTA. Indeed, Linder and Goode (1988; Goode et al., 1989) have now shown that ^{67}Cu can be released from ceruloplasmin into ultrafiltrates in the presence, but not in the absence, of plasma membrane, at 0°C. (Release of freed ^{67}Cu from the copper receptor was affected by addition of excess "cold" Cu-NTA, which does not release ^{67}Cu from ceruloplasmin.)

As with nonradioactive ceruloplasmin, addition of nonradioactive copper (as Cu-NTA) reduced total binding of ^{67}Cu to a low plateau (Fig. 3-10A). [The same occurred when Cu-NTA was added after membrane receptors had first been "loaded" with ^{67}Cu-ceruloplasmin (Linder and Goode, 1988).] Based on the data with and without Cu-NTA, the binding of ceruloplasmin (labeled with ^{67}Cu) approached saturation and was half-maximal at about 90 nM. Scatchard analysis (Fig. 3-10B) gave evidence of positive cooperativity and an apparent K_D of 150 nM. These K_D values are most likely too high, as the (unknown) concentration of the receptor–ceruloplasmin complex (which would be quite low), rather than

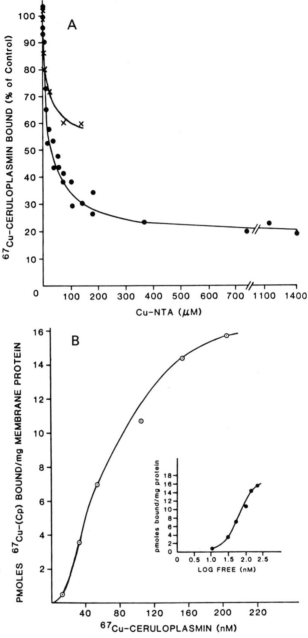

Figure 3-10. Binding kinetics of ^{67}Cu from ceruloplasmin to brain membranes in the presence and absence of excess Cu(II)-NTA. (A) Effect of increasing Cu-NTA concentrations on binding. Data for heart membranes are also shown (X). (B) Specific binding of ^{67}Cu from ceruloplasmin at varying ^{67}Cu-ceruloplasmin concentrations. Reprinted, with permission, from Orena *et al.* (1986).

that of free ceruloplasmin copper in solution, should be used for the calculation. Another uncertainty (working in the opposite direction) is the number of copper atoms labeled with ^{67}Cu and the number transferred (per ceruloplasmin molecule). In any event, these findings imply that there is not just a receptor that recognizes and binds the protein moiety of ceruloplasmin, but also one that receives ceruloplasmin copper released during interaction with the membrane.

The site receiving copper from ceruloplasmin appears to be involved in copper transport, since uptake of ceruloplasmin ^{67}Cu (like specific membrane binding) was blocked by the addition of an excess of Cu(II) (as the NTA complex) in the case of the cultured Chinese hamster ovary (CHO) cells examined (Orena et al., 1986). What all this also indicates is that copper is released from ceruloplasmin at the cell surface and that uptake occurs (at least primarily) by a mechanism not involving endocytosis of ceruloplasmin.

Summarizing the evidence: (1) Ceruloplasmin and ceruloplasmin-copper receptors are found on the cell surface and in isolated plasma membrane preparations; (2) in these membranes, copper can be released from ceruloplasmin to a copper binding site; (3) externally administered Cu(II)-NTA blocks uptake of ceruloplasmin Cu by cultured cells; (4) monensin, an inhibitor of receptor recycling (Tartakoff and Vassalli, 1978), has no effect on copper uptake from ceruloplasmin in cultured cells (Orena et al., 1986); and (5) metabolic energy is probably not involved (Harris and Dameron, 1987). (Endocytosis requires energy.) Finally, the very recent work of Harris and Percival (1988) with cultured K562 cells demonstrates that the ^{67}Cu from ceruloplasmin is taken up much more rapidly than the protein itself (labeled with ^{125}I), which may not be taken up at all by these hematopoietic tumor cells.

It is uncertain whether the ceruloplasmin-copper binding site on the cell surface is the same one discussed previously as a transporter for ionic copper (Schmitt et al., 1983; Darwish et al., 1984b; McArdle et al., 1987a, b; Weiner and Cousins, 1980), although this remains an intriguing possibility. Two factors that may not fit are the lack of competition of Zn(II) and the much lower K_m in the case of the receptor for ceruloplasmin Cu. Whether the release of copper from ceruloplasmin at the cell membrane also involves reduction and reoxidation, as previously suggested by Frieden (1980) in a review on ceruloplasmin, and by analogy with the work of Morley and Bezkorovainy (1985) on release of Fe(III) from transferrin in the plasma membrane, must still be examined, although Harris and Percival (1988 and 1990) have reported that ascorbate enhances uptake; and N_3^- pretreatment of ceruloplasmin (which would bind to the "blue" type 1 copper) prevents cellular uptake from ceruloplasmin.

Figure 3-11 summarizes what might be the mechanism(s) by which ceruloplasmin delivers copper to cells, based on what has been discussed. Ceruloplasmin attaches to specific cell surface receptor proteins on most cells, the abundance of which can vary dramatically with cell type. This receptor is capable of inducing a conformational change in ceruloplasmin, possibly involving reduction and reoxidation of the copper, resulting in release of one or more copper atoms previously tightly bound and buried in the protein. In most cases, an adjacent protein (that could also be considered a subunit of the ceruloplasmin receptor) receives the released copper [in the form of Cu(II)] and feeds it into the cell. This mechanism

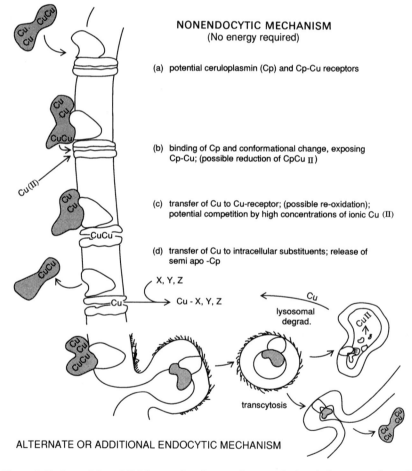

Figure 3-11. Potential model(s) for uptake of copper from ceruloplasmin by mammalian cells.

would account for most of the observations cited but not for the internalization of ceruloplasmin protein, which appears to occur, but at a slower rate than uptake of ceruloplasmin copper, at least in some cells (Campbell *et al.*, 1981). To account for the endocytosis of some ceruloplasmin protein, one could postulate that there are (in some or most cells) other receptors, or the same kinds of receptors but lacking the copper receptor, that gather in clathrin-coated pits and are endocytosed, with degradation of ceruloplasmin and concurrent intracellular copper release. In some special cases, such as liver endothelial cells (Tavassoli *et al.*, 1986) and perhaps the blood–brain barrier, "transcytosis" across two cell membranes may also take place, releasing the protein on the other side. (During this process, some chemical modifications may occur, such as desialylation in the case of the liver.)

Extracellular Copper Substituents and Mammalian Copper Transport

4.1 Components of the Blood Plasma and Other Extracellular Fluids Associated with Copper

At present, there are at least five major components of the blood plasma with which we know copper to be associated. These are ceruloplasmin, a fairly abundant α_2-glycoprotein, long recognized as containing the largest proportion of copper in most animal and human plasma; albumin, containing (except in the dog, pig, and chicken) a single high-affinity Cu-binding site at the N-terminus; two larger, nonceruloplasmin, nonalbumin proteins, one called ferroxidase II, and the other as yet less defined and named transcuprein; and a fraction of the plasma comprising small molecules (especially amino acids and small peptides), among which histidine may be the most important ligand. Other potential copper ligands include a histidine-rich glycoprotein and α_2-macroglobulin, both of which (like albumin) are also associated with zinc. It is thought that most of these components are present not just in plasma but also in interstitial fluid and in other fluids involved with copper transport, such as the cerebrospinal fluid in the brain (on the other side of the blood–brain barrier) and fluids in specialized areas (testes, amniotic fluid). This chapter reviews what is known about these components, their structure, properties, and potential function in copper transport.

4.1.1 Ceruloplasmin

Ceruloplasmin (also known as ferroxidase I) has been recognized as the major copper-containing component in mammalian plasma/serum since its isolation by Holmberg and Laurell in the 1940s (Holmberg and Laurell 1947, 1948). This blue α_2-glycoprotein contains multiple copper atoms that are not directly available to binding substances, or exchangeable with

↦ **67KDal**

1
Lys-Glu-Lys-His-Tyr-Tyr-Ile-Glu-Thr-Thr-Trp-Asp-Thr-Ala-Ser-Asp-His-Ser-Ile-Ser-Val-Asp-Thr-Glu-Lys-Lys-Leu-Ile-Gly-His-Ser-Asn-Tyr-Leu-Gln-Asn-Gly-
            20   40

Pro-Asp-Arg-Ile-Gly-Arg-Leu-Tyr-Lys-Lys-Ala-Leu-Tyr-Val-His-Leu-Gln-Tyr-Asp-Glu-Thr-Phe-Arg-Thr-Ile-Glu-Lys-Pro-Val-Trp-Leu-Gly-Phe-Leu-Gly-Ile-Ile-Ala-
            60   ★ 80 (GlcN)

Glu-Thr-Gly-Asp-Lys-Val-Val-His-Leu-Lys-Asn-Leu-Ala-Ser-Arg-Pro-Tyr-Thr-Phe-His-Ser-Gly-Ile-Thr-Tyr-Lys-Gly-Ala-Ile-Tyr-Pro-Asp-Asn-Thr-
            100   120

Thr-Asp-Phe-Gln-Arg-Ala-Asp-Asp-Lys-Val-Tyr-Pro-Gly-Glu-Gln-Thr-Tyr-Met-Leu-Ala-Thr-Glu-Gln-Ser-Pro-Gly-Glu-Gly-Asp-Gly-Asn-Cys-Val-Thr-Arg-Ile-Tyr-
            140   160

His-Ser-His-Ile-Asp-Ala-Pro-Leu-Ile-Gly-Leu-Ile-Cys-Lys-Lys-Asp-Ser-Leu-Ala-Lys-Glu-Lys-His-Ile-Asp-Arg-Glu-Phe-Val-Val-
            180   200

Met-Phe-Ser-Val-Val-Asp-Glu-Asn-Phe-Ser-Trp-Tyr-Leu-Glu-Asp-Asn-Ile-Lys-Thr-Tyr-Cys-Ser-Glu-Pro-Glu-Lys-Val-Asp-Lys-Asn-Gly-Asp-Phe-Gln-Ser-Asn-Arg-Met-
            220   240

Tyr-Ser-Val-Asn-Gly-Tyr-Thr-Phe-Gly-Ser-Leu-Pro-Gly-Leu-Met-Cys-Ala-Glu-Asp-Arg-Val-Leu-Trp-Tyr-Leu-Phe-Gly-Met-Gly-Asn-Glu-Val-Asp-Val-His-Ala-Phe-Phe-
            260   280

His-Gly-Gln-Ala-Leu-Thr-Asn-Lys-Asn-Tyr-Arg-Ile-Asp-Thr-Ile-Asn-Leu-Phe-Pro-Ala-Thr-Leu-Phe-Asp-Ala-Tyr-Met-Val-Ala-Gln-Asn-Asp-Pro-Gly-Leu-Trp-Met-Leu-Ser-Cys-Gln-
            300   320

Asn-Leu-Asn-His-Leu-Lys-Ala-Gly-Leu-Gln-Ala-Phe-Phe-Gln-Val-Gln-Gln-Cys-Asn-Lys-Ser-Ser-Ser-Lys-Asn-Ile-Ala-Gln-Asn-Pro-Gly-Leu-Tyr-Met-Val-Arg-His-Tyr-Ile-Ala-Glu-
           (GlcN)   360

Glu-Ile-Trp-Asn-Tyr-Ala-Pro-Ser-Gly-Ile-Phe-Lys-Asn-Leu-Thr-Ala-Pro-Gly-Asn-Ala-Pro-Gly-Glu-Gly-Thr-Gly-Thr-Arg-Gly-Ile-Gly-Ser-
           (GlcN) 380   400

Tyr-Lys-Lys-Leu-Val-Tyr-Arg-Gly-Tyr-Thr-Ala-Ser-Phe-Thr-Asn-Arg-Lys-Gly-Arg-Gly-Pro-Gly-Glu-His-Leu-Gly-Pro-Val-Ile-Trp-Ala-Gly-Ile-Leu-Gly-Ser-
            420   440

Thr-Ile-Arg-Val-Thr-Phe-His-Asn-Lys-Gly-Ala-Tyr-Pro-Leu-Ser-Ile-Glu-Gly-Val-Arg-Phe-Asn-Asn-Glu-Gly-Thr-Tyr-Tyr-Ser-Pro-Asn-Tyr-Asn-Pro-Gln-Ser-
  ★   460   480

↦ **50KDal**   **67KDal** ↤

Arg-Ser-Ala-Val-Pro-Pro-Ser-Ala-Ser-His-Val-Ala-Pro-Thr-Glu-Phe-Thr-Phe-Val-Pro-Thr-Val-Pro-Gly-Val-Gly-Val-Pro-Thr-Asn-Ala-Asp-Pro-Val-Cys-Leu-Ala-Lys-Met-Tyr-
1   20   520

Tyr-Ser-Ala-Val-Asp-Pro-Thr-Lys-Ile-Gly-Phe-Thr-Gly-Leu-Ile-Gly-Pro-Met-Lys-Ile-Cys-Lys-Lys-Gly-Leu-His-Ala-Asn-Gly-Arg-Gln-Lys-Ser-Leu-Lys-Gly-Lys-Asp-Val-Asp-Phe-Tyr-
40   60   560

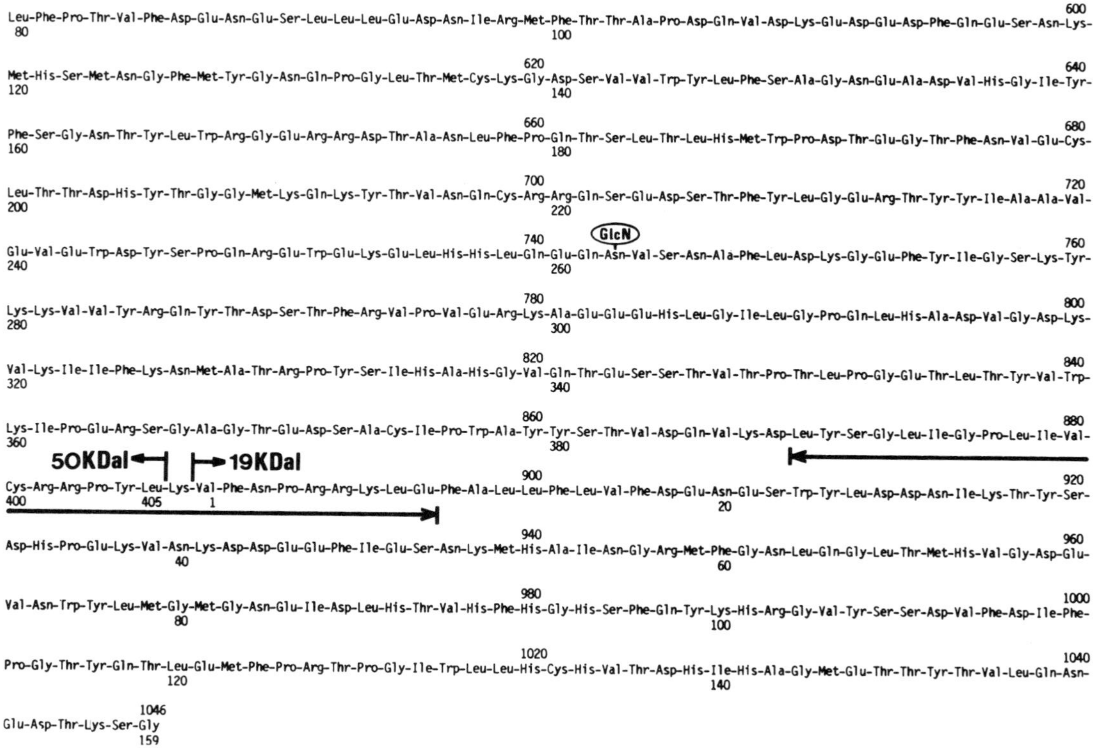

Extracellular Copper Substituents and Mammalian Copper Transport

Leu-Pro-Thr-Val-Phe-Asp-Glu-Asn-Gly-Ser-Leu-Leu-Glu-Asp-Asn-Ile-Arg-Met-Phe-Thr-Thr-Ala-Pro-Asp-Gln-Val-Asp-Lys-Glu-Asp-Glu-Asp-Phe-Gln-Ser-Asn-Lys-
80   100   600

Met-His-Ser-Met-Asn-Gly-Phe-Met-Tyr-Gly-Asn-Gln-Pro-Gly-Leu-Thr-Met-Cys-Lys-Gly-Asp-Ser-Val-Val-Trp-Tyr-Leu-Phe-Ser-Ala-Gly-Asn-Glu-Ala-Asp-Val-His-Gly-Ile-Tyr-
120   140   640

Phe-Ser-Gly-Asn-Thr-Tyr-Leu-Trp-Arg-Gly-Glu-Arg-Arg-Asp-Thr-Ala-Asn-Leu-Phe-Pro-Gln-Thr-Ser-Leu-Thr-Leu-His-Met-Trp-Pro-Asp-Thr-Glu-Gly-Thr-Phe-Asn-Val-Glu-Cys-
160   180   680

Leu-Thr-Thr-Asp-His-Tyr-Thr-Gly-Gly-Met-Lys-Gln-Lys-Tyr-Thr-Val-Asn-Gln-Cys-Arg-Arg-Gln-Ser-Glu-Asp-Ser-Thr-Phe-Tyr-Leu-Gly-Glu-Arg-Thr-Tyr-Tyr-Ile-Ala-Ala-Val-
200   220   720

Glu-Val-Glu-Trp-Asp-Tyr-Ser-Pro-Gln-Arg-Trp-Pro-Arg-Glu-Tyr-His-Glu-His-Leu-Gln-Leu-Gln-Asn-Val-Ser-Asn-Ala-Phe-Leu-Asp-Lys-Gly-Glu-Phe-Tyr-Ile-Gly-Ser-Lys-Tyr-
240   260   760

Lys-Lys-Val-Val-Tyr-Arg-Gln-Tyr-Thr-Asp-Ser-Thr-Phe-Arg-Val-Pro-Val-Glu-Arg-Lys-Ala-Glu-Glu-His-Leu-Gly-Ile-Leu-Gly-Pro-Gln-Leu-His-Ala-Asp-Val-Gly-Asp-Lys-
280   300   800

Val-Lys-Ile-Ile-Phe-Lys-Asn-Met-Ala-Thr-Arg-Pro-Tyr-Ser-Ile-His-Ala-His-Gly-Val-Gln-Thr-Glu-Ser-Ser-Thr-Val-Thr-Pro-Thr-Leu-Pro-Gly-Glu-Thr-Leu-Tyr-Val-Trp-
320   340   840

Lys-Ile-Pro-Glu-Arg-Ser-Gly-Ala-Gly-Thr-Glu-Asp-Ser-Ala-Cys-Ile-Pro-Trp-Ala-Tyr-Tyr-Ser-Thr-Val-Asp-Gln-Val-Lys-Asp-Leu-Tyr-Ser-Gly-Leu-Ile-Gly-Pro-Leu-Ile-Val-
360   380   880

50KDal◄─┐ ┌─►**19KDal**
Cys-Arg-Arg-Pro-Tyr-Leu-Lys-Val-Phe-Asn-Pro-Arg-Arg-Lys-Leu-Glu-Phe-Ala-Leu-Phe-Asp-Glu-Asn-Glu-Ser-Trp-Tyr-Leu-Asp-Asn-Ile-Lys-Thr-Ser-
400   405 1   20   920

Asp-His-Pro-Glu-Lys-Val-Asn-Lys-Asp-Asp-Glu-Phe-Ile-Glu-Ser-Asn-Lys-Met-His-Ala-Ile-Asn-Gly-Arg-Met-Phe-Gly-Asn-Leu-Gln-Gly-Leu-Thr-Met-His-Val-Gly-Asp-Glu-
                              40   60   960

Val-Asn-Trp-Tyr-Leu-Met-Gly-Met-Gly-Asn-Glu-Ile-Asp-Leu-His-Thr-Val-His-Phe-His-Gly-His-Ser-Phe-Gln-Tyr-Lys-His-Arg-Gly-Val-Tyr-Ser-Ser-Asp-Val-Phe-Asp-Ile-Phe-
                         80   100   1000

Pro-Gly-Thr-Tyr-Gln-Thr-Leu-Glu-Met-Phe-Pro-Arg-Thr-Pro-Gly-Ile-Trp-Leu-Leu-His-Cys-His-Val-Thr-Asp-His-Ile-His-Ala-Gly-Met-Glu-Thr-Tyr-Thr-Val-Leu-Gln-Asn-
                      120   140   1040

Glu-Asp-Thr-Lys-Ser-Gly
                    1046
                    159

Figure 4-1. Complete amino acid sequence of human ceruloplasmin. Sequences involving the 67-, 50-, and 19-kDa fragments due to plasmin are indicated, along with four asparagines thought to anchor the glycan chains (GlcN) and locations of amino acid substitutions (marked with asterisks). Reprinted, with permission, from Takahashi *et al.* (1984).

copper ions, in the mammalian plasma environment: already Holmberg and Laurell (1947) found that the copper was "not dialyzable", which has many times since been confirmed; it was not available to diethyldithiocarbamate, added under endogenous (plasma) conditions (Holmberg and Laurell, 1947; Gubler et al., 1953; Jackson 1978); radiocopper on ceruloplasmin is not exchangeable with Cu(II) in serum (Gubler et al., 1953; Sternlieb et al., 1961), or in phosphate pH 7.0 (Orena et al., 1986), and so on. There has also been a notable lack of success in removing most of the copper from ceruloplasmin or reconstituting it after copper removal, without irreversible denaturation of the protein (see more below). These observations imply that ceruloplasmin copper is deeply buried within the tertiary structure of the protein and fueled the early view (already discussed in Chapter 3) that ceruloplasmin is not involved in copper transport. This view now seems untenable (see Chapter 3). However, the likelihood that ceruloplasmin is involved in other kinds of activities cannot be dismissed, either. From the current vantage, ceruloplasmin must continue to be viewed as a multifuncional protein, as stressed by Frieden already some time ago (see Frieden, 1981). Reviews on ceruloplasmin structure and function include those of Frieden and Hsieh (1976), Frieden (1979, 1981), Goldstein and Charo (1982), Fee (1975), and Scheinberg and Morell (1973).

Ceruloplasmin is composed of a single polypeptide chain to which several oligosaccharide units are attached. The total amino acid sequence of the major form of human ceruloplasmin has been known for some time (Fig. 4-1) and consists of 1046 amino acids, for a total molecular weight of 120,085 (Takahashi et al., 1984). Within the molecule, there are three highly homologous units (of about 350 amino acids), each of which can be subdivided into domains, with differing structures (A-1, A-2, and B) (Ortel et al., 1983). The molecule also has little α-helix but a great deal of β-sheet and unordered secondary structure (Noyer and Putnam, 1981). Based on primary and secondary structural considerations and hydrophilicity/hydrophobicity, a molecular model of the overall structure of human ceruloplasmin has been proposed by Putnam and co-workers (Fig. 4-2) (Ortel et al., 1983). Only two apparent sites for amino acid interchange (Gly/Lys) were identified (Takahashi et al., 1987), although there are at least five genetically determined forms encountered in humans at or near polymorphic frequencies (Morell and Scheinberg, 1960; Shreffler et al., 1967; Poulik and Wiess, 1975). The structural basis for the latter heterogeneity is still unknown, although (apart from primary structure) there are several possibilities, including different numbers of glycan units and copper atoms (see below). [Strawinski et al. (1979) also report polymorphism in lowland cattle.]

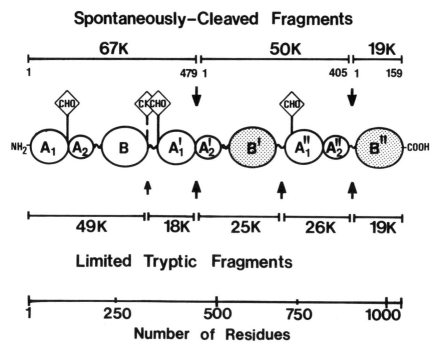

Figure 4-2. Model of the structure of human ceruloplasmin deduced from the primary structure. Shaded Bs indicate the most likely location of the type 1 Cu atoms. B" is thought to contain three Cu atoms. Reproduced with permission from Ortel et al. (1983).

The gene for ceruloplasmin has been assigned to human chromosome 3, arm q (Weitkamp, 1983; Naylor et al., 1985), in the region 21 to 24 (Royle et al., 1987), about 10 centimorgans from the gene for transferrin (Bowman and Yang, 1987). Genes for the rat (Aldred et al., 1987), mouse (J. Mercer and D. Danks, personal communication), and human proteins (Bowman et al., 1985; Mercer and Grimes, 1986; Kunapuli et al., 1986; Koschinsky, 1988; Shvartsman et al., 1990) have been cloned and have been, or are currently being, sequenced. A nonfunctional human pseudogene (Koschinsky et al., 1987) has been assigned to chromosome 8 (21.13–23.1) by in situ hybridization (Wang et al., 1988). The functional human gene appears to have nucleotides for a 19-amino acid signal sequence (Met-Lys-Ile-Leu-Ile-Leu-Gly-Ile-Phe-Leu-Cys-Ser-Thr-Pro-Ala-Tyr-Ala) (Koschinsky et al., 1986). An additional (200 kDa) form of human ceruloplasmin has also been reported (Sato et al., 1990) in normal adult, normal newborn, and Wilson's disease patient sera. This may be explained on the basis that the ceruloplasmin mRNA in human liver comes in two sizes (about 3.7 and 4.2 kb). Both are expressed by the liver, but only the smaller one is expressed by hepatoma cells (Koschinsky et al., 1986). [Messenger RNA of only the smaller size is found in liver of rats, hamsters, and mice (Gitlin, 1988a).] Yang et al., (1990) have obtained evidence that both forms of human ceruloplasmin are made from the same gene transcript by alternative splicing. Ceruloplasmin is expressed primarily, but not exclusively, in liver, from which it is secreted into the plasma (see also Chapter 7). Other tissues expressing it are brain (especially the choroid plexus), testis (especially Sertoli cells), yolk sac, and uterus (Aldred et al., 1987), and apparently also peripheral blood monocytes, macrophages and fetal lung (Gitlin, 1988b; personal communication) [and also a few cancer cell lines, especially hepatomas (see Chapter 9)]. There is a recent Russian report that the protein is expressed in, but not secreted from, the heart and kidney in rats (Gaitskhoki et al., 1990), which does not agree with findings of some others (Aldred et al., 1987; Fleming and Gitlin, 1990).

At least three glycan chains ending in sialic acid (Yamashita et al., 1981; Endo et al., 1982), two biantennary, plus one or more triantennary, are attached to the N-terminal 67,000-Da portion of human ceruloplasmin, via asparagine residues numbers 119, 339, and 378 and 743 (Fig. 4-1) (Takahashi et al., 1984). Their structure and sequence is given in Fig. 4-3. There is nothing particularly remarkable about these glycans. They begin with the usual branched N-acetylglucosamine plus mannose backbone, and

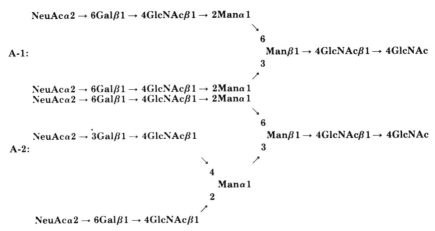

Figure 4-3. Structure of glycan chains of human ceruloplasmin. These structures agree with those obtained by Endo et al. (1982), for human ceruloplasmin and transferrin, although transferrin also contains quadriantennary chains. Abbreviations: NeuAc, N-acetylneuraminic acid; Gal, galactose; GlcNAc, N-acetylglucosamine; Man, mannose. Reprinted, with permission, from Yamashita et al. (1981).

there are mannose, N-acetylglucosamine, and galactose residues (working away from the protein) before the N-acetylneuraminic (sialic) acid units at the ends. At least for humans, they appear to be identical to the oligosaccharides of transferrin (involved in transport of Fe and some other metals) (Yamashita et al., 1981; Endo et al., 1982; Irie and Tavassoli, 1988). One or two additional (similar) glycan chains are usually attached elsewhere on the molecule (Yamashita et al., 1981), and the average total number is probably four, contributing 12,000 Da to a total molecular weight of 132,000 for ceruloplasmin (Takahashi et al., 1984). Fucose may also be present in some units, contributing to microheterogeneity (Kolberg et al., 1983).

Based on the behavior of pure human and rat ceruloplasmins in nondenaturing gel electrophoresis (M. C. Linder, J. H. Dawson, H. B. Gray, unpublished results; D. Baer, C. A. Goode, and M. C. Linder, unpublished results) (Fig. 4-4) and in lectin affinity chromatography (Irie and Tavassoli, 1988) (although this study was done with commercial ceruloplasmin not characterized as to integrity), there appears to be considerable microheterogeneity. This may be ascribable to variations in sialylation of the glycan chains and/or variations in the number of bi-and triantennary glycan units on the protein. Variations in copper content are also implicated (Fig. 4-4). Whether variations in glycosylation will explain the apparent heterogeneity of total molecular weight, apparent in sodium dodecyl sulfate (SDS) gel electrophoresis, is still unclear. DiSilvestro and Harris (1985), for example, showed that chicken ceruloplasmin contained four somewhat faster moving bands in addition to the major component of M_r 120,000. Similar results have been obtained for pure human

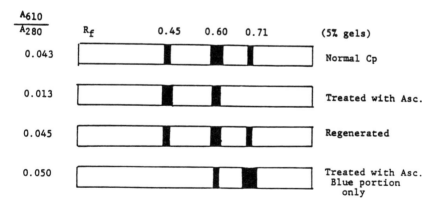

Figure 4-4. Behavior of pure human ceruloplasmin, before and after ascorbate treatment, in nondenaturing electrophoresis (M. C. Linder, J. Dawson, and H. B. Gray, unpublished results). All three bands had M_r 132,000, 124,000, and 111,000 in SDS electrophoresis.

ceruloplasmin (M. C. Linder, J. Dawson and H. B. Gray, unpublished results).

Crystallization and X-ray diffraction studies of human ceruloplasmin have been initiated by Russian investigators (Oimatov et al., 1986), who reported trigonal crystals to have 3, 2, and 32 symmetry (at 6-Å resolution) and give overall (very asymmetric) ceruloplasmin dimensions of 214 by 85 Å (length by width). This is consistent with the hydrodynamic behavior of ceruloplasmin as a molecule of M_r 150,000 rather than 132,000 and the axial ratio of 3.6 determined by sedimentation (see Fee, 1975). Indeed, until the careful X-ray and sedimentation equilibrium work of Magdoff-Fairchild et al. (1969), which established a molecular weight of 132,000 (now confirmed by the amino acid and carbohydrate sequence work), it was considered that ceruloplasmin had a molecular weight of 150,000, with eight Cu atoms (in analogy to laccase, which has four Cu atoms; M_r 58,000) (see more below). The confusion about molecular weight contributed to the still existing confusion about the exact copper content of the ceruloplasmin molecule. Also contributing has been the lability of some of the peptide bonds within the structure, at least two of which are readily hydrolyzed by plasmin (Ryden, 1971a,b). Plasmin is a serine protease activated by proteolysis during blood clotting. This protease (and other serine proteases) is inhibited by ε-aminocaproic (ε-amino hexanoic) acid, now routinely added to plasma and solutions during ceruloplasmin isolation (Ryden, 1971a,b; Kingston et al., 1977; Moshkov et al., 1979). Other proteases may also be involved in the scission of up to five labile bonds (Prozorovski et al., 1982). This lability explains earlier reports of ceruloplasmin "subunits" (Kasper and Deutsch, 1963a; Poulik, 1962; Poillon and Bearn, 1966; Freeman and Daniel, 1973), of 65–67, 50–55, and 15–18 kDa (Takahashi et al., 1984).

Much of the information on the structure of ceruloplasmins is summarized in Table 4-1. The bovine protein may be quite similar to the human one. The rat protein (data not shown) appears to be less homologous, as is emerging from sequencing of the cDNA clones (Aldred et al., 1987; Fleming and Gitlin, 1990) and from its observed lack of reactivity to some polyclonal antibodies against human ceruloplasmin (M. C. Linder, unpublished results). The chicken (DiSilvestro and Harris, 1985; Calabrese et al., 1988) (Table 4-1) and goose (Hilewicz-Grabska et al., 1988) proteins (present in lower concentrations overall) also may be somewhat different. They appear to be of smaller size, with a much lower tyrosine content, which gives them a much lower extinction coefficient at 280 nm and thus a high maximum 610:280 absorbance ratio (0.07). (Despite some overall similarities, the chicken and goose proteins nevertheless also display considerable differences in amino acid composi-

Table 4-1. Summary of Molecular Properties of Holoceruloplasmins[a]

	Ceruloplasmin source		
	Human	Bovine[b]	Chicken[c]
Molecular weight			
Total	132,000	124,000	124,000
			(Sephadex G-150)
Polypeptide	120,000 (91%)		
Carbohydrate	12,000[h] (9%)[d]		7.3% or more
Sedimentation coefficient (S20, w)	7.1 S		
Axial ratio	3.6		
N-terminus; C-terminus	Lysine; glycine		
pI	4.4		
No. of sialic acid residues	9 (also 6, 7)[e]	7[h] (1.6% wt.)	4[h] (0.6% wt.)
No. of glycan chains	3–5		
E (1%/1 cm; max)	0.69 (610 nm)	NA (614 nm = max)	0.73 (602 nm)
E (1%/1 cm; 280 nm)	15.0		10.6 (less Tyr)
Maximum ratio, $A_{610}:A_{280}$	0.045	0.045	0.070
Copper content			
Percent	0.28–0.34		0.17[f]–0.20
Atoms/molecule	5.8–7.0	6[h]	3.3–3.9; 5[g]

[a] Modified from Frieden (1981).
[b] Calabrese et al. (1981), Sakurai and Nakahara (1986).
[c] DiSilvestro and Harris (1975).
[d] Assuming four glycan chains (Takahashi et al., 1984).
[e] S. Irie and M. Tavassoli (1988).
[f] After Chelex-100 treatment.
[g] Calabrese et al. (1988).
[h] Value is the best guess from information currently available.

tion.) The exact sizes of these proteins are also still uncertain. They were determined by gel chromatography and SDS electrophoresis. As already explained, the former technique gives a 14% overestimate of the actual molecular weight for human ceruloplasmin, because of the asymmetric nature of the molecule. (On the basis of this assumption, the actual molecular weight of the chicken and goose proteins would be about 110,000.) Despite these uncertainties, it seems that the chicken protein has about the same proportion of carbohydrate but much less sialic acid and less copper than the human protein. The goose protein also has the same amount of carbohydrate and less sialic acid, but 0.32% by weight of Cu (Hilewicz-Grabska et al., 1988), which is high (Table 4-1). Reptilian ceruloplasmin, from the turtle, has also been isolated, and retains similar overall characteristics: an apparent molecular weight of 145 kDa (including carbohydrate), 5 Cu atoms/mol (50% of which are detectible by EPR), absorbance at 603 nm, but a very low oxidase activity (except towards Fe(II)), and a higher resistance to proteolysis (Musci et al., 1990).

While the copper content for the human and bovine proteins is fairly well established as being six or, at most, seven atoms per molecule (see below), that of the chicken and turtle proteins is probably five and that of the goose protein could be eight or nine. [The Cu content of the chicken protein, after removal of loosely bound Cu by Chelex-100, was 0.17% according to DiSilvestro and Harris (1985) but was estimated as five atoms per molecule by Calabrese et al. (1988).] All of this needs confirmation. With regard to glycan content, it seems likely that there are similar numbers of chains (number adjusted for an overall 17% lower molecular weight relative to that of the human protein), but that these are less sialylated.

The exact number of copper atoms and their nature and ligands in ceruloplasmin are still not entirely certain. Table 4-2 summarizes what is

Table 4-2. Types of Copper in Human Ceruloplasmin[a]

No. of atoms per molecule	Designation	Detectability	Other properties
1 Fast type 1	(Blue) Cu(II) (also known as Cu_A)	A_{max} at 610 nm; EPR detectable	Fast reoxidation; involved in enzyme activity
1 Slow type 1	(Blue) Cu(II) (also known as Cu_B)	A_{max} at 610 nm; EPR detectable	Slow reoxidation; not involved in enzyme activity
1 Type 2	Permanent Cu(II)	EPR detectable	Binds anions; involved in enzyme activity; may be artifact of isolation[b]
2 Type 3	Cu(II)	A_{max} at 330 nm; EPR silent	Postulated to be spin coupled, in analogy with other multicopper oxidases
1 Type 4	Cu(I or II)	EPR silent	Required by total Cu content
1 Additional	Cu(II)	EPR silent	Exchangeable, removed by treatment with Chelex-100; may be contaminant acquired during protein isolation[c]

[a] Modified from Frieden (1981).
[b] Evans et al. (1985).
[c] McKee and Frieden (1971).

probably the currently most accepted view on human ceruloplasmin among the investigators working in this field (Frieden, 1981; Sakurai and Nakahara, 1986; Evans et al., 1985; Gray and Solomon, 1981). It may be summarized as follows. There are six copper atoms tightly bound or buried in the molecule. They are nondialyzable and not available to diethyldithiocarbamate (DDC). They include at least two atoms with an absorbance maximum at 602–614 nm (depending on the species source), responsible for the blue color of the protein (Fig. 4-5) and designated as type 1. These atoms are reduced and reoxidized at different rates (Deinum and Vanngard, 1973; Sakurai and Nakahara, 1986), and only the atom that is reoxidized rapidly is thought to be involved in the oxygenase activity of ceruloplasmin (see more below). The type 1 blue copper atoms are the ones most easily removed and perhaps also restored (Dawson et al., 1979), although Scheinberg and Morell (1973; Morell and Scheinberg, 1958) as well as Aisen and Morell (1965) reported that half of the total Cu in ceruloplasmin is exchangeable, upon reduction, and that all of it may be removed, at low pH. Treatment with cyanide or ferrocyanide or ascorbate plus DDC effects reduction and release of Cu (Morell and Scheinberg, 1958; Kasper and Deutsch, 1963a). These coppers can be partly reconstituted by readding copper, in the presence of ascorbate. Recently, Harris and Dameron (1987a,b) have used incubation of ceruloplasmin with ^{67}Cu(II) in the presence of 0.05% ascorbate (30 min, at room temperature) to label the protein with ^{67}Cu, followed by treatment with Chelex-100 (Dameron and Harris, 1987a,b). This procedure results in a radiolabeled ceruloplasmin free of loosely bound copper but uncharacterized with regard to the position(s) of the ^{67}Cu on the protein (E. D. Harris, personal communication). [Recent work by Goode et al. (1989) suggests that the added copper may not be bound in the normal way.]

The blue copper center in ceruloplasmin is analogous to that in other blue copper proteins (such as ascorbate oxidase, laccase, and azurin), with a high molar absorptivity (at about 600 nm), an EPR spectrum with very small $A_\|$ values, and a reduction potential in the range of 300–500 mV [much higher than for aqueous Cu(II)] (Gray and Solomon, 1981). Cysteines are thought to be actively involved as ligands (probably nos. 680 and 1021; Figs. 4-1 and 4-2). The amino acid sequences thought to surround the blue copper-binding cysteines in ceruloplasmin and analogous proteins are given in Fig. 4-6 and show what appears to be significant homology. This agrees with recent findings, based on gene cloning, that the C-terminal portion of neurospora laccase bears a striking resemblance to that of the third homologous unit of ceruloplasmin and has homology with the small (single blue) copper proteins plastocyanin (abundant in plants) and azurin (used in electron transport by bacteria) (Ryden, 1988). It also

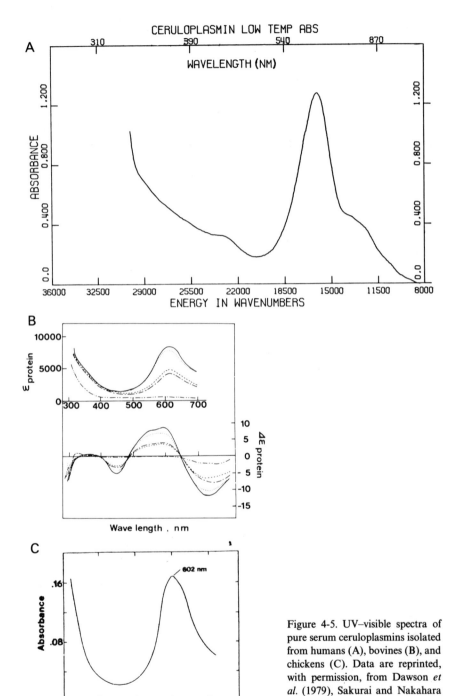

Figure 4-5. UV–visible spectra of pure serum ceruloplasmins isolated from humans (A), bovines (B), and chickens (C). Data are reprinted, with permission, from Dawson *et al.* (1979), Sakurai and Nakahara (1986), and DiSilvestro and Harris (1985), respectively.

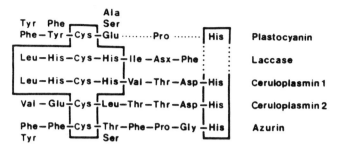

Figure 4-6. Sequence of amino acids and location of cysteine thought to be involved in the binding of type 1 Cu in ceruloplasmin and other blue copper proteins. Reprinted, with permission, from Reinhammer and Malmström (1981).

agrees with the assignment by Takahashi *et al.* (1984) of both blue Cu atoms in ceruloplasmin to the third homologous unit of the protein (Fig. 4-1) and with evidence that three of the six (or seven) Cu atoms in ceruloplasmin are associated with the C-terminal portions of the third homologous sequence ("B," Fig. 4-2). Additional homologies of potential copper binding regions to those of other (smaller) Cu binding proteins have been described by Moshkov *et al.* (1988); but Messerschmidt and Huber (1990) suggest that one (blue) copper is associated with the C-terminal end of *each* of the three homologous units of ceruloplasmin (Figure 4-2), and that the trinuclear copper cluster (also seen in laccase and ascorbate oxidase) is associated with both the N- *and* C-terminal parts of the total molecule. (Further information on laccase and ascorbate oxidase is found in Chapter 10.)

In addition to two type 1 Cu atoms, it is fairly well accepted that there is one atom of type 2 Cu in isolated ceruloplasmins, detectable in low-field EPR. What is not entirely clear is whether this form of copper (recognized in other blue or multicopper oxidases; Reinhammar and Malmström, 1981) is present *in vivo*, in the native state. Originally, Vanngard (1967) found no type 2 copper in a concentrated human serum sample, but, later, Andreasson and Vanngard (1970) reported that there were two type 2 copper atoms in the isolated protein (assuming a molecular weight of 150,000). They suggested that the earlier failure to observe it might be due to a line-broadening effect induced by the serum environment. More recently, Dooley *et al.* (1981), for the pure bovine protein, and Dawson *et al.*, 1979, for the human protein suggested a ratio of 1 : 2 for type 2 to type 1 copper, which would be more consistent with the findings for laccase and ascorbate oxidase (Reinhammar and Malmström, 1981). However, Evans *et al.* (1985) confirmed the absence of a low-field EPR signal in fresh, partially purified human and porcine serum samples (Fig. 4-7) and

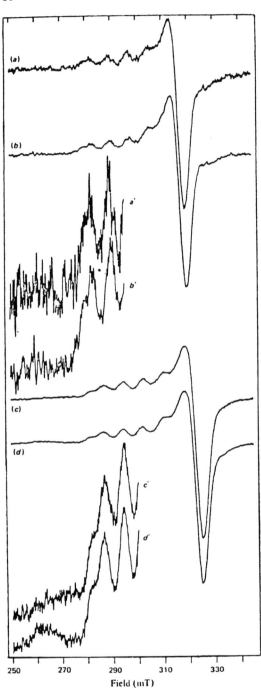

Figure 4-7. EPR spectra of fresh, partially purified human and porcine ceruloplasmins. Fresh human serum is shown after (a) Cibacron Blue–agarose chromatography (to remove albumin) or (b) ethyleneimine–agarose chromatography; a' and b' show low-field portions at a ten fold higher gain, with peaks corresponding to two type 1 copper atoms. Fresh porcine serum is shown (c) after ethyleneimine, ion-exchange chromatography and (d) subsequent concentration and gel filtration; d' shows the appearance of a type 2 Cu signal in the 260-mT field region (ten fold higher gain than d). Reprinted, with permission, from Evans et al. (1985).

found that the EPR signal for type 2 copper appeared after exposure of the porcine ceruloplasmin extract to gel chromatography on Sephadex G-200 or Sephacryl S-200 (Fig. 4-7). Fish ceruloplasmin has also been subject to the appearance and disappearance of type 2 copper (Syed et al., 1982).

There is general agreement on the presence of two type 3 Cu atoms, assumed to be spincoupled and thus EPR silent, in human, bovine, and porcine ceruloplasmins, again in analogy with other blue copper oxidases. These copper atoms are thought to be responsible for the presence of a 330-nm shoulder in the UV-visible absorption spectrum of the protein (Fig. 4-5 A, B). It is noteworthy that the spectrum for the chicken protein appears to have no 330-nm shoulder, although absorption in that region appears after reduction of the type 3 form (Calabrese et al., 1988).

At least one additional atom of copper per mammalian ceruloplasmin molecule is needed to fulfill a stoichiometry of six in total, the most likely number of nondialyzable copper atoms present. Although the values reported for the copper contents of pure isolated ceruloplasmin have ranged from 0.28 to 0.34%, figures have tended to center on 0.30%, which translates to 6.2 atoms per molecule. The occasional higher values may be due to the presence of contaminating Cu(II) loosely bound to low affinity, Chelexable sites (K_D 10^{-10} M or higher), identified by McKee and Frieden (1971) and Zgirski and Frieden (1990). It seems unlikely that this site is occupied in vivo in the blood plasma, as albumin (and transcuprein), with much higher affinities for Cu(II), should easily compete for it. (See more later in this chapter.) At the same time, it is generally agreed that some of the copper on ceruloplasmin is labile to storage and to various of the purification procedures that may be used. This would tend to result in an underestimate of the total copper present. Not only that, but evidence (Chapter 3) indicates that ceruloplasmin delivers copper to cells, which implies that ceruloplasmin from which some copper has been removed should also be present in plasma and serum. Indeed, there have been periodic reports of the presence of low-copper ceruloplasmin serum extracts (Carrico et al., 1969; Milne and Matrone, 1970; Kasper and Deutsch, 1963a; Holtzman and Gaumnitz, 1970b; Calabrese et al., 1981). Carrico et al. (1969) estimated that 10–20% of human ceruloplasmin is in the apo form, although it is not entirely clear whether this means it is partially or fully devoid of copper. [A similar estimate has been made by Mallet and Aquaron (1983).] Holtzman and Gaumnitz (1970b) reported the presence of low-copper ceruloplasmin in copper-deficient rats, estimating the remaining copper content at 10% of that for holoceruloplasmin. Goode et al. (C. A. Goode, F. Alemzadeh, and M. C. Linder, unpublished results) have recently confirmed that a large proportion of rat serum ceruloplasmin is of the low-copper variety, with no blue absorbance, and elutes from ion-exchange columns at a lower salt concentration than the typical blue material that has *p*-phenylenediamine oxidase activity. [Low-copper human ceruloplasmin eluting with a higher salt concentration had previously been reported by Kasper and Deutsch (1963a).] If apo- and holoceruloplasmin even partly copurify (which they appear to do), the latter facts would suggest that the existing values for maximal copper content may be underestimates.

In any event, it seems likely that mammalian ceruloplasmins contain a unit of four Cu atoms, made up of one type 1 and two type 3 Cu atoms, as well as one atom in another form that becomes type 2 on isolation, all of which are involved together in oxidation/reduction reactions involving this protein, and in analogy with the multicopper oxidases laccase and ascorbate oxidase (not found in vertebrates). [The two type 3 and the "type 2" form are in the form of a trinuclear cluster (see later, and also Chapter 10).] At least two other Cu atoms (one type 1 and one or more "type 4") are present elsewhere in the molecule (perhaps in another domain) and are perhaps engaged in other (nonenzymatic) functions of the protein. The case of chicken ceruloplasmin is less clear. A total of five Cu atoms per molecule would suggest one blue Cu (type 1) and one type 2, plus two type 3, as in the basic oxidizing unit postulated for the other ceruloplasmins, especially as chicken ceruloplasmin has full *p*-phenylenediamine and amine oxidase activities (E. D. Harris, personal communication). This leaves one additional (blue) Cu for other functions. The goose protein (with a higher copper content) might have two oxidizing units, or more extrafunctional copper atoms (for example for transport). This would need to be examined further.

The presence of a basic, four-Cu oxidative unit is precisely analogous to what has been found for the blue multicopper oxidases, laccase and ascorbate oxidase, which have four and eight Cu atoms, respectively (Reinhammer and Malmström, 1981). These two proteins, found in plants (and mushrooms in the case of laccase), have very similar spectral and EPR properties to those of ceruloplasmin (Table 4-3), and there is sequence homology of the blue Cu cysteine site(s) of laccase with the C-terminal portion of ceruloplasmin (Lerch and Germann, 1988), as has already been mentioned. A hypothetical model of the environment of the four Cu atoms in laccase is given in Fig. 10-10A (Chapter 10) and may be relevant to what occurs in ceruloplasmin (and ascorbate oxidase). The most recent data for ascorbate oxidase suggest the four-Cu unit is actually composed of a single trinuclear cluster (of one type 2 and two type 3 Cu atoms) and that the more distant type 1 copper atoms (one in each of two subunits, at either end of the trinuclear centers) might be the initial recipient(s) of incoming electrons (then transferred to the trinuclear cluster)

Table 4-3. EPR Parameters and Oxidation/Reduction Potentials of Ceruloplasmin, Laccase, and Ascorbate Oxidase[a]

Parameter	Ceruloplasmin		Laccase		Ascorbate oxidase
			Polyporus	*Rhus*	
EPR parameters[b]					
Type 1 Cu					
g_x			2.033	2.030	2.036
$g_\perp$	2.06	2.06			
g_y			2.051	2.055	2.058
g_z	2.215	2.206	2.190	2.300	2.227
A_z	9.5	7.4	9.0	4.3	5.8
Type 2 Cu					
$g_\perp$	2.060		2.036	2.053	2.053
$g_\parallel$	2.247		2.243	2.237	2.242
$A_\parallel$	18.9		19.4	20.6	19.9
Oxidation/reduction potentials (mV)					
Type 1 Cu	490	580	785	394	—
	(pH 5.5)		(pH 5.5)	(pH 7.5)	
Type 2 Cu				365	
				(pH 7.5)	
Type 3 Cu	—		782	434	
			(pH 5.5)	(pH 7.5)	

[a] Summarized from Reinhammar and Malmström (1981).
[b] Hyperfine coupling constants, in 10^{-3}/cm.

(Messerschmidt *et al.*, 1988). The same is most likely the case for ceruloplasmin (Calabrese *et al.*, 1989; Messerschmidt and Huber, 1990). It has the same trinuclear copper cluster and probably receives electrons from one or both of the (unequal) blue copper centers. (See also Chapter 10, Sections 10.5.3 and 10.5.4, Table 10-5, and Figs. 10-9 and 10-10B.)

As already alluded to, ceruloplasmin is well known for its capacity to catalyze a variety of oxidative reactions involving molecular oxygen. Presumably, the four Cu atoms cited earlier are all involved as mediators of electron flow from substrate to dioxygen in these reactions. Frieden (1981) has classified them as falling into three categories, based on the substrate (Fig. 4-8): (a) the reaction with Fe(II), designating ceruloplasmin as a ferro-O_2-oxidoreductase (EC 1.16.31), or as ferroxidase I, Fe(II) being the substrate for which at least the human protein has the lowest K_ms (0.6 and 5.0 mM; Huber and Frieden, 1970), the kinetics (biphasic K_ms) being explained on the basis of substrate activation; (b) the reaction with a broad variety of bifunctional aromatic amines and phenols (substrate K_ms

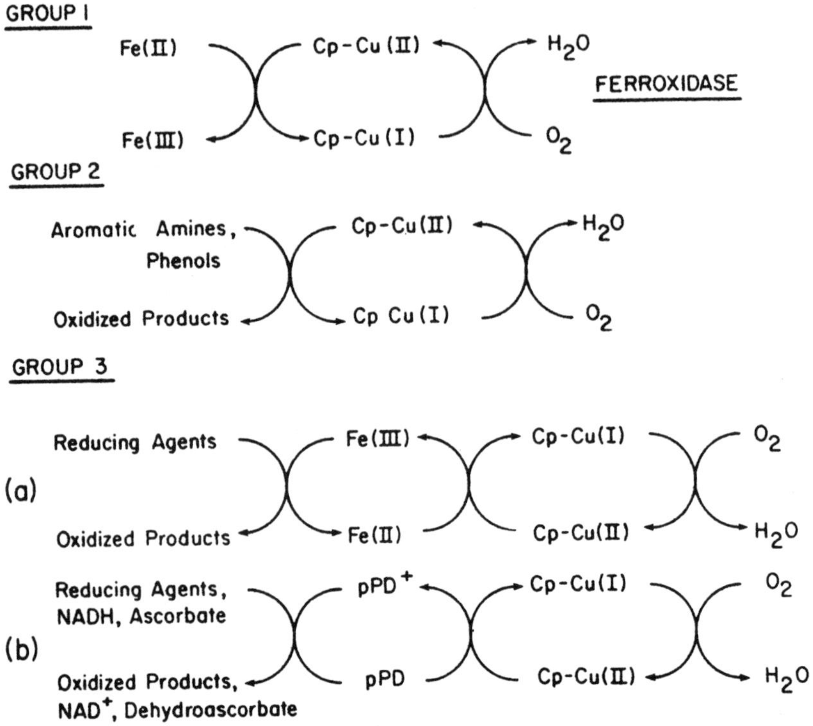

GROUP 1

```
Fe(II)      ⟶ Cp-Cu(II) ⟵  ⟶ H₂O
         ╳              ╳              FERROXIDASE
Fe(III) ⟵   ⟶ Cp-Cu(I)    ⟶ O₂
```

GROUP 2

```
Aromatic Amines,  ⟶ Cp-Cu(II) ⟵  ⟶ H₂O
Phenols           ╳              ╳
Oxidized Products ⟵  ⟶ Cp-Cu(I)   ⟶ O₂
```

GROUP 3

(a)
```
Reducing Agents   ⟶ Fe(III) ⟵   ⟶ Cp-Cu(I) ⟵  ⟶ O₂
                 ╳            ╳            ╳
Oxidized Products ⟵ ⟶ Fe(II)    ⟶ Cp-Cu(II) ⟵ ⟶ H₂O
```

(b)
```
Reducing Agents,   ⟶ pPD⁺ ⟵      ⟶ Cp-Cu(I) ⟵  ⟶ O₂
NADH, Ascorbate   ╳            ╳            ╳
Oxidized Products, ⟵ ⟶ pPD       ⟶ Cp-Cu(II) ⟵ ⟶ H₂O
NAD⁺, Dehydroascorbate
```

Figure 4-8. Types of enzymatic reactions involving catalysis by ceruloplasmin. Substrates falling into group 2 include p-phenylenediamine (K_m variously reported as 21–3000 μM); aminophenols ($K_m > 180\ \mu M$); 5-hydroxyindoles ($K_m > 900\ \mu M$); phenothiazines ($K_m > 900\ \mu M$); and catecholamines ($K_m > 2.6\ \text{m}M$) [possibly also ascorbate (K_m 5.2 mM)]. Group 3 substrates would include many reducing agents, such as O_2 and H_2O_2. Details have been reviewed by Lovstad (1978) and Frieden and Hsieh (1976). Reprinted, with permission, from Frieden (1981).

ranging from 21 to 3000 mM; Frieden, 1981); and (c) reactions with pseudo substrates that can rapidly reduce Fe(II) or free-radical intermediates. Frieden (1981) believes that these pseudo substrates will react with ceruloplasmin indirectly via Fe(II) or the free radicals, and not directly. The basis for this view was the fact that traces of iron are usually found in isolated ceruloplasmin preparations (removable with apotransferrin) and that (in the presence of low concentrations of ascorbate) this iron stimulates oxidation of various reducing agents (including ascorbate) (Huber and Frieden, 1970). Nevertheless, Curzon and Young (1972) had concluded that ascorbate is a true substrate for the enzyme (K_m 5 mM), on the basis of the assumption that addition of 100 mM

EDTA eliminated the involvement of contaminating Fe(II) in the reaction. [In this connection, Rice (1962) showed that *p*-phenylenediamine oxidation by serum was the product not just of direct ceruloplasmin catalysis but also of oxidation by traces of metal ions, presumed to be Fe(III), that could be chelated (and prevented from acting) by addition of EDTA. This has become part of the protocol for ceruloplasmin assays involving phenylamines.] Whether EDTA is indeed effective in inhibiting mediation of oxidation of other "substrates" via contaminating Fe(II)/Fe(III) is not certain. However, ceruloplasmin can inhibit the oxidative reactions leading to deoxyribose damage induced by Fe(III)-EDTA + H_2O_2 or O_2^-, or by Fe(II)-bipyridyl (Gutteridge, 1985).

In this connection, another way of viewing category 3 reactions involving ceruloplasmin is to focus on its direct catalytic involvement with free-radical substrates generated *in vivo* or *in vitro*, that is, as an antioxidant (Fig. 4-8, 3b). There has been considerable interest in the free-radical-scavenging potential of this extracellular protein, beginning with the report of Al-Timini and Dormandy (1977) that pure ceruloplasmin prevented autoxidation of lipids induced by low concentrations of ascorbate or Fe(II)/Fe(III) and reports of Gutteridge (1978, 1985; Gutteridge *et al.*, 1980) that it could inhibit lipid and DNA peroxidation induced by iron or copper salts *in vitro*. [An extensive review of the antioxidant capacities of ceruloplasmin is that of Goldstein and Charo (1982).] The paper by Gutteridge (1985) is also especially informative, and more recent data confirm the protective capacity of ceruloplasmin to prevent red cell lysis (Caffrey *et al.*, 1990) and the release of oxygen radicals from polymorphonuclear leukocytes actively engorging (and inactivating) foreign substances (Broadley and Hoover, 1989).]

Goldstein *et al.* (1979) reported that, during purification of ceruloplasmin, plasma superoxide dismutase activity coincided with ceruloplasmin oxidase activity (measured with dianisidine) and that ceruloplasmin with an $A_{610} : A_{280}$ ratio of 0.040 (about 90% pure) also had this activity. For these studies, the same (indirect) dismutase assay was used as that originally employed by McCord and Fridovich (1969) when they demonstrated that a major Cu- and Zn- containing protein of erythrocytes, "erythrocuprein" (also found in other tissues), was an intracellular superoxide dismutase. In this assay, the inhibition of ferricytochrome *c* reduction by dianisidine dihydrochloride is measured, as mediated by xanthine oxidase. Other (indirect) inhibition assays used to measure superoxide dismutase confirmed the results for ceruloplasmin. Superoxide dismutases catalyze the production of peroxide from superoxide (Table 4-4). However, when this was examined directly for ceruloplasmin, it was found that ceruloplasmin actually inhibited peroxide formation

Table 4-4. Reactions Involving Generation and Disposal of Oxygen-Derived Species That Can Damage Organic Molecules in Cells

Fenton reaction

$$H_2O_2 + Fe(II) \rightarrow OH\cdot + OH^- + Fe(III)$$
Inhibited by ceruloplasmin

Superoxide dismutation[b]

Overall reaction

$$2O_2^- + 2H^+ \rightarrow H_2O_2 + O_2$$
Inhibited by ceruloplasmin

Steps, involving Fe (as an example):

$$X-Fe(III) + O_2^- \rightarrow X-Fe(II) + O_2$$
$$X-Fe(II) + O_2^- \rightarrow X-Fe(II)\text{-}O_2^{2-}$$
$$X-Fe(III)\text{-}O_2^{2-} + 2H^+ \rightarrow X-Fe(III) + H_2O_2$$
Ceruloplasmin inhibits these reactions

Damage via generation of hydroxide radicals[c]

$$Fe(II)/Fe(II)\text{-dipyridyl} + O_2 + dR \xrightarrow[OH\cdot]{via} \text{oxidized dR}$$

Not inhibited by SOD
Inhibited by ceruloplasmin
Slightly inhibited by Cu-His (1:1)

Damage via generation of superoxide[c]

$$Fe(III)\text{-EDTA} + O_2 + dR + \text{reducing agent} + \xrightarrow[O_2^-]{via} \text{oxidized dR}$$
$$(H_2O_2 \text{ or } O_2^-)$$

Inhibited by SOD
Inhibited by ceruloplasmin
  with H_2O_2, boiling only partly destroys protection
  with O_2^-, boiling prevents inhibition
Inhibited partly by Cu-His (O_2^-)
Slightly inhibited by Cu-Alb (H_2O_2)

---

[a] Abbreviations: dR, deoxyribose; SOD, superoxide dismutase (enzyme); Cu-His, Cu–histidine complex; Cu-Alb, Cu–albumin complex (1:1 molar).
[b] Based on Fee (1981).
[c] Gutteridge (1985).

from xanthine oxidase-generated superoxide (Goldstein and Charo, 1982). This was also shown by Bannister *et al.* (1980), who noted that superoxide did nevertheless react directly with the ceruloplasmin, causing a reduction of some of its Cu(II). [This fits the group 3b reaction of Frieden (Fig. 4-8).]

The view of Gutteridge on these matters, and further insight into the antioxidant properties of ceruloplasmin, is provided by the data in a recent publication (Gutteridge, 1985), also summarized in Table 4-4. In these studies, hydroxide radicals (OH ·) were generated directly or indirectly by Fe(II)/Fe(III), in the presence and absence of chelators and superoxide (generated through a xanthine oxidase system) or peroxide. Production of damaging radicals was assessed by following net production of thiobarbituric acid (TBA)-reactive material from deoxyribose. In all cases (except Fe-EDTA alone), ceruloplasmin (1.18 mM) inhibited the oxidative degradation of deoxyribose by 40–96%. In systems involving generation of superoxide, ceruloplasmin was almost as inhibiting as bovine (Cu/Zn) superoxide dismutase (mole for mole). (All enzymes were purchased commercially.) A 1 : 1 Cu–histidine complex (5.1 mM) was also somewhat inhibitory [especially in the Fe(III)-EDTA + superoxide system], while Cu–albumin (5.9 mM) tended to be ineffective [except in the case of Fe(III)-EDTA + peroxide]. Of additional interest was the finding that preboiling ceruloplasmin eliminated its effects, except in the Fe(III)-EDTA + H_2O_2 system (where it reduced inhibition from 83 to 58%).

Almost all of the effects cited fit into the group 3 category of Frieden (1981) (Fig. 4-8). Clearly, some of the inhibition was due to the oxidation of Fe(II) by ceruloplasmin (ceruloplasmin's ferroxidase activity; group 1), which inhibits interaction of reduced iron species with peroxide, for example, in the Fenton reaction, that would otherwise produce destructive OH radicals (Table 4-4). However, ceruloplasmin also directly reacts with free-radical intermediates (group 3b; Fig. 4-8), thus harmlessly dissipating the extra electrons via formation of water from dioxygen. Recent studies of Calabrese and Carbonaro (1986) indicate that ceruloplasmin is also capable of reacting with H_2O_2, via the blue copper atom that is "slowly reoxidized" after reduction with ascorbate. Whether ceruloplasmin also directly interacts with superoxide (rather than indirectly, via Fe, as suggested in Fig. 4-8, group 3a), is not entirely settled. Nevertheless, investigators agree that ceruloplasmin is not a superoxide dismutase and that it is a scavenger of free radicals (Goldstein and Charo, 1982; Gutteridge, 1985). They also agree that the presence of ceruloplasmin in serum (and in interstitial or other body fluids) is important in protecting against peroxidative damage that might or would otherwise occur under various physiological conditions. Indeed, ceruloplasmin would seem to be an "all-purpose" neutralizer of reactive oxygen species. This likely to be of particular importance in connection with the formation of extracellular reactive oxygen forms by white blood cells, when they are activated by inflammatory processes and/or during their digestion of bacteria or immune complexes (Babior, 1978a,b; Broadley and Hoover, 1990). Further aspects of this part of copper function and metabolism are covered in Chapter 9.

Returning to the matter of ferroxidase activity, this activity has been a focus of interest as a way of linking copper and iron metabolism (Frieden, 1970). It is thought that iron leaves storage sites in cellular ferritin after reduction to Fe(II) and chelation. Frieden and co-workers have postulated that ceruloplasmin is normally employed in oxidizing this ferrous iron, so that it may attach to plasma transferrin for transport. (This matter is addressed further in Chapter 7, Section 7.2.1.2a.) Apart from allowing the normal flow of stored iron to sites where it may be needed, ceruloplasmin ferroxidation will work to minimize peroxidative damage that might otherwise occur with Fe(II) (Gutteridge, 1985, 1986).

4.1.2 Albumin

Albumin is well known as the most abundant protein in vertebrate blood plasma and interstitial fluids. A rather acid protein (pI 4.7) of 68,000 Da, it is involved in the transport of numerous substances, from tryptophan, fatty acids, and bilirubin to drugs and metal ions. These various substrates mostly occupy different positions on the protein. It is also well recognized that albumin binds copper and has a role in copper transport. The interactions of Cu with albumin and amino acids (relating to transport) have been reviewed in detail by Sarkar *et al.* (1983).

In contrast to the copper in ceruloplasmin, copper on albumin is readily exchangeable with that on other ligands, particularly amino acids, that may be present in serum and other body fluids (Lau and Sarkar, 1971; Sarkar and Wigfield, 1986). Albumin has one high-affinity binding site (Breslow, 1964), with a dissociation constant of about 10^{-17} M in the case of the human protein (Lau and Sarkar, 1971). It also binds additional copper atoms, but with lower affinity. In the presence of histidine, which itself forms a tight His_2-Cu complex (see Chapter 3) (K_D 10^{-17} M; Lau and Sarkar, 1981a,b), His_2-Cu is in equilibrium with albumin-Cu through an intermediary, ternary, albumin-Cu-His complex,

$$Cu\text{-}His_2 \leftrightharpoons Cu\text{-}His \leftrightharpoons albumin\text{-}Cu\text{-}His \leftrightharpoons albumin\text{-}Cu$$

which may in fact be the predominant *in vivo* species. The dissociation constant for the ternary complex has been measured as being 10^{-22} M (Lau and Sarkar, 1971). [Recently, Suzuki *et al.* (1989; Suzuki and Karasawa, 1990) report that a ternary complex between albumin, copper, and cysteine may also be formed *in vivo*.]

The high-affinity copper binding site is situated at the amino terminus (Breslow, 1964; Peters and Blumenstock, 1967; Bradshaw and Peters,

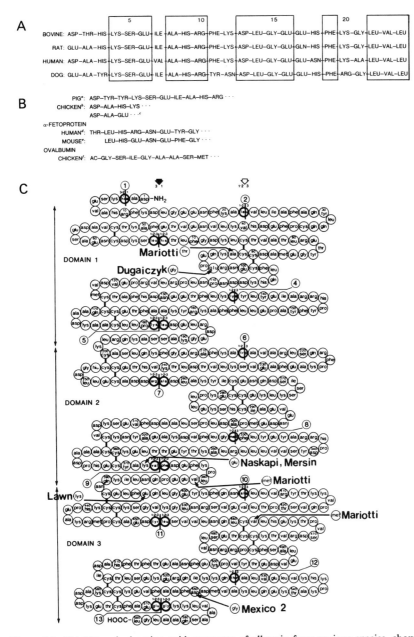

Figure 4-9. (A) N-terminal amino acid sequences of albumin from various species, showing the location of residues involved in the high-affinity Cu site and the absence of histidine in dog (and some other) albumin. Reprinted, with permission, from Dixon and Sarkar (1974). (B) Additional sequences from other literature: [a] Peter (1977); [b] Sorkina et al. (1976); [c] B. Sarkar, personal communication; [d] Bowman and Yang (1987); [e] Gorin et al. (1981); [f] Gagnon et al. (1978) and Nisbet et al. (1981). (C) Amino acid sequence of human albumin, showing the locations of some rare substitutions or mutations. Reprinted, with permission, from Takahashi et al. (1987).

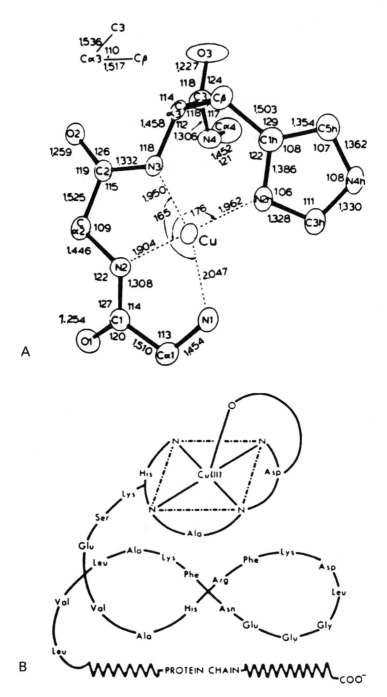

1969) and involves primarily the histidine in position 3 (Fig. 4-9). This histidine is essential for tight binding, a matter that was largely proved by nature's own experiment, the dog. In this species (and the pig; Peters, 1977), the crucial histidine has been replaced by tyrosine, and the albumin does not have the same (high) affinity for copper (Appleton and Sarkar, 1971). It does have several lower affinity sites, however (as do other albumins). [In the chicken, the third amino acid is glutamate, according to B. Sarkar (personal communication), although Sorkina *et al.* (1976) reported it as histidine. Differences in strains?] Owen (1980) has proposed that the lower affinity of canine albumin for copper may contribute to the unusually high levels of Cu in dog liver (through more rapid uptake occurring in the absence of high-affinity albumin binding that would slow down Cu uptake. This would be consistent with the data of Goresky *et al.* (1968), who showed that livers of mongrel dogs had copper concentrations of 82 ± 29 mg/g wet weight ($N = 10$) (more than 10 times that of most other species; other tissues were in the "standard" range). Dog liver rapidly took up copper administered intravenously, but dogs were unable to tolerate the kinds of repeated doses of this element that can be given to rats. This suggests that albumin in blood functions as a buffer for excess copper that might otherwise cause intracellular (organ) toxicity. The only problem with this logic is that the pig and chicken, both of which have "standard" liver copper concentrations (Table 1-6), also have albumins with no histidine in the third position (i.e., with no high-affinity copper binding site). This implies that another factor in the dog, such as the ability to excrete excess copper in the bile, is also limiting (see Chapter 5).

Returning to the high-affinity site, further elegant work by Sarkar and colleagues has used peptides with structures identical or analogous to that of the N-terminal end of human serum albumin to characterize the nature of the copper ligands and the binding architecture, using titration combined with visible and circular dichroism (CD) spectroscopy, as well as ^{65}Cu-EPR and ^{13}C-NMR spectroscopy. The result (summarized in Fig. 4-10) combines the picture obtained from X-ray crystallography of the glycylglycylhistidine-N-methylamide analogue (A) with that obtained from

---

Figure 4-10. Mechanism of binding of Cu to the amino-terminal amino acids of albumin. (A) Binding to the N-terminal analogue glycyl-glycyl-histidine-N-methylamide; X-ray structure determined on material crystallized at physiological pH, showing binding distances in angstroms (to four significant figures). Reprinted, with permission, from Camerman *et al.* (1976). (B) Binding to the N-terminal 24-residue peptide fragment of human albumin, including the carboxyl side chain of aspartic acid at the N-terminal itself. Reprinted, with permission, from Laussac and Sarkar (1984).

titration/spectroscopy of the peptide comprising the last 24 amino acids of human albumin (B). The upper drawing (A) shows the coordinate bonding of Cu(II) with four nitrogens (the N-terminal amino nitrogen, two peptide nitrogens, and one imidazole nitrogen) in a slightly distorted square planar arrangement. Similar chelation was observed for the more physiological (albumin-derived) peptides, with the additional involvement of carboxylate oxygen (from the aspartate at the N-terminal position; Fig. 4-10B). The high-affinity copper binding capacity of albumin can be fully explained by interaction of the metal with just the last N-terminal amino acids and does not require the allosteric involvement of other portions of the albumin molecule. This also implies that other substances binding to the albumin will not usually influence copper binding. [Zinc goes to another site. Only Ni(II) binds in the same position, and with similar stabilities (Glennon and Sarkar, 1982).] Just how widespread among vertebrates is the "proper" sequence for high-affinity copper binding is unknown, although the pig and chicken (like the dog) seem to be missing the crucial histidine (Peters, 1977), while the bovine sequence is like that for the rat and human (Fig. 4-9). α-Fetoprotein (which serves a similar function to albumin in the fetus) does again appear to have histidine as the third residue at the N-terminus (at least in humans) (Yang et al., 1985; Lau et al., 1989), and it does bind copper (Aoyagi et al., 1978), as does ovalbumin (egg albumin) (Chikvaidze, 1990), giving the same ESR spectrum. (Like albumin, α-fetoprotein is synthesized as a preproprotein, whose signal sequence and a few other amino acids must be removed before copper binding can occur.) Apart from the single high-affinity specific binding site, albumin can also bind multiple copper ions more loosely, as in the case of dog albumin (Appleton and Sarkar, 1971; Mohanakrishnan and Chignell, 1984). This loose copper can be removed by exposure to Chelex-100, a metal ion-binding resin with high affinity for Cu(II). Thus, *in vivo*, excess copper in dog serum may still preferentially bind to albumin [and at least partly to the N-terminus (Mohanakrishnan and Chignell, 1984)] versus to other proteins; at least, this possibility cannot now be excluded.

4.1.3 Low-Molecular-Weight Copper Ligands in Vertebrate Plasma

The most studied additional factors thought to be involved in copper binding are in the amino acid and small peptide fraction of the blood. Indeed, for some time, a leading concept has been that this fraction is the one most directly involved in delivery of copper to cells, as it is considered to be largely in equilibrium with the exchangeable, nonceruloplasmin copper of the plasma.

In a seminal paper on this topic, Neumann and Sass-Kortsak (1967) showed that plasma amino acids (especially histidine) were able to release copper from, and establish an equilibrium of copper with, the exchangeable human serum protein pool (see also Chapter 3). For this, they added specific (mostly above physiological) amounts of Cu, as ^{64}Cu(II), to serum and determined the amounts of radioactivity in the top portion of supernatants obtained by sedimenting for 24 h at $115,000 \times g$. [This sedimentation brings albumin and most other proteins (and protein-bound copper) to the lower half to third of the tube.] Using this procedure on fresh serum predialyzed for 48 h (in four to six changes of isotonic saline adjusted to pH 7.4), they found $0.2 \pm 0.1\%$ of the added tracer in the top supernatant (or low-molecular-weight) fraction (mean $\pm$ SD; $N+12$), with Cu : albumin molar ratios in the physiological range (0.002–0.007 : 1). In the case of undialyzed serum, the value was $0.6 \pm 0.2\%$, and for predialyzed serum to which maximal concentrations of 21 amino acids expected in the blood had been readded, the percentage was $0.75 \pm 0.2\%$. (Similar data were obtained for Cu : albumin molar ratios of up to 0.6 : 1, and much larger percentages resulted when ratios approached 1 : 1 or more.) What this showed was that in the absence of most small metabolites normally found in the blood, only a tiny fraction of the copper attached to proteins dissociates to free copper, in equilibrium with the protein-bound pool. However, a significantly larger portion is in the low-molecular-weight fraction of serum, which contains nonprotein ligands. The latter can be mimicked by adding back amino acids at physiological concentrations. This implies, first, that other metabolites in the plasma are not involved in binding copper and, second, that Cu–amino acid complexes are in equilibrium with (exchangeable) Cu bound to proteins. With a higher Cu : albumin ratio of 2 : 1, Neumann and Sass-Kortsak (1967) examined the specificity of the amino acid effect, concluding that, *on an equimolar basis*, histidine, cystine, asparagine, and threonine were most important (in that order). However, *on the basis of maximum physiological concentration*, the order was histidine, glutamine, threonine, and cystine (again in decreasing order of importance). [Cysteine (versus cystine) is unlikely to be involved, as it would reduce Cu(II) to Cu(I), thus lowering its binding affinity.] Their evidence also suggested that other amino acids, and interactions of the most important amino acids, synergistically enhanced copper release from the protein fraction.

The conclusions from these original findings still largely hold. Sarkar and Kruck (1966) investigated the copper associated with ultrafiltrates of human serum with thin-layer chromatography, concluding that it was associated largely with histidine and threonine (as a His$_2$ and His-Thr complex). Data on the stability constants of Cu(II) with various amino acids (Table 4-5) also support the overriding importance of histidine, especially as the bis complex. The studies of Ettinger and his colleagues (1986; see Chapter 3) on uptake of Cu(II) by cultured cells in the presence of albumin with and without histidine and those of Lau and Sarkar (1971) on dissociations of the ternary albumin-Cu-His complex confirm the importance of this amino acid in bringing albumin-bound copper into equilibrium with that of the low-molecular-weight fraction.

May et al., (1976, 1977) carried out computer simulations aimed at quantitating the percentage of copper in the low-molecular-weight fraction that is associated with various plasma/serum amino acids and other metabolites, work that again supported the importance of histidine and the

Table 4-5. Trends of Stability and Stability Constants Observed for Copper–Amino-Acid-Anion–Metal Complexes[a]

Amino acid	Trend[c]	log Stability constant	
		CuL	CuL$_2$
Simple amino acids[d,e]	0	8.0	14.7
Alanine	0	8.0	14.6
Aminobutyric acid	−	7.7	14.0
Arginine	−	7.4	13.7
Asparagine	−	7.7	13.7
Glutamic acid	+	8.7	14.9
Glutamine	−	7.2	13.4
Glycine	−	8.0	14.7
Histidine	0	9.8	17.5
Isoleucine	0	8.0	14.7
Leucine	0	8.0	14.7
Lysine	0	9.3	14.6
Methionine	−	7.7	14.1
Ornithine	+	9.8	14.8
Phenylalanine	−	7.7	14.4
Proline	−	8.7	16.0
Serine	−	7.6	14.0
Threonine	−	7.6	14.0
Tryptophan	+	8.1	15.3
Tyrosine	+	9.1	15.1
Valine	0	7.9	14.6

[a] Modified from May et al. (1977).
[b] Cumulative stability constants measured at 37°C; $l = (0.15 \text{ mol/dm}^3)$ [KNO$_3$].
[c] "Trend" means the strength of binding of the particular amino acid relative to the average for the series of similar ligands (L).
[d] Alanine, glycine, leucine, isoleucine, and valine.
[e] It should be noted that (while not present so much in blood plasma) sugar alcohols and amino sugars also form strong complexes with Cu(II) (Micera et al., 1989) (see also Chapter 7).

other amino acids mentioned (Table 4-6). For these studies, the best available information on stability constants and concentrations of 7 metal ions and 40 potential ligands, interacting and forming up to about 5000 potential complexes, was considered. Where missing, necessary constants were estimated by reasonable or accepted procedures. It may be seen that at least 92–94% of the total Cu estimated to be attached to low-molecular-weight (amino acid-sized) ligands was computed to be in the form of diamino acid complexes. Also noteworthy (and agreeing with earlier conclusions) is that little or no copper was estimated to be associated with non-amino acid substituents, such as citrate, lactate, salicylate, or ascorbate. Naturally, computer simulations will not necessarily provide correct answers with regard to what is occurring in the complex microenvironment of the blood, with its may still unrecognized components, but in this case some of the answers are consistent with other data.

Further evidence for the existence of copper associated with low-molecular-weight material comes from gel permeation chromatography of whole serum, (a) after addition of tracer radiocopper (copper components detected by radioactivity) or (b) with no additions (copper detected by fur-

Table 4-6. Distribution of Exchangeable Copper(II) among Low-Molecular-Weight Ligands in Blood Plasma[a]

Cu(II) complex	Charge	Percent of the total metal in the low-molecular-weight fraction		
		pH 7.2	pH 7.4	pH 7.6
Cystinate histidinate	−1	16	21	28
Protonated cystinate histidinate	0	20	17	14
Bis (histidinate)	0	12	11	10
Histidinate threoninate	0	9	8	7
Histidinate valinate	0	5	5	5
Protonated histidinate lysinate	+1	5	5	4
Alanate histidinate	0	4	4	4
Histidinate serinate	0	4	4	4
Histidinate phenylalanate	0	3	3	3
Glycinate histidinate	0	3	3	3
Histidinate leucinate	0	2	2	2
Glutamate histidinate	−1	2	2	2
Glutaminate histidinate	0	2	2	2
Protonated histidine ornithinate	0	2	2	1
Histidinate prolinate	0	2	1	1
Histidinate isoleucinate	0	1	1	1
Histidinate tryptophanate	1	1	1	1

[a] Modified from May et al. (1977).

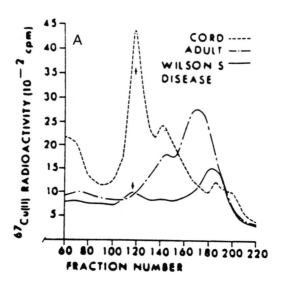
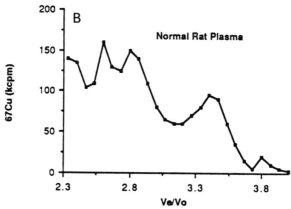
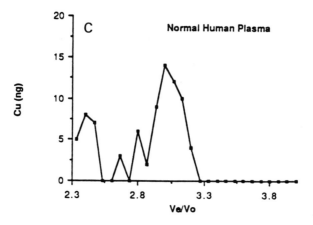

nace atomic absorption or activation analysis). As illustrated by the data in Fig. 4-11A, a significant but tiny portion of the added radioactivity is associated with components eluting near the column volume (fraction 180). Even more interesting is the apparent presence of several components with molecular weights between 1000 and 15,000. [Cytochrome *c*, which is 13,000 Da, would elute at about fraction 110 in this system; Lau and Sarkar, 1984.] The data for umbilical cord blood serum are particularly impressive, with about four components. These data are consistent with those of Weiss (1983) and Weiss *et al.* (1985) using rat plasma labeled *in vivo* with ^{67}Cu (Fig. 4-11B), where multiple components are again evident. They are also, in principle, consistent with the data of Wirth and Linder (1988) for samples of normal and cancerous human serum, chromatographed on Sephadex G-150 and assayed directly for Cu by furnace atomic absorption (Fig. 4-11C). Several peaks, with specific elution volumes, usually appeared at the end of the bed volume and beyond (V_e 80–120% of the column volume). In samples from some individuals (data not shown), one or more of these peaks became prominent (Wirth and Linder, 1985). Again, we are here not just seeing copper associated with amino acids and small metabolites, but with some larger substituents. Lau and Sarkar (1984) have suggested that polyamine conjugates with small peptides may be involved and have reported on work in progress to isolate one of these components from cord blood.

Apart from amino acids, the identities of copper-binding species that could be present in blood plasma or other intracellular vertebrate fluids, with a molecular weight in the range well below that of albumin, are essentially unknown. One exception is histamine (Berthon and Kayali, 1982); it has also been shown that administration of copper (but not zinc) can decrease the likelihood of lethal anaphylactic shock (Walker *et al.*, 1975) (anaphylactic shock being due mainly to the release of histamine from mast cells and basophils). [The formation constant for the histamine$_2$–Cu complex is on the order of 10^{21} (Kayali and Berthon, 1980).] Another exception is a peptide, probably glycylhistidyllysine (GHL), reported to be with

---

Figure 4-11. Elution of low-molecular-weight copper components in gel permeation chromatography of serum. (A) Elution of radioactive copper after addition of tracer ^{67}Cu(II) and fractionation on Sephadex G-100. Only the elution of low-molecular-weight components is shown, for normal human serum, cord serum, and serum from a patient with Wilson's disease. Reprinted, with permission, from Sarkar *et al.* (1983). (B) Elution of normal rat plasma, labeled with ^{67}Cu(II) by intraperitoneal injection, on Sephadex G-150 in 0.9% NaCl (Weiss, 1983). (C) Elution of copper, measured by furnace atomic absorption spectrophotometry, from normal serum applied to Sephadex G-150, in 20 mM potassium phosphate, pH 7.0 (P. L. Wirth and M. C. Linder, unpublished results).

Cu(II) [and Fe(III)] in human plasma by Pickart and colleagues (Pickart and Thaler, 1973, 1980; Pickart *et al.*, 1979; Schlesinger *et al.*, 1977). This peptide was apparently most abundant in the albumin-rich fraction, suggesting that it might be in the form of a ternary complex with albumin and Cu, as was previously found for histidine and albumin (Lau and Sarkar, 1971). The Cu–GHL complex was shown to have growth-modulating properties in culture, apparently prolonging the survival of hepatocytes and stimulating hepatoma cell growth and intercellular adhesion (Pickart and Thaler, 1980). More recently, Pickart has named the Cu–GHL complex "iamin" and has provided some evidence that its application to wounds and burns enhances their rate of healing (as compared with that of control wounds to which vehicle alone was applied) (Pickart, 1987). Evidence was also put forward that GHL might be having its effects by stimulating uptake of copper. Using HTC_4 hepatoma cells, Pickart *et al.* (1980) showed that 10–200 ng/ml concentrations of GHL stimulated uptake of copper, but not iron, in a serum-free, amino acid-containing medium. However, in more extensive work with rat hepatocytes in culture (see Chapter 3), Darwish *et al.* (1984b) found that, at low ratios of GHL to Cu, GHL inhibited uptake in the same manner as albumin, and the effect was additive with that of albumin, at 10 mM concentrations of each. Since the normal concentration of the tripeptide has been estimated at 200 ng/ml, or about 0.6 μM, while that of albumin is about 0.65 M (10^6-fold higher), it seems unlikely that the quantitative effect of GHL is significant with regard to Cu uptake inhibition compared with that of the (much larger) protein albumin.

The presence of an additive inhibitory effect also implies that Cu is not in a ternary complex with GHL and albumin, although the careful potentiometric, spectral, and equilibrium dialysis studies of Lau and Sarkar (1981a, b) allow for such a possibility. These latter inverstigators also showed that the stability constants of complexes of copper with GHL are greater than those of complexes of copper with histidine ($10^{16.4}$ and 10^{21}–10^{38} versus $10^{9.8}$ and $10^{17.5}$, for the mono and bis complexes, respectively). At high concentrations, GHL was also more effective than histidine in removing Cu(II) from Cu(II)-albumin [42% bound to GHL at equimolar concentrations with albumin, in equilibrium dialysis (Lau and Sarkar, 1981a)]. That this is relevant at physiological ratios (which are 10^6-fold lower) is not supported by the uptake work already cited. This does not mean that GHL cannot have growth modulatory functions. The X-ray crystallographic work of Pickart's group (Pickart *et al.*, 1980) has identified tridentate bonding of Cu(II) to the imidazole, α-NH_2, and peptide nitrogens of GHL in a polymeric structure. The assignment of these nitrogens is consistent with the potentiometric work of Lau and Sarkar

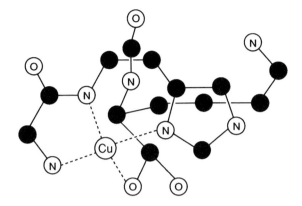

Figure 4-12. Possible structure of the glycylhistidyllysine (GHL) tripeptide–Cu(II) complex. Based on the X-ray structure of Pickart *et al.* (1980) and data and suggestions of Sarkar *et al.* (1983). Peak maximal absorbance is at about 610 nm.

(1981a, b). However, as the stability constant is much higher than expected for this tridentate complex, Sarkar *et al.* (1983) suggested that a carbonate oxygen is also involved, for an overall square planar configuration in solution. This would agree with their proton displacement and spectroscopic findings. Such a structure is shown in Fig. 4-12.

4.1.4 Other Copper-Binding Proteins in Blood Plasma/Serum

4.1.4.1 Histidine-Rich Glycoprotein

As already partly recounted (Chapter 3), the α_2-glycoprotein known as histidine-rich glycoprotein (HRG) was first isolated from human serum by Heimburger *et al.* in 1972 and was subsequently isolated by others from human and rabbit serum (Morgan, 1978, 1981), human plasma (Lijnen *et al.*, 1980; Koide *et al.*, 1982), and platelets (Leung *et al.*, 1983), where it is released during thrombin stimulation. It tightly binds heparin (and less tightly to plasmin and thrombospondin) and has a N-terminal amino acid sequence homologous with that of antithrombin (Koide *et al.*, 1982). It also interacts with the lysine binding site of plasminogen (Lijnen *et al.*, 1980), inhibits T-cell expression of IL-2 receptors (Saigo *et al.*, 1989), and binds heme and metal ions (Morgan, 1985). Human HRG has a single gene located on chromosome 3 (Van den Berg *et al.*, 1990). It has been cloned (Koide *et al.*, 1986) and consists of a sequence of 507 amino acids (with an 18-amino acid leader sequence), more than half of which contains five different types of internal repeats (Fig. 4-13). The last three, larger repeats (type V; between amino acids 320 and 400) have 12 tandem sub-repetitions of the sequence Gly-His-Pro-His. This histidine-rich portion of the protein accounts for more than half of its overall histidine content and most probably its metal-binding properties. Recent work of Morgan *et al.* (1989) indicate that a sequence (gly-his-phe-pro-phe-his-try) in the C-terminal part of the protein is critical for its metal (and low-spin heme) binding. The five types of amino acid, tandem repeat sequences cited also show considerable homology to the histidine-rich region of high-molecular-weight kininogen (Koide *et al.*, 1986), and thus probably to the superfamily of "cystatins" (cysteine protease inhibitors) (Rawlings and Barrell, 1990).

The total molecular weight of human HRG is about 67,000, which includes 14% carbohydrate, probably attached to asparagines 45, 69, 107, and 326 (and/or 327), with a sedimentation coefficient of 3.8 S. There are 64 His residues in human HRG (12.6% of the total amino acids), 35 of which are located between residues 320 and 400 (already indicated), along with considerable proline and glycine. This would make it elute in about the same position as albumin in the typical gel permeation columns that have been used to fractionate plasma substituents binding copper.

The rabbit protein (Morgan, 1978; Guthans and Morgan, 1982) as well as a 30,000-Da fragment (containing at least part of the histidine-rich region) was found to bind multiple Zn(II), Ni(II) and Cd(II) ions, with considerable avidity and positive cooperativity and more strongly than albumin. Dissociation constants for these metals were in the range of 10^{-8} M. Copper(II) seemed to have the same effect as Zn(II), totally inhibiting heme binding at a molar ratio of 1:2 (metal ion:heme). HRG is present in normal human serum at levels in the range of 11–13 mg/dl (110–130 µg/ml) or a little less, about one-third the normal concentration of ceruloplasmin. There is a slight increase with heart disease and a decrease during pregnancy, toward parturition (Morgan *et al.*, 1978; Failla *et al.*, 1982). Concentrations are low at birth (about 20% of adult levels) but almost reach mature levels (of about 86 µg/ml) by 2 years of age (Corrigan *et al.*, 1990).

The question then is whether HRG normally binds some of the copper in plasma/serum. As the K_D for Cu(II) of rabbit HRG is on the order of 10^{-6} M (Morgan, 1981), with about ten metal-ion-binding sites per molecule, it seems unlikely that HRG could compete with albumin (or histidine) for copper ions. [These substituents have binding constants of 10^{17} or more (Table 4-5).] It does compete with albumin for zinc binding (Morgan, 1981). HRG may nevertheless bind some serum copper in certain circumstances, as, for example, in the case of the dog (which has no high-affinity Cu site on albumin). All this remains to be explored.

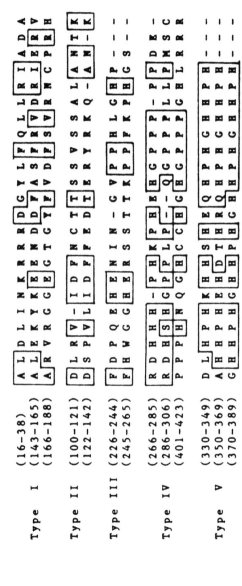

Figure 4-13. Internal histidine-rich repeats in the primary structure of histidine-rich glycoprotein (HRG). Numbers of the amino acid residues, from the N-terminus, are indicated on the left. Identical residues in each repeat are enclosed in boxes. Gaps were inserted for maximal homology in the amino acid sequences. The single-letter code for amino acids is as follows: Ala, A; Asn, N; Asp, D; Cys, C; Gln, Q; Glu, E; Gly, G; His, H; Ile, I; Leu, L; Lys, K; Met, M; Phe, F; Pro, P; Ser, S; Thr, T; Trp, W; Tyr, Y; Val, V. Reprinted, with permission, from Koide et al. (1986).

4.1.4.2 Ferroxidase II

Ferroxidase II was originally detected by Topham and Frieden as a nonceruloplasmin (non-azide-inhibitable) Fe(II) oxidizing activity in mammalian plasma (Johnson et al., 1967; Topham and Frieden, 1970; Sung and Topham, 1973; Topham and Johnson, 1974) and may serve as a means for oxidizing Fe(II) to Fe(III) (water is the other product of the reaction) to allow attachment to transferrin in species (such as birds) or conditions (such as Wilson's disease) in which ceruloplasmin levels are low. (See Chapter 7, Section 7.2.1.2a for a discussion of these aspects.) In normal human serum, it accounts for 5% or less of total ferroxidase activity, the rest being due to ceruloplasmin (ferroxidase I) (Topham and Frieden, 1970; Gutteridge et al., 1985). Ferroxidase II has an apparent total molecular weight of about 800,000 (Topham and Frieden, 1970) and is comprised of two very large subunits (M_r 320,000 and 220,000), the smaller of which has the copper and catalytic activity (Lykins et al., 1977). One copper atom per molecule appears to be necessary for activity (Garnier et al., 1981), although the protein has been isolated with much more copper loosely attached. The enzyme appears also to be a lipoprotein, with about 15% phospholipid and 11% cholesterol (by weight) attached to the smaller subunit. [The larger subunit appears necessary for the stability and full activity of the enzyme (Lykins et al., 1977).] The lipid (and copper) is removable with phospholipase C (Sung and Topham, 1973; Topham et al., 1975). The copper is partly removable at pH 4–5, with 0.1 M EDTA. (The copper is bound but not removed by diethyldithiocarbamate.)

No direct estimates of the potential contribution of this enzyme to the copper content of the plasma appear to be available in the literature. However, from the activity of the enzyme in rat plasma (about 190 mmol/min per ml) and the specific activity of the purified enzyme (about 190 mmol/min per mg) (Topham et al., 1975), the concentration of the enzyme in rat serum can be calculated as about 1 mg/ml. At 0.07% Cu, this would mean about 700 ng of copper on ferroxidase II per ml. This cannot be correct and may be due to adhesion of additional copper picked up during enzyme purification. Indeed, the isolated human and rat enzymes have been reported to contain 12 nmol of Cu/mg, which translates into about seven atoms per molecule. On the basis of one (tightly bound) Cu per molecule (required for activity), the contribution to rat plasma or serum would then only be 100 ng/ml, which is much more reasonable. Calculations from human data give a much lower value: with a specific activity of the isolated ferroxidase II of 4.9 ΔA_{460}/10 min per mg of protein (Topham and Frieden, 1970), a serum activity of 0.0105 ΔA_{460}/10 min per mg of protein, and assuming a serum protein concentration of 65 mg/ml, the ferroxidase concentration is calculated to be 0.14 mg/ml of serum. Assuming one copper atom per molecule, this comes to about 15 ng of Cu on human ferroxidase II per ml of serum or plasma. As will be seen later in this chapter, Topham *et al.* (1975) have shown that ferroxidase II activity (like ceruloplasmin) markedly decreases in rats placed on a copper-deficient diet, confirming the importance of copper for enzyme activity. Because at least the essential copper is not labile to diethyldithiocarbamate, it seems unlikely that ferroxidase II is part of the labile copper pool of the blood plasma, although this cannot be entirely ruled out.

4.1.4.3 Transcuprein and α_2-Macroglobulin

Transcuprein was discovered as a radioactive copper peak eluting in the void volume of Sephadex G-150, upon application of ^{67}Cu-labeled plasma from rats given ionic radiocopper (shortly before) by intragastric or intraduodenal intubations or by intravenous or intraperitoneal injection (Linder, 1983; Weiss *et al.*, 1985; Linder *et al.*, 1987). Labeling could also be obtained instantaneously *in vitro*. Lau and Sarkar (1984) had previously also found a portion of radiocopper in the void volume when they fractionated human plasma [with added ^{64}Cu(II)] on Sephadex G-100. There appeared to be rapid equilibration between the copper in this peak and in albumin (Weiss *et al.*, 1985). This is exemplified by the data in Fig. 4-14, showing the distribution of added ^{67}Cu(II) between albumin and transcuprein in mixtures containing different proportions of these proteins (Fig. 4-14A) and in serum samples to which different amounts of actual copper had been added (Fig. 4-14B) (Linder *et al.*, 1988; H. M. Vu and M. C. Linder, unpublished). As shown, transcuprein competed very well for Cu against albumin, although even under the "best" of circumstances (A-2), it was present in amounts vastly lower than albumin (albumin is 65% of plasma protein). This implies that transcuprein has a greater affinity for copper than does albumin (in this case, in the absence of histidine). The implication is corroborated by the data for whole serum (containing histidine) (Fig. 4-14B), which show that the capacity of transcuprein for binding added copper is saturated long before that of albumin, at their respective lopsided physiological concentrations (concentrations that should greatly favor albumin). Also bearing this out are data on the off-rates of ^{67}Cu from partially purified fractions of transcuprein and albumin (Linder *et al.*, 1988; see Fig. 4-15). In these studies, ^{67}Cu-transcuprein and ^{67}Cu-albumin (obtained from whole rat serum gel filtration) were dialyzed at pH 7.0 in 20 mM phosphate–0.9% NaCl for several days. Off-rates for ^{67}Cu from both proteins were linear and similar, amounting to about 1.4% per 24 h. Assuming that on-rates are diffusion controlled (10^{10} s^{-1}), this

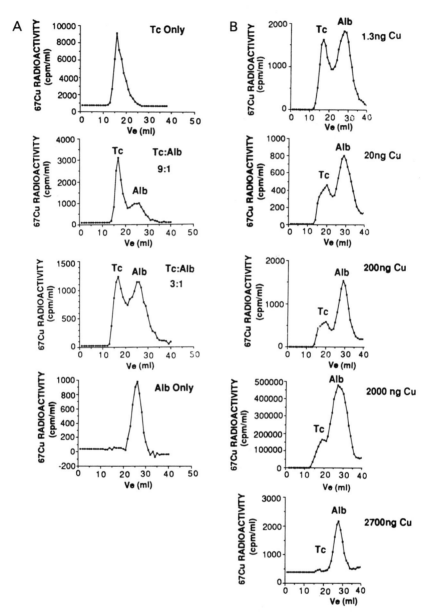

Figure 4-14. Distribution of copper between transcuprein and albumin fractions in various mixtures and under various conditions (H. M. Vu and M. C. Linder, unpublished results). (A) Trace (10-ng) quantities of ^{67}Cu(II), as the 1:1 nitrilotriacetate (NTA) complex, were added to different mixtures of rat plasma transcuprein and albumin, isolated by gel chromatography, in 20 mM phosphate, pH 7.0, and fractioned on Sephadex G-150. The elution of radioactivity is shown, with ratios of transcuprein to albumin decreasing from top to bottom. (B) Increasing quantities of Cu(II), as Cu-NTA (as indicated), were added to whole rat serum prior to fractionation on Sephadex G-150. The corresponding changes in distribution of eluted ^{67}Cu in the void volume (transcuprein) and albumin fractions are shown.

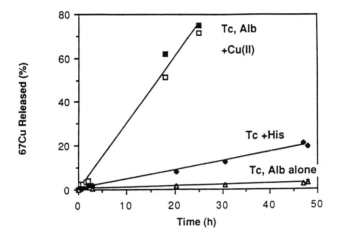

Figure 4-15. Off-rate of ^{67}Cu from rat transcuprein and albumin, during dialysis in 0.15 N NaCl, 20 mM potassium phosphate, pH 7.0, at 4°C, in the presence and absence of excess nonradioactive 100 μM Cu(II)-NTA or His (M. C. Linder, unpublished results).

indicates that K_Ds for both rat proteins are in the range of 10^{-17} M. [This is similar to the value obtained by Lau and Sarkar (1971) for human albumin and Cu(II).] Transcuprein thus emerges as a protein with a high affinity for newly absorbed or added Cu(II) in human and rat serum or plasma, Cu(II) that is in equilibrium with the copper on albumin.

Other known properties of transcruprein are summarized in Table 4-7, including an apparent molecular weight of 270,000, which distinguishes it from ferroxidase II (as well as from ceruloplasmin and albumin). Due to its low concentration, it has only just been isolated in pure form (Goode et al., 1989) and is currently being characterized (M. T. Tsai, and M. C. Linder, unpublished results). It is noteworthy that the protein does not react with antiserum against ceruloplasmin or albumin (Weiss et al., 1985) and has a higher molecular weight than that of most known plasma proteins. It behaves as a globulin, precipitating with $(NH_4)_2SO_4$, especially in the range of 30–50% saturation (4°C). Because of the lability of transcuprein copper to the typical high-pH buffers used in nondenaturing electrophoresis (Table 4-7; Linder et al., 1987b), it has not yet been established into what globulin subgroup it falls. Also of importance is the observation that a significant amount of the actual copper in serum (detected by furnace atomic absorption spectroscopy) elutes in the same void volume fractions of Sephadex G-150 or G-100 as transcuprein (see more in Section 4.2). What portion of this copper is actually on transcuprein will not be clear until an assay for the protein has been developed and its

Table 4-7. Summary of Properties of Rat Transcuprein

Behavior in purification procedures	
Procedure	Result
Ammonium sulfate fractionation	Precipitates between 30 and 50% saturation
Sephadex G-150 chromatography	Elutes in the void volume
Ultrogel AcA 34 chromatography	Elutes with M_r 270,000
DEAE-cellulose chromatography	Elutes with about 0.15 M NaCl in 0–0.5 M gradient, phosphate, pH 7.0
Pseudo affinity chromatography (Affigel Blue)	Elutes with 0.5 M KSCN
Metal affinity gel chromatography	Binds tenaciously

Properties of purified transcuprein (^{67}Cu labeled *in vivo* or *in vitro*)

Nondenaturing disc PAGE, pH 8.8 (5% separating gel): One major band, R_f 0.25
SDS electrophoresis: Three major bands appear to have a subunit M_r of about 200,000, 100,000, and 66,000
K_D about 10^{-17} M (rat protein)

stoichiometry of copper binding determined. (Some of this copper is attached to ferroxidase II, which also elutes in the void volume.)

In the process of purifying transcuprein, it appeared that at least a small portion of the void volume ^{67}Cu (on Sephadex G-150) was with a very large plasma protein precipitating with 0–35% saturation ammonium sulfate. Nondenaturing polyacrylamide gel electrophoresis (in 5% acrylamide) of the partially purified radioactive component revealed a major and two minor bands, barely entering the gel (consistent with the behavior of α_2-macroglobulin). In large-pore (Ultrogel AcA 34) gel chromatography, a portion of the serum radioactivity was with a component in the void volume ($M_r > 700,000$). (The alternative is that it is on ferroxidase II.) The immunoreactivity of this copper component and the stability constant of α_2-macroglobulin for copper need to be measured and will be compared with a value $3 \times 10^7 – 8 \times 10^7$ M obtained by several investigators for multiple Zn(II) sites (Parisi and Vallee, 1970; Adham *et al.*, 1977; Pratt and Pizzo, 1984). [α_2-Macroglobulin is considered a major plasma ligand for Zn (Parisi and Vallee, 1970; Giroux, 1975). In fact, the portion of plasma zinc bound to this protein is not considered exchangeable (May *et al.*, 1977).] The "exchangeable" (nonceruloplasmin) serum copper pool is probably about half as great as that for zinc (40 versus 65%, respectively, of a similar total plasma concentration). All in all, more work is needed to determine whether a small portion of the exchangeable serum Cu is indeed associated with this protein, *in vivo*, although about 10% of the radioactivity of tracer ^{67}Cu(II) added to rat serum is associated with the 0–35% saturation ammonium sulfate fraction (M. C. Linder, unpublished results).

4.1.4.4 Amine Oxidase, Superoxide Dismutase, and Blood Clotting Factor V

Small amounts of a copper-containing *amine oxidase* (and diamine oxidase) are also present in the blood plasma of vertebrates, as well as in vertebrate tissues, where they are thought to be important in the inactivation and catabolism of physiologically active amines, such as histamine, tyramine, and polyamines. Various copper-containing amine oxidases are found in the tissues of most living organisms, and their structure and function are reviewed more extensively in Chapters 6 and 10. Reviews of the serum amine oxidases include those of Pettersson (1985) for structure and reactions and Bombardieri *et al.* (1985) for clinical aspects. Briefly, the plasma proteins, as exemplified by the well-studied pig amine oxidase and beef diamine oxidase enzymes, come as two-subunit molecules, of about 190,000 Da, with 7–10% carbohydrate. Each subunit contains one Cu(II) atom bonded to two N and four O ligands which are detectable by EPR spectroscopy. The copper is tightly bound but can be removed (at lower pH?) by dialysis against diethyldithiocarbamate. Visible absorbance in the 400–500-nm region is due to the presence of another cofactor, the structure of which has just recently been elucidated for amine oxidase (Jane *et al.*, 1990) but which as long thought to be related to pyridoxal, and more recently, pyrroloquinoline quinone (Moog *et al.*, 1988; Hartman and Klinman, 1988). This cofactor, 6-OH dopa, is involved in the oxidation/reduction reactions of the enzyme.

The serum amine oxidases, though similar in structure, display species differences in substrate specificity. Enzymes from all sources will oxidize benzyl- and heptylamine, by the reaction $RCH_2NH_2 + O_2 + H_2O \rightarrow RCHO + H_2O_2 + NH_3$. However, the enzyme from nonruminants (human, horse, pig, and rabbit) is active mainly with primary (biogenic) amines, such as histamine, tyramine, dopamine, and serotonin, and may be important for their inactivation. At the same time, the enzyme from these sources is not very active with diamines (such as spermidine or spermine). In contrast, the bovine and sheep enzymes are much more active against the latter (versus the former) substrates and are thus mainly diamine oxidases. They are often even referred to as spermine oxidase. [The flavin amine oxidases (containing no copper) can oxidize primary, secondary, and even tertiary amines (Hamilton, 1981; Mondovi, 1985).]

It is unclear just how much amine (or diamine) oxidase protein is in vertebrate plasma, though it falls into the class of trace enzymes used in clinical diagnosis, thought to be leaked by tissues. Conditions that result in elevated serum activity have the common denominator of connective tissue activation/deposition: liver fibrosis, congestive heart failure, hyperthyroidism, childhood, and senescence, for example (Bombardieri et al., 1985). This had led to the hypothesis that the serum enzyme originates from fibroblasts, although plasma-type amine oxidase activity has not been detected in this cell type. As very little enzyme is present in the plasma, it seems unlikely that (at least normally) more than a few nanograms of Cu in this fluid would be attributable to amine oxidase. Since the enzyme is important for hormone inactivation, it also may not be part of the exchangeable serum copper pool and is not likely to be involved in copper transport.

Small amounts of a copper-containing *extracellular superoxide dismutase* (EC-SOD) are also found in blood plasma and extracellular fluids (Marklund, 1982) (Table 4-8). This enzyme is not the same as the well known Cu/Zn SOD found in most vertebrate (and many other) cells. It is a somewhat hydrophobic, 132,000-Da tetramer, composed of subunits of

Table 4-8. Distribution of "Extracellular" Superoxide Dismutase in Blood Plasma and Tissues of Three Species[a]

Tissue	Superoxide dismutase (U/g or U/ml)		
	Human	Rat	Mouse
Blood plasma	26 ± 4 (51)[b]	332	400
Liver	69	20	301
Skeletal muscle	75	32	55
Spleen	89	16	110
Heart	223	35	193
Kidney	315	79	416
Intestine	395[c]	19	127
Lung	545[c]	117	3225
Thyroid	1210	—	—
Uterus	1260[c]	—	—
Brain			
Gray matter	63[c]	—	—
White matter	125	—	—
Lymph	61 ± 5 (3)[b]	—	—
Cerebrospinal fluid	44 ± 7 (8)[b]	—	—

[a] From Marklund (1984a, c). Mean values for two determinations. Most values were consistent.
[b] Karlsson and Marklund (1987).
[c] Highly variable (more than 100% difference between two values).

222 amino acids, of which the sequence from 96 to 193 has about a 50% homology with the final two-thirds of that for eucaryotic Cu/Zn SODs (Tibell et al., 1987). (Residues 194–222 are involved in the characteristic heparin binding of this enzyme.) Each subunit contains one copper atom and probably also one zinc atom. Although, in humans, EC-SOD activity appear to be much more concentrated in tissues than in blood plasma (Table 4-8), the enzyme is a glycoprotein and is translated with an 18-amino acid leader sequence, both of which facts suggest that it is a secreted protein and should be extracellular. [Indeed, it is absent from a variety of cells grown in culture (Marklund, 1984c).] The reason for its high tissue concentration may be that it attaches to (extracellular) glycosaminoglycans of connective tissue, such as heparan sulfate. In this connection, Marklund (1984b) has observed that administration of heparin increases plasma concentrations of EC-SOD, suggesting that this glycosaminoglycan allows release of the enzyme from heparin-like sites.

The tissue/cell origin of EC-SOD is still unclear, although the liver should probably be ruled out, due to its low activity (Table 4-8). The data (Table 4-8) for tissue versus blood plasma concentrations in three species are not very enlightening as to the cell source. Whereas, in the human, activity in plasma is low relative to that in other tissues, this is not the case for the rat, for example. Similarly, lung concentrations of EC-SOD are very high in the human and the mouse, but not nearly so high in the rat. (Are there differences in enzyme specific activity?) Nevertheless, generally, concentrations in the lung (and in the thyroid and uterus) are high. This may reflect the function of EC-SOD (with ceruloplasmin) as a scavenger of superoxide and in the general protection of this organ against oxidative damage. (The lung with its direct air contact may be particularly susceptible to this type of damage.) Marklund (1986) has estimated that, at least in humans, the EC-SOD activity of blood plasma is such that it will scavenge twice as much superoxide as ceruloplasmin. However, EC-SOD may not be an acute phase reactant and may therefore be proportionately less important than ceruloplasmin in the inflammatory process.

The potential/actual contribution of EC-SOD to the copper content of plasma and extracellular fluids in different species is still far from clear, but it may be significant. Based on its molecular weight, it would elute in the same position as ceruloplasmin and thus may contribute to the copper commonly associated with this peak in gel permeation chromatography (see below). From the specific activity of the pure enzyme (Marklund, 1982) and the activity in blood and other extracellular fluids (Marklund et al., 1982), it can be estimated that human plasma contains about 9 μg/ml of the enzyme. With a copper content of 0.19% (four Cu atoms per 132,000 Da), this would indicate that about 18 ng of copper are present per ml. Clearly, much remains to be learned about this interesting copper protein and its role in metabolism.

Blood clotting factors Va and VIII have both been found to have amino acid sequences with significant homologies to repeated portions of ceruloplasmin (Church *et al.*, 1984). This lead to the finding that these factors are also cuproteins (Mann *et al.*,1984; Walker and Fay, 1990), containing one atom of Cu per molecule for the human and bovine species. Factors V and VIII are both nonenzymatic substituents of the coagulation process and are probably produced by endothelial cells, such as those lining the blood vessels (Cerveny *et al.*, 1984). Both are activated (to factors Va and VIIIa) by thrombin, which in turn has been formed from prothrombin by factor Va in conjunction with Ca^{2+}, serine protease Xa, and cell/platelet membrane (as part of a positive feedback loop). [VIIIa in turn forms Xa (from X) with the help of serine protease IXa.] A common decapeptide sequence on ceruloplasmin allows the latter to inhibit interaction of Factors V and VIII with activated protein C, at least *in vitro* (Walker and Fay, 1990). Factors V and VIII (330 and 240 kDa, respectively) circulate in trace amounts, until activated by cleavage to active peptides of 94 + 74 and 82 + 69 kDa, respectively, for the human system. Human plasma contains about 10 μg of factor V per ml (Nesheim *et al.*, 1982) (and perhaps an equal amount of factor VIII). (The pig has less than 1 μg/ml of factor V per ml.) At 0.019% Cu (by weight), this would translate into 2 ng of Cu/ml of plasma, in humans. Thus, factor V is a very small, but nevertheless real, contributor to the total copper in the blood plasma.

4.2 Amounts of Copper Associated with Various Substituents in Blood Plasma and Extracellular Fluid

For a long time, it was considered that ceruloplasmin accounted for 90–95% of the copper in blood plasma and serum and that the remaining 5–10% was on albumin and amino acids. This conclusion was based largely on measurements of copper available to diethyldithiocarbamate (DDC) and lack of evidence that added copper would associate with anything but albumin (and amino acids, as already discussed; see Section 4.1.3). As regards DDC, Gubler *et al.* (1952a) had reported that little or no ceruloplasmin copper was available to this chelating agent at pH 7, although half or all could be removed at acid pH (Morell *et al.*, 1964). Using addition of DDC to human serum samples, in the presence of a high concentration of sodium pyrophosphate, Gubler *et al.* (1952) then reported that about 5% of the total copper present could be detected as the colored DDC complex. At least in retrospect, there are two problems with these findings. First, use of DDC to detect quantities of copper below 0.5–1 μg is highly unreliable. Second, it is not certain that some of ceruloplasmin-bound copper is not available also to DDC and other such chelating agents. Using commercially supplied ceruloplasmin (admittedly not ideal but perhaps not any worse than some of the original protein preparations used by others), Jackson (1978) reported the release of considerable ceruloplasmin copper to EDTA, penicillamine, and tetra-ethylenetetramine (at concentrations of 25 mM), with similar findings for fresh human serum and plasma, an effect that was enhanced by DDC.

With regard to the matter of albumin (and amino acids) as the only other large serum copper ligand, this conclusion was based (a) on evidence (already reviewed) that albumin does indeed have a high-affinity binding site and (b) on studies in which relatively large doses of copper, as ^{64}Cu(II) were added to samples *in vitro* or injected into animals. Under such conditions, very little radioactivity would be able to associate with the less abundant binding components, as they would be saturated with copper at low concentrations. As increasing radioactivity Cu is added, the proportion attaching to albumin (which has a huge excess of unoccupied sites) increases and that on other components must diminish. (This is illustrated also by the data in Fig. 4-14B.)

Additionally, the conclusions about 90–95% serum copper on ceruloplasmin (the rest on albumin) were drawn in the days before the development of current, much more sensitive, detection methods for copper, namely, furnace atomic absorption spectroscopy and activation analysis. With the use of the latter techniques, a new picture of the quantities of copper with various serum components is emerging, exemplified by the data in Tables 4-8, 4-9, and 4-10 and Fig. 4-16. Sections A and B of Fig. 4-16 show elution of nonradioactive, endogenous copper from human serum upon fractionation on Bio-Gel P-150 or Sephadex G-150, respectively. In both cases, a significant amount of copper is evident in the void volume. It was suggested that this might be contamination on the column, swept off by the void volume protein fraction (Evans and Fritze, 1969; Gardiner *et al.*, 1981). However, precleaning of gel with borohydride did not alter the results, and, of course, a significant radioactive peak (which could not have been from the column) was also found in fractions from samples labeled with ^{67}Cu *in vivo* (Weiss *et al.*, 1985) and *in vitro* (Lau and Sarkar, 1984; Weiss, 1983; Weiss *et al.*, 1985). L. Barrow and M. S. Tanner (1988) (Fig. 4-16C) have recently confirmed the presence of a significant percentage of serum copper in the transcuprein/void volume fraction of Sephadex G-150, for normal and diseased individuals. A summary of the reported results is given in Table 4-9. From the direct measurements made

Table 4-9. Estimates of the Amounts of Copper in the Void Volume Fractions of Human and Animal Serum Separated in Gel Chromatography

Serum	Copper concentration ng/ml	nM	Method	Reference
Normal human adult blood	4.4[a] 20[b]	69 310	% of exchangeable tracer radiocopper added to whole serum; Sephadex G-100	Lau and Sarker (1984)
Human cord blood	18	280	7.1% of 25.4 μg/dl	Sarkar et al. (1983)
Normal adult rat blood	33[b]	520	33% (max.) of ^{67}Cu tracer (1–2 ng of Cu/ml) added to whole serum exchangeable pool; Sephadex G-150	Linder et al. (1987)
Normal human adult blood	120	1900	Direct measurement; furnace atomic absorption; Sephadex G-150	Wirth and Linder (1985)
Normal human adult blood	74[c]	1160	Direct measurement; neutron activation analyses; Bio-Gel P-150	Evans and Fritze (1969)
Normal human adult blood	250[c]	3900	Direct measurement; electrothermal atomic absorption; Sephadex G-150	Gardiner et al. (1981)
Normal human adult blood	77	1200	Direct measurement; furnace atomic absorption; Sephadex G-150	Barrow and Tanner (1988)

[a] Assuming that the exchangeable Cu pool is 10 μg/dl.
[b] Assuming that the exchangeable Cu pool is 45 μg/dl (Wirth and Linder, 1985).
[c] Approximated, from a graph.

by four different laboratories, it appears that, per ml, about 80 ng of Cu are associated with the transcuprein fraction. Assuming that any copper on α_2-macroglobulin (also in this fraction; see Section 4.1.4.3) is freely exchangeable with that of transcuprein, then about 25–30% of this amount (or 20–25 ng) might be on the macroglobulin and the rest (about 60 ng) on transcuprein, at least in the plasma of the rat and the human. It cannot, however, be entirely excluded that some of the copper in this fraction is nonexchangeable and associated with an additional component (or components). Indeed, a small percentage (perhaps more in the rat) will also be associated with ferroxidase II (see Section 4.1.4.2.).]

Returning to the contributions of ceruloplasmin (as well as those of other components) to the total copper in human serum, Wirth and Linder (1985) fractionated serum from two groups of normal adults on Sephadex

Table 4-10. Distribution of Copper among Components of Normal Human Serum, Determined by Gel Filtration and Furnace Atomic Absorption

Source of serum	No. of samples	Copper content of fraction[a] (ng/ml serum)			
		Transcuprein fraction	Ceruloplasmin	Albumin	Low-molecular-weight fraction[b]
Adults[c]					
61 ± 3 years	14	120 ± 30	600 ± 90	170 ± 70	110 ± 40
24 ± 3 years	6	120 ± 30	570 ± 100	120 ± 40	70 ± 60
Adults[d]	8	77 ± 19	790 ± 110	217 ± 55	23 ± 33
Neonates[d]	4	63 ± 15	280 ± 83	310 ± 70	—[e]

[a] Mean ± SD for number of samples indicated.
[b] Includes material below 40 kDa.
[c] Wirth and Linder (1985).
[d] Barrow and Tanner (1988).
[e] Too low to measure. Henkin et al. (1973) reported about 40–100 ng of Cu/ml in ultrafiltrates made with filters having a cutoff of 50 kDa (and about 110 ng/ml in adults) (see Chapter 8).

G-150 (to obtain "profiles" such as these in Fig. 4-16B), the data for which are summarized in Table 4-10. From these, it is clear that ceruloplasmin is still the major binding component, but the percentage of total copper accounted for by ceruloplasmin is more like 60% versus the 90–95% estimated 30 years earlier. The potential objection that 32 mg of ceruloplasmin protein per dl (the normal standard widely used for human ceruloplasmin; see Campbell et al., 1981) will account for about 90% of the copper in plasma is based on the assumption that all ceruloplasmin contains 0.28–0.32% copper (32 mg/dl × 0.02% = 96 µg/dl vs. a total of about 115 µg/dl, or 83%). As already discussed (see Chapter 3), this assumption is unlikely to be true, in view of evidence that ceruloplasmin is a copper source and that low-copper ceruloplasmin may comprise a substantial portion of the immunoreactive protein present (see Section 4.1.1 and Fig. 4-4).

Earlier estimates of 5–10% of serum copper on albumin are a little low, but not far off the mark, based on current procedures (Table 4-10). These values still need to be refined, because the gel permeaton medium used often did not effect a sufficient separation of albumin from ceruloplasmin (Fig. 4-16) [also, possibly because the albumin fraction might be masking a portion of copper associated with HRG (see section 4.1.4.1).] With regard to the separation of albumin and ceruloplasmin, the values gathered for two groups of normal adults were based on standard elutions of ceruloplasmin (measured as p-phenylenediamine oxidase activity) and pure human serum albumin in the same kinds of

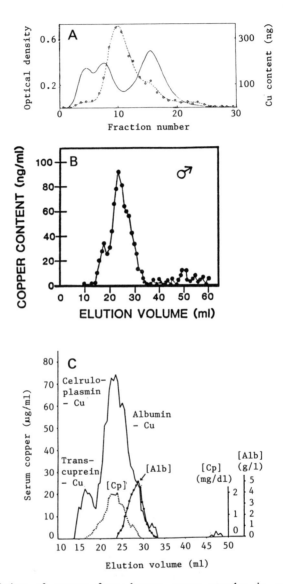

Figure 4-16. Elution of copper from human serum samples in gel permeation chromatography. (A) Serum (2.0 ml) applied to Bio-Gel P-150, eluted with 0.15 M ammonium acetate. Copper was measured by neutron activation analysis. Reprinted, with permission, from Evans and Fritze (1969). (B) Serum applied to borohydride-treated Sephadex G-150, eluted with 20 mM phosphate, pH 7.0 (P. L. Wirth and M. C. Linder, unpublished results). (C) Serum, eluted as in (B); location of ceruloplasmin (Cp) and albumin (Alb) peaks, determined immunologically, is shown. Reprinted, with permission, from Barrow and Tanner (1988).

columns. Elution of the latter protein also coincides with the second major UV-absorbing peak eluting from serum samples.

With regard to copper attached to ligands of low molecular weight, the values obtained by several investigators, using reasonable (trace) amounts of radiocopper added *in vitro*, tend to agree that amino acids and components below 15,000 Da comprise 0.5–0.6% of the total exchangeable pool (Table 4-11). The furnace atomic absorption data, however, for material below 15,000 Da from human serum profiles (Fig. 4-16) suggest that actual values are considerably higher (Table 4-10). To account for these differences, there are two possibilities. Either the data from direct analysis are wrong and reflect contamination or the assumptions about the size of the exchangeable pool are wrong, or both (to some extent). With regard to the former, the lack of an effect of precleaning gel columns to remove contaminating copper and the consistency of elution of specific low-molecular-weight peaks (Wirth and Linder, 1985) argue against

Table 4-11. Estimates of Copper in Low-Molecular-Weight (LMW) Fractions of Human Serum

Total copper concentration of LMW components		Method	Reference
ng/ml	nM		
0.6[a]	9	Tracer ^{64}Cu addition;	Neumann and
2.7[b]	42	ultracentrifugation	Sass-Kortsak (1967)
0.3[a]	5	Tracer radiocopper addition;	Earl et al. (1954)
1.4[b]	23	0.2–0.4% ultrafiltrate	
0.5[a]	8	0.5% of added ^{67}Cu tracer;	Lau and Sarkar (1984)
2.3[b]	36	Sephadex G-100	
0.8[a]	13	Computer simulation of plasma	May et al. (1977)
6[b]	90	conditions	
90[c]	1400	Direct measurement, furnace atomic absorption; Sephadex G-150	Weiss et al. (1985)
23[c]	360	Direct measurement, furnace atomic absorption; Sephadex G-150	Barrow and Tanner (1988)
78[c,d]	1200	Direct measurement, neutron activation analysis; Bio-Gel P-150	Evans and Fritze (1969)

[a] Assuming the total exchangeable (nonceruloplasmin) copper pool is 10 μg/dl.
[b] Assuming the total exchangeable (nonceruloplasmin) copper pool is 45 μg/dl.
[c] Includes material up to M_r 40,000.
[d] Approximated, from a graph.

contamination as an explanation. With regard to the assumptions on exchangeable pools, it seems increasingly reasonable to assume that, *de facto*, only a few specific components that bind copper may be part of a rapidly exchanging and equilibrating pool. From the radiocopper evidence, this might include just albumin, transcuprein, and perhaps α_2-macroglobulin, in equilibrium with a very small amount of amino acid-bound and free copper. Thus, copper attached to GHL, ferroxidase II, and some other potential ligands (as yet undefined) might not be part of the exchangeable pool. Such other components might have "constitutive" (tightly bound or buried) Cu necessary for specific regulatory (or enzymatic) functions (such as growth promotion), functions that are independent of copper transport. Indeed, it would seem important that regulatory or other functions of copper components not overlap too much with the need for transport and distribution of ionic copper entering the blood from the diet or on the way to excretion.

Taking all of these data into account, a current reasonable estimate of the amounts of copper in various human serum compartments might be those given in Table 4-12. This would be for blood plasma or serum of normal human adults. The picture will be somewhat different for samples from subjects with various diseases or from infants and young children or of umbilical cord blood, etc., in which it is known that there are major differences in the ceruloplasmin content. (Some of the other components will also vary, although much of this remains to be examined.) Changes in ceruloplasmin copper alone (which is not part of the exchangeable copper pool) will perhaps not change the copper economy of the rest of the components in principle but may do so indirectly through alterations in the

Table 4-12. Summary of Current Best Estimate of the Distribution of Copper among Serum Compartments of Normal Human Adults

Component	Approximate copper concentration		
	ng/ml	nM	% of total
α_2-Macroglobulin or ferroxidase II	(20)[a]	(310)	2
Transcuprein	80	1,250	8
Ceruloplasmin	650	10,200	65
Albumin	180	3,000	18
(His-rich glycoprotein)	(0)	(0)	(0)
15–60-kDa components	30–50	630	4
Amino acids and small peptides	20–50	550	4
Free copper	0.0001	0.0002	~0

[a] Values in parentheses are best guesses from information currently available.

Table 4-13. Total Copper Content and Ceruloplasmin Content of Human, Nongastrointestinal, Nonblood Body Fluids

Fluid	Total copper concentration		Ceruloplasmin concentration	
	μg/ml $(N)^a$	μM	μg protein/ml	μg Cu/ml^b
Cerebrospinal	5 ± 2 (4)	78	0.9^c	0.002
Lymph	1.2 (1)	19	$1-2^d$	0.003
Synovial	0.2, 0.5 (2)	5	284^e	0.6
Amniotic	0.1–0.25 (1)^f	3		
Seminal	(0.5, 1.5) (2)^g	16		

^{a} Mean, or mean of means, ±SD (N = number of studies) (see Appendix B).
^{b} Calculated based on the assumption that ceruloplasmin has an average Cu content of 0.20%. This is based on plasma values of 320 μg of ceruloplasmin protein per ml and 650 ng of ceruloplasmin Cu per ml for the normal human adult.
^{c} Schreiber (1987).
^{d} Estimated from the data of Trip et al. (1969).
^{e} From patients with rheumatoid arthritis (Gutteridge et al., 1984).
^{f} From Chez et al. (1978) throughout gestation.
^{g} Includes cells (values are for semen); values are thus less certain (parentheses).

economy of copper transport. Thus, in Wilson's disease, for example, in which there is less copper incorporated into ceruloplasmin, larger amounts of copper are associated with nonceruloplasmin components (see Chapter 9), not just in terms of the percentage of total copper but in terms of absolute concentrations. In other species, and in other body fluids of vertebrates, there will be additional variations in the properties of copper associated with these various components. In this connection, Table 4-13 summarizes what is currently known about the total copper content of various internal (nongastrointestinal) nonblood fluids of the adult human, as well as their ceruloplasmin content (where known), measured not as copper but as enzyme activity or immunoreactivity.

4.3 Origins of Extracellular Copper Components

Clearly, most of the proteins in blood plasma and interstitial fluid are products of liver secretion, with the exception of immunoglobulins and trace quantities of enzymes, hormones, etc. The liver origin of ceruloplasmin and albumin has recently been verified by the molecular biological studies of Schreiber, Mercer, and collaborators (Aldred et al., 1987), in the rat. Of additional interest, however, have been their observations that the genes for plasma proteins are expressed not just by hepatocytes, but also by certain portions of the brain, testes, placenta, and yolk sac. The work on testes corroborates an earlier finding of Skinner and Griswold (1983) that Sertoli cells secrete ceruloplasmin (and transferrin), and the general tissue distribution of ceruloplasmin mRNA has been corroborated by the recent work of Gitlin (J. D. Gitlin, personal communication), for rats, mice, and hamsters. As brain, testes, and yolk sac each have their own "circulation" separate from that of the blood, it is tempting to think of the respective cells synthesizing plasma proteins in these special regions as the "hepatocytes" of these respective regions, in charge of the special composition of fluids required there, as proposed by Schreiber (1987). Thus, each fluid within the body may have its own complement of specific plasma proteins, amino acids, etc., and its own particular complement of Cu-binding components that have roles in Cu transport and in other functions for which extracellular Cu or Cu components may be required. The cerebrospinal fluid of rats, for example, contains relatively less albumin and ceruloplasmin, but more prealbumin (Schreiber, 1987). Yolk sac, on the other hand, would appear to express no albumin, significant amounts of ceruloplasmin, and very large amounts of α_2-macroglobulin (Schreiber, 1987). Gene expression may also change in disease, as discussed in Chapters 7 and 9.

4.4 Copper Transport and Delivery to Organ Systems

In this section, the roles of plasma/serum and interstitial fluid substituents in delivery of copper to organs and their cells, in the whole vertebrate organism, will be considered in the light of our knowledge about these components and their potential capacities for copper delivery, already reviewed in this and the previous chapter. Here, the stress will be on the delivery and distribution of incoming copper to organs and cells. Transport of copper for ultimate excretion will be reviewed in Chapter 5.

We have examined what is known about components in the blood plasma that bind copper (this chapter) and what has been done to study their capacity to deliver copper to specific cell and organ systems at the cellular and molecular level (Chapter 3). In some cases, much is known both about their structure and biochemical characteristics or about aspects of the mechanisms whereby copper delivery may be operating at the cellular level. In other instances, few of the molecular and mechanistic details are as yet available. Bearing this in mind, we can build a picture of the most likely, normal ways in which incoming dietary copper is distributed throughout the body, both in the initial phases and in the longer term, during which all cells must ultimately be supplied with this element to carry out oxidative phosphorylation and other essential functions.

For this, we must consider what happens to incoming copper from the time it first enters the circulation to the time that it is fully distributed to the tissues of the mammal. Owen (1965, 1971) did the first extensive studies on the distribution with time of radioactive copper to most of the tissues of the rat, at various times after its intravenous administration as $^{64}Cu(II)Cl_2$ or $^{67}CuCl_2$. As shown in Fig. 4-17, one sees an initial influx of radioactivity especially into the liver and kidney and then a slower influx into other organs. The 1965 organ data have not been corrected for radioactivity of residual blood present (which will of course be substantial, especially early on). The data for blood plasma alone are biphasic (a matter

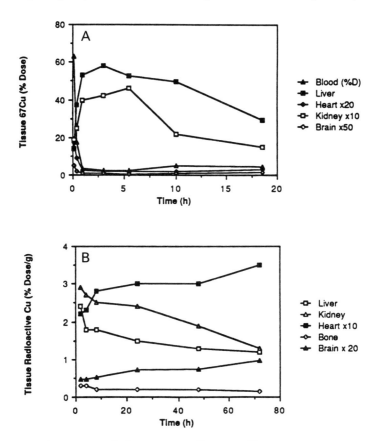

Figure 4-17. Time course of distribution of ^{64}Cu and $^{67}CuCl_2$ to rat tissues after its intravenous injection. Note that data for brain and heart have been scaled up for visibility. (A) Mean data of Owen (1965) as percent of dose per organ, for early times, with no subtraction of residuel blood radioactivity. (B) Data of Owen (1971) for later times, with subtraction of some residual blood radioactivity.

reconfirmed several times since then). They show an initial rapid drop to a low (at 1–2 h), followed by a resurgence of blood radioactivity starting at 4 to 6 h. This reflects the characteristic influx of nonceruloplasmin copper into the liver and subsequent secretion of copper on ceruloplasmin back into the plasma. Liver and kidney show a parallelism in uptake, with maximum incorporations at 2–4 h. Indeed, liver and kidney alone accounted for about half the administered radioactivity at 2 h after injection. With the possible exceptions of intestine and bone, the other tissues showed relatively little radioactivity and little change in radioactivity until after 8 h, following which they increased their contents of radioactive copper, as that of the blood and the other tissues decreased. This sequence of events has since been confirmed by Marceau and Aspin (1973) and Campbell *et al.* (1981) and, most recently, by Weiss *et al.* (1985). In the latter study, the amounts of radioactivity associated with various plasma components were also followed (Fig. 4-18). In the earliest studies, unphysiologically high amounts of copper were usually administered (for example, 105 μg per rat; Owen, 1965) versus only 30 to 300 ng per rat, in the case of Campbell *et al.* (1981) and Weiss *et al.* (1985). If so, this would be expected to load proportionately more copper than normal onto albumin. [At least in humans, the albumin in normal serum has a capacity to hold 40 μg of Cu per ml, although it normally holds only about 150 ng (see Sections 4.1.2 and 4.2).] Apparently, this did not markedly affect the distribution of radioactivity obtained in terms of the percent dose absorbed by various tissues. At least for the first 24 h (where lower doses could be used), the results were qualitatively similar, over a 300-fold different dose range (Fig. 4-18 versus Fig. 4-17), although the times for maxima and minima were different. An increase in exchangeable copper pools would, however, be expected to enhance the rates of uptake of copper from the plasma by mass action effects on the available uptake systems [used by free Cu(II) and/or transcuprein copper, for example].

Returning to Fig. 4-18, one sees that radioactivity in the blood plasma was immediately in the transcuprein and albumin fractions. [Indeed, this is how transcuprein was discovered, and the same initial distribution was found when ionic radiocopper was added to plasma or serum, *in vitro* (see Section 4.1.4.3) (Fig. 4-14).] Radioactivity associated with these fractions decreased in parallel and became very low or virtually nil by 6 h. The decrease in the specific radioactivity of copper on transcuprein and albumin was matched by an increase in specific radioactivity of total liver and kidney copper, which achieved a peak at around 6 h. Beginning with its detection already by 2 h but achieving a peak at about 18 h at the doses of copper used (500–800 ng/rat; M. C. Linder, unpublished results), is ceruloplasmin copper, the specific activity of which increases as that of the

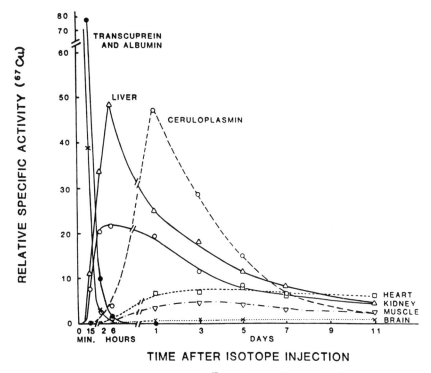

Figure 4-18. Time course of distribution of $^{67}CuCl_2$ to rat tissues and plasma components, following intraperitioneal injection. Data of Weiss et al. (1985) for specific activity of ^{67}Cu in plasma components and tissues have been replotted on a nonlog scale of percent dose. It should be noted that intravenous and intrapertoneal injections produced the same results on Sephadex G-150, at various times after injection into whole animals, as did intragastric or intraduodenal administration (tempered by a less than total absorption and much broader time influx of copper from the gut).

liver declines. Subsequently, the specific radioactivity of ceruloplasmin copper declines also, and that of peripheral tissues, such as the brain and heart, continues to rise. It is noteworthy again that there is very little radioactivity in these peripheral tissues until ceruloplasmin copper begins to appear.

From the standpoint of blood plasma components alone, Fig. 4-19 shows the distribution of radioactivity at various times after administration. There is little or no radioactivity eluting in the position of ceruloplasmin early on, and only traces eluting in one to three components at close to the bed volume of the column. Radioactivity then appears in the ceruloplasmin region. Eventually, this becomes dominant, and, in fact, ceruloplasmin becomes virtually the only labeled substituent detectable in

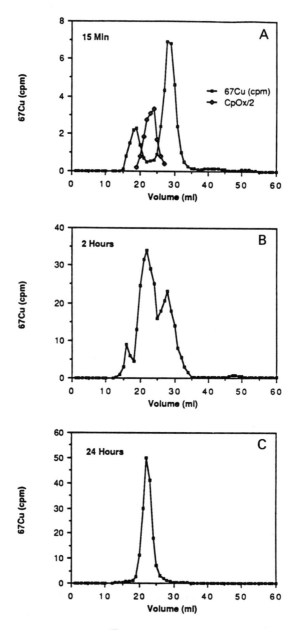

Figure 4-19. Distribution of injected ^{67}Cu among plasma components at various times after injection into rats. Data are from Weiss *et al.* (1985), Weiss (1983), and Weiss and Linder (unpublished results), at (A) 15 min, (B) 2 h, and (C) 24 h. In (A), the elution of ceruloplasmin oxidase activity (◇) is also shown.

the plasma, although traces may be persisting in low-molecular-weight components. What is interesting here is that we do not see the development of an equilibrium between the radioactive copper that has recently entered and the nonradioactive copper of the various plasma substituents; that is, the radioactivity profile never began to resemble the profile for chemically/ physically determined copper in serum compartments, over the 10-day period in which the radioactivity of the short-lived ^{67}Cu isotope was followed ($t_{1/2}$ 61.8 h; $t_{1/2}$ for ^{64}Cu is 13 h). Of course, this partly reflects the fact that ceruloplasmin copper is not part of the exchangeable plasma copper pool, and we can detect equilibration of copper within this pool, comprised of transcuprein, albumin, and probably the amino acid (and free copper) fractions. However, newly absorbed copper appears to be virtually gone from the exchangeable pool by 6 h after it has entered the blood, in the case of the rat. Presumably, this reflects its replacement by additional incoming dietary copper and perhaps also by copper released from cells, as it may be that certain intracellular copper compartments "equilibrate" with the exchangeable plasma/interstitial pool (Figs. 3-1 and 3-6). As already mentioned and as suggested by the available data (Section 4.2), this leaves room for the existence of other nonceruloplasmin plasma components, components that are not part of the exchangeable pool but contribute to the overall copper content of this fluid and have other functions.

Two major points that can be made from these various data are that incoming copper is rapidly incorporated into ceruloplasmin synthesized in the liver, and that there may be two stages in the distribution of newly absorbed copper to tissues, the first involving transport by albumin, transcuprein, and possibly amino acids and uptake mainly by liver and kidney, and the second involving liver secretion of ceruloplasmin copper and its uptake by other tissues, as well as perhaps some reabsorption by the liver (see Chapter 5).

With regard to the first stage, there is no dispute that the liver and kidney are the organs most abundant in copper (Table 1-6) and that they initially absorb the "lion's share" of copper that has entered the blood. The liver is also in a good physical/physiological position to accept most of the dietary copper, by virtue of the fact that the incoming blood in the hepatic portal circulation is that draining the gastrointestinal tract (including the stomach). (The liver does, of course, perform similar bulk removal of other incoming nutrients, such as glucose, fructose, and amino acids.) At the same time, all of the entering copper will not be removed during its first pass through the liver, as demonstrated repeatedly by examining the radioactivity of peripheral blood samples within minutes of intragastric or intraduodenal administration of ^{64}Cu or ^{67}Cu. Indeed, as already noted, the kidney is also a major initial recipient of incoming copper, and it is much further away (in terms of circulation) from the original (gastrointestinal) sources. Initially, then, portal and peripheral blood will contain newly absorbed copper bound to albumin, amino acids, and the transcuprein fraction, all in rapid equilibration with a minute (undetectable) portion present as free Cu(II). The various cells of the liver (and kidney), comprised (in the former case) of Kupffer and endothelial cells lining the sinusoids, as well as rows of hepatocytes partly shielded by those cells, will interact with these components, with the net result that copper is rapidly absorbed, presumably primarily into the hepatocytes. [The hepatocytes make up about 85% of the liver cell volume (Weibel et al., 1969) and are thought to be the most important sites for expression of most plasma protein genes, for which the liver is the source.]

Assuming that portal blood plasma is not greatly different from peripheral blood plasma (see more below), we will have at least three significant potential sources of newly absorbed copper for the hepatocytes and other cells of the liver (and kidney). These are albumin, transcuprein, and His$_2$-Cu or free Cu(II). As already shown by Ettinger and colleagues and by McArdle et al. (Chapter 3) and as summarized in Fig. 3-1, albumin is not a direct source, but holds copper in equilibrium with that of histidine and free Cu(II). The latter forms are the immediate source for a transport system with K_m of 11 mM. In most normal physiological conditions, it is not clear how active this transport system may be, because the concentrations of amino acid and free copper (Table 4-12) are several orders of magnitude lower than the transport K_m. Nevertheless, some uptake by this high-capacity system will occur, and, under certain conditions, such as high dietary intake of following injection, this would also be a safety valve for the quick disposal of excess copper in the blood. [Excess copper might otherwise lead to peroxidative cell membrane damage, one of the most accepted mechanisms for toxicity of copper in vertebrates (see Chapter 9).]

The presence of transcuprein, clearly also a part of the *in vitro* exchangeable copper pool in plasma, provides the possibility for an additional more rapid mechanism by which exchangeable copper might be funneled into these cells. This is illustrated in Fig. 4-20 (bottom right, "A"), in which copper from albumin (and amino acids) might be transferred via the attachment of transcuprein to specific transcuprein receptors on the cell surface. [Whether mediation of uptake by α_2-macroglobulin is also involved rather than (as illustrated) its participation as another site for binding, transport, and sequestration of the exchangeable plasma copper pool is, of course, currently unknown.] It seems probable that similar mechanisms are operative for other body cells (Fig. 4-20; "B"), especially in the case of fibroblasts, in which this has been studied. However, the relative capacities of these transport systems for copper uptake are likely to be

Extracellular Copper Substituents and Mammalian Copper Transport

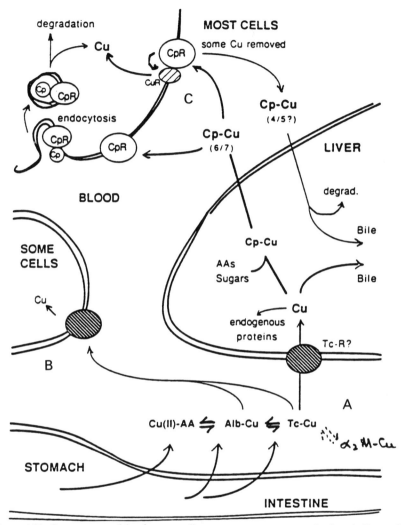

Figure 4-20. Summary model of mammalian copper transport and tissue/cell uptake. (A) Possible mechanism of delivery of dietary Cu to hepatocytes by a transcuprein receptor (Tc-R) (see text). (B) Delivery by a similar mechanism to some other cells. (C) Proposed mechanisms for delivery of ceruloplasmin copper (and ceruloplasmin protein) to most cells.

lower than they are for the liver (as has already been shown for some cells). This is demanded by the observations (Figs. 4-17 and 18) that other tissues of the body take up most of their newly absorbed copper after uptakes by liver (and kidney) have reached a peak and during the period when ceruloplasmin has become the most important or only plasma carrier of the copper that has entered (Fig. 4-18).

It should be noted that Gordon *et al.* (1987) have challenged the hypothesis that transcuprein is involved in liver copper uptake on the basis of what they describe as their inability to detect radiocopper associated with such a fraction in the portal blood of rats given oral (or intraduodenal) ^{67}Cu. This is in contrast to the findings of Linder *et al.* (1958), who reported substantial labeling of portal transcuprein fractions upon intraduodenal tracer administration or *in vitro* addition. Part of this discrepancy may lie in the interpretation of the data (Fig. 4-21), which indeed show labelling of components larger than albumin early on, especially at 30 min. The investigators have assumed that this labeling is of ceruloplasmin, although the ceruloplasmin peak (prominent in Fig. 4-21C and D) does not coincide with the peak in Fig. 4-21B. Indeed, Campbell *et al.* (1981) and then Weiss *et al.* (1985) also first assumed that the excluded volume peak was ceruloplasmin, until they investigated it further (Fig. 4-19). As already noted in Section 4.1.4.3 of this chapter, the dose of labeled copper administered and its specific activity upon entering the

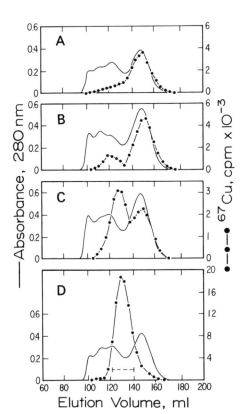

Figure 4-21. Distribution of ^{67}Cu among components of portal blood serum after intragastric intubation. Gel permeation chromatography was on Sephadex G-150, in phosphate-buffered saline, pH 7.5. Samples taken at (A) 10 min, (B) 30 min, (C) 90 min, and (D) 8 h. Reprinted, with permission, from Gordon *et al.* (1987).

blood will influence the proportion of radioactivity detected in the transcuprein fraction (see Fig. 4-14B). Assuming that Gordon *et al.* used 100 μCi of ^{67}Cu(II) with a specific activity of 3000 mCi/mg, they gave 33 ng of Cu, which was diluted by an unspecified (but certainly substantial) amount of endogenous gastric, duodenal, and mucosal copper before entering the blood. Nevertheless, what is likely to be a peak of radioactivity on transcuprein appeared to be present.

The next phase of copper transport (beyond the liver) certainly also involves ceruloplasmin (Fig. 4-18), for which specific cell surface receptors have been detected and which has also been shown to be a major source of copper for tissues and cultured cells, even for incorporation into specific cellular proteins (see Chapter 3, Section 3.2.4). As previously discussed, the mechanism of uptake may not primarily involve endocytosis of ceruloplasmin protein, but transfer of some of its copper at the cell surface (Fig. 3-11; Fig. 4-20, "C"), though some ceruloplasmin protein also appears to be absorbed. Receptors for ceruloplasmin and its copper appear to be more abundant on nonliver (and nonkidney) cells, again in agreement with the two-stage absorption concept, the first involving liver and kidney uptake, but not ceruloplasmin. Thus, secretion of ceruloplasmin copper might be viewed as a second (and for some cells, more important) means for distribution of copper to cells, cells which might contain, in addition to the transport mechanisms already mentioned, receptors for ceruloplasmin and ceruloplasmin copper (Fig. 4-20, "C"). Existing data on the affinity of such receptors, in conjunction with known ceruloplasmin concentrations, would indicate that transport systems for ceruloplasmin copper uptake should indeed be operating under physiological conditions.

As a final twist, one should briefly consider how copper might be delivered to cells bathed by fluids not directly derived from the blood plasma, such as those of the brain or amnion. To gain access, copper must first cross specific cellular barriers, such as the blood–brain barrier (made up of endothelial cells with very tight intercellular junctions; Gjedde, 1986) and the placenta, respectively. The question is then how copper crosses these barriers. This is currently unknown. However, Fishman *et al.* (1987) have reported "transcytosis" of transferrin and insulin across the blood–brain barrier, meaning that the proteins are taken up from the plasma via receptors on one side of these barrier cells and delivered on the other side, to the cerebrospinal fluid. However, since transferrin and ceruloplasmin are both also produced by the choroid plexus (and some other specific portions of the brain itself) and by the yolk sac (Aldred *et al.*, 1987), one wonders about the need for transcytosis. Delivery of copper to the brain by ceruloplasmin would fit data on the time course of distribution to that tissue (Figs. 4-17 and 4-18), but whether this delivery is just to the blood–

brain barrier or across it into the brain remains to be established. Other mechanisms of delivering copper to the brain must also be operating, at least under certain conditions, since during gestation the developing fetus has a ceruloplasmin-poor blood, and this is also the case in Wilson's disease (in which an overabundance of copper accumulates in the brain).

In summary, there are clearly several ways in which the copper of extracellular fluids can be delivered to and taken up by cells. Some of these will normally predominate in certain cells, others in others. All may contribute to a greater or lesser degree to the copper economy of cells. An initial assessment is that there is a distinction between the mechanisms favored by hepatocytes (and certain kidney cells), involving albumin, transcuprein, and amino acids, and those for cells in other tissues, involving ceruloplasmin. This picture will need to be refined as new information becomes available from additional biochemical and physiological work.

Excretion of Copper in the Mammal

5.1 Routes of Excretion

In considering by what routes metals may be excreted from the bodies of animals and humans, interest generally focuses on the urine and the bile. Other routes to consider include salivary, gastric, and pancreatic secretions, direct intestinal secretion, and the sloughing-off of cells along the gastrointestinal tract and on outside body surfaces (skin), as well as hair loss. It seems highly unlikely that exhalation plays a role, so this will be ignored.

Starting with the gastrointestinal tract, studies mainly in the human (and the rat) have shown that the various secretions, especially the bile, have significant concentrations of copper (Table 5-1). Except for bile, concentrations are nevertheless well below those of the blood plasma (which are about 1.2 µg/ml in the adult man or rat; Table 1-6). Much lower levels are found in the urine. With regard to the sloughing-off of cells, the intestinal mucosa contains about 2 µg of Cu/g in both species (Appendix B). Assuming that, for the average 60-kg adult person, 1.8% of body weight is stomach plus intestine (Magnus-Levy, 1910), half of which may be mucosa, and assuming that about 20% of mucosal cells are sloughed off per day, calculations suggest that about 220 µg of Cu may be removed with gastrointestinal cells per day. All of the estimates of losses from the gastrointestinal tract must address the matter of potential reabsorption (of which more below).

Losses from the body exterior would include skin, sweat, hair, and nails. Human skin has about 0.8 µg of Cu/g and regenerates at a rate of about 1–2% per day (Linder, 1985). With the assumption that 4.5% of body weight is skin (Magnus-Levy, 1910), 30–60 µg of Cu might be lost with skin cells per day. The potential contributions of hair (and nails) to daily copper losses are harder to estimate but will be greater for men than women; copper concentrations are about 20 µg/g for both sexes, but hair

Table 5-1. Copper Secreted in Gastrointestinal Fluids and Urine[a]

Fluid	Copper concentration (μg/ml)	Average daily secretion[b] (μg)	Reference
Saliva	0.22 ± 0.08 (4)[c]	(330)	Owen (1982a)
		~450	Brewer (1986)
	0.2–3[d]		Owen (1982a)
Gastric juice	0.39	(975)	Gollan et al. (1971b)
		~1000	Brewer (1986)
Bile	4.0 ± 1.9 (5)[c]	(2500)	Owen (1982a)
	1.8 ± 0.8 (3)[c,d]		Owen (1982a)
Pancreatic fluid	0.26[e]	390	Ishihara et al. (1987)
	0.34–0.9	500–1300	Underwood (1977)
Duodenal juices[f]	0.17	1181 ± 495 (74)[c]	Frommer (1974)
		1180 ± 1060 (6)[c]	Salaspuro et al. (1981)
Feces		334	Owen and Hazelrig (1968)
		110–230[d]	Hill et al. (1984)
Urine	>0.02	>30	Butler and Newman (1956)
	~0.05	75	Ishihara and Matsushiro (1986)
	(0.27)[d]	(3–4)[d]	Gibbs and Walshe (1977); Hill et al. (1984)

[a] Data for adult humans, unless otherwise indicated.
[b] Volumes assumed for daily human secretions were 1500 ml for saliva, 2500 ml for gastric juice, and 500–750 ml for bile (Shils and Randall, 1980); also, 1500 ml each for pancreatic fluid and for urine (Ishihara et al., 1987). Values in parentheses were calculated from the copper concentration data indicated.
[c] Mean ± SD for no. of means indicated.
[d] Data for rats.
[e] 5 min after administration of secretin.
[f] May include biliary and pancreatic secretions.
[g] Calculated from value for average daily secretion, assuming a volume of 1500 ml (Ishihara et al., 1987).

grows faster in men. [Deeming and Weber (1978) report much higher values for women (and 23 μg/g for men), but for a small number of subjects, so the sex difference in loss rate may be very small.] It seems unlikely that, averaged over time, much more than about a gram of hair (plus nails) a day (20 μg) would be lost. With regard to sweat, women may again lose about the same amount as men. The concentration of Cu in sweat obtained from an armbag, when subjects were placed in a sauna for 15 min, was three times higher for the women (1.6 versus 0.6 μg/ml) (Hohnadel et al., 1973; Sunderman et al., 1974), but the women sweated three times less volume than the men. Total body washdowns yielded about 150 μg of copper (Cohn and Emmett, 1978), which might give a good indication of daily losses by that route. Heavy exercise would increase these losses. Combined daily losses of copper from hair, nails, skin, and sweat are thus usually probably no more than about 200 µg for Western human adults, a small fraction of the 1200–1800 µg consumed by them daily (Linder, 1985b). (Daily intakes of at least 2 mg are suggested by the National Research Council of the U.S. National Academy of Sciences.)

For women, losses of copper through menstrual bleeding were estimated at about 0.5 mg per period (range 0.2–0.7 mg; Ohlson and Daum, 1935; Leverton and Binkley, 1944) long ago, while more recent studies gave values of 0.11 mg for adolescent girls (Greger and Buckley, 1977). Considering that, typically, menstrual blood flow does not exceed 80 ml per period, this suggests that more than whole blood (Cu content of about 1 µg/ml) is contributing to these losses. However, the recent values are much lower, so the earlier ones may be overestimates. Either way, menstrual losses are only a small fraction of daily losses in women and girls.

Daily losses via the urine are also very small. Estimates have been on the order of 30–80 µg per day (Table 5-1) for humans, or about 4% of average intake. (This does not reflect intestinal absorption which is usually 50–60%; Chapter 2.) Urinary excretion by the adult rat may be even a smaller percentage of the typical normal intake of 200 µg (Hill et al., 1984).

Based on the available data, the bulk of copper excretion is thus clearly via the gastrointestinal tract. Here, there still are some questions about the relative importance of the various secretions to the process. As summarized by Owen (1982a), various reports agree that the copper content of the saliva is variable, centering around 0.2 µg/ml (Table 5-1). Shils and Randall (1980) have estimated that the average adult human secretes 1–2 liters of saliva. If that is the case, 200–400 µg of endogenous body copper would be entering the digestive tract by this route. The estimate of Brewer (1986; Brewer et al., 1988) (Table 5-1) is at the upper end of this range. With regard to gastric secretions, Gollan et al. (1971b) determined a copper concentration of 0.4 µg/ml for normal adult humans, which (at 2.5 liters secreted) amounts to almost 1000 µg per day (Table 5-1). Then there is the bile, which is uniformly touted as the main excretory route for this element (beginning with Judd et al, 1935), and for good reason (Table 5-1). Nevertheless, it should be noted that determinations of concentration in bile have given highly variable results, ranging from 0.2 to 7.5 µg Cu/ml, in studies from the 1950s through 1971 (Owen, 1982a). This suggests that biliary copper losses in humans might range from 135 to more than 4600 µg per day. This may depend on diet (including fat and copper consumptions) and, indeed, on copper status. For example, in this regard, a direct relationship between liver and biliary copper concentrations has been shown in pigs (Skalicky et al., 1978), which is not surprising, considering that liver is the source of bile. Iyengar et al. (1988), who reported an average copper content of 1.2 μg/ml in normal volunteers, found that copper concentration of (cholecystokinin stimulated) bile increased to 3.0 μg/ml in the same individuals upon dietary supplementation with 5–6 mg of copper (as sheep liver). Very low values for hepatic bile (0.10–0.15 μg Cu/ml) have also recently been obtained from patients with cholelithiasis or cholecystolithiasis (duct obstructions), following surgery. Whether these values are simply at the low end of the normal range or represent an adaptation of the excretory route in this condition remains to be clarified. [Data of Owen (1980) in rats suggest that surgery enhances biliary copper secretion.]

Pancreatic fluid probably also makes a quite variable contribution to the copper content of intestinal fluid (Table 5-1), as suggested by the data on copper concentration and rates of secretion, perhaps amounting to 500 μg per day.

Measurements have also been carried out on duodenal juices, using perfusion. These should be a result of the mixing of biliary, pancreatic, and intestinal mucosal juices. Shils and Randall (1980) estimated that, overall, intestinal mucosal secretions (for the upper to middle small intestine) amount to 2–3 liters and might in themselves (if low in copper) explain the overall mean values of 1180 μg/day obtained by two independent groups (Table 5-1). No direct measurements of the copper content of non-biliary, nonpancreatic-derived intestinal juices appear to have been made. However, mucosal cells are known to secrete some copper (see Chapter 2), and ligation of the common bile duct (in the rat) only cuts total body copper excretion rates in half (Linder et al., 1986). Also, Terao and Owen (1974) have shown (in rats) that intravenously administered ^{67}Cu(II) is taken up in some quantity by the stomach and small intestine (where it may be bound for excretion). The projected biliary contribution alone would also account for considerably more copper than was measured as output into the duodenum in perfusion studies. This poses a bit of a dilemma. One possibility is that the perfusion studies (with fasted individuals) really only measured direct intestinal mucosal copper output, with little contribution from either pancreas or bile. (Secretions from the latter are stimulated by food intake, especially fat in the case of the bile.) If so, the duodenal juice measurements must be added to projected pancreatic and biliary values in estimating total intestinal copper output. Another possibility is that copper resorption is so efficient that, even with intestinal perfusion, a good portion returns to the blood. A less satisfactory alternative is that values are so variable that such estimates are relatively meaningless. [The 95% confidence limits for the duodenal juice values do indeed range from 191 to 2200 µg (based on Frommer, 1974), but, even at the high end, these are still less than the supposed "average" value for secretion of biliary copper alone.]

In summary, we are left with a clear indication that copper is considerably more concentrated in bile than it is in other gastrointestinal fluids. Nevertheless, because of the volumes secreted, other fluids together are at least of equal potential import. Adding it all up for the adult human, it would appear that, on average, something like 5000 µg of copper may be released into the digestive tract from various sources (also including sloughed cells). About half of that may be derived from the bile, with the rest mainly from gastric and direct intestinal outputs.

The amounts of copper secreted into the digestive tract are much higher than the net fecal losses obtained from balance studies. For zero balance in humans, an intake of more than 2000 µg is recommended, suggesting that net fecal losses may amount to 1000–1500 µg. This would indicate that about two-thirds of the copper secreted from the mouth through the intestine is reabsorbed. The question is thus whether copper derived from particular secretions is more or less available for reabsorption, or, in other words, how do we get net excretion of copper from the body, when so much is reabsorbed? (Urine is a source of net excretion, but its contribution is small.)

In contrast to rats and humans, sheep do not appear to excrete the bulk of their copper in the form of biliary substituents (Soli and Rambaek, 1978). Indeed, biliary Cu concentrations are as low as concentrations in the urine. Overall excretion is low [only about 1.5% of an injected dose of $^{64}Cu(II)$ in 60 h], and gastrointestinal routes other than bile are thought to be most important. Also, biliary copper concentrations do not reflect liver concentrations in this species (Skalicky et al., 1978). The reduced capacity for biliary copper excretion would explain the special sensitivity of sheep and lambs to copper toxicity. (Further information on these matters, for humans and rats, follows below.)

5.2 Absorbability of Endogenously Secreted Copper

Effects of digestive secretions on the availability of dietary (mostly ionic) copper for absorption in the digestive tract have already been reviewed (Chapter 2), but without stressing what is know about the reabsorbability of endogenous forms of copper secreted into the digestive tract. Here, one important fact is quite clear. The copper found in the bile is largely unavailable for reabsorption by the small intestine. As summarized in Table 2-3, there are studies in the rat showing that bile inhibits the uptake of ^{64}Cu(II). Already in 1964, Owen found that there was a greater percentage of ^{64}Cu in the feces (over the first 24 h) when radioisotope was given orally with bile, at low as well as higher doses. At least at the doses of bile employed, however, the effects were not dramatic (48–56% with bile versus 39–50% without). Later, using ^{64}Cu placed in the intestine, Gollan (1975) confirmed that carcass uptake was less in the presence of bile but that salivary and gastric juices made no difference under the same conditions, and Jamison *et al.* (1981) reported that the proteins of pancreatic juice inhibited intestinal uptake of ^{64}Cu(II) in the rat about as much as did bile (Table 2-3).

Inhibitory actions of bile and perhaps also pancreatic juice would explain the less than totally efficient absorption of ionic copper observed with test doses of ionic radioisotope or with normal forms of dietary copper in balance studies. [Overall absorption in the human is about 60% and a bit less in the rat (Chapter 2).] However, it is also well established that the copper in bile itself is also largely unavailable for reabsorption. In the 1964 studies, Owen showed that with doses of less than 1 μg of Cu, 22–36% of biliary copper was reabsorbed (did not appear in the feces within 24 h, after oral administration to rats) in comparison to 50–61% of ionic ^{64}Cu. At higher doses, apparent absorption was considerably lower in the case of biliary copper (6-16% versus 39–47% for the ionic form). Farrer and Mistilis (1968; Mistilis and Farrer, 1968) carried out more extensive studies, summarized in Table 5-2, on dialyzed and non dialyzed

Table 5-2. Dialyzability and Intestinal Absorption of Biliary Copper[a]

Time of bile collection (h)	Dialyzable ^{64}Cu (%)	^{64}Cu precipitable with TCA (%)	Absorption, after intrapyloric intubation[b] (%)	
			Undialyzed bile	Dialyzed bile
0–4	50 ± 7 (9)	15	15 ± 2 (3)	16 ± 3 (5)
4–8	28 ± 6 (9)	32	13 ± 2 (2)	12 ± 5 (6)
8–24	16 ± 3 (9)	55	8 (1)[c]	10 ± 3 (3)

[a] Data of Farrer and Mistilis (1968) and Mistilis and Farrer (1968) for rats. ^{64}Cu-labeled bile was from animals injected intravenously with 100 μg of Cu as ^{64}Cu(II) acetate and collected for various periods. Absorption of Cu from bile was determined 24 h after intrapyloric administration of dialyzed or undialyzed bile containing 2.5 μg of radioactive copper. Dialysis of bile was by repeated exposure to 100 volumes of 0.9% NaCl (0°C); trichloroacetic acid precipitation was with a 5% final acid concentration (0°C). Absorption of control ^{64}Cu acetate was 39%.
[b] Mean ± SD (N).
[c] Bile from a Cu-EDTA-injected rat.

bile obtained from rats at various times after injection of ^{64}Cu (mostly very large doses), with quite similar results. The data show several interesting things. First, they reveal that a considerable portion of the biliary copper incorporated into bile immediately after injection is dialyzable, whereas very little is dialyzable when bile is taken at later times. One might think that the early incorporation may result from a need to process the large (100 μg) unphysiological doses of copper injected. However, the same trends were observed with lower doses (5–10 μg) (Farrer and Mistilis, 1968a). So it may be that a certain fraction of biliary copper is alway dialyzable, and probably this fraction may derive from ionic copper incorporated directly into the bile after absorption from the diet (or after injection) and transport to the liver (see further, below).

The second observation of interest is that, with decreasing dialyzability, the copper may be increasingly tightly bound to protein (even after acid denaturation) (Table 5-2). More importantly, perhaps, the copper of "older" bile is less absorbable than earlier bile and is much less absorbable than ionic ^{64}Cu (as the acetate) given in the same manner to similar rats. It should be noted, however, that even with "older" bile, absorption is not zero, at least in these studies. It "bottoms out" at about 25% of that for Cu(II) acetate (or 10% overall). The data shown also suggest that the tightly bound, dialyzable copper in the bile is no more available for reabsorption than the nondialyzable forms, although this has not been as consistent a finding (see more below). The relative unavailability of copper from human or rat bile for intestinal absorption has been confirmed by others (Owen, 1964; Lewis, 1973; Gollan, 1975) and is considered a well-established fact (see also Chapter 2). Gollan (1975) compared the absorption of *in vitro* labeled biliary copper with that in other fluids (Table 5-3). Although the method used for the labeling (see footnote to Table 5-3) may, in retrospect, not have been ideal and ^{64}Cu, as $Cu(OH)_2$, might have precipitated to varying degrees within the samples, there was a clear difference between the absorbability (and dialyzability) of copper from gallbladder bile and that from other fluids. Not only did the copper appear to associate much more loosely with saliva and gastric juice, but also the copper in these fluids was as absorbable as the (highly absorbable) Cu-acetate complex (Table 5-3). Unfortunately, pancreatic juice was not tested (and has, so far as can be ascertained, never been tested). Biliary copper, whether taken directly from the liver or from the gallbladder, appeared to be half as available for absorption. It is noteworthy that copper added to hepatic bile appeared to be much more readily dialyzable than that binding to gallbladder bile, although the absorbability of both was low. (Might "aging" in the gall bladder make copper increasingly less available?)

Table 5-3. Dialyzability and Absorbability of Copper Associated with Various Gastrointestinal Fluids[a]

^{64}Cu source (in vitro labeled)	Dialyzable copper[b] (%)	Retention 76 h after intraduodenal administration to rats[b] (%)
Cu(II) acetate	—	13 ± 4 (28)
Cu(II)-histidine	—	13 ± 2 (6)
Saliva	98 ± 2 (6)	13 ± 3 (6)
Gastric juice	97 ± 2 (6)	12 ± 6 (10)
Hepatic bile	85 ± 5 (6)	6 ± 1 (6)
Gallbladder bile	12 ± 12 (5)	6 ± 1 (21)
Gallbladder bile plus saliva	12 ± 7 (3)	—
Gallbladder bile plus gastric juice	20 ± 10 (3)	—

[a] Data of Gollan (1975) for bile and other fluids taken from the human. Absorbability was measured as whole-body retention of radioactivity in rats 76 h after intubation of the various solutions. Labeling with ^{64}Cu occurred by addition of excess ^{64}Cu(II) acetate, then bringing to pH 8 (with NaOH), and centrifugation, to remove unbound Cu as precipitated hydroxide. Dialysis was against 0.9% NaCl–10 mM Tris, pH 8, or running water (comparable results) over 48 h.
[b] Mean ± SD (N); recalculated.

These various findings point to the bile as a major (or perhaps the only major) source of endogenously secreted copper that is largely unavailable for reabsorption in the human and the rat and also as the major source (perhaps with pancreatic juice) of inhibitors of dietary copper absorption in the digestive tract. All this implies that bile is the major source of copper destined for net excretion from the digestive tract in these species, although, again, pancreatic and other intestinal juices have not been tested. The rat studies suggest that only 10–15% of biliary copper is reabsorbed. If this is the case also for humans, then 85–90% of biliary copper secreted daily should be destined for fecal excretion. This would, for the average human adult, amount to about 2200 µg per day, which seems reasonable.

5.3 Nature of Copper Components in Urine, Bile, and Other Alimentary Secretions

Little has been done to characterize the copper found in urine. Suzuki et al. (1981; Suzuki and Yoshikawa, 1981) have fractionated human urine on Sephadex G-75 after its 10-fold concentration by ultrafiltration with an Amicon UM-2 membrane (M_r cutoff about 2000). In normal individuals with a urinary Cu concentration of 8–24 µg/liter, more than 80% was in a low-molecular-weight fraction, eluting with the column volume and A_{280}. A small amount was also in the excluded volume ($M_r > 50,000$). In industrial workers suffering from cadmium toxicity or in Cd-injected rats, total urinary copper concentrations increased, and the proportions of copper eluting in the void volume and with the metallothionein fraction (M_r 10,000) increased. (Cadmium tends to result in kidney tubular damage.) The nature of the (most abundant) low-molecular-weight urinary copper form(s) detected in these studies is unknown, although, in general, copper tends to associate preferentially with N-containing ligands, and some small peptides would be present in urine that would be candidates for binding. Indeed, there are several peptide-sized copper-binding components in blood plasma (Wirth and Linder, 1985; Chapter 4) that might be filtered out by the kidneys. (Amino acids and Cu–diamino acid complexes would largely have been lost during the ultrafiltration.) It is also unclear whether all of the copper in urine was retained by the UM-2 filter, although the majority appeared to have been, based on recoveries relative to the initial (variable) total concentrations (Suzuki et al., 1981b). So we do not know whether the small amounts of Cu–amino acid complexes in blood plasma are reabsorbed by the kidney tubules or not. If not, loss of these complexes would, of course, contribute to the total Cu losses in the urine. (Indeed, a certain portion of the plasma amino acids does enter the urine.)

We have already alluded to the apparent dialyzability of copper associated with various secretions (Section 5.2), the copper associated with bile being strikingly less dialyzable (80–90%) than copper associated with saliva or gastric juice (more than 95%). Gollan (1975) showed that all the ^{64}Cu(II) in saliva eluted as material of low molecular weight, in the same position as amino acids, on Sephadex G-25 (eluted with Tris–NaCl buffer, pH 8). The material in which ^{64}Cu was found in gastric juice [brought to pH 8 to remove excess unbound ^{64}Cu(OH)$_2$] also had a low molecular weight (less than 400), but the elution of copper did not coincide with that of amino acids. The copper components of human and rat bile were also studied by Gollan and by others before and after. What emerges from these studies is that, under conditions in which bile is fractionated on gel columns with buffered isotonic saline solutions, two peaks of radioactive copper are always found (Fig. 5-1). These are there whether bile has been labeled with radioactive copper in vivo or in vitro (Gollan and Deller, 1973; Terao and Owen, 1973; Gollan, 1975; Frommer, 1971), giving additional credence to the studies with in vitro labeled bile already discussed. Moreover, the two peaks are obtained in various gel systems, from Sephadex G-200 to G-75. Both Terao and Owen (1973) and Cikrt et al. (1974) found that injected radioactivity migrated from the smaller to the larger component over time. With Sephadex G-50 or G-25, an additional

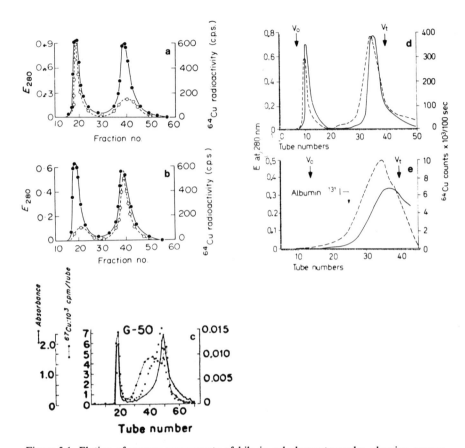

Figure 5-1. Elution of copper components of bile in gel chromatography, showing copper radioactivity (---) and absorbance at 280 nm. (A) Human gallbladder, labeled *in vitro*, separated on Sephadex G-75. (B) Human hepatic bile, labelled *in vitro*, separated on Sephadex. (C) Rat bile, labeled with $^{67}CuCl_2$ *in vivo*, separated on Sephadex G-50, also showing elution of copper (×) determined by atomic absorption. (D) Human T-tube bile separated on Sephadex G-200 in 10 mM Tris–0.9% NaCl, pH 8. (E) As for (D), in the same buffer with 15 mM glycocholate (close to the concentration of this detergent in human hepatic bile). Reprinted, with permission, from Gollan (1975), Terao and Owen (1973), and Frommer (1977) for (A) and (B), (C), and (D) and (E), respectively.

low-molecular-weight peak was separated, its elution coinciding with that of much of the absorbance at 280 nm (Fig. 5-1B) (Terao and Owen, 1973; Gollan, 1975). Nonradioactive copper was associated with both low-molecular-weight components (Fig. 5-1C).

A potentially important discovery of Frommer was that the presence of *bile acid* abolished the large component and displaced all the copper (and A_{280}) into components of low molecular weight (Fig. 5-1, E versus D). Using 15 mM glycocholate, Frommer (1977) found that the majority of the ^{64}Cu was included in Sephadex G-50 (M_r about 4000), but was excluded by Sephadex G-25, suggesting an apparent molecular weight a bit greater than 5000. (A second, minor ^{64}Cu component of M_r 8000 was sometimes also seen.) It is noteworthy, however, that, in the presence of detergent, the elution of the main (broad) radiocopper peak coincided with a shoulder eluting later than the main A_{280} peak (Fig. 5-1E); also, tryptic digestion of bile failed to alter the behavior of biliary copper components in gel chromatography (Gollan, 1975). Both of these findings would suggest that most biliary copper either is not associated with protein or is associated with a small protein or peptide(s) not digestible by trypsin or by other digestive proteases. Otherwise biliary copper should be reabsorbable. (Unfortunately, this has not been followed up by other workers.) More recently, Martin et al. (1986) have tested bile ultrafiltrates for copper binding, using "modified gel chromatography" (in which $CuNO_3$ was included in the buffer) on Sephadex G-10 and G-15, and ultrafiltration though Amicon membranes. Three copper-binding components of bile were identified. Two were excluded from UM-5 membranes but filtered through YM-10, thus having apparent molecular weights between 5 and 10 kDa. A third filtered through UM-5 and was thus smaller. All eluting peaks contained ninhydrin-reactive materials and amino acids but were highly impure. All of these studies leave us with the concept that the copper in bile is mainly associated with a component, complex, or micelle of M_r 5000 that tends to aggregate, is at least partly dissociable with bile acids, and is largely or completely immune to proteolytic digestion. Another, smaller portion is associated with much smaller ligands having absorbance at 280 nm (perhaps amino acids); its size is still quite unclear (amino acids tend to adsorb to the gel).

This picture does not entirely fit with a report by Samuels et al. (1983) of the existence of a 50,000-Da protein, binding five Cu atoms (mostly EPR silent), in the bile of rats given an azo dye carcinogen. The high apparent molecular weight reported was obtained by denaturing electrophoresis in SDS and does not fit with the late elution of this protein from columns of Sephadex G-75. Therefore, these authors may actually have been dealing with the same 5000-Da protein seen by others. It was noteworthy that this protein did not react with antiserum against ceruloplasmin or albumin, had a high pI (of about 7), and had a rather distinctive amino acid composition, high in Gly, Asx, and Glx and relatively low in Leu. Most recently, Iyengar et al. (1988) reported that the copper peak of human bile that is excluded from Sephadex G-75 (in the absence of detergent) reacted with antibody against ceruloplasmin. Indeed, they proposed that a trypsin-resistant fragment of ceruloplasmin may be the major form in which copper is excreted from the body. Although there is good evidence that ceruloplasmin (or its fragments) are in the bile (see below), the question is how much copper in the bile does it account for (and how big is the ceruloplasmin component?). The high-molecular-weight peak accounts for about 35% of the total copper in the bile of normal people and a somewhat larger proportion after copper supplementation, but the effect of detergent on this peak was not examined.

The apparent properties of the major biliary copper component are summarized in Table 5-4. Of particular interest may be the work of Lewis (1973), which showed that synthetic bile made with lecithin, electrolytes, and taurochenodeoxycholic acid (versus other common bile salts) had similar ^{64}Cu-binding properties and similar behavior to those of authentic radiocopper-labeled human or rat bile in agar gel electrophoresis and in terms of extractability with chelating agents and availability for intestinal resorption. [Similar electrophoretic results were obtained for rat bile by Cikrt and Tichy (1972).] The presence of lecithin also was important in producing the synthetic bile. Lewis (1973) noted that the pattern of secretion of copper over time, after surgical (T-tube) cannulation, closely paralleled that of taurodeoxy- and taurochenodeoxycholate secretion. (In both cases, secretion dropped dramatically between 6 and 8 h after surgery.) Taurochenodeoxycholate (a small proportion of the total bile acid in normal human bile) with lecithin (Fig. 5-2) may thus be responsible for some of the copper binding in bile. Unless in the form of a large aggregate, lecithin plus taurochenodeoxycholate together (in a potential copper

Table 5-4. Properties of the Major Copper Component Associated with Human or Animal Bile

Property	Finding	Reference
Apparent size	~5000 Da	Frommer (1977)
	(+ second, much smaller component)	Terao and Owen (1973)
Dialyzability	Low	Various (see earlier tables)
Stability to pH	High at pH 5+	Gollan (1975)
	Lower at pH 4 or less	Gollan and Deller (1973)
Availability to chelating agents	Mostly unavailable to 0.1 M EDTA or penicillamine	Gollan and Deller (1973)
Electrophoretic migration	Rapid, to anode, at alkaline pH; associated with pigment, not with albumin	Cikrt and Tichy (1972) Lewis (1973)
Bile acid complex	With taurochenodeoxycholate, not taurodeoxycholate	Lewis (1973)

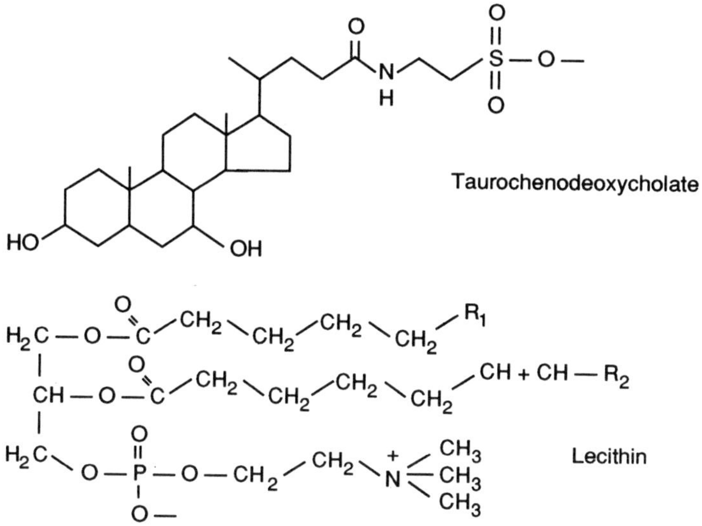

Figure 5-2. Components of bile that may play a role in copper binding.

complex) cannot alone account for the apparent size of the major biliary copper component identified by gel chromatography. This implies that an aggregate or micelle of these components is formed or that additional factors (maybe even of a protein nature) are bound as well. Indeed, several groups had reported coprecipitation of biliary copper with protein (Owen, 1964; Farrer and Mistilis, 1968; Frommer, 1972a,b). The question is what this or these proteins/peptides might be.

The possibility that *metallothionein* is (or is a major part of) this protein is not supported by studies of Sato and Bremner (1984): In copper-injected rats, only 1–2% of the copper in bile was bound to metallothionein, and less than 10% of liver metallothionein appeared to end up in the bile intact. The possibility that *ceruloplasmin* is in some way a source of this protein (and of at least some of the biliary copper) has been investigated several times with mixed results but remains a significant prospect. In this regard, there is little argument that ceruloplasmin is at least an indirect source of biliary copper (see more below). In early studies, Aisen et al. (1964) found less than 5% of ceruloplasmin-^{64}Cu entering rabbit bile over 67 h, and Jeunet et al. (1962) also found low percentages in rat bile during liver perfusion with ceruloplasmin. However, Jeunet et al. (1962) and, most recently, Iyengar et al. (1988) (with human bile) did detect immunologically some ceruloplasmin (or its fragments) in bile. Kressner et al. (1984) did more extensive studies in cannulated rats, with

^{67}Cu-and ^{64}Cu-labeled human ceruloplasmin, before and after desialylation. They showed that asialoceruloplasmin was the best source of biliary ^{67}Cu (8% of radioactivity being excreted in 3 h) but that both sialoceruloplasmin and copper administered i.v. as cupric acetate were significant sources (with 2 and 3% appearing in the bile by 3 h). (The doses of copper involved are unclear.) More importantly for the present question, they reported that about 75% of the ^{67}Cu in the bile of rats given ^{67}Cu-sialo- or asialoceruloplasmin precipitated with ceruloplasmin antiserum (and only a few percent in the case of rats given ^{67}Cu acetate). This does *not* mean that a similar proportion of *total* biliary copper was associated with ceruloplasmin (or its fragments). Simultaneous administration of another asialoglycoprotein (orosmucoid, or α_1 acid glycoprotein) inhibited the appearance in bile of ^{67}Cu from asialoceruloplasmin. As desialylated plasma proteins (including ceruloplasmin) are thought to enter hepatocytes by galactose-receptor-mediated endocytosis, this finding was not unexpected and was confirmatory of the concept that endocytosis is at least the major route taken by asialoceruloplasmin-copper on its way to the bile. [Van den Hamer *et al.* (1970) have reported that removal of only 2 sialic acid residues exposing galactose is sufficient to promote the rapid uptake of ceruloplasmin by the liver, and specifically the hepatocyte (Morell *et al.*, 1968).] The data of Tavassoli (1985) on sialo- and asialoceruloplasmin receptors on different liver cell types are also compatible with this idea and suggest that holoceruloplasmin copper enters the bile after desialylation within liver endothelial cells, during transcytosis (see Chapter 3). The size or sizes of the ^{67}Cu-labeled protein in bile reacting with ceruloplasmin antiserum are still unknown. The protein may at least partly be intact, since there is some evidence that other whole plasma proteins enter the bile (Schiff *et al.*, 1984; Renston *et al.*, 1980).

Consistent with the concept that whole ceruloplasmin or a copper fragment (or fragments) may be present in the bile are the early studies of Waldman *et al.* (1967), in which the gastrointestinal absorbability of ceruloplasmin- copper was examined in several species and shown to be very low. Specifically, orally administered, *in vitro* labeled human ^{67}Cu-ceruloplasmin in humans, dogs, and rats was largely recovered in the stool over the following days ($88 \pm 18\%$, $92 \pm 10\%$ and $98 \pm 13\%$, respectively, mean $\pm$ SD, for 8 humans, 4 dogs, and 6 rats). Thus, ceruloplasmin-bound copper (produced by a procedure that probably does not label all forms of copper in the molecule) seems to be remarkably indigestible and to be similarly unavailable for absorption/reabsorption as copper in bile. It is tempting to think that a portion of the ceruloplasmin molecule, high in Cu and known to be indigestible by trypsin, is contributing to the net excretion of copper from the body by exiting through the bile. Indeed, this has recently been suggested by Brewer (1986; Brewer et al., Iyengar et al., 1988). If so, the most likely fragment is the histidine-rich portion in part B″ of the molecule (Fig. 4-2), thought to carry three Cu atoms and to be unaffected by most proteolytic enzymes.

5.4 Sources of Copper for Excretion from the Body

With regard to urine, low-molecular-weight peptide and perhaps also amino acid complexes have been identified as the likely forms by which copper is excreted by this route. Since urine derives from the blood plasma, it seems likely that one or more of the low-molecular-weight copper components identified in plasma (Chapter 4) may be the direct source(s) for urinary excretion. As reviewed in Chapter 4, these components are still largely uncharacterized, with the exception of amino acid complexes and GHL. Since the vertebrate organism does not waste amino acids and reabsorbs them through the kidney tubules after glomerular filtration, it seems possible that the traces of copper lost by this route are mainly in forms other than amino acid complexes, and one wonders what these might be. In Wilson's disease, in which there is a defect in biliary copper excretion, the amounts of copper associated with low-molecular-weight components in blood are usually increased (Sarkar et al., 1983), and urinary excretion is enhanced (see Chapter 9). This is consistent with there being a reciprocity between the rate of urinary copper excretion and amounts of copper associated with low-molecular-weight components in the blood plasma. It is also well know that oral or parenteral administration of chelating agents with a high affinity for copper will increase the fraction of plasma copper associated with small molecules and enhance urinary copper excretion (Gibbs and Walshe, 1977; see Chapter 9, Section 9.1.1). This is the main basis for the therapeutic treatment of patients with Wilson's disease, in which copper accumulates to toxic levels in liver, brain, and other tissues.

The sources of copper for excretion/secretion in alimentary fluids other than bile also are poorly defined at this time. The initial observations of Gollan (1975), made for gastric fluid and saliva (Table 5-3), suggest (as for urine) that copper is bound to small dialyzable substances. This again suggests the involvement of peptides and/or amino acids that could be derived from the blood plasma or interstitial fluid. In the case of pancreatic fluid, the studies of Jamison et al. (1981) on absorption inhibition in inverted gut sacs implicated proteins as responsible for the binding (and inhibition). The status and contributions (of any) of mucosal cell (or plasma) copper to direct small intestinal secretion is unknown.

With regard to bile, it is clear that a major proportion of an administered ionic radiocopper dose finds its way to this fluid within 24 h and that much more is excreted by this route than through the urine (Table 5-5), at least in the rat and in the human. Many of the studies were done over relatively short times, directly after injection of copper radioisotope, and it is noteworthy that some of the administered radioisotope makes a very rapid initial appearance in the bile. The discrepancies between the percent dose results obtained for rats with cannulated versus uncannulated bile ducts probably reflect enhanced biliary copper secretion induced by surgery (Owen, 1980). As it has been shown that dietary copper initially binds to the same blood plasma components as injected ionic copper (Weiss et al., 1985; Linder et al., 1988; Chapter 4) and appears to follow the same distribution route, the data obtained for excretion of injected ionic ^{64}Cu (in Table 5-5) should be directly relevant to what is found for the disposition of copper entering the blood from the diet.

This is borne out more directly by the results of studies with the perfused liver. Owen and Hazelrig (1968) followed the appearance of radiocopper in the bile after addition of ^{64}Cu(II) acetate to the blood circulating through isolated livers of rats. The added radiocopper in the circulation would, under these conditions, be associated with albumin and

Table 5-5. Loss of Injected ^{64}Cu(II) by Various Routes in Humans and Rats

Species	Loss per first 24 h (% of dose)	Route	Dose given (μg)	Reference
Human	6–18[a]	Gallbladder	500 μCi (~100–500)	Gollan and Deller (1973)
Rat	24–75[a]	Bile (cannulated)	1–100	Farrer and Mistilis (1968)
	31			Cikrt (1972
	11	Bile (not cannulated)		Owen et al. (1975)
	10–14	Total Gl tract	0.01–0.2	Linder et al. (1986)
	2	Urine	100	Owen (1965)
				Owen et al. (1975)
	2–7	Urine	0.01–0.2	Linder et al. (1986)
	3–4	Urine	5	Owen and Hazelrig (1968)
Rat				
Neonate	0.07–0.18[b]	Total loss	1	Mearrick and Mistilis (1969)
Weanling	0.28	Total loss	1	Mearrick and Mistilis (1969)
	51[a]	Bile (cannulated)	5	Owen and Hazelrig (1968)

[a] Extrapolated from a 4–5-h test period.
[b] Extrapolated from a 12-day test period.

transcuprein (Chapter 4). As shown in Table 5-6, one-fifth of the radiocopper was secreted into the bile within 5 h. About half as much ended up in circulating ceruloplasmin, and most of the rest was still in the liver. In the case of copper-loaded livers (from rats on very high copper diets), smaller proportions of the radiocopper dose ended up in bile and ceruloplasmin, reflecting its dilution by excess copper in the tissue. (Probably, much larger absolute amounts of copper entered the bile compartment.) In the case of copper-deficient livers, very little of the radiocopper went into the bile, reflecting the need to retain body copper, and increasing proportions went into ceruloplasmin, presumably to make up for the decrease in ceruloplasmin synthesis (and blood concentration) that accompanies copper deficiency. [Linder et al. (1979b) have shown that copper administration to copper-deficient, but not to copper-sufficient, rats enhances the rate of ceruloplasmin synthesis, measured with [^{3}H] leucine.] These results indicate that, depending on copper status, different proportions of newly absorbed copper entering the liver are shuttled into processes occurring in parallel: production and secretion of bile and production and secretion (into the blood) of ceruloplasmin. This also suggests that at least part of the copper in bile does not come from ceruloplasmin. One factor that may be involved in this shuttling to the bile is glutathione, as suggested by studies with diethyl maleate (in rats) in which this compound decreased liver glutathione concentrations 20–25% and similarly decreased biliary Cu secretion (Alexander and Aaseth, 1980).

As already indicated, ceruloplasmin is a source of copper entering the bile and has long been implicated as being on the route to excretion. Here, the most significant points have been as follows. First, in Wilson's disease, in which there is a toxic accumulation of copper in liver and other organs,

Table 5-6. Disposition of Copper Infused as Ionic Copper into Rat Liver[a]

Rat diet	N	Distribution at the end of 5 h[b] (% of dose)		
		Liver	Bile	Ceruloplasmin
Normal	18	38 ± 7	21 ± 7	8 ± 3
Excess copper				
29 weeks	8	62 ± 9	9 ± 4	2 ± 6
Copper-deficient				
3–14 weeks	5	59 ± 8	2 ± 0.3	10 ± 3
23–32 weeks	5	50 ± 7	0.3 ± 0.1	17 ± 4

[a] Modified from Owen and Hazelrig (1968). Perfusate blood contained 5 μg of added Cu, as ^{64}Cu(II) acetate.
[b] Mean ± SD.

due to a decreased capacity for copper excretion, there also is a relative deficiency in liver ceruloplasmin synthesis (Scheinberg and Sternlieb, 1967). [On average, patients with Wilson's disease have about one-quarter the normal levels of circulating ceruloplasmin (see Chapter 9).] Second, when the bile duct is obstructed or ligated, there is a decrease or cessation of bile (and biliary copper) secretion, and this is accompanied by an increase in the circulating levels of ceruloplasmin (Smallwood et al., 1968; Worwood et al., 1968; Linder et al., 1986), as if there was an accumulation of ceruloplasmin destined for removal from the circulation by the liver for infusion into the bile. Other studies linking ceruloplasmin to copper excretion are those in which the protein labeled with radiocopper was infused intravenously and the apparent half-life of the copper in plasma (and its total loss from the body) were followed over time (Table 5-7). In most of the species examined, including the human, a significant percentage of copper deriving from ceruloplasmin was excreted each day.

Direct evidence for ceruloplasmin as a source of biliary copper and biliary protein may be taken from the data of Kressner et al. (1984), already summarized. Here again, radiocopper added in ionic form (attaching mainly to albumin and transcuprein) was an immediate source of some biliary copper. However, on the basis of percent dose over 3 h (though different doses were used), sialo- and especially asialoceruloplasmin were equal or better sources. As already noted, Tavassoli et al. (1986) have reported that endothelial cells separating hepatic sinusoids from the Space

Table 5-7. Turnover and Excretion of Ceruloplasmin Copper in Different Species, as Measured with Radioactive Copper Given Intravenously as Ceruloplasmin

Species	Apparent half-line in plasma (days)	Daily loss (% of dose)	Reference
Human	4.5[a]	11–22 (GI tract)	Waldman et al. (1967) Audran et al. (1974)
Dog	6.1	7–11 (GI tract)	Waldman et al. (1967)
Rabbit	1.8–2.8	>5 (bile)	Aisen et al. (1964)
Rat	5[b]	10–15 (total)	Marceau and Aspin (1972)
	6.5	8 (total)	Owen (1971)

[a] The apparent half-life for Cu-deficient ceruloplasmin (measured with radiocopper) is about 1 day in humans (Audran et al., 1974).
[b] The half-life of Cu-deficient ceruloplasmin in the rat (determined in two rats by labeling the protein moiety) is 6 h, and that of the holoprotein 12 h (also measured in two animals) (Holtzman and Gaumnitz, 1970a).

of Disse (that leads to the hepatocytes) appear to transport and desialylate ceruloplasmin. Desialylation appears to occur only to molecules that have triantennary carbohydrate units (at least in the rat) (Irie et al., 1989) and involves transcytosis (cross-cell endo- and exocytosis) that can be inhibited by monensin and NH_4Cl (Omoto and Tavassoli, 1989), as shown with *in vitro* endothelial cell preparations. This would explain why ceruloplasmin with its sialate residues initially intact was also a significant source of biliary copper (and ceruloplasmin). The hepatocyte, on the other hand, may be absorbing only the desialylated form of the protein, and this might be the form destined for the bile. Indeed, liver uptake (and plasma clearance) of asialoceruloplasmin is well known to be more rapid than that of the fully sialylated protein (Gregoriadis et al., 1970; Van den Hamer et al., 1970; Ashwell and Morell, 1974; Stockert et al., 1980). [Removal of only two sialic acid residues is sufficient to effect ceruloplasmin removal (van den Hamer et al., 1970)]. It is also clear that ceruloplasmin partially denuded of its copper (as in the case of treatment with ascorbate or cyanate) is removed from the circulation much more rapidly than the holo protein (Holtzman and Gaumnitz, 1970a; Audran et al., 1974) (Table 5-5). (Whether low-copper ceruloplasmin also contributes to biliary copper is unknown.)

Uptake of desialylated plasma proteins by hepatocytes is thought to occur via cell membrane receptors that have a high affinity for carbohydrate residues (on glycoproteins) ending in galactose, which is the end result of desialylation (Ashwell and Morell, 1974; Stockert et al., 1980). This may be linked to the bile pathway in a more general way. Dive (1971) has suggested that up to 1 g of plasma proteins may enter human bile by such a route every day. Immunoglobulin A has long been known to enter the bile (and digestive tract) more or less intact (Birbeck et al., 1970; Schiff et al., 1984). Thomas and Summers (1978) showed that small proportions of α_1 acid glycoprotein, fetuin, and carcinoembryonic antigen (CEA) entered the bile within an hour after their intravenous infusion as radioiodinated protein into rats, as determined by radioactivity and also by radioimmunoassay (RIA) of the proteins in bile. However, here, desialylation prior to infusion did not make a great deal of difference. [Indeed, even proteins (or protein fragments) with no carbohydrate (such as insulin and horseradish peroxidase) have been shown to enter bile canaliculi, after fusion with endocytic vesicles receiving the proteins by liver perfusion (Renston et al., 1980).] A comparison of the results of the ^{125}I and RIA assays suggested that, at least for CEA, part of the protein in bile was fragmented, especially at later times. (There was ^{125}I radioactivity yet little immunoreactivity.) In this connection, Gregoriadis et al. (1970) had found that immunoprecipitable ^{64}Cu, in *lysosomal–mitochondrial fractions* of liver homogenates derived from rats given ^{64}Cu-ceruloplasmin, went from 90% at 3 min to about 20% after 30 min, implying that rapid degradation of absorbed ceruloplasmin might occur along the way. This suggests either that a portion of ceruloplasmin may go more directly into the bile or that the relative indigestibility of at least a portion of the molecule or the rapidity of its movement into bile allows entry in only partially degraded form.

In summary, the route that ceruloplasmin takes to the bile from the blood may involve the following: uptake (perhaps especially of copper-deficient sialoceruloplasmin) by liver endothelial cells and desialylation; transfer across the Space of Disse to hepatocytes; uptake by hepatocytes, along with other asialoglycoproteins, by galactose-receptor-mediated endocytosis, followed perhaps by fusion with lysosomes whose contents, after partial digestion, are destined for the bile (Scheinberg and Sternlieb, 1967), the end result being copper associated with fragments (or a particular fragment) of ceruloplasmin in the bile. Bile is then secreted into the digestive tract, and the ceruloplasmin-fragment-bound copper, not being readily reabsorbed, is excreted from the body.

We are left to reconcile the findings that both ionically administered copper and ceruloplasmin copper are sources of biliary copper with findings of one major and one or more minor copper-binding components in the bile. The direct copper assays of Terao and Owen for rat bile (Fig. 5-1C) suggest that about 40% is in the component of lowest molecular weight (x) and the rest in the 5000-Da component (or excluded peak that dissociates into this component). One highly speculative possibility is that the 5000-Da component contains a fragment of ceruloplasmin and that the other smaller component derives from ionic copper delivered to hepatocytes by albumin and transcuprein (see Chapter 4). Perhaps the latter copper ends up with lecithin and taurochenodeoxycholate (Lewis, 1973). There may also be an additional small fraction of biliary copper, derived from ionic copper delivered to the hepatocyte, perhaps the 10–15% that is dialyzable, that was not recovered from the gel permeation columns [where recovery was 70–80%, at least in the studies of Frommer (1977)].

5.5 Overall Turnover of Copper in the Body

Very few studies have been carried out to ascertain the half-life of whole body copper in vertebrate animals or man. In 1971, Owen (1971) made the first attempts in normal and copper-deficient rats, using ^{67}CuCl$_2$, and followed whole body retention of the isotope for more than two weeks.

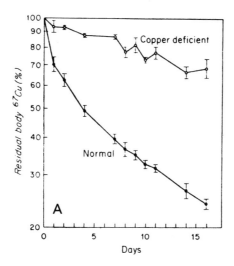

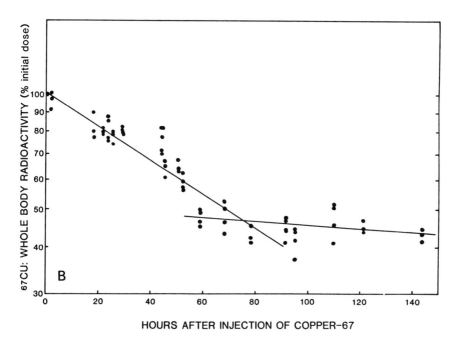

Figure 5-3. Turnover of whole-body copper in rats, determined with ^{67}Cu. (A) Data of Owen (1971), for normal and copper-deficient rats injected intravenously with ^{67}CuCl$_2$; reprinted with permission. (B) Data of Linder et al. (1986), from rats injected intraperitoneally with ^{67}CuCl$_2$; reprinted with permission from the publisher.

At least in the normal animals, loss of total body radioactivity was not a simple exponential function (Fig. 5-3A), suggesting the involvement of more than one body copper pool, from which copper might be lost at different rates. Linder et al. (1986) have recently confirmed and expanded these findings (Fig. 5-3B), showing that loss of whole body copper is at least biphasic, drawing from two different compartments which have half-lives of 67 h and more than 200 h, respectively. These results are entirely compatible with half-life estimates that may be drawn from Owen's data. The question is then whether ceruloplasmin and ionic copper (associated with albumin, transcuprein, and amino acids; Chapter 4) are the two pools in question or intermediates for the two pools. This is not easy to say, first because there are considerable discrepancies between apparent half-lives obtained for rat ceruloplasmin (and its copper) by various laboratories. The data of Owen (1971) and Marceau and Aspin (1972) for rats give values (based on radiocopper) somewhat longer than those for humans (Table 5-7) and much longer than those obtained for the protein by Holtzman and Gaumnitz (1970a). This suggests that copper is being removed from and restored to ceruloplasmin many times over, while the protein itself is removed and replaced at a different rate. This idea is not entirely untenable and perhaps even has merit. Ceruloplasmin is a source of copper for excretion. It is also a source of cellular copper (Chapter 4). As a source of cellular copper, some of its copper will be removed, and at least partially-apo ceruloplasmin is found in the plasma. This deficient ceruloplasmin might perhaps receive copper from cells to return it to the liver, since we do not know how else copper may be returned by peripheral tissues to the blood and liver, for excretion. In this regard, all but traces of radioactivity present in blood plasma from 1 to 10 days after injection of ^{67}Cu were attached to ceruloplasmin, and not to albumin, transcuprein, small peptides, or amino acids, in the case of the rat (Fig. 5-4) (Weiss et al., 1985), a period in which about 65–70% of the injected dose of ^{67}Cu was excreted. Although it is clear that removal and restoration of copper from/to ceruloplasmin cannot occur by simple exchange or dialysis, removal is mediated by specific membrane receptors (Chapter 3) and so might be reloading.

The apparent half-life values (5–7 days) for ceruloplasmin-copper from Owen (1971) and Marceau and Aspin (1972) do not correspond to the apparent half-lives of either the fast-or slow-turnover whole body compartment. However, Weiss et al. (1985) determined an apparent half-life of only 2.4 days, using *in vivo* labeling of ceruloplasmin. These values for half-life would nevertheless be the result of several factors, including dose of copper or ceruloplasmin administered, losses of copper from ceruloplasmin to nonhepatic as well as to hepatic cells (and bile), perhaps also return of

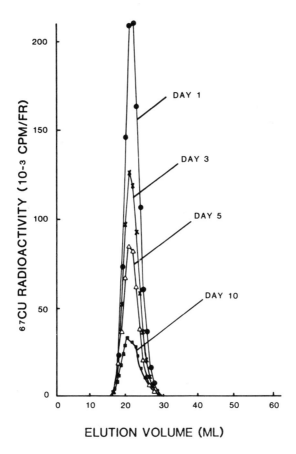

Figure 5-4. Distribution of ^{67}Cu among components of rat plasma at longer times after ingestion of ^{67}CuCl$_2$. Sephadex G-150 chromatography of plasma samples from rats taken 1–10 days after intraperitoneal injection. Elution position of radioactive peak is the same as for ceruloplasmin (measured as oxidase activity). Reprinted, with permission, from Weiss *et al.* (1985).

some copper from nonhepatic cells to apoceruloplasmin, and even perhaps recycling of copper through ceruloplasmin re-formed and secreted by the liver. Thus, it may be that the slow-turnover compartment is the one involving liver extraction and biliary excretion of ceruloplasmin fragments with copper. The faster turnover compartment might then be that replenished in the liver (and perhaps elsewhere) from nonceruloplasmin plasma sources. For the slow-turnover compartment, desialylation of ceruloplasmin would be a factor promoting return of this copper to the liver, for excretion (van den Hamer *et al.*, 1970) (where it would enter hepatocytes through the galactose receptor).

Similar whole body turnover studies have not been done in humans or other vertebrates, although it seems likely that the results for the human would be similar to those for the rat (or dog). The apparent half-lives of ceruloplasmin and percentages of ceruloplasmin copper appearing in the bile are not dissimilar (Table 5-7) nor are the percentages of whole body copper appearing in the bile or gastrointestinal tract (Table 5-5).

5.6 Regulation of Copper Excretion

It is clear that the rates of copper excretion from the body will vary in different physiological states. More importantly, excretion (which is largely nonurinary) is geared to copper status, being greater when excess copper is present in the body and much lower (or almost absent) in states of copper deficiency (Fig. 5-3A; Table 5-8). Indeed, excretion is the primary factor regulating copper homeostasis in vertebrates. (As indicated in Chapter 2, copper status has less of an effect on copper absorption.) Evidence available for copper excretion in various physiological states is summarized in Table 5-8 (and also Table 5-5). Most of the data do not distinguish between rates of loss from the fast-and slow-turnover compartments. One sees that excretion is slower in neonates (Table 5-5), in prepuberty (Table 5-8), in copper deficiency, and upon ligation of the bile

Table 5-8. Rates of Copper Excretion in Different Physiological States in Rats[a]

State	Half-life (days)			References
	Fastest rate	Average rate	Slowest rate	
Normal adult	~2	~5	~12 (8–16)	Owen (1971)
Normal adult	2.8		9.2	Linder *et al.* (1986)
Cu-deficient adult		~24[b]		Owen (1971)
Prepuberty	3.3[c]		4.2[c]	Linder *et al.* (1986)
Estrogen treated	4.6[c]	ND[d]	ND	Karp *et al.* (1986)
Tumor bearing	3.8[c]	ND	ND	Karp *et al.* (1986)
Ligated bile duct	4.8[b]	ND	ND	Linder *et al.* (1986)
Sham operation	2.3	ND	ND	Linder *et al.* (1986)

[a] Rats injected with $^{67}CuCl_2$ or $^{64}CuCl_2$ and followed by whole-body counting. All rats were adult (3.5–5 months of age), unless otherwise indicated.
[b] $p < 0.01$ for difference from appropriate (adult) controls.
[c] $p < 0.05$ for difference from appropriate (adult) controls.
[d] ND: Not determined.

duct. With regard to the latter, it has already been noted that this only cuts in half the rate of total copper excretion from the body, implying that the rat either uses other gastrointestinal routes for net excretion or can adapt to use them out of necessity. (The sheep is thought to use other gastrointestinal routes.) Indeed, Terao and Owen (1974) showed that the stomach and intestine were appreciably labeled when ^{67}Cu(II) was administered intravenously. Excretion is also decreased in the case of rats injected with large doses of estradiol or in those implanted subcutaneously with fast-growing tumors. Both of the latter treatments induce increases in the concentration of ceruloplasmin, measured as ceruloplasmin oxidase activity (Karp et al., 1986). The tumor effect is via an incrase in the rate of ceruloplasmin synthesis (Linder et al., 1979b). That of the estrogen must be by a different mechanism, as the tumor and estrogen effects are additive (Karp et al., 1986). Most of this regulation of excretion may occur through actions of the liver, involving the shuttling of copper from different liver and blood compartments to the bile, as suggested by liver perfusion studies (Owen and Hazelrig, 1968) (Table 5-6). Indeed, normal liver function is necessary for this process, and injury of hepatocytes from administration of CCl_4 results in decreased biliary output of copper, without an apparent decrease in the rate of bile flow (Cikrt et al., 1975). Alexander and Aaseth (1980) have shown that injection of dimethyl maleate, which depresses hepatocyte and bile concentrations of glutathione, or diethyldithiocarbamate (DDC), which accumulates in liver and has a high affinity for copper, both suppress biliary copper excretion in the rat. This suggests that glutathione might be involved in the excretory mechanism and that a form of copper available to DDC in hepatocytes is bound for the bile. The rate of ceruloplasmin desialylation, in the plasma or interstitial fluid, should also affect the rate of biliary copper excretion as it would influence the rate of ceruloplasmin uptake by hepatocytes via the galactose receptor (van den Hamer et al., 1970). [It is not clear where desialylation of ceruloplasmin may occur, although some occurs in liver endothelial cells (see earlier).]

Finally, the adrenal cortex (and hypophysis) plays at least a permissive role in biliary copper excretion in the case of the rat, as demonstrated by Evans et al. (1970a) and Evans and Cornatzer (1971). Adrenalectomy or hypophysectomy results in an elevation of the copper content of the liver, a decreased rate of bile flow (from 1.0 to 0.5 ml/h in cannulated animals), and a decreased loss of copper by this route, at least in this species. Similarly, corticosteroids injected directly or released in surgery and anesthesia enhance biliary copper excretion (without altering bile flow) in sheep (which do not use the bile as a major excretory route for copper). In rats, there is also a significant positive relationship between body temperature and biliary copper secretion, over a range of 30 to 40°C

(Klaassen, 1973). Thus, fever and probably also surgery (Owen, 1980) enhance copper excretion. These latter phenomena would be expected also to involve an acute phase response, in which ceruloplasmin formation by liver (Gitlin, 1988a) and probably also reticuloendothelial cells (such as macrophages) (Gitlin, 1988b) would be increased, resulting in elevated plasma ceruloplasmin concentrations. (Ceruloplasmin might then return increasingly to the liver, for exit in the bile.)

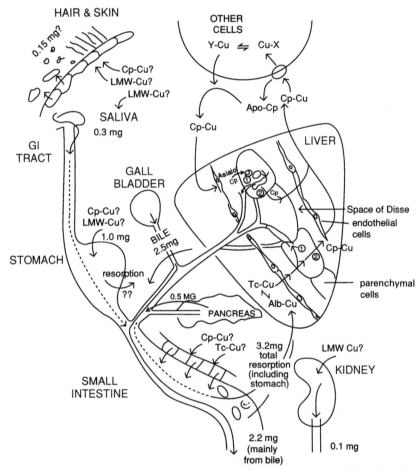

Figure 5-5. Overview of copper and secretions responsible for daily losses of this mineral from the body. Amounts indicated are daily secretions or losses. Areas of the figure show hair and skin; gastrointestinal tract; the enterohepatic pathways and organs, and the interchanges between liver and other tissues. Cp, Ceruloplasmin; ApoCp, apoceruloplasmin; AsialoloCp, ceruloplasmin without sialic acid; LMW, low molecular weight; Tc, transcuprein; Alb, albumin; X and Y, intracellular copper binders.

5.7 Summary

The most salient conclusions about sources and mechanisms used by vertebrates to excrete copper from the body are summarized in Fig. 5-5. Large amounts of copper are secreted into the digestive tract and traces into the urine. Except for biliary copper, most of the copper may come directly or indirectly from the nonceruloplasmin (perhaps even the nonprotein) portion of the blood plasma. Most of this copper secreted into the digestive tract is reabsorbed, contributing to a constant influx of copper, probably ionic, from the intestine (and stomach?) into the blood. Biliary copper, on the other hand, derives from two sources processed independently by the liver, resulting in two forms of biliary copper compounds highly unavailable for resorption. Both the nonceruloplasmin and ceruloplasmin fractions of the plasma contribute copper (and protein) to this pathway, as probably does one (or more) of the copper pools of the hepatocyte. Biliary copper is associated mainly with a component of about 5000 Da, which may include an indigestible ceruloplasmin fragment, and with a smaller component, which may contain lecithin and taurochenodeoxycholic acid. These are the main forms in which endogenous copper destined for actual excretion is lost from the body every day. However, at least when the bile duct is blocked, other gastrointestinal routes, such as the pancreatic fluid, and direct secretions from mucosal cells may become important for ridding the body of excess copper.

Copper within Vertebrate Cells

6.1 Distribution of Copper among Cell Organelles

Beginning with Thiers and Vallee (1957), a number of research group have examined the subcellular distribution of copper among organelles of *liver* cells, with fairly consistent results. Some of the data are presented in Table 6-1 and show that the largest portion of copper, in livers of normal adult rats or humans, is in the soluble (supernatant) fraction obtained by centrifuging for 60 min at $100,000 \times g$. This would represent copper binding to components of the cell cytosol, possibly excluding very large molecules that might have sedimented out. The next most abundant compartment is probably the nucleus, although, in these kinds of studies, unbroken cells will contaminate the nuclear fraction and add to its total copper content. The mitochondrial and microsomal fractions have lower, but significant, amounts of copper as well. Livers of rats and humans differ primarily by having different percentages of total tissue copper in the cytosolic fractions, about 65% in the rat and about 80% in the human. Pig liver may have a lower percentage in the cytosol (about 40%) and a larger percentage (35%) in the nuclear fraction (Owen *et al.*, 1977).

The distribution of copper within liver cells is altered to some extent in the neonate and in the case of a deficiency or excess of copper within the body, induced by dietary restriction or excess administration. It is well recognized that the liver becomes a repository for copper during the latter part of gestation, presumably in most or all mammalian species (Linder and Munro, 1973). (See also Chapter 8.) The rationale for this may be to provide the element for the early growth of the newborn which is not receiving much copper from its milk diet. [The same rationale holds for iron and zinc (Linder, 1895b).] As shown in the fifth column of Table 6-1, most of the extra copper in the newborn rat liver appears in the nuclear fraction. Indeed, it has been shown by immunohistochemistry that much of this metal really is within the nuclei, where it associates especially with

Table 6-1. Subcellular Distribution of Copper (and Incoming Radiocopper) in Liver Cells under Various Physiological Conditions

	Distribution of copper (%)[a]							Distribution of radioactive copper (%)		
Fraction	Rat liver (Thiers and Vallee, 1957)	Rat liver (Owen et al., 1977)	Adult human liver (Owen et al., 1977)[b]	Fetal human liver (Nartey et al., 1987b)	Neonatal rat liver (Owen et al., 1977)	Copper-loaded rat liver (Owen et al., 1977)	Copper-deficient rat (Owen et al., 1977)	Isolated rat hepatocytes (0.5 min) (Schmitt et al., 1983)	Rat liver (60 min) (Campbell et al., 1981)	Rat heart (60 min) (Campbell et al., 1981)
Connective tissue	2	—	—	—	—	—	—	—	—	—
Nuclei and whole cell residue	20 (0.95)	12 (0.59)	16 (0.97)	33	53 (8.0)	38 (8.7)	23 (0.6)	12	42	38
Heavy mitochondria	⎱ 8 (0.39)	9 (0.45)	6 (0.34)	18	9 (1.3)	12 (2.8)	24 (0.7)	⎱ 26	⎱ 10	⎱ 6
Light mitochondria	⎰ —	6 (0.27)	7 (0.41)	5	5 (0.7)	8 (1.8)	10 (0.3)	⎰	⎰	⎰
Heavy microsomes	⎱ 5 (0.24)	4 (0.19)	3.5 (0.21)	4	2 (0.4)	4 (0.6)	7 (0.2)	17 ⎱	⎱ 49	⎱ 55
Light microsomes	⎰ —	3 (0.13)	3.5 (0.21)	—	1 (0.2)	3 (0.4)	6 (0.2)	6 ⎬	⎰	⎰
Supernatant	64 (3.0)	67 (3.2)	64 (4.9)	22	27 (4.0)	43 (9.9)	30 (0.8)	40 ⎰	—	—

[a] Values in parentheses are actual concentrations in fraction ($\mu g/g$ wet wt. of starting tissue).
[b] Percentages markedly increased in nuclear fraction, decreasing in the supernatant, in biliary (but not micronodular) cirrhosis, with liver Cu accumulation.

metallothionein (Banerjee et al., 1982). (The whole matter of metallothionein and its function in copper metabolism is addressed later in this chapter.) The other particulate fractions of the liver also appear to have more copper, while the cytosol has about as much as in the normal adult rat.

In the liver of copper-loaded adult rats, the proportion of copper in the nuclei is also higher, but the absolute amounts in all fractions are increased to some extent (Table 6-1). This is also illustrated by the early data of Milne and Weswig (1968) (Fig. 6-1). In contrast are the data of Feldmann et al. (1972), for rats injected daily for 7 days with 500 μg of $CuSO_4 \cdot 5H_2O$, which suggest that the largest proportion of the excess copper is deposited in the mitochondrial rather than the nuclear or cytoplasmic fractions. [Here, marker enzymes were used to identify the organelles, with about 75% of cytochrome oxidase activity being in the mitochondrial fraction (and 10% in the nuclear fraction). About 55% of acid phosphatase was in the lysosomal fraction (with 30% in the mitochondrial fraction), and about 70% of glucose-6-phosphatase was in the microsomes (8% in the supernatant).] Although the Feldmann studies were distinguished by their use of enzyme markers, it seems more likely that the accumulation of extra copper is really in the nuclei and cytosol, as indicated by other studies and in view of evidence that metallothionein can accumulate in the nuclei (Banerjee et al., 1982). [It is generally agreed that metallothionein is the site for deposition of most excess intracellular copper (see Section 6.3.1).]

The data from the studies of Milne and Weswig and of Feldmann do, however, both illustrate another point, namely, that neither diets high in copper nor injections of large amounts of the element result in much copper retention by liver, at least in the rat (Fig. 6-1B). In the case of the diets, intakes of 100 to 1000 μg per day gave identical liver values for total copper content, while injection of more than 1000 μg (over 7 days) only increased liver Cu concentrations five- to six-fold (with an accumulation of about 400 μg of Cu) (Feldmann et al., 1972). As the liver is the main site for deposition of excess copper, this illustrates that excess copper is easily lost from the body. (See also Chapter 5.) Copper-*deficient* rats, on the other hand, retain about the same amounts of copper as normal rats in all liver cell fractions, *except* the cytosol/supernatant, where the amount is much lower (Table 6-1). In other words, excess copper preferentially accumulates in nuclear (as well as cytosolic) fractions, but copper is preferentially lost from the cytosol during deficiency, at least in the liver. In conditions of biliary cirrhosis in humans, in which there also is copper accumulation, there again appears to be deposition of the excess copper primarily in nuclear and large particulate fractions (Owen et al., 1977).

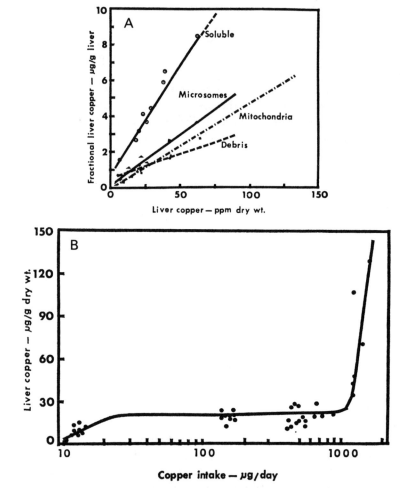

Figure 6-1. Distribution of copper among subcellular fractions of liver, with variations in dietary copper intake and total liver copper concentrations in rats. (A) Intracellular distribution at different liver copper concentrations. (B) Total liver copper in relation to dietary intake. Reprinted, with permission, from Milne and Weswig (1968).

This may not hold for copper accumulated in micronodular liver cirrhosis, in which the percentage in the supernatant may remain very high.

The pattern of distribution and the pathway that incoming copper takes to enter liver cell organelles has not been studied systematically with regard to changes over time. The relatively scanty data available nevertheless suggest that the largest fraction of newly absorbed copper first enters the cytosol, as measured with ^{64}Cu(II) or ^{67}Cu given to cultured

Table 6-2. Effects of Feeding Low- and High-Copper Diets on Concentrations of Copper in the Cytosol Fractions of Different Rat Tissues[a]

	Supernatant Cu concentration (μg/g)		
Tissue	Normal diet[b]	Copper-deficient diet	Excess copper diet
Kidney	4.5 ± 1.2 (60)	1.6 ± 0.4[c]	13.2 ± 7.1[c]
Liver	2.8 ± 0.7 (68)	1.2 ± 0.3[c]	8.0 ± 2.7[c]
Stomach	1.5 ± 0.6 (71)	0.9 ± 0.2[d]	2.3 ± 0.8
Intestine	1.2 ± 0.3 (74)	1.1 ± 0.3	2.7 ± 0.7[c]
Heart	1.8 ± 0.7 (34)	1.0 ± 0.4[d]	1.4 ± 0.4
Lung	1.5 ± 0.3 (84)	1.0 ± 0.3	1.9 ± 0.6
Testis	1.5 ± 0.6 (81)	1.1 ± 0.2	1.4 ± 0.4

[a] Data of Terao and Owen (1976) for groups of 3–12 rats (mean ± SD).
[b] Values in parentheses are amount of supernatant copper as percent of tissue copper.
[c] Probably a significant change from the normal ($p < 0.01$).
[d] Possibly a significant change from the normal (no exact N values provided with the mean ± SD).

hepatocytes as Cu(II)-histidine (10 mM) (Ettinger *et al.*, 1986) or intravenously as ^{67}CuCl$_2$ (Campbell *et al.*, 1981). The studies of Schmitt *et al.* (1983) indicate that a substantial portion also becomes rapidly associated with the liver cell mitochondria, although the data of Campbell *et al.* (1981) (taken after 60 min rather than immediately) show a major association with the nuclear fraction. More recent data of Ettinger and

Table 6-3. Distribution of Copper between Organelles and Cytosol of Different Rat Tissues[a]

		Distribution of incoming ^{67}Cu to supernatant (%)[c]			
		At 60 min		At 24 h	
	Supernatant Cu concentration[b]	Cu(II)	Cp	Cu(II)	Cp
Heart	1.6 ± 0.5	31 ± 8	80 ± 7	42 ± 3	71 ± 5
Lung	1.7 ± 0.4	55, 67	78 ± 3	79 ± 14	53 ± 2
Testis	1.5 ± 0.5	23 ± 3	64 ± 5	62 ± 7	50 ± 3
Stomach	1.5 ± 0.6	31 ± 3	48 ± 4	31, 25	57 ± 5
Intestine	1.2 ± 0.3	63 ± 7	79 ± 5	33 ± 3	51 ± 9
Kidney	4.5 ± 1.2	72 ± 4	70 ± 3	89	75 ± 6
Liver	2.8 ± 0.7	44 ± 8	55 ± 4	89	51 ± 13

[a] Data of Terao and Owen (1976).
[b] Data for tissues of groups of 3–6 rats on normal copper diets; mean ± SD ($N = 3$–12).
[c] In most cases, proportions of incoming copper (^{67}Cu) in supernatants did not change with copper-deficient or excess copper diets, so means of means were calculated for low-, normal-, and high-Cu diets (mean ± SD; $N = 3$). Copper was administered as CuCl$_2$[Cu(II)] or ceruloplasmin (Cp), as indicated.

co-workers (M. Ettinger, personal communication) on mouse and rat hepatocytes in culture confirm that the cytosol comes to equilibrium rather rapidly with regard to incoming copper but that the nuclear and mitochondrial fractions continue to accumulate the trace element over longer time periods. The potential meaning of these observations is difficult to interpret at this time, but the data indicate that there are probably fast- and slow-turnover compartments within the cell, as would be expected. Moreover, the mitchondrial compartment (and its cytochrome oxidase) may be turning over rather slowly.

The percentages of total cellular copper found in $100,000 \times g$ supernatants have also been studied in a variety of other rat tissues (Tables 6-2 and 6-3). While liver and kidney have about two-thirds of their copper in the supernatant, the percentages are even higher in other tissues, except the heart. Thus, the amounts of copper in the cytosol generally reflect the total copper content of the tissue. In the case of liver and kidney (which have the highest tissue concentrations), but not so much in other organs, levels of cytosol copper change with copper status induced by changes in dietary intake. This is the basis for the concept that the liver and kidney are the sites for storage of most of the excess copper that may accumulate (at least temporarily) within the body. [Part of the liver "accumulation" may also reflect the role of the liver in providing copper for its excretion through the bile (see Chapter 5.)]

Very little has been done to examine the further subcellular distribution of copper in other organs and species. However, the available data on rat heart, fractionated by a procedure modified for this organ (see Campbell *et al.*, 1981), suggest that the entering copper distributes itself to the various heart cell organelles in a very similar fashion to that in hepatocytes (Table 6-1). Some data are also available for kidney (versus liver) cells of weanling rats, given a normal or copper-deficient diet for 6–18 weeks (Alfaro and Heaton, 1974). Here, the nuclear fractions of kidney cells had less copper (and a smaller proportion of the cellular copper) than those of liver cells, while the proportion (and amounts) of copper in the mitochondrial and microsomal fractions was much higher. (Supernatants had similar proportions of total copper.) In deficiency, the largest decreases in the kidney copper (as in liver) were again seen in the soluble fraction, and the next largest in the nuclear fraction, with little change in mitochondrial and microsomal fractions. In the brain, cerebral cortex studies of Matsuba and Takahashi (1970) in rats indicated that half the copper was in the mitochondria, and about 30% in the supernatant fraction of the brain homogenates, leaving much smaller amounts for other cell organelles. The high mitochondrial copper content emphasizes the quantitative importance of energy metabolism to the brain. [The brain accounts for about 20% of the caloric expenditure of basal matabolism (BMR).]

Distribution of copper among subcellular compartments has also been examined in the trout and quahog. In trout liver, with a total copper concentration close to that of mammals (7.9 µg/g), more of the copper was in the mitochondria and lysosomes than in the cytosol (Julshamn et al., 1988). With a copper concentration in the feed of 100 µg/g, a larger proportion was in the cytosol, but with even higher copper feeding, more went into the lysosomes. The kidney of the bivalve *Mercenaria mercenaria*, taken from "pristine" waters, but with a very high total Cu concentration (60 µg of Cu/mg of protein), seemed to deposit the copper mainly within characteristic intra- and extracellular granules, with high concentrations also in the nuclear fraction and the cytosol (Sullivan et al., 1988).

With regard to incoming copper measured as ^{67}Cu, the source may make a difference in terms of the percentage entering the cell supernatant/cytosol (Table 6-3). If copper was given intravenously as ^{67}Cu-ceruloplasmin, rather than as ^{67}CuCl$_2$ (attaching to albumin and transcuprein in the plasma), a much larger percentage initially appeared to be in the cytosol (at 60 min) in the case of heart, stomach, lung, and testes, whereas a much smaller percentage was in the cytosol of these tissues when it was given as ionic copper(II). The source of the copper made little difference in liver, kidney and intestine. Thus, the source of the copper appears to be a determinant of the intracellular path copper will take in some cells, of which more will be said later in this chapter. The discrepancy between the apparent paths for copper from ceruloplasmin versus transcuprein and albumin within certain cells, especially the heart, may also be related to the preferential uptake of ceruloplasmin copper versus other forms of plasma copper observed for this and some other organs and the relative abundance of ceruloplasmin receptors in these tissues versus in liver or kidney (see Chapter 3).

6.2 Copper Components Associated with Different Fractions of the Cell

6.2.1 The Nuclear Fraction

As already implied, evidence points to the small cysteine-rich, metal-binding protein metallothionein as one of the "players" in the accumulation of copper (and perhaps also zinc) in the nucleus of liver cells. In this regard, Banerjee et al. (1982) showed that cadmium injection of rats, which induced metallothionein accumulation in liver and kidney, resulted in positive immunohistochemical staining for metallothionein in the nuclei, while cells from untreated animals had metallothionein mainly in the cytosol (in small amounts). As we know that metallothionein has a higher affinity for copper than cadmium and also that metallothionein is the main repository for excess copper in the liver of the newborn, where metallothionein also accumulates in the nuclei (Panemangalore et al., 1983), it seems likely that, under conditions of metallothionein accumulation, copper will be held by the nuclei, if it is present in amounts above those needed for more vital cell functions.

It also seems likely, however, that copper is in the nuclei for other purposes as well, although what they might be is difficult to say at this time. Certainly, even in Cu deficiency, nuclear copper is retained much more than copper in the cytosol. One possible nucleus-related function in which copper has been implicated is in the stabilization of chromosomes. Using HeLa cells, Lewis and Laemmli (1982) found that Cu(II) [and not Mn(II), Zn(II), Co(II), or Hg(II)] specifically stabilized isolated metaphase chromosomes that had been depleted of histones, by generating a scaffolding structure involving two large proteins of about 135 and 170 kDa. Nondividing chromosomes were also stabilized by copper (in nucleated cells), via such nonhistone proteins (Laemmli et al., 1981). There are also reports of copper-binding proteins associated with washed (calf thymus) chromatin (Hardy and Bryan, 1975; Bryan et al., 1981) and with specific chromatin fragments (Bryan et al., 1985) that may be related to the findings of the Laemmli group but may also be different. Chromatin isolated from frozen calf thymus, repeatedly extracted with nucleosome isolation buffer, was reported to contain 25 ng of tightly bound Cu (and about 100 ng of tightly bound Zn) per mg of DNA. [This translates to about 75–100 ng of Cu per g of tissue and would approximate perhaps 10% of the Cu found in normal nuclei (Table 6-1).] In these and subsequent studies, chromatins from frozen and fresh sources were treated with micrococcal nuclease and/or DNase I (which preferentially digest transcriptionally active regions). This appeared to release most of the copper that had been associated with a particular fraction of nucleoprotein, separated in gel chromatography. Copper may thus be involved in transcription and/or in specific transcription-stimulatory events effected by certain hormones. The angiotensin II system is being examined in this regard (Re, 1987).

6.2.2 The Mitochondrial Fraction

The mitochondrial fraction contains the well-known copper enzyme cytochrome c oxidase, which is the terminal oxidase of electron transport directly involved in the reduction of elemental oxygen in the mitochondrial cristae. (Further details on this enzyme are given later in the chapter.) In liver cells, which have been studied the most, the mitochondria normally contain 10–30 ng of Cu/mg of protein, as calculated from the rat data of Feldmann et al. (1972) and Milne and Matrone (1970). The values reported by Williams et al. (1976) for pig liver are higher (Table 6-4), which may reflect generally higher levels of copper in the tissues of this species (Bremner, 1976) (Table 1-6). Levels of copper in heart mitochondria appear to be considerably lower, averaging about 1.4 ng per mg of protein (Page et al., 1974), while those in brain may be much higher (see below and Table 6-15). The tissue differences suggest that cytochrome c oxidase may not be the only copper-associated factor in mitochondria. This concept is strengthened by the fact that, although mitochondrial copper concentrations are very much lower in heart than liver, cytochrome c oxidase activities (per mg of mitochondrial protein) are the same or higher (see later; Table 6-11). Additional support comes from the data in Table 6-4 for the copper and cytochrome c oxidase content of liver mitochondria of normal and copper-deficient pigs. Here, the fall in oxidase activity (in deficiency) is not nearly as great as the apparent drop in mitochondrial copper. In contrast, the drop in heme synthesis rate is close to that for cytochrome c oxidase activity (Table 6-4), consistent with heme synthesis being an energy-dependent process. Therefore, particularly in liver (which has a relatively high mitchondrial copper concentration), it would appear that other copper-binding factors are also present in the mitochondria, and that these factors lose copper more readily than cytochrome c oxidase when less copper is available.

Table 6-4. Copper Content, Cytochrome c Oxidase Activity, and Heme Synthesis in Normal and Copper-Deficient Mitochondria and Reticulocytes from Pigs[a]

	Normal[b]	Copper-deficient[b]	% Decrease
Hematocrit (%)	42 ± 1	25 ± 3	40
Serum Cu (µg/g)	1.8 ± 0.1	0.2 ± 0.0	89
Liver mitochondria			
Cu content (ng/mg protein)	61 ± 30	4 ± 1	94
Cytochrome c oxidase (10^{-3} IU/mg protein)	244 ± 15	72 ± 14	70
Heme synthesis (nmol/mg protein per h)	2.6 ± 0.2	1.3 ± 0.1	50
Iron content (µg/mg) protein	1.2 ± 0.3	5 ± 1	—
Reticulocytes			
Iron uptake (nmol/10^9 cells per 2 h)	6.7 ± 0.6	3.5 ± 0.4	48
Heme synthesis (nmol/10^9 cells per 2 h)	4.2 ± 0.4	1.4 ± 0.2	67

[a] From Williams et al. (1976).
[b] Mean ± SE.

In the heart, a close correspondence has been found between the total copper content in that tissue and variations in mitochondrial mass induced by postnatal growth, ventricular hypertrophy (due to aortic constriction), and treatment (of hypothyroid rats) with thyroxine (Page et al., 1974). This is despite the fact that mitochondria contain only one portion of the total cellular copper of the heart. These authors attributed the changes in mitochondrial copper to changes in the amount of mitochondrial cristae, as verified morphologically. They also suggested that, among different species, the total concentration of copper in the heart (especially in the left ventricle) is positively related to the adult resting heart rate (which may, of course, be related to the density of mitochondrial cristae in the tissue). Rat and mouse hearts appeared to have the highest copper concentrations (from 0.43 down to 0.39 μmol/g dry weight), rabbits and guinea pigs were intermediate (at 0.33 μmol/g), and larger animals and the human had the lowest values (from 0.29 down to 0.23 μmol/g dry weight) (Page et al., 1974).

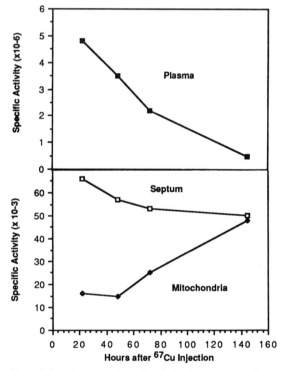

Figure 6-2. Specific activity of copper in cells of the heart ventricular septum and their mitochondria, at various times after intravenous administration of $^{67}CuCl_2$ to rats. Replotted from Page et al. (1974).

The respiratory apparatus and electron transport system seem a particularly important area of copper metabolism and function. At the same time, at least in animals with sufficient copper, the mitochondrial compartment does not appear to be the site for rapid copper uptake and turnover. Examining just the uptake of *ceruloplasmin* copper by heart, which appears to be its preferred source (Campbell *et al.*, 1981; Weiss *et al.*, 1985), most of the entering ^{67}Cu probably first appears in the cytoplasm (Table 6-3; 60 min) after intravenous administration. Moreover, as shown in Fig. 6-2, even with administration of ionic ^{64}Cu, the specific activity of heart mitochondrial copper is initially much lower than that of total heart tissue, reaching equilibration with it only after 140 h (in the rat) (Page *et al.*, 1974). Even earlier studies by Dallman (1967) indicated that restoration of cytochrome *c* oxidase activity in tissues of copper-depleted rats refed the element was slow, except in rapidly proliferating cells such as those of the intestinal mucosa. This implies that the copper component(s) of mitochondria may be part of the slow-turnover compartment of whole-body copper (see Chapter 5). Cederbaum and Wainlio (1972) showed that, in contrast to what happes *in vivo*, isolated (liver) mitochondria rapidly bound Cu(I) or Cu(II) with high affinity, to a maximum of about 5 µmol (or 320 µg) per mg of protein, but it seems unlikely that this has much significance physiologically.

6.2.3 The Lysosomal and Microsomal Fractions

The concept that lysosomes may contain metallothionein and metallothionein aggregates complexed with copper, zinc, and/or cadmium has prevailed for a long time. This concept arose in part from observations that neonatal liver accumulated copper in particulate organelles and histochemical observations on the copper staining of sections of tissue from patients or animals with copper overload due to Wilson's disease, biliary cirrhosis, or excess Cu intake/administration. Nevertheless, *neonatal* accumulation of excess liver copper is not lysosomal (Table 6-1) but nuclear and associated with accumulation of metallothionein in that organelle, as demonstrated by the extensive immunohistochemical work of Cherian's group (Panemangalore *et al.*, 1983; Nartey *et al.*, 1987b). (See also Chapter 8.) (Some accumulation is also in the cytosol.) The same appears to be true for metallothionein (and perhaps copper) also in thyroid tumors (Nartey *et al.*, 1987a). In contrast are disease conditions of the adult, in which large copper accumulations are found in liver and kidney. As shown most recently by X-ray probe analysis, using unfixed frozen as well as glyceraldehyde-fixed sections of human liver, there was a good coincidence between the distribution pattern for areas concentrated in copper and sulfur with the patterns for lysosomes (Hanaichi et al., 1984). Phosphorus showed no such pattern. A particular and consistent ratio of Cu to S was found for the lysosomal particles, identified electron microscopically by their high density and peribiliary canalicular location in the hepatocytes. As metallothionein is well known for its extraordinary high cysteine content, the relationship strongly implies that in abnormal circumstances of copper accumulation, at least in the liver (and kidney), excess copper accumulates in lysosomes attached to metallothionein. This has also been well documented histochemically in rats given a large excess of copper in the diet (Haywood et al., 1985), as well as in humans with alcoholic liver cirrhosis (Lesna, 1987). In normal circumstances, however, there appears to be little copper or metallothionein in the lysosomes, and, at least in the adult, the metallothionein-bound copper is largely in the soluble/cytosol cell fraction.

The microsomal fraction is comprised principally of fragments of plasma and endoplasmic reticular membranes, as well as polyribosomes and other large soluble cell components (for example, cytosol ferritin). Data on the copper content of isolated plasma and reticular membranes (Table 6-5) are extremely variable but suggest that small amounts of the metal are associated with them. It is presently unknown what the copper-binding components involved might be. Certainly, it is possible that some of the copper is attached to copper transport proteins, such as ceruloplasmin, bound to cell surface receptors in the process of delivery or while protecting the outer membrane surface. Also, transport systems for shuttling the copper across the cell surface may be binding it. (For a discussion of these transporters and transport mechanisms, see Chapter 3.) All of this remains to be further explored.

Table 6-5. Reported Copper Contents of Various Cell Membranes[a]

Membrane	Copper content (ng/mg protein)	Reference
Pig erythrocyte plasma membrane	29	Mackellar and Crane (1982)
Rat liver plasma membrane	134	Vassiletz et al. (1976)
Rabbit liver plasma membrane	<0.6	Ichikawa and Yamano (1970)
Rabbit liver microsomes	24	Mason et al. (1965)
Rat liver endoplasmic reticulum	128	Vassiletz et al. (1976)
Rabbit liver endoplasmic reticulum	6	Ichikawa and Yamano (1970)
Rabbit liver Golgi membranes	13	Ichikawa and Yamano (1970)

[a] Modified from Mackellar and Crane (1982).

6.2.4 The Cytosol Fraction

As already mentioned (Table 6-2), the largest fraction of cellular copper in most cells (except perhaps the heart and brain) is in the cytosol, as represented by the $100,000 \times g$ supernatant obtained from cell homogenates. Consequently, in tissues of the normal adult mammal, the level of cytosol copper reflects the level of total copper in the tissue, and changes in dietary intake are thus also reflected in changing levels of copper in components of the cytosol. This is particularly the case for liver and kidney, the two organs in which copper levels tend to be the highest and in which excess copper accumulates (Table 6-2). Kidney and liver supernatants have considerably higher concentrations of the element than those of heart, intestine, and other organs. While most organs appear to reduce their levels of cytosol copper in deficiency (with the exception of the small intestine, with its already low copper content), the reductions would appear to be much greater in the liver and kidney than in other tissues, both absolutely and on a percentage basis. Similarly, the accumulations in these organs are greater when extra copper enters from the diet, and here the intestine and stomach also show a substantial elevation (which is not surprising, as they are in direct contact with the high-copper diet).

The liquid portions of many kinds of cells have been fractionated by many investigators, who have separated and in part identified several copper-binding components. Figure 6-3 shows the distribution of chemically/physically determined (stable) copper in fractions of cytosol separated by gel permeation chromatography on Sephadex G-100 or G-75. Three components are generally evident in tissue supernatants, as illustrated here for rat liver and kidney, and as shown previously for small intestine (Chapter 2, Fig. 2-5). The first component elutes in the void volume, indicating an apparent molecular weight exceeding 80,000. The second component elutes between albumin and cytochrome c and has an apparent molecular weight in the range of 35,000. This is close to the size of superoxide dismutase (SOD), and enzyme found in most (or all) cells and known to contain copper and zinc. (Further details on this enzyme are provided later in the chapter.) This component would appear to be much more abundant in the liver than the kidney (Fig. 6-3A versus B). The third component (more prevalent in kidney than liver) elutes in the position of Cd-thionein, implying that this fraction of copper is attached to cytosolic metallothionein. Traces of copper also appear to be present in some of the fractions eluting with or after the small peptide and amino acid fraction (represented by absorbance at 280 and 254 nm at or beyond the column volume). It has generally been accepted, although not fully documented experimentally, that the second and third copper peaks are indeed largely superoxide

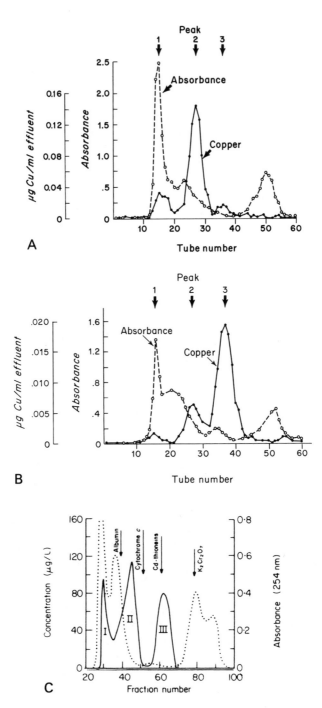

dismutase and metallothionein. Whether other (coincident) copper components are also present is neverthesess an open question, and there are preliminary reports suggesting that the story is much more complex (Ettinger, 1987; Waldrop et al., 1988, 1990; Palida et al., 1990) (see later). Indeed, the skewed shape of the second peak in Fig. 6-3C suggests that there is a smaller amount of copper attached to something a bit larger than SOD, as also reported by Alfaro and Heaton (1974) (M_r about 65,000, although no standard curve was shown). This is not evident in the data of Terao and Owen (Fig. 6-3A and B).

The nature of the largest component is much more obscure, although ceruloplasmin may account for a portion of it. Terao and Owen (1973, 1974), as well as Linder and Moor (1977), have demonstrated that tissue supernatants (obtained from homogenates centrifuged at 100,000 x g for 60 min) contain p-phenylenediamine (pPD) oxidase activity. This activity is characteristic of ceruloplasmin but is also exhibited by the respiratory electron transport system of mitochondria (which should sediment out completely at 100,000 x g). The gel fractionation data of Terao and Owen show that pPD oxidase activity elutes in the void volume of Sephadex G-100, as it should if due to ceruloplasmin (M_r 132,000). The question is whether this activity is actually due to ceruloplasmin and, if so, whether the ceruloplasmin is a contaminant or truly intracellular and in the cytosol. Arguing for the intracellular nature of the activity is the fact that it was present in liver supernatants even after extensive preperfusion of the tissue with saline to remove the blood (Terao and Owen, 1973) and also the finding that the amount of activity present in liver and a variety of tissues could not be accounted for by residual blood contamination (Linder and Moor, 1977). Arguing in favor of it being ceruloplasmin is its elution as a larger protein (Fig. 6-3A, B) and its copurification with plasma ceruloplasmin in ion-exchange chromatography (Linder and Moor, 1977). This remains to be confirmed and further investigated. While there is evidence of ceruloplasmin uptake by tissues (Campbell et al., 1981; Kressner et al., 1984), one wonders how it would get into the cytosol (as opposed to endosomes or lysosomes). At the same time, there is evidence that ceruloplasmin is not normally made by cells in the body other than

---

Figure 6-3. Fractionation of cytosolic copper-binding proteins by gel permeation chromatography. (A) and (B) Fractionation on Sephadex G-100 of rat liver and kidney supernatants, respectively, showing elution of copper (——) and absorbance at 280 nm (— —). Reprinted, with permission, from Terao and Owen (1973 and 1974, respectively). (C) Rat liver supernatant on Sephadex G-75, showing elution of copper (——), absorbance at 254 nm (···), and the elution positions of molecular weight standards. Reprinted, with permission, from Sharma and McQueen (1981).

those of the liver (and a few others that have a separate circulation) (Aldred et al., 1987). One possibility is that extracellular ceruloplasmin has been released from cell membrane receptors during homogenization and other treatments of the tissue.

Even if ceruloplasmin is present, it is unlikely that this is the only copper-binding component in the void volume fractions of the cytosol (Fig. 6-3). Indeed, more recent data of Norton and Heaton (1980), for supernatants of hepatocytes of liver, skeletal muscle, and small intestine homogenates fractionated on a larger pore gel, suggested that there are two components of M_r 150,000 or more, and also two more between M_r 40,000 and 75,000 (Fig. 6-4). (In the latter studies, no standard curve was shown, although a series of standard proteins was used to established apparent molecular weights, but only two fractionations were performed.) This suggests that a number of additional components with copper may be present. Indeed, we know that there are a number of other copper-containing enzymes in some tissues (see below).

A small proportion of soluble Cu may also be bound to carbohydrates of low molecular weight, as suggested by the work of Templeton and Sarkar (1985, 1986) for kidney. Working with Ni(II), these authors have partially characterized several sulfated oligosaccharide complexes, where Cu(II) can readily be substituted. It remains to be seen whether similar complexes are found in other tissues.

Another approach to the identification of soluble intracellular copper-binding components has been to use radioactive copper. Figure 6-5 shows three components, eluting in the same positions as the three consistent nonradioactive peaks already described, that were rapidly labeled with incoming radioactive copper and separated on Sephadex G-100: one in the void volume, one eluting close to superoxide dismutase, and the third in the position of metallothionein. Again, all three peaks appeared to be present in a variety of tissue supernatants (Fig. 6-5; Terao and Owen, 1974), although exact standardization of each column was not carried out, and there appeared to be some variations in the elution volumes of peaks II and III (Fig. 6-5). Assuming for the moment that the three components are always identical, it is evident that the labeling of the soluble components changes with time and also that the relative prominence of the labeling in one versus another component seems to be tissue specific. *In the case of liver*, the third peak (metallothionein?) is the most prominent initially (at 30 min) (Fig. 6-5A), and the second peak (near SOD) increases with time. (The label in the void volume also decreases with time.) By 24 h after intravenous administration of ^{67}Cu(II), the elution profile of the radioactivity had merged with that for nonradioactive copper (Fig. 6-3A), implying full intracellular equilibration. *In the case of the large intestine*

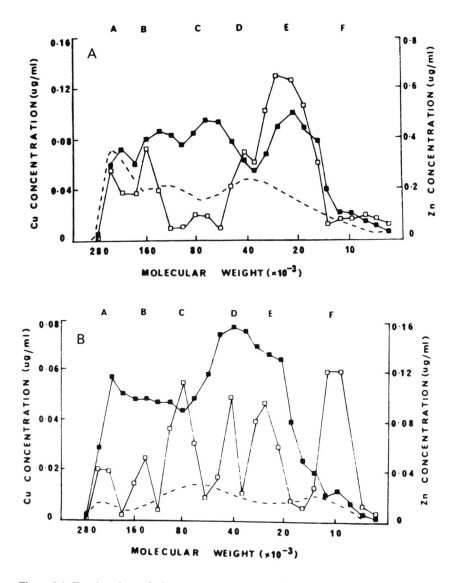

Figure 6-4. Fractionation of tissue supernatants on Sephadex G-150: copper-binding components. Data are for $100,000 \times g$ supernatants and show Cu (□) or Zn (■) concentrations of fractions eluting at different apparent molecular weights. It should be noted that Sephadex G-150 excludes proteins above about 150 kDa. (A) Adult male rat liver. (B) Adult male rat skeletal muscle. Reprinted, with permission, from Norton and Heaton (1980).

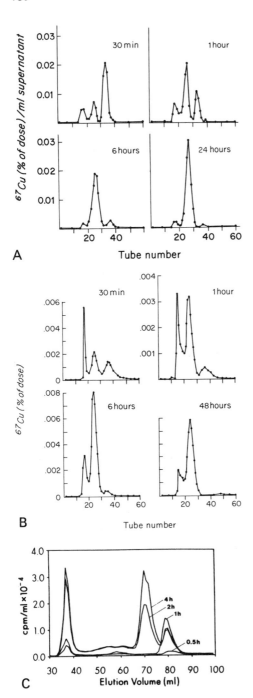

Figure 6-5. Changes in the distribution of ^{67}Cu among intracellular cytosolic components at various times after intravenous administration as ^{67}Cu(II)Cl$_2$ to rats. Samples of $100,000 \times g$ supernatants were fractionated on Sephadex G-100. (A) Results for liver. Reprinted, with permission, from Terao and Owen (1973). (B) Results for large intestine. Reprinted, with permission, from Terao and Owen (1974). (C) Results for HAC hepatoma cells, showing tracer ^{67}Cu associated with void volume components, MT, and GSH, respectively (peaks from left to right) at various times after the start of incubation with the metal (in hours, as indicated). Separation of cytosolic components was on Superose 12B FPLC. Reprinted, with permission, from Freedman et al. (1989).

(Fig. 6-5B), the void volume peak and peak II were initially most prominent, again changing solely to peak II (near SOD) with time. [Here the third peak (metallothionein?) also decreased with time, and its elution position seemed to vary.] These two patterns of response appeared to be prototypic for almost all of the tissues examined by Terao and Owen (1973, 1987). In all cases but that of the testis, labeling of the reputed SOD peak was initially low and increased with time. The labeling of the third (metallothionein?) peak decreased with time, as was generally the case for the void volume component(s). In the case of heart, skeletal muscle, lung, spleen, and small intestine, labeling of the void volume peak was initially much more prominent than that of the third peak, while the reverse was the case for liver, kidney, and brain. (In stomach, large intestine, and testis, labeling of both peaks was initially about the same.) This suggests that the copper newly entering the cell sap from the blood initially binds to void volume components (or a void volume component), as well as to metallothionein, and that SOD (or something in that fraction) gradually becomes the major recipient of that copper over time, while the copper associated with peaks I and III tends to turn over more rapidly. The degree of initial metallothionein labeling probably depends upon the concentration of this protein already present, which is largely regulated by other metal ions and hormones, depending on the tissue (see more below). [Cu easily displaces Zn (or Cd) from this protein.] Thus, liver and kidney, known to be sites of abundant metallothionein production, have what has been assumed to be a prominent ^{67}Cu-metallothionein peak, 30 min after *in vivo* administration of ^{67}Cu (Fig. 6-5A). [It is noteworthy that the labeling of intestinal cytosol components with ^{67}Cu(II) entering from the *blood* side (Fig. 6-5B) differs from the picture obtained when ^{67}Cu(II) is given from the side of the intestinal lumen (see Chapter 2; Fig. 2-7), where only the metallothionein peak appears to be labeled.] The data of Marceau and Aspin (1973), where liver supernatants were compared in the rat 30 min after giving ^{64}Cu(II)-serum and 56 h after giving ^{67}Cu-ceruloplasmin, essentially confirm the time pattern of labeling changes seen by Terao and Owen (1973), except that there was little or no labeling of a void volume component.

The very recent studies of Freedman *et al.* (1989) add an important step to this picture, namely, that glutathione (GSH)-bound copper may mediate binding of incoming copper to metallothionein and may, in fact, be the first major cytosolic substituent to which copper binds. Comparing a copper-resistant line of hepatoma cells (HAC600) with the wild type (HAC), they discovered that a major difference in the two was a high level of GSH peroxidase in the resistant line. This led them to investigate levels of GSH *per se* and the effects of GSH depletion on copper resistance and storage (which also fell). Further, they showed that most of the tracer ^{67}Cu entering cells (from serum-free RPMI medium) initially bound to GSH (Fig. 6-5C), and, with time, it went to the metallothionein fraction. They also found evidence of a ternary GSH–Cu–metallothionein intermediate. These findings may prove crucial to our understanding of the steps involved in entry and distribution of intracellular copper.

The current work of Ettinger (1987; Palida et al., 1990) in rat liver and kidney, in which supernatants from ^{64}Cu(II)-treated cells are fractionated on Sephadex G-150 or in Superose FPLC, suggests that the apparent SOD component seen by previous workers is not SOD. In this connection, Reed et al., (1970) isolated a pink Cu(II) protein from bovine erythrocytes [Initially detected by Shields et al. (1961)]. It had an apparent molecular weight of 32,000 but was distinct from SOD in terms of its spectral, electrophoretic, chromatographic, and catalytic properties. (It also had no amine oxidase activity.) One wonders whether a similar protein might be present in other cells and account for some of the labeling (of a component in the SOD molecular weight range) seen in most tissues. Ettinger's group is also finding that two larger proteins are involved in the initial binding of copper absorbed by cells, one of which may possibly be a heat shock protein (M. Ettinger, personal communication). The component of M_r 38,000 in Sephadex G-150 (peak II in the earlier studies; Fig. 6-5) has an M_r of 46,000 in Superose but is about 55,000 Da in SDS electrophoresis. (It is still unclear whether the latter is actually the same component and associated with the copper.) Another less prominent component has an M_r of 88,000 in G-150, 65,000+ in Superose, and apparently 48,000 in SDS electrophoresis. Still another (80,000 Da) ^{64}Cu-binding peak in G-150 is also about 65,000 Da in Superose and 80,000 Da in SDS electrophoresis. Elucidation of the nature and function of these components and any role(s) they might play in copper transport will be a major challenge.

6.3 Specific Copper-Binding Components in Most Vertebrate Cells

There appear to be at least three proteins that contain/bind copper in almost all vertebrate cells. These are metallothionein(s), superoxide dismutase, and cytochrome c oxidase. What is known about the structure, function, and metabolism of these is summarized below. More extensive reviews are available elsewhere, as indicated.

6.3.1 Metallothioneins

Metallothioneins were first isolated by Margoshes and Vallee (1957) from equine kidney. These small, cysteine-rich metal-binding proteins have since been found in a wide variety of human and animal organs, as well as in most eucaryotic species, including birds, amphibia, fish, lizards, yeasts, and also plants. Reviews (or volumes) on the structure, genetics, function, and regulation of these proteins include those of Kägi and Nordberg (1979), Foulkes (1982), Kägi et al. (1984), Karin (1985), Hamer (1986), Kägi and Kokima (1987), and Bremner (1990).

Metallothioneins (MTs) in all mammalian species are comprised of a single polypeptide chain of about 61 amino acids, 20 of which are cysteine and none of which are aromatic amino acids or histidine, for a total molecular weight just above 6000 (Table 6-6). The lysine (plus arginine) and serine contents are also high, and these residues tend to be intercalated between the cysteines (Kägi et al., 1984) in the amino acid sequence (Fig. 6-6). Most of the cysteines are found either as Cys-X-Cys or adjacent to each other. Among species, the cysteine sequences tend to be the most conserved (Hamer, 1986). Methionine is at the N-terminus and is usually

Figure 6-6. Amino acid composition of vertebrate metallothioneins and related invertebrate proteins. One-letter codes: a, Ala; C, Cys (upper-case C for emphasis); d, Asp; e, Glu; f, Phe; g, Gly; h, His; i, Ile; k, Lys; l, Leu; m, Met; n, Asn; s, Ser; t, Thr; v, Val. Reprinted, with permission, from Nemer et al. (1985).

Table 6-6. Amino Acid Composition of Metallothioneins and Metallothionein-Like Proteins[a]

Amino acid	Metallothionein I			Metallothionein II			Sea urchin[b]	Nonvertebrate proteins		
	Human	Equine	Mouse	Mouse	Equine	Human		Drosophila[c]	Neurospora	Saccharomyces[d]
Asp	1	2	3	3	2	3	3	1	1	2
Asn	2	1	1	1	1	1	3	1	2	6
Thr	3	3	5	1	1	2	5	1	0	3
Ser	9	9	9	10	9	8	2	6	7	8
Glu	2	1	0	0	1	1	4	1	0	6
Gln	1	1	1	3	2	1	1	1	0	5
Pro	2	3	2	2	3	2	1	2	0	3
Gly	5	6	5	4	4	5	9	7	6	5
Ala	5	5	5	6	7	7	5	2	1	0
Val	0	1	2	1	3	1	3	0	0	0
Cys	20	20	20	20	20	20	20	10	7	12
Met	2	1	1	1	1	1	1	1	0	0
Ile	1	0	0	1	7	1	1	0	0	0
Lys	8	6	7	8	1	8	4	5	1	1
Arg	0	2	0	0	0	0	0	0	0	0
Total	61	61	61	61	62	61	62	38	25	51

[a] From Kägi and Nordberg (1979), modified mainly from Nemer et al. (1985).
[b] Nemer et al. (1985).
[c] Lastowski-Perry et al. (1985) for Mtn; another form of MT (MTo) with 43 amino acids and 12 cysteines has also been cloned from Drosophila (Mokdad et al., 1987).
[d] Karin et al. (1984). Note that this composition reflects the gene sequence, but that the processed protein is missing 8 amino acids from the N-terminus (Winge et al., 1984; Butt et al., 1984).

acetylated (Winge et al., 1984). The protein is quite asymmetric, which accounts for its elution in gel permeation chromatography with an apparent molecular weight of 10,000–11,000. The asymmetric structure consists of two domains, with differential affinities for various numbers of bivalent metal ions (Zn, Cu, Co, Cd, Hg, Ag, Pb, Ni, and even Au). The cysteine residues are directly involved in the metal binding. Winge and colleagues (Winge and Miklossy, 1982; Boulanger et al., 1982) have demonstrated that the C-terminal domain (A or α) of mammalian MT binds 4 Zn and/or Cd ions with the help of 22 cysteine sulfhydryl groups, while the β domain (at the N-terminus) only binds 3, with 9 cysteine residues. All seven Zn or Cd ions achieve an approximately tetrahedral geometry of binding (Vasak and Kägi, 1981), as part of oligonuclear clusters in the α or β domains. Neutralization of the excess negative charge [about three SHs for every Zn(II) or Cd(II)] appears to be a function of the intercalated basic amino acid residues, as inferred from ^{1}H-NMR data (Vasak et al., 1985) and from protection of lysine side chains from arylation by trinitrobenzenesulfonic acid in the presence of metal ions (Pande et al., 1985) and, also, from effects of lysine prearylation on the Cd spectrum of the Cd-thiolate clusters. Zinc and cadmium ions bind preferentially to the α domain. In contrast, the same metallothioneins appear to bind 11 or 12 Cu atoms, preferentially, adding 6 to the β-domain and 5 or 6 to the α during reconstitution of the apoprotein (Nielson and Winge, 1984). The isolated yeast protein with 53 amino acids (8 missing from the N-terminus in the β domain) has also been studied. It binds 8 Cu(II) ions through coordination with 12 cysteine residues (Winge et al., 1984; George et al., 1988). This implies that the two additional Cu(I) ions in the mammalian Cu-thionein would depend upon the cysteines in the N-terminal 8-amino acid sequence. Data from EXAFS studies of several Cu thioneins are compatible with trigonal coordination to the cysteine thiolate sulfurs (George et al., 1986; Abrahams et al., 1986; George et al., 1998) (Fig. 6-7B). [Ag(I) is thought to bind like Cu(I), while Hg(II) may bind like Cd or Zn.] The likely configuration for the yeast Cu-thiolate cluster is shown in Fig. 6-7C.

The X-ray structure of the Cd/Zn form of crystallized rat liver MT II has been characterized at 2.3-Å resolution (Fig. 6-7A) by X-ray diffraction (Furey et al., 1986, 1987). It confirms the two-domain globular nature of the protein, each domain with a diameter in the range of 20 Å and showing a similar (but sequentially opposite) folding of the polypeptide chain around the metal clusters. The β domain comprises the first 29 amino acids and is separated from the α domain at the junction between a cysteine and two lysine residues (30 and 31) (Fig. 6-6). There is little physical and chemical interaction between the two domains, and each can in fact bind

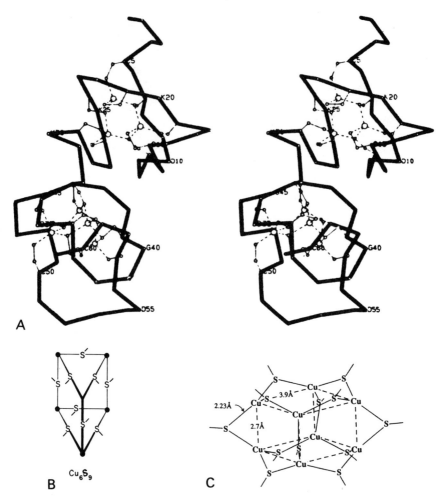

Figure 6-7. (A) Crystal structure of rat Cd/Zn metallothionein, determined by X-ray crystallography. Reprinted, with permission, from Furey et al. (1986). (B) Potential structure of the Cu-thiolate cluster in rat liver metallothionein, based on EXAFS. Reprinted, with permission, from George et al. (1986). (C) Potential structure of the Cu-thiolate cluster in yeast metallothionein. Reprinted, with permission, from George et al. (1988).

metal ions entirely independently, as demonstrated *in vitro* with isolated β and α fragments. The configuration of Cu-thionein should be different *but retain some similarities* (see also Fig. 6-7B). Overall, it would appear that Cu binds preferentially to the β domain, and Zn to the α domain and that *in vivo* this metal specificity may be important, perhaps even for MT function (D. R. Winge, personal communication). Naturally occurring hybrid molecules with Cu and Zn so distributed have been isolated. Also, Cu binding may elicit conformational changes in the Zn-bound domain.

The genes for human, mouse, and yeast metallothioneins have been most studied. Most mammalian species appear to have at least two distinct functional genetic cistrons, expressing two forms of the protein that differ by a few amino acids (Table 6-6). The designation of these two categories of metallothioneins is based purely on their elution in ion-exchange (DEAE) chromatography (Hamer, 1986). The mouse probably has only one functional gene each for metallothioneins I and II, and both are on chromosome 8 (Cox and Palmiter, 1982), only six kilobases apart (Searle et al., 1984). In the human and the horse, one functional gene for MT II and several functional MT I genes (as well as several pseudogenes) have been cloned (Karin and Richards, 1982; Richards et al., 1984). All or most are on chromosome 16 (Schmidt et al., 1984). In other words, MT I appears to have more variants than MT II. It is also noteworthy that the three exons composing the mammalian MT genes cover sequentially most of the highly invariant N-terminal region of the β domain (through residue 10) (exon 1), the rest of the β domain (exon 2), and the entire α domain (exon 3) (Winge et al., 1984). Exons 2 and 3 are spliced at the junction of codons for amino acid residues (Lys) 30 and 31.

In contrast to the mammalian metallothioneins are the MT-like proteins of very different species, many of the genes for which have also been cloned (Hamer, 1986). The sea urchin (Nemer et al., 1985) and crab (Lerch et al., 1982) proteins contain 62 and 57–58 amino acids, respectively, and have 20 and 18 cysteines, but much less homology (Fig. 6-6; Table 6-6). Their capacity for Cd binding is probably also less: Crab MT binds six rather than seven Cd(II) ions. (There is no information available for the sea urchin proteins.) The *Neurospora crassa* protein is much more divergent, having only 25 amino acids, but maintaining exactly the spacing of the seven cysteine residues in the 25-residue sequence of the β domain of mammalian MTs (residues 3–27; Fig. 6-6). The metallothionein-like protein of yeast (*Saccharomyces cerevisiae*) is 53 amino acids (12 cysteines) and has even less sequence conservation except for the hexapeptide segment seen in residues 30–35 (Fig. 6-6) of the mammalian MT consensus (Winge et al., 1985). It is capable of binding four Zn or Cd ions and eight Cu or Ag ions.

Other metallothionein-like, small metal-binding proteins, containing some aromatic amino acids and especially a lower percentage of cysteine residues, were reported in the early days of work on these proteins (Table 6-6) (Premakumar et al., 1975a,b; Riordan and Gower, 1975a,b; Winge et al., 1975). Most of the findings of low cysteine content can now be ascribed to the lability of Cu metallothionein to $-SH$ oxidation during its isolation (Madapallimatam and Riordan, 1977; Geller and Winge, 1982a). It is nevertheless not entirely to be excluded that some small proteins of lower cysteine content are involved in Cu binding, especially within nonmammalian organisms and in the intestinal mucosal cells of mammals (Mason *et al.*, 1981). (See Chapter 2, Section 2.5.) The 5' untranslated region of the MT genes (in humans and mice) contains elements responsive to metal ions and often also has elements regulated by interferon, glucocorticoids, and other hormones (Fig. 6-8). In contrast, the yeast gene is regulated by copper (and not by hormones) (Hamer, 1986) and may be feedback regulated by MT itself (Gorman *et al.*, 1986) by virtue of binding the inducing copper ions. These findings imply that expression of multiple MT genes is differentially regulated. In vertebrates, one MT may be regulated mainly by metal ions, others by hormones that coordinate functions within the body.

Regulation by metal ions is part what would appear to be a major function of the protein, namely, its capacity to sequester metals that might otherwise bind elsewhere and interfere with normal cell processes. The list of metal ions known to bind to metallothioneins is long and includes the following (in the order to their decreasing affinities): $Hg > Cu > Cd > Ag > Zn$ (Nielson and Winge, 1985). Co(II) and Ni(II) and even Au can also be substituted (Vasak *et al.*, 1981). [Stability constants have been estimated at 10^{17}–10^{19}, 10^{15}–10^{17}, and 10^{10}–10^{14} for Cu, Cd, and Zn, respectively (Hamer, 1986).] Clearly, excesses of free ionic Cu or Hg could produce a host of detrimental effects in cells, the former leading to oxidative damage (as, for example, to cell membranes; see Chapter 9), the latter to occupation of vital sulfhydryl groups on innumerable enzymes, for example. Binding and detoxification of such metal ions can therefore be viewed as a helpful and health-promoting function of these proteins. In general, it has been found that Cd is the most successful MT-inducing agent in whole organisms (Table 6-7). Zn is next best, requiring high concentrations, while copper is a relatively weak inducer, except in cultured cells (Table 6-7) (Mayo and Palmiter, 1981; Sadhu and Gedamu, 1989). Induction involves the binding of regulatory proteins to metal (or hormone)-responsive elements in the promoter region of the MT genes, resulting in increased transcription.

The role of MT in detoxification is supported by work from Hamer's group with yeast, in which it was shown conclusively that deletion of the MT gene rendered the cells hypersensitive to Cu poisoning (Hamer *et al.*, 1985). [It also rendered the Cu MT regulatory region (CUPI) more sensitive to activation, since there was less MT to bind any added Cu (Thiele *et al.*, 1986).] The importance of MT for metal detoxification (and storage) is also supported by data showing that, at least in rats, feeding large

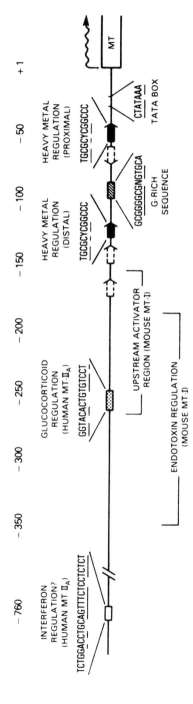

Figure 6-8. Regulatory sequences of mammalian metallothionein genes, in the 5' untranslated region. Solid arrows indicate the positions of elements controlled by metals [Cd(II), Zn(II), Cu(II)]. A possible consensus sequence is given, with invariant nucleotides underlined. Hollow arrows indicate partial repeats of this sequence in the mouse MT-I gene. The putative human interferon (□) and glucocorticoid (▥) (controlled) elements are also indicated. Reproduced from Hamer (1986), with permission.

Table 6-7. Regulation of Metallothionein Gene Expression by Metals and Hormones

Factor	Target organs cells	Selected references
Copper	Liver, kidney?	Ryden and Deutsch (1978)
		Yagle and Palmiter (1985)
	Cultured cells	Mayo and Palmiter (1981)
	Yeast cells	Butt et al. (1984)
Cadmium	Liver, kidney, intestine, heart, muscle, brain, spleen	Piscator (1964)
		Durnam and Palmiter (1981)
		Richards et al. (1984)
Zinc	Liver, kidney, small intestine, cultured cells	Mayo and Palmiter (1981)
		Yagle and Palmiter (1985)
		Searle et al. (1984)
		Schmidt and Hamer (1983)
Glucocorticosteroids (mammals)	Cultured hepatocytes, fibroblasts, and erythroid and cancer cells	Failla and Cousins (1978)
		Mayo and Palmiter (1981)
	Primarily liver; also possibly kidney, skeletal muscle, spleen (rats and mice)	Hager and Palmiter (1981)
Glucagon (fasting), Angiotension II, α- and β-adrenergic agents	Rat liver	Cousins (1985)
Inflammatory response (turpentine, endotoxin, CCl_4)	Liver, kidney	Oh et al. (1978)
		Sobocinski et al. (1977)
	Liver (adrenalectomized rats)	Sobocinski et al. (1981)
Interferon	Neuroblastoma cells	Friedman et al. (1984)

amounts of ionic Zn, Cu, Cd, Hg, or Ag resulted in accumulation of the metals in the MT fractions of liver and kidney (Whanger and Ridlington, 1982). This was not the case with Pb, however.

Temporary binding and long-term *storage* of metals are also part of the repertoire of metallothionein function, a role that is sometimes hard to distinguish from that of preventing metal toxicity. Here, two examples are perhaps of special interest. Liver metallothionein appears to be the site for temporary storage of much plasma zinc during the acute phase of infectious disease, when levels of circulating Zn and Fe are decreased. [It is thought that this decrease occurs to keep the metals from infectious organisms that need them for profileration (Weinberg, 1974; Webster et al., 1981).] In acute inflammatory conditions, such as induced experimentally by turpentine or endotoxin (Table 6-7), there is induction of MTs in liver, and, to some extend, also in kidney, which cannot be ascribed to glucocorticosteroids. (It occurs in adrenalectomized animals.) Interferon release, also associated with infections and inflammations, can induce increased synthesis of MT, an involvement initially deduced from the presence of interferon regulatory sequences upstream of the human MT II_A gene (Fig. 6-8) (Friedman *et al.*, 1984). A second example is the long-term sequestration of Cd by MT, particularly in the kidney. As there is no well-developed route for Cd excretion from the body, it tends to accumulate in the kidney with age, where it has no known essential functions but can cause damage when present in excess amounts. Presumably less damage will occur if it is sequestered by metallothionein. Increased synthesis of metallothioneins can be induced by cadmium itself, involving increased transcription (and processing) of MT mRNAs via as yet poorly understood interactions of hormone–receptor complexes, second messengers, and/or metal complexes with the regulatory regions of the genes. In the case of the metal-regulatory elements (MRE), a core concensus sequence (TGCRCNG) appears to be involved in the binding of a metal-regulated protein recently identified in nuclear extracts from the livers of rats (Searle, 1990) and mice (Imbert *et al.*, 1989). [The mouse protein (MBF-I) is about 14 kDa.] The various MT genes have different numbers of MREs and HREs in their promotor regions and are differentially regulated by metals and glucocorticosteroids (Jahroudi *et al.*, 1990).

While copper is not nearly as good an inducer of MT as cadmium or zinc and (in the case of liver) the glucocorticosteroids, Cu(I) can easily displace Zn(II) and Cd(II) in metallothioneins (Holt *et al.*, 1980; Day *et al.*, 1981), thus allowing its immediate sequestration in existing MT when entering in potentially harmful amounts. From the data in Table 6-7, it is evident that hormonal control occurs mainly in the liver and that liver, intestine, and kidney tend to respond both to Cd and Zn, but to Cu only when present in very high concentrations. For example, Bremner and Davies (1974a) found that injections of a 300-μg dose of copper into rats caused a cycloheximide-inhibitable accumulation of copper in the liver MT fraction (separated on Sephadex G-75), while doses of 100 μg had relatively little effect. [Both doses are very high and can be toxic in some strains of rats (Linder *et al.*, 1979b).]

Another potential function of metallothionein may be in the intracellular as well as extracellular transport of metals. Udom and Brady (1980) and Li *et al.* (1980), as well as Winge and Miklossy (1982), have shown that Zn-thioneins readily function *in vitro* as a source of Zn for apocarbonic anhydrase. In the case of Li *et al.* (1980), the apoenzyme of red blood cells was able to rapidly extract Zn(II) from equine kidney or rat liver metallothionein and at rates several order of magnitude greater than those achieved by EDTA. (Reconstitution with Zn-thionein was as rapid as with ionic Zn.) Winge and Miklossy (1982) showed that MT I was the preferred donor of the metal to carbonic anhydrase and that the fully Zn-loaded thionein was a better donor than a Cd/Zn mixed thionein. Geller and Winge (1982a) also demonstrated *in vitro* transfer of copper from rat liver Cu-thionein to apo-superoxide dismutase (SOD) from rat liver and erythrocytes. However, metal transfer (either to EDTA or apoenzyme) was much more successful in the case of *partially oxidized* Cu-thionein. As already noted, Cu-thionein is especially unstable and is easily oxidized during and after its isolation (Geller and Winge, 1982a), unless stored anaerobically or with $-SH$ reducing agents. (This accounts for some of the variations in MT reactively, apparent function, and even composition reported in the literature.) Oxidation of some of the cysteine thiols also can lead to polymerization of the protein, and this may occur when storage of copper in liver is high, as in copper overload. [Here, extraction of the protein requires addition of $-SH$ reducing agents (Winge *et al.*, 1984)]. Whether Cu-thionein is a direct source of Cu for SOD, *in vivo*, is therefore unclear, although the time course of changes in radioactive copper distribution (first to MT, later to SOD?) (Fig. 6-5) would be consistent with this concept. Recent studies by Brouwer *et al.* (1989) on three lobster MTs indicate that those with structural homology to MT I were unable to transfer Cu to apohemocyanin (the O_2 carrier in lobster "blood"), although a third form did so, *in vitro*. Studies with yeast have demonstrated that lack of MT in no way impairs the normal growth and function of this organism under conditions of normal copper availability (Hamer *et al.*, 1985). This does not mean that MT cannot still function as a direct (or indirect) source of copper for copper-dependent enzymes if it is present, but it is probably not essential for this.

As regards extracellular transport, Garvey and Chang (1981) were the first to show (by radioimmunoassay) that traces of metallothionein are present in the blood plasma. Since then, several studies from this laboratory and that of Bremner have established that normal adult rats have about 1-2 ng of MT/ml of plasma but that the MT concentration in plasma is markedly higher in animals injected with Cu(II) or Cd(II) [and to a lesser extent, when injected with Zn(II) (Garvey and Chang, 1981; Mehra and Bremner, 1983).] This is also the case in neonatal rats, where liver Cu-thionein levels (especially in the nucleus) are very high (Mehra and Bremner, 1984). As shown in Fig. 6-9, levels of plasma MT I decreased rapidly with age, and this change paralleled (though not in perfect synchrony) changes in intracellular liver and kidney MT concentrations. (It should be noted that the values for plasma are in ng/ml as opposed to $\mu g/g$ for the tissues.) These data indicate that the plasma levels of MT I may reflect tissue levels of MT. As tissue levels of the protein are largely

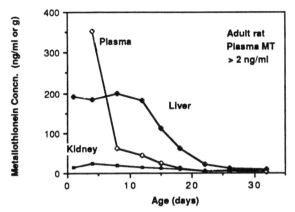

Figure 6-9. Changes in the concentrations of plasma and tissue metallothionein during the first month after birth, in the rat. Data of Mehra and Bremner (1984). Metallothionein was determined immunologically.

regulated by factors other than copper (although there can also be some copper regulation at the extremes of availability), plasma MT probably does not play much of a role in normal Cu homeostasis or transport. Nevertheless, Bremner has proposed that it may serve to shuttle some Cu from the liver to the kidneys of copper-poisoned animals (I. Bremner, personal communication, cited by Camakaris, 1987).

MT has also been found in the urine, especially in the case of animals or humans exposed to cadmium (which tends to accumulate in the kidney MT) (Tohyama *et al.*, 1981; Shaikh and Hirayama, 1979). Most metallothionein filtered by the kidney would appear to be resorbed (Kägi and Nordberg, 1979, p. 41), but some still escapes into the urine, especially when there is kidney damage (as in Cd overexposure). Traces have also been found in the bile (see Chapter 5). Thus, MT may play a small but limited role in interorgan and even excretory copper transport.

Turnover of MTs has been studied by several researchers, but the results do not necessarily allow for straightforward conclusions. Clearly, the apoprotein (which takes the form of a random coil; Vasak *et al.*, 1980) is relatively unstable *in vitro* and may be so also *in vivo*. Indeed, the apoprotein has not been isolated from tissues. Both *in vitro* (Mehra and Bremner, 1985) and *in vivo* (Bremner *et al.*, 1986), it was shown that the half-life of rat MT I is shorter than that of MT II. In the *in vivo* or cell work, however, one cannot always be certain that some of the disappearance of labeled protein is not due to its uptake into lysosomes or other particulate fractions. Potential oxidation and polymerization are also problematic for consistent recoveries of representative MT samples (for radioactive analysis). The stability of the protein also depends to some extent upon the metal ions that are bound. Specifically, Feldman et al. (1978) and Mehra and Bremner (1985) showed that Zn-thionein was readily hydrolyzed by liver lysosomal enzymes, but this did not appear to be the case for Cu-thionein (Bremner et al., 1986). In agreement with a lack of biological function for Cd, it has also been shown that Cd-thionein appears to turn over much less rapidly than Zn-thionein, in cultured liver and kidney cells (Kobayashi et al., 1982). Recent studies of Laurin and Klasing (1990) in chicken macrophages confirm these relations of Zn and Cd to MT synthesis and degradation; while data of Sadhu and Gedamu (1989) in HepG$_2$ cells indicate a lasting effect of copper but not zinc on induction of MT mRNAs.] Although Cu-thionein may be more resistant to lysosomal degradation than Zn-thionein, it nevertheless does not bind Cu so tenaciously that Cu does not easily find its way to copper-dependent enzymes, such as SOD, where it is needed (Fig. 6-5) or to factors that ensure its secretion from cells (Chapter 3) and excretion in the bile or by other routes (Chapter 5).

In summary, metallothioneins would appear to function as sites for temporary storage and detoxification of excess amounts of intracellular copper, as potential (though probably not crucial) sources of Cu for other (Cu-dependent) proteins within the cell, and as a means of shuttling traces of Cu between kidney and liver and out of the body. Apart from inhibiting intestinal Cu transport in conditions of extremely high zinc intake (where it is induced to high concentrations), it would seem that the role of metallothionein in overall copper metabolism is otherwise quite passive, adjusting to influx and efflux primarily by displacing zinc in favor of copper (which, in turn, may induce more of the protein).

6.3.2 Superoxide Dismutase

The copper-containing protein we now call copper/zinc superoxide dismutase (SOD) was first isolated from bovine red blood cells and liver by Mann and Keilin in 1938 and later also from brain (Porter and Folch, 1957). In 1970, Carrico and Deutsch reported that these various "cupreins" (as they were then called) also contained Zn, while McCord and Fridovich (1969) discovered that they had enzyme activity against superoxide anions, catalyzing their dismutation to peroxide and dioxygen: $2O_2^- + 2H^+ \rightarrow O_2 + H_2O_2$. Since then, it has become apparent that most cells in most eucaryotic organisms probably contain this Cu/Zn enzyme and that it may play an important role in shielding intracellular components from oxidative damage. Numerous reviews on the structure and function of this rather unique metalloprotein have appeared, including more recently those of Bannister *et al.* (1987), Valentine and Mota de Freitas (1985), Valentine and Pantoliano (1981), Fee (1981), and Vignais *et al.* (1980).

Within animal cells, Cu/Zn-containing SOD is mainly in the cytosol. Early biochemical studies with chicken liver suggested that a small portion was also in the intermembrane space of mitochondria (Peeters-Joris *et al.*, 1975; Weisiger and Fridovich, 1973). This was not confirmed in rat liver by immunocytochemistry (Slot *et al.*, 1986). (A less abundant manganese-containing, cyanide-insensitive SOD is in the mitochondrial matrix.) As shown biochemically (Weisiger, 1973) and now by immunospecific gold labeling of ultrathin rat liver cryosections, Cu/Zn SOD is also evenly distributed within the *nuclear* matrix (Slot *et al.*, 1986), and some (at least in the liver) is present in the lysosomes, as shown previously also by Geller and Winge (1982b). (It was found absent from endoplasmic reticulum channels, Golgi elements, secretory vesicles, and mitochondria.)

Copper (and zinc) deficiency reduces the levels of Cu/Zn SOD in animal tissues, especially in the liver, kidney, and red cells (Paynter *et al.*, 1979), and also in the fungus *Dactylium* (Shatzman and Kosman, 1978, 1979). In fungi and yeast at least, Cu can induce and regulate formation of the protein (Shatzman and Kosman, 1978, 1979; Gregory *et al.*, 1974). Thus, Cu availability appears to have some affect on synthesis and/or turnover of the protein, as tends to be the case with cofactors in biological systems. This apparent regulation does not, however, seem to hold at the other extreme, that is, under conditions of copper excess (Paynter *et al.*, 1979), at least when given in the diet. [Under these conditions, tissue copper contents also tend *not* to be markedly increased (see earlier).] However, injection of rats with Cu(II) as $CuSO_4$ (in very large doses) has been reported to cause a 35% increase in cytosolic SOD activity in hepatocytes (with a supposed greater percentage increase of mitochondrial intermembrane Cu/Zn SOD) (Asada *et al.*, 1973). In contrast to iron and manganese SODs, there is no evidence that Cu/Zn SOD is induced by oxidative stress (Bremner *et al.*, 1987). Potential mechanisms for copper/zinc regulation would include stabilization of the protein against denaturation and/or degradation (i.e., turnover should be increased upon imposition of a Cu or Zn deficiency) and regulation of SOD gene transcription (and/or SOD mRNA processing or translation) by the metal ions. The *in vivo* aspects of these phenomena have not yet been explored, although Roe *et al.* (1988) have demonstrated the importance of metal ion binding for thermal stability of SOD, *in vitro*.

Cu/Zn SODs have been isolated from a number of different eucaryotic organisms, and the amino acid sequences of the bovine (Steinman *et al.*, 1974), human (Jabusch *et al.*, 1980), and yeast (Johanson *et al.*, 1979) proteins have been known for some time. Most recently, the human and yeast proteins have been cloned. Levanon *et al.* (1985) have identified an 11-kb gene on human chromosome 21, with five exons and four introns. This gene appears to be processed into mRNAs of at least two sizes, 0.5 and 0.7 kb, with different lengths of the 3' untranslated region (UTR), and also variations in the 5' UTR (Sherman *et al.*, 1984). The full amino acid sequences of the human, bovine, horse, and yeast enzymes are given in Fig. 6-10. Four nonfunctional human pseudogenes have also been identified, and the yeast gene has also been cloned (Bermingham-McDonogh *et al.*, 1988). The monkey and pig genes would appear to be on chromosome 9 and the mouse gene on chromosome 16 (see Bannister *et al.*, 1987).

The protein is a homodimer, comprised of two subunits, each with about 150 amino acids, giving the whole a molecular weight near 32,000. There is considerable homology in the amino acid sequence for the Cu/Zn

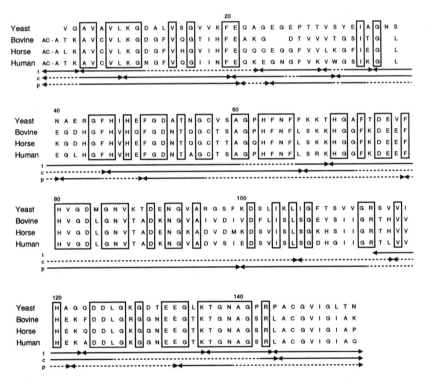

Figure 6-10. The amino acid sequence of human, bovine, horse, and yeast Cu/Zn superoxide dismutases. Redrawn, with permission, from Barra *et al.* (1980). One-letter amino acide code: A = ala; C = cys; D = asp; E = glu; F = phe; G = gly; H = his; I = Ile; K = lys; L = leu; M = met; N = asn; P = pro; Q = gln; R = arg; S = ser; T = thr; V = val.

SODs from 11 species that have so far been examined (Fig. 6-10), especially among the mammals (where percentages range from 78 to 84, for the human, rat, pig, cow, and horse, from closest to farthest from the human, respectively) (Bannister et al., 1985). The homology with the fish enzyme is 66% and that with SOD from flies, cabbage, yeast, and fungi 52–58%. (The rare Cu/Zn SOD of a few bacteria is only 25% homologous.) The amino acids directly or indirectly involved in metal binding appear to be conserved throughout. Because the tryptophan and tyrosine contents of the protein are low or nil (see Albergoni and Cassini, 1974), there is a relatively low absorbance in the 280-nm region (Table 6-8). The content of glycine, on the other hand, is very high, at 22–25%. Each of two subunits contains one intramolecular disulfide bond and one Cu(II) and one Zn(II) atom in the fully active native protein (Table 6-8). The presence of the Cu ions (Fee, 1981) greatly enhances the absorbance of the protein at the absorption maximum (for the bovine protein) of 258 nm. The metal-bound protein is also faintly blue, with an absorbance peak at 680 nm.

The structure of bovine erythrocyte Cu/Zn SOD in monoclinic cyrstals has been deduced from X-ray diffraction studies at 2-Å resolution (Tainer et al., 1982, 1983) and is pictured in Fig. 6-11. It shows an elongated symmetrical molecule, with one disulfide bridge per subunit and interactions between those portions of the subunits that contain the C- and N-termini. Apparently, these (hydrophobic) interactions are extremely tight and, in some cases, do not even yield to 4% SDS or 8 M urea (see Valentine and

Table 6-8. Properties of Bovine Cu/Zn Superoxide Dismutase

Property	Erythrocyte enzyme[a]
Molecular weight (from AA sequence)	31,200
Subunits	2 identical
Intrasubunit S—S bonds	1 per subunit
Metal ion content (native protein)	1 Cu(II), 1 Zn(II) per subunit
Color	
Native	Blue-green
Apo	Colorless
λ_{max} (nm) (UV–vis spectrum)	
Native	258, 250–270, 680 nm
Apo	258, 250–280 nm
EPR spectrum (native)	
9 GHz 77K	$g_{\parallel} = g_z = 2.265$
9 GHz 30K	$g_x = 2.03$
35 GHz 70K	$g_y = 2.11$, $A_{\parallel} = 130$ G
Redox potential (native), pH-dependent	$E^{0'} = 0.42$ or 0.28 V

[a] Rearranged, with permission, from Valentine and Panteliano (1981).

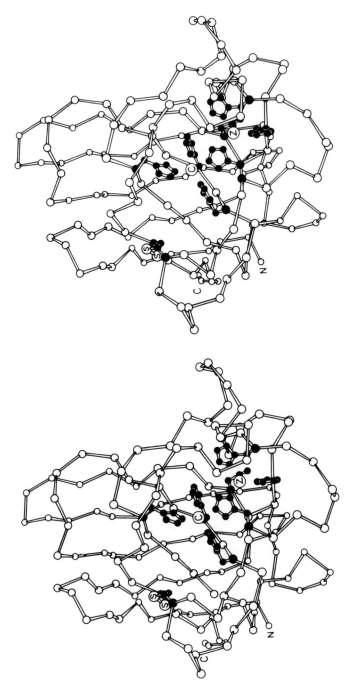

Figure 6-11. The crystal structure of bovine erythrocyte Cu/Zn superoxide dismutase, determined by X-ray crystallography. Only the polypeptide backbone of the two subunits of the molecule is indicated. Reprinted, with permission, from Valentine and Pantoliano (1981), based on results of Tainer *et al.* (1982).

Pantoliano, 1981). Although the Zn- and Cu-binding portions of the two subunits are about 34 Å apart, the metal occupancies of each subunit are sensed by the other, as demonstrated with coulometric titrations (Lawrence and Sawyer, 1979) and during metal reconstitution (Rigo et al., 1977a,b).

In each subunit, the Cu ion is bound to four imidazole side chains (of histidines 48, 50, 135, and 77 for the human; bovine sequence 44, 46, 118 and 61) and one water ligand (Fig. 6-12) in a distorted five-coordinate arrangement. The Zn ion is, in turn, bound to two additional histidines, an aspartate carboxylate oxygen, and the deprotonated histidine imidazole that bridges to Cu (human His 77, bovine His 61), all with approximate tetrahedral geometry. The latter (77 or 61) histidine imidazole ring is thought to be completely deprotonated and serves as a bridge between the two metal ions, which are 6.4 Å apart in this rather unusual arrangement. The ligand state of the Cu is also unusual in having a sterically blocked axial site facing a hydrophobic region of the protein, while the opposite axial site faces a narrow hydrophilic channel and is occupied by H_2O, anions, or other ligands, including the substrate O_2^- (see more below.) (Anions like CN^- and N_3^- also bind.) The Zn is completely buried in the protein and not directly involved in substrate binding.

Removal of the metal ions from SOD does not occur easily under most normal physiological conditions but can be effected at pHs below 4, in the presence of chelating agents (such as EDTA). Moreover, reconstitu-

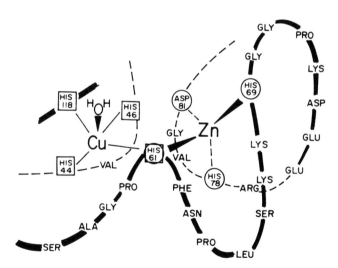

Figure 6-12. Schematic rendering of the active site of bovine red cell Cu/Zn superoxide dismutase, showing the metal–ligand interactions and histidine bridge between Cu and Zn. Reprinted, with permission, from Valentine and Pantoliano (1981).

tion from the apoprotein is slow and begins with binding of the metal ions to sites not involved in the native protein, followed by rearrangements to achieve the native state (Rigo *et al.*, 1978). This suggests that the metal ions are incorporated during synthesis and folding of the protein and also that Cu and Zn ions may not turn over separately from the protein itself or that some other factor would be necessary for rapid entry or exit of the metals, if that occurs within cells. In other words, it seems unlikely that the protein is constantly losing and rebinding its metal ions.

As already indicated, the Cu at the active site of the enzyme is at the bottom of a long, narrow, hydrophilic channel, made up of highly conserved amino acid residues that help direct the very negative O_2^- substrate to the highly positive catalytic binding site (Getzoff *et al.*, 1983; Tainer *et al.*, 1983). Glutamate 131 and lysine 134 appear to be involved in directing the substrate approach, while arginine 151 (bovine 141) is directly involved [with Cu(II)] at the catalytic site (otherwise occupied by a water ligand). [See also Valentine and Mota de Freitas (1986).] [This arginine is also involved in the interaction of SOD with phosphate (Mota de Freitas *et al.*, 1987).] Despite wide differences in the overall pI of the protein from different species (sheep 8.0, pig 6.2, ox 5.2), the electrostatic characteristics of the substrate channel are strictly maintained (Desiderio *et al.*, 1988).

For superoxide dismutation in general, one or more coordination sites on the Cu ion should be available for interaction with O_2^- (Fee, 1981), as is the case for Cu/Zn SOD. Studies by Fielden *et al.* (1974), Bray *et al.* (1974), and Klug-Roth *et al.* (1972, 1973) have indicated that the first step in the SOD reaction involves reduction of Cu(II) by O_2^-, with formation of O_2. This is followed by interaction of the reduced Cu(I) protein with another superoxide anion (and two protons) to form H_2O_2 and the reoxidized enzyme. Since the copper of SOD is so deep within the protein structure, and access to it is only through a narrow channel that would exclude most other substrates, cells are able to perform selective dismutations (of O_2^-) without catalyzing potentially deleterioius reactions with larger substrates (such as might be catalyzed by simple Cu-amino acid complexes) (Fee, 1981). Dismutation is possible over a rather wide range of pH. With regard to mechanism, the Cu(II)-His(141)-Zn(II) bridge (Fig. 6-12) is thought to break and protonate reversibly after reduction by one electron from superoxide (Fee, 1981), in the first part of the dismutation reaction.

$$Cu(II)-N-N-Zn \rightarrow Cu(I) \; HN \quad N-Zn(II)$$

This important catalytic property of SOD (superoxide dismutation) would seem to be at the center of its functioning in living systems. As such, it is part of a network of proteins, involving Cu, Fe, and Se and S amino acids, designed to minimize oxidative and peroxidative damage to intra and extracellular substituents, including cell membranes (Fig. 6-13). It is placed to protect cells from the inside, while ceruloplasmin (a superoxide anion scavenger, but not a dismutase) may protect the outside cell surface and extracellular fluids (see Chapter 4). Without Cu/Zn SOD, there is little or no catalytic activity available to convert O_2^- to H_2O_2 (which can then be removed by glutathione peroxidase and/or catalase). Production of superoxide anions is a natural consequence of many internal cellular reactions, involving or involved in phagocytosis (Bannister and Bannister, 1985), mitochondria (Loschen et al., 1974), or oxidative enzymes such as xanthine oxidase, aldehyde oxidase, and dihydroorotic acid oxidase (Fridovich, 1975). [(See also Bannister et al. (1987).] Autoxidation of hemoglobin is also a potential source and would explain the importance of SOD to erythrocytes. Without dismutation, the alternative is to produce hydroxyl (or related radicals) that can initiate chain reactions leading to destruction of membrane lipids and other intracellular structures. It has been estimated that steady-state erythrocyte O_2^- concentrations might normally be 10^{-12} M (Bannister et al., 1987) and that SOD concentrations are normally sufficient to prevent most interactions of O_2^- with other molecules. [Moreover, O_2^- itself is not that reactive in aqueous solution (Sawyer and Valentine, 1981).] The extremely broad distribution of SOD among different cell types (Table 6-9) and its very high activity in liver (a tissue extremely active in metabolism and thus likely to produce more superoxide) and red blood cells (especially vulnerable and exposed to oxygen) support the concept that superoxide dismutation is indeed the primary function of Cu/Zn SOD. Moreover, an increased fragility of erythrocytes accompanies copper deficiency (and subnormal levels of SOD) in animals, as has been known for some time (Bush et al., 2965a). Finally, Gregory et al. (1974) showed that oxygen exposure induced a large increase in SOD and increased oxygen resistance in yeast, while Bermingham-McDonogh et al. (1988) showed recently that replacement of the Cu/Zn SOD gene in oxygen-sensitive mutants that had lost it restored oxygen tolerance. These findings clinch the importance of SOD in defending against the consequences of oxygen exposure intracellularly. Small amounts of a copper-containing, extracellular SOD in plasma (and extracellular fluids) have also been reported (Table 6-9), an SOD that appears to bear little structural resemblance to the Cu/Zn enzyme (see Chapter 4). Cu/Zn SOD, injected intravenously, has a very short half-life (Huber and Saifer, 1977) and would therefore not normally be expected to contribute to extracellular SOD activity.

Another potential function of Cu/Zn SOD to be considered is the

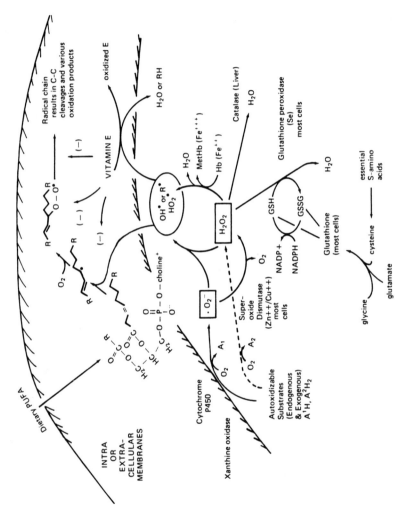

Figure 6-13. Interaction of SOD with other enzymes and other factors in the protection of oxidizable intracellular soluble and membrane substituents against oxidation and free radical damage. Reprinted, with permission from Linder (1985c).

Table 6-9. Distribution of Cu/Zn SOD (and Extracellular Cu SOD) in Different Organs of the Rat and the Human

Organ	Cu/Zn SOD activity (relative units/g)			Extracellular SOD (relative units/g)
	Rat			
	I[a]	II[b]	Human[c]	Human[c]
Liver	100	100	100	0.08
Adrenal	35	—	—	—
Kidney	50	69	33	0.37
Spleen	9.8	21	15	0.10
Heart	9.4	20	16	0.26
Pancreas	5.8	—	—	—
Brain	6.7	17	12	0.76
Lung	4.4	11	8.9	0.64
Intestine	2.7	11	10	0.46
Thyroid	—	—	15	1.4
Skeletal muscle	—	5.6	15	0.09
Blood	17			0.03[d]

[a] Relative to liver Cu/Zn SOD activity (4480 u/g); Peeters-Joris et al. (1975).
[b] Relative to liver Cu/Zn SOD activity (74,000 u/g); Marklund (1984).
[c] Relative to liver Cu/Zn SOD (intracellular) activity (85,000 u/g); Marklund (1984a). (See Chapter 4 and Table 4-8 for more information on extracellular Cu SOD.)
[d] Karlsson and Marklund (1987) ($=26\pm 4$ U/ml; $N=51$).

storage of copper and zinc. Reasons for considering it include the fact that apo-SOD is an excellent ligand for a wide variety of metal ions (Valentine and Mota de Freitas, 1985). Perhaps more importantly, it has not been clear on a theoretical basis (and on an actual basis until recently) that superoxide is very toxic to the cell and that its dismutation (and thus removal) could not be accomplished sufficiently rapidly either by spontaneous disproportionation or by other copper complexes (most of which have SOD activity). Even now, the exact reactions involved in rendering O_2^- toxic to cells are mainly surmised, although (as already indicated) the importance of Cu/Zn SOD for the prevention of oxygen toxicity has been established with yeast mutants (Bermingham-McDonogh et al., 1988). Reasons for *not* considering SOD a storage or metal detoxification protein have already been touched upon: (1) the metal ions are deeply buried, not making for easy entry or exit from the protein; (2) there are only two Cu and two Zn atoms per molecule (32,000 Da), which would be rather inefficient for metal storage (in contrast to metallothionein, for example, which holds seven Zn or ten Cu atoms per 6100 Da of protein); (3) upon influx of Cu(II) into cells, SOD tends not to be a major initial recipient of the metal, only gradually incorporating it over time* (Fig. 6-5); and (4) the protein is an efficient enzyme, with a special arrangement of the metal ions and substrate channels seemingly designed for dismutation of O_2^-, which does provide cells protection against oxygen toxicity. Finally, although SOD activities decrease in copper deficiency, the decreases are usually not as severe as those encountered by cytochrome c oxidase (Prohaska, 1983). This strongly suggests that SOD activity is just as vital to cells as the activity of the terminal enzyme of electron transport.

6.3.3 Cytochrome c Oxidase

The terminal enzyme of the electron transport chain of mitochondria, cytochrome c oxidase (EC 1.9.2.1), performs one of the most ubiquitously essential reactions in living cells, namely, the direct reduction of O_2 with four electrons (from cytochrome c) to form (with four protons) two molecules of H_2O. In conjunction with the other components of the electron transport chain, this permits the concomitant formation of ATP and uses Krebs cycle intermediates as the original sources of the electrons. The electrons for oxygen reduction are carried by four metal clusters, involving heme a, heme a_3, and Cu_A and Cu_B. Inhibition of activity by anions (such as CN^- or N_3^-) that bind to the reduced heme a_3 or heme a units on the enzyme vastly reduces the cell's capacity for ATP formation and is commonly used as a means of identifying energy-dependent processes (which will be strongly limited).

Clearly, the electron transport chain, including especially this enzyme, is fundamental to life processes as they have evolved on Earth, and as we know them. Thus, it is no surprise that a core portion of this protein appears to have been highly conserved and is prevalent in most (or all) species, from mammals, birds, fish, and other vertebrates to yeast, fungi, plants, and bacteria. The number of subunits of the protein appears to be inversely related to the complexity of the organism, with 1–3 in procaryotic bacteria; 5–7 in higher plants (Nakagawa *et al.*, 1987); 8–10 in fungi, protozoa, and yeasts; and 12 or 13 in vertebrates. Recent reviews of the structure and function of the protein include those of Capaldi (1990), Malmström (1990), Malatesta *et al.* (1990), Chan and Li (1990), Boer and Gray (1988), Kadenbach *et al.* (1987), Denis (1986), Wikstrom and Saraste (1984), Freedman and Chan (1984), and Capaldi *et al.* (1984); for topology, see also Jarausch and Kadenbach (1985) and Millett *et al.* (1983). The

---

* As already indicated, there is some very rapid labeling of a cytosol component eluting in a position similar to that of SOD, but evidence suggests that this is another protein.

book on *Oxidases and Related Redox Systems* (edited by King *et al.*, 1988) also contains several reviews of different aspects, and there is a New York Academy of Sciences volume, *Cytochrome Oxidase* (Brunori and Chance, 1989) as well as the article of Friere (1990) proposing a mechanism of assembly for the yeast enzyme.

Bolstered by recent information on the simpler forms of cytochrome *c* oxidase in bacteria and plants, the notion is arising that the basic "core" unit of the enzyme is composed of one or two subunits, corresponding to subunit I with or without II (Table 6-10), the first and third largest of the mammalian 12- (or 13-) subunit enzyme (Denis, 1986). Subunits I, II and III are the only ones coded for by the mitochondrial DNA and synthesized in eucaryotic mitochondria. Generally, the bacterial enzyme appears to be comprised of two or three subunits, corresponding to mitochondrial subunits I and II or I, II, and III, respectively, with sequence homology to the vertebrate subunits (Steffens *et al.*, 1983; Fink *et al.*, 1987; Steinbrucke *et al.*, 1987). Mitochondria and bacteria have a variety of other similarities, including their form of DNA, tRNAs, and ribosomal subunits, which distinguish them from the corresponding nuclear and cytoplasmic elements of eucaryotic cells. The general impression has been that the bacterial enzymes (and even the complex multisubunit forms of eucaryotic

Table 6-10. Properties of and Nomenclature for Subunits of Bovine Heart Cytochrome *c* Oxidase[a]

Nomenclature[b]				M_r of bovine heart subunits	Site of synthesis	N-terminal amino acid sequence (bovine heart enzyme)
Kadenbach	Buse	Capaldi	King			
I	I	I	I	56,993	Mitochondria	fMet-Phe-Ile-Asn-Val-Val
II	II	II	II	26,049	Mitochondria	fMet-Ala-Tyr-Pro-Met-Gln
III	III	III	—	29,918	Mitochondria	(Met)Thr-His-Glu-Thr-His
IV	IV	IV	III	17,153	Cytosol	Ala-His-Gly-Ser-Val-Val
Va	V	V	V	12,436	Cytosol	Ser-His-Gly-Ser-His-Glu
Vb	VIa	a	IV	10,670	Cytosol	Ala-Ser-Gly-Gly-Gly-Val
VIa	Vb	b	—	9,419	Cytosol	Ala-Ser-Ala-Ala-Lys-Gly
VIb	VII	c	VI	10,068	Cytosol	Ac-Ala-Glu-Asp-Ile-Gln-Ala
VIc	VIc	VI	—	8,480	Cytosol	Ser-Thr-Ala-Leu-Ala-Lys
VIIa	VIIIc	—	VII	6,244	Cytosol	Phe-Glu-Asn-Arg-Val-Ala
VIIb	—	—	VII	5,900[c]	Cytosol	Ser-Gly-Tyr-Ser-Val-Val[d]
VIIc	VIIIa	VII-Ser	VII	5,541	Cytosol	Ser-His-Tyr-Glu-Glu-Gly
VIII	VIIIb	VII-Ile	VII	4,962	Cytosol	Ile-Thr-Ala-Lys-Pro-Ala

[a] Modified from Kadenbach *et al.* (1987) and Wikström and Saraste (1984).
[b] Kadenbach *et al.* (1987); Buse *et al.* (1982); Capaldi *et al.* (1983); King *et al.* (1978).
[c] Apparent molecular weight average.
[d] Pig heart.

cytochrome c oxidase) have subunits I and II as their basic functional unit, containing all four of the metal clusters that are directly involved in the electron transfer and oxidation/reduction reactions of the protein. This may need to be reassessed. At least one bacterium, *Thermus thermophilus*, appears to have a fully functional, one-subunit form of the enzyme (M_r 55,000) (Yoshida et al., 1984). Also, Muller et al. (1988a,b) have shown that *Paracoccus denitrificans* subunit I is fully functional after subunit II has been removed. It turns out that, at least in this organism, subunit I contains *both* hemes, which fits with information for subunit II indicating that it has no heme *a* binding sequence (Benne et al., 1986; Muller et al., 1988a). Steffens and Buse (1988) have arrived at the same conclusion based on their discovery of an additional Cu ion on subunit I in the vertebrate as well as in bacterial (two-subunit) cytochrome oxidases (Steffens et al., 1987; Bombelka et al., 1986). [The vertebrate (beef heart) enzyme also binds one zinc and one magnesium ion.] Steffens and Buse (1988) proposed that the third Cu (on subunit II) helps to funnel electrons from ferrocytochrome c to subunit I.

The rest of the subunits in the various forms of the eucaryotic enzyme are less than 20,000 Da (Table 6-10), encoded by nuclear DNA, and synthetized in the cytosol. Recent work with monospecific antibodies suggests that different forms of subunits VIa, VIIa, and VIII may be important in establishing the tissue specificity of cytochrome c oxidase. For example, the capacities to bind cardiolipin, and the turnover numbers are not the same for cytochrome oxidases of beef heart and liver. Merle and Kadenbach (1982) have found that the activity of the liver enzyme, per molecule, is always greater, and the cardiolipin content (per mole) after isolation higher (two versus one mole per mole) than that for the heart. Differences in subunits also may occur in relation to growth conditions or phases, as seen for *Dictyostelium discoidum* (Bisson et al., 1984).

The amino acid sequences of 12 subunits of the bovine heart enzyme (Buse et al., 1982) and of many subunits of the enzymes from other species have been determined or deduced from cloning studies. The topology of the mitochondrial enzyme had until recently been conjectured solely from kinetic, spectrophotometric, and chemical modification/cross-linking work. Electron microscopy has also been helpful (Fuller et al., 1979; Fuller, 1981; Deatherage et al., 1982; Frey et al., 1982), resulting in the summary models given in Fig. 6-14. Attempts to analyze the crystal structure of this complex bovine molecule have begun as well (Yoshikawa et al., 1988). The models show that the 12–13-subunit mammalian enzyme has a lopsided branched "Y" shape, with the "stem" of the molecule (containing parts of most subunits) protruding into the intermembrane space of the mitochondrion, and the branches, containing in one case (domain M_1) hydrophobic parts of subunits I and II (with subunit IV at the aqueous surface; Wikstrom and Saraste, 1984) and in the other case (domain M_2) subunit III, penetrating and bridging the inner mitochondrial membrane. Coupled, high-spin, ferric heme a_3 and Cu_B (3.75 Å apart) form a binuclear center (in subunit I) perhaps in the M_1 domain but nearer the intermembrane space (Fig. 6-14). Cu_B seems to be bound by three N or O ligands and one S (which may be a methionine) (Mark et al., 1987), in a very aromatic region of the protein, where heme a_3 is also found (probably linked to a histidine). [Recent EXAFS data indicate two types of Cu-S (or Cu-Cu) interactions for the whole molecule (covering distances of 2.3 and 2.6 Å), as well as Cu-N or Cu-O bonds of 1.97 Å (Graham et al., 1989).] The other two metal clusters, with magnetically isolated, low-spin ferric heme a and Cu_A, are either both attached to subunit II or to subunits I and II, respectively, probably about 20 Å from the binuclear metal cluster (heme a_3-Cu_B). The iron ion in heme a may be bound to two histidine imidazole nitrogens, and Cu_A (responsible for the absorbance at 830 nm) is bound to at least two histidines, and probably two cysteines (Li et al., 1987). The two cysteines in subunit II are absolutely conserved among species (Wikström and Saraste, 1984), and the spectral properties of Cu_A resemble those of type 3 copper in laccase and ceruloplasmin (Chapter 4, Table 4-3). Recent work of Rich et al. (1988) confirms that the Cu_A is very close to the cytosolic surface.

The C-terminal end of subunit II, in domain M_1, at the cytoplasmic end of the inner mitochondrial membrane (Fig. 6-14), is the high-affinity site to which cytochrome c attaches and near where Cu_A is also thought to be located. The concentration of cytochrome a near the inner membrane surface and its diffusion rate are limiting for the (uncoupled) electron transport activity of the enzyme, and it has been determined that the interaction (collision) of cytochrome oxidase with cytochrome c is a highly efficient process in mitochrondria (Gupte and Hackenbrock, 1988). In bacteria, at least in T. thermophilus (Yoshida et al., 1984), the cytochrome may be much more tightly bound. Due to artifacts of the isolation procedure, cytochrome c oxidase initially appeared to be a dimer (of all subunits), which allowed proximity of cytochrome c also to subunit III (Denis, 1986). This does not appear to be the case. The M_2 domain, with subunit III, would appear to be involved in another function of the mammalian protein, namely, that of pumping protons through the membrane, at least in the case of the mammal (Azzi et al., 1984; Prochaska and Reynolds, 1986; Prochaska and Fink, 1987). However, the two-subunit bacterial proteins also pump protons, and some investigators have detected partial proton pumping activity in preparations of the mammalian enzyme deficient in subunit III (Thompson and Ferguson-Miller, 1983; Puettner et al., 1985), suggesting

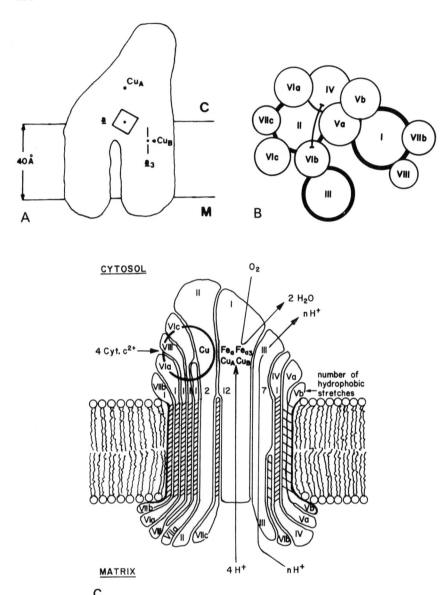

Figure 6-14. Model of the configuration of mammalian cytochrome *c* oxidase, as it sits in the inner mitochondrial membrane. (A) Configuration deduced from image reconstruction studies of two-dimensional "crystals" (from Wikström and Saraste, 1984). (B) and (C) Configuration deduced for the rat liver enzyme subunits by cross-linking studies (nomenclature of Kadenbach) by Jarausch and Kadenbach (1985) and Kadenbach et al. (1988), respectively. (Figures reprinted with permission of the authors and publishers.)

that the "core" oxidation/reduction unit (comprised of subunits I and II) indeed also comes with this capacity.

As might be expected, there are conformational changes in protein structure during oxidation/reduction, which are reflected in spectral changes in the Soret and other regions of the UV–visible spectrum, and these have been intensively studied (see, e.g., Antonini *et al.*, 1977; Brunori *et al.*, 1983; Kumar *et al.*, 1984). It is thought that reduction of the Cu_A site (on subunit II) may initiate the major conformational changes observed (Nilsson *et al.*, 1988). The interaction of the protein with the lipid membrane and specifically with cardiolipin, phosphatidylethanolamine, and phosphaltidylcholine (Seelig and Seelig, 1985) is also important. Not only is bound cardiolipin (1:1, mole:mole) essential for oxidase activity (Wikström *et al.*, 1981; Azzi, 1980; Goormaghtigh *et al.*, 1982), but the phospholipids may also be helpful in binding cytochrome *c* (Speck *et al.*, 1983).

Electron transfer rapidly occurs from reduced cytochrome *c* (or from comparable cytochromes in lower organisms) attached to subunit II, in the "stem" of cytochrome oxidase that protrudes between the inner and outer mitochondrial membranes. The molar ratio of cytochrome *c* to cytochrome oxidase (complex IV) content in most mitochrondria is about 1, although it may be double that in heart mitochondria (Wikström and Saraste, 1984). Transfer of electrons is to Cu_A and heme *a* in subunit II (or subunits II and I), perhaps in that order. (There is very rapid electron equilibration between these sites.) Transfer of electrons continues to the enzyme's complex heme a_3-Cu_B O_2 reduction center in subunit I. It is thought that dioxygen is then reduced to water, in two concerted two-electron steps, at the reduction site (Wikström and Saraste, 1984). How the Cu and heme iron ions may be involved in the other (proton pumping) function of the enzyme is unknown, although heme *a* (in subunit I) is thought to be involved (and, of course, subunit III), while Cu_A probably is not (Mueller and Azzi, 1986). The third Cu in subunit I may substitute for Cu_A (in subunit II) in some circumstances or species (Steffens and Buse, 1988). It is expected that discussion and debate over the exact mechanisms of both reactions will continue for some time. However, it is unquestionable that this enzyme, with two to three vital Cu atoms, is central to life processes on this planet and is one of the most conserved proteins known (subunits I and II) (Fig. 6-15).

The mechanism by which synthesis of the subunits encoded by the nucleus (and, for that matter, cytochrome *c*) is coordinated with that of the mitochondrion-derived subunits is also still largely unclear, as are the mechanisms by which these polypeptides cross the inner mitochondrial membrane. It is thought that the N-terminal peptide sequence of the

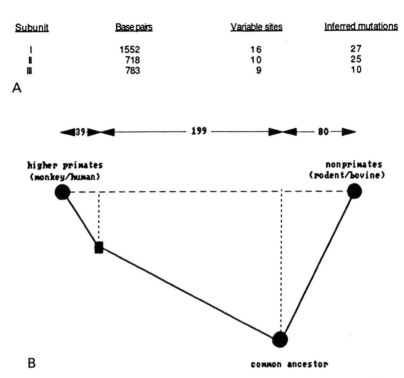

Figure 6-15. Evolution of cytochrome c oxidase subunits. (A) Data on variable sites and inferred mutations in cytochrome oxidase codons in mitochondria from 112 humans all over the world. From Cann et al. (1984). (B) Aspects of the evolution of the more rapidly evolving subunit II of mammalian cytochrome oxidase. Reprinted, with permission, from Ramharack and Deeley (1987).

nuclear-coded subunits may carry the signal to cross into the mitochondria (Nakagawa et al., 1987). As regards gene expression of the nuclear-coded subunits, heme concentrations (and perhaps also oxygen tension) play a role (Saltzgaber-Muller and Schatz, 1978; Gollub and Dayan, 1985) and may have differential effects on isologues of the nuclear-coded subunits, as studied mainly in yeast (Trueblood et al., 1988). Schatz and Mason (1974) reviewed the evidence some time ago and concluded that the main regulation appears to be via expression of nuclear genes. During inhibition of cytoplasmic protein synthesis, mitochondrial synthesis initially continues but eventually stops. In the reverse case (when chloramphenicol is used to block mitochondrial synthesis), synthesis and accumulation of the cytoplasmically derived subunits in the mitochondria still occurs, at least in yeast (in which most of these studies were carried out). It is thought that the cytoplasm usually supplies mitochondria with an excess of subunits to be combined with the mitochrondrially derived subunits, until they run out, at which point synthesis of the mitochondrial proteins also stops. How copper is added to the protein is also unknown, although this must occur within the mitochondrion itself, since attachment is to subunits synthesized therein. Cytochrome oxidase activities of tissues probably normally reflect their metabolic and respiratory functions but will be reduced when the availability of copper (and iron) becomes limiting. In complex organisms in which most iron is in the form of hemoglobin and myoglobin, sufficient iron should always be available for the enzyme, as it requires only a tiny fraction of the total iron in the body. Copper deficiency is thus much more likely than iron deficiency to affect the rate of oxygen metabolism. Cytochrome oxidase activities of various tissues in rats and mice (mostly early in life) and their response to dietary copper deprivation are given in Table 6-11. The data show that the heart (and intestine) have the highest respiratory capacities, as judged by maximal cytochrome oxidase activities (measured *in vitro*). Liver, brain, and kidney are next, and other tissues considerably below. Upon imposition of severe dietary copper deprivation, plasma ceruloplasmin oxidase activity decreases dramatically [from 14 to 1.3 U/ml in mice (Prohaska, 1981a) and from 54 to 0.3 U/ml in rats

Table 6-11. Total Copper Concentrations and Copper-Enzyme Activities in Tissues of Normal and Copper-Deficient Mice and Rats

Tissue	Cytochrome oxidase activity (U/mg)		Tissue Cu (μg/g)		SOD activity (U/mg)	
	Normal	Cu deficient	Normal	Cu deficient	Normal	Cu deficient
Mice (11–12 days of age)[a]						
Liver	0.39	0.16	46	1.3	170	95
Kidney	0.33	0.19	3.2	1.2	47	36
Brain	0.40	0.08	1.3	0.3	84	57
Heart	1.08	0.15	4.0	1.1	31	22
Small Intestine	0.66	0.26	3.1	0.8	90	73
Spleen	0.17	0.04	3.1	0.7	52	20
Thymus	0.18	0.05	—	—	47	
Bone marrow	0.17	0.03	—	—	18	3
Rats (3 weeks)[b]						
Liver	0.47	0.13	20	0.7		
Brain	0.45	0.14	2.2	0.5		

[a] Prohaska (1983b).
[b] Prohaska and Smith (1982).

(Prohaska and Smith, 1982)], reflecting a similar drop in liver (but not in other tissue) copper concentrations. Total Cu concentrations in nonliver tissues of the same animals were reduced only three- to fourfold. In most cases, cytochrome oxidase activities also fell to the same degree as total copper, with less of a drop in liver and kidney, and more of a drop in heart (which normally has very high levels). Superoxide dismutase activities were not nearly as affected as cytochrome oxidase, suggesting that all tissues (and the heart) are greatly overendowed with cytochrome oxidase (from the point of view of maintaining life). This is not surprising, as the rates of respiration and metabolic pathways used to generate ATP are only used to their maximum (by heart, for example) under conditions of strenuous exercise. [Strenuous exercise might not be necessary (or possible) in such copper deficiency.] Intestine, liver, and kidney, however, use most of their energy for other purposes, involving uptake and disposition of nutrients, urine production, etc., which must be maintained at all costs. This would explain their higher residual cytochrome oxidase activity in copper deprivation (Table 6-11). Tissues are thus normally endowed with a content of cytochrome oxidase well above what they tend to need in everyday circumstances, and this excess capacity is lost in prolonged and severe copper deficiency.

6.4 Specific Copper Components in Special Tissues

A number of other copper-dependent enzymes known in vertebrates appear to be confined to specific tissues. These include the lysyl oxidase of connective tissue, necessary for the maturational cross-linking of elastin and collagen, and dopamine β-hydroxylase of nervous tissue, long known as essential for the synthesis of catecholamines. A summary of our knowledge of these (and a few other) copper proteins follows.

6.4.1 Connective Tissue and Lysyl Oxidase

Lysyl oxidase (EC 1.4.3.13) is an essential enzyme for connective tissue function in all areas of the vertebrate organism. [The only major reviews of this enzyme are those of Siegel (1979) and Kagan (1986).] Lysyl oxidase catalyzes the oxidative deamination of lysine ε-amino groups (Fig. 6-16) to form (peptidyl) aminoadipic semialdehyde (also called allysine), with the release of ammonia and peroxide. Although a number of amine oxidases in vertebrate and nonvertebrate organisms can catalyze the same reaction (and are also Cu-dependent enzymes), lysyl oxidase is a unique amine

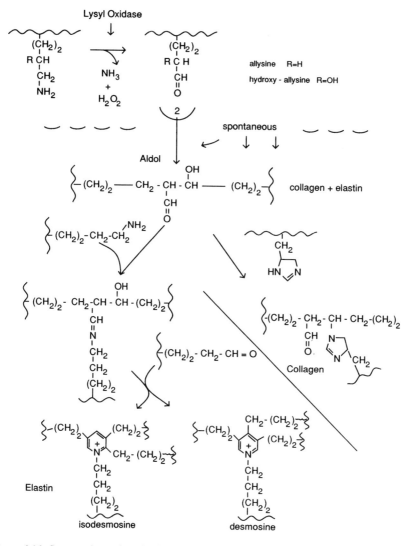

Figure 6-16. Steps and reactions in the cross-linking of collagen and elastin. Modified from Siegel (1979).

oxidase in having a marked preference for lysine side chains on immature collagen and elastin, and not reacting with free lysine *per se*. Another characteristic is its irreversible binding to (and inhibition by) β-aminopropionitrile (BAPN) (Pinnell and Martin, 1968; Trackman and Kagan, 1979) (at least 50%, at 10 μM). However, some preparations of the bovine aorta enzyme have been reported to oxidize some simple primary amines and diamines (Trackman and Kagan, 1979), as well as other lysine-containing peptides and basic proteins (Kagan et al., 1980, 1984).

Lysyl oxidase acts preferentially on newly formed collagen and to a lesser extent on partially converted procollagen, but probably not on procollagen *per se* (Siegel, 1979; Kuivaniemi et al., 1984). It also acts on soluble (tropo-) elastin and has been found at the interface between nascent amorphous elastin and microfibrils (that surround the elastin bundles), by ultrastructural immunolocalization (Kagan et al., 1986). Both lysine and hydroxyllysine side chains are oxidized, and there is a preference (at least in cartilage) for hydroxyllysine and for α_2 rather than α_1 collagen chains (Siegel, 1979). [Siegel has suggested that the enzyme attaches to a specific repeat sequence of 11 amino acids (Gly-Met (or Ile)-Hyl-Gly-His-Arg-Gly-Phe-Ser-Gly-Leu).] The more recent data of Cronlund et al. (1985) with collagen type I fibrils with and without pepsin digestion (to remove telopeptide regions) indicates that binding of lysyl oxidase is primarily to the triple-helical portions of the collagen. Binding may be proximal to lysines 9^N and 17^C (at the N- and C-terminal ends), positioning the enzyme to oxidize these lysines in adjacent quarter-staggered tropocollagen units. Cronlund et al. (1985) have also confirmed that initiation of cross-linking must occur before fully formed collagen fibrils are laid down, perhaps at the 4-D staggered dimeric or trimeric nuclear stages, or soon after lateral aggregation of these precursor units. They also suggest that, while procollagen may not appear to be an optimal substrate for the enzyme, its oxidation by lysyl oxidase has been demonstrated *in vitro* (Leung et al., 1979) and it cannot be ruled out that this also occurs *in vivo*.

Current data (Williamson and Kagan, 1986) suggest that the lysyl oxidase reaction begins with the attachment and oxidation of the ε-amino group, releasing the oxidized product (peptidyl allysine) (Fig. 6-16) and with retention of the N moiety on the reduced enzyme. Oxygen then binds to the enzyme, receives electrons, and converts the N moiety so that it can be hydrolyzed from the protein. Ammonia and peroxide are the products. Following formation of allysines (or hydroxyallysines), a series of spontaneous reactions occurs between the oxidized products of different peptide chains, to form simple and more complex cross-links between them. Formation of these cross-links appears to be nonenzymatic (after lysyl oxidase initiation) and does not require molecular oxygen (Narayanan et al., 1978). In the case of collagen, there is also reaction with histidyl side chains (Siegel and Lian, 1975), while, in elastin, complex cross-links form desmosine and isodesmosine (Fig. 6-16). For the latter, three allysines condense with a lysine into 1,2-dihydropyridine, which is then oxidized (via allysine aldol, dehydronorleucine, and dehydromerodesmosine) to isodesmosine and desmosine (Piez, 1968; Davis and Anwar, 1970). "Coacervation," which inolves the concentration of tropoelastin substrate molecules into a localized space (during their alignment into a fibril), is necessary for desmosine formation (Narayanan et al., 1978) (and does not occur at 15°C). Lysyl oxidase and its substrates are released from connective tissue fibroblasts (and smooth muscle cells; Ferrera et al., 1982), and all these reactions occur extracellularly. (Usually about 10% lysyl oxidase is intracellular, the rest extracellular, in a tissue.)

Lysyl oxidase has not as yet been cloned, but it has been purified from a variety of tissues. Recent data obtained by extraction with (and in the presence of) 6 M urea (which does not alter enzyme activity) indicate that it has an apparent molecular weight near 32,000 and comes in four isoforms that can be separated by ion-exchange chromatography (Kuivaniemi et al., 1984; Kuivaniemi, 1985; Iguchi and Sano, 1985). The basic subunit of the enzyme is also about 32,000 DA, but, in the absence of urea (presumably also in vivo), larger aggregates may form. This would account for earlier reports of the enzyme M_r being 60,000 (or much higher). It is currently unclear whether the isoforms actually differ in amino acid composition or sequence. The data of Kagan et al. (1979) would indicate that there are small but significant differences among the amino acid compositions of the bovine aortic species (Table 6-12), while those of the human placental enzyme (Kuivaniemi et al., 1984) would appear to be indistinguishable (although here the first two and last two species are pooled for the amino acid analysis). In the latter case, all species also reacted identically with different substrates and were equally inhibited by specific antibodies. Supporting the concept of real differences (in composition and/or sequences) are the clear separation of the isoforms in ion-exchange chromatography, in the absence of differences in internal disulfide bonding (three such bonds per 32,000 Da in the bovine enzyme; Williams and Kagan, 1985) or carbohydrate content (Kagan et al., 1979). It seems likely that the different isoforms (if they exist) would have different specificities for elastin and/or collagen substrates that may not as yet have been detected. This is also suggested by the affinity chromatographic data of Ferrera et al. (1982), in which various and differing portions of the cellular and secreted enzymes bound to immobilized elastin.

Using a specific ELISA assay for lysyl oxidase protein and methods of purification less dependent on urea (after the initial urea extraction), Burbelo et al. (1985) the available data suggest that most of the lysyl oxidase (at least from bovine aorta) is aggregated and inactive. In the purification, immunoreactive aggregated material was obtained that was enzymatically inactive and could not be dissociated by reduction and boil-

Table 6-12. Amino Acid Composition and Tissue Distribution of Lysyl Oxidase

Amino acid[a]	Amino acid composition in lysyl oxidase from[a]:							Connective tissue (human)[b]	Lysyl oxidase activity (relative U/g)
	Bovine aorta				Human placenta		Chick cartilage		
	I	II	III	IV	I	II			
Ala	71	81	83	85	77	75	66	Skin	100
Arg	56	61	56	49	52	57	59	Vena	83
Asx	123	125	122	94	125	123	136	Fascia	73
Cys	27	24	15	17	ND[d]	ND	30	Pleura	47
Gly	120	87	108	177	114	111	97	Cartilage	38
Glx	113	136	136	124	133	130	106	Lung	37
His	39	25	27	23	27	29	29	Placenta	7
Ile	30	27	31	29	33	33	40	Teeth (bovine)[c]	
Leu	64	86	78	65	73	77	67	Odontoblast	
Lys	31	36	36	29	46	47	31	—predentine	
Met	15	16	15	17	ND	ND	15	layer	432–505
Phe	27	30	26	25	31	34	27	Subodontoblast	
Pro	60	51	50	51	57	58	58	layer	93–121
Ser	104	86	104	122	98	101	82	Pulp layer	36–48
Thr	57	55	56	48	59	55	53		
Tyr	25	31	21	18	24	22	65		
Val	39	42	35	32	51	48	39		

[a] Residues per thousand; modified from Kuivaniemi et al. (1984). Data for isoforms (I–IV) of the bovine aortic enzyme (data of Kagan et al., 1979) and pools of isoforms (I + II) = I and (III + IV) = II from human placenta (Kuivaniemi et al., 1984), along with data for unfractionated enzyme from chick cartilage (Stassen, 1976).
[b] Data from Kuivaniemi (1985).
[c] Numata and Hayakawa (1986).
[d] ND: Not determined.

Table 6-13. Purification of Lysyl Oxidase Activity and Immunoreactive Protein (Bovine Aorta)[a]

Step	Lysyl oxidase activity		Lysyl oxidase antigen	
	Total activity ($U \times 10^{-6}$)	Recovery (%)	Total antigen (mg)	Recovery (%)
4 M Urea extract	4.5	100	18.4	100
Gelatin affinity chromatography	4.5	100	9.4	51
α-Elastin affinity chromatography	2.8	61	5.2	28
Ultrafiltration (UM-10)	0.98	22	4.5	24
Sephacryl S-200 gel chromatography	0.67	15	0.49	2.7

[a] Data of Burbelo et al. (1985).

ing in SDS. The 15% yield of enzyme activity (Table 6-13) was ascribable to 1.7% of the original lysyl oxidase protein quantitated by the ELISA, and the active (32-kDA) enzyme accounted for only about 10% of the total enzyme protein in the purified concentrate, when separated in Sephacryl S-200. The question is whether the aggregated material exists *in vivo* or is (at least partly) an artiffact of purification, including urea extraction. Some of the inactivation of the enzyme occurred during the ultrafiltration (concentration) step in the purification (Table 6-13), while a major portion of inactive protein seems to have been lost during the gelatin affinity step. The low-molecular-weight form(s) of the lysyl oxidase certainly appears to be active. Use of isolation procedures not involving urea also produces active forms of low molecular weight, as in the affinity purification of lysyl oxidase from the medium of cultured aortic smooth muscle cells, via elastin immobilized on Hydrogel (plus size-exclusion HPLC) (Ferrera *et al.*, 1982), but we do not know whether larger (inactive) forms were also present in these preparations.

The existence of Cu as a cofactor for the enzyme/cross-linking reaction was deduced long before lysyl oxidase activity was first measured and the enzyme purified. One of the hallmarks of Cu deficiency is a defect in elastic tissue formation, leading to aortic aneurysms and spinal curvature, increased proportions of soluble collagen, and other defects of connective tissue (collagen and elastin) formation. Defective development of elastin structure and alveolar space in the lung also occurs (O'Dell *et al.*, 1978). These phenomena can be ameliorated or reversed by copper treatment (see, e.g., Coulson and Carnes, 1967; Waisman *et al.*, 1969). This implies the existence of a copper-dependent factor or enzyme necessary for normal connective tissue formation. [The same defects occur in lathyrism produced by the lathyrogens of certain pea plants that cause irreversible, noncompetitive inhibition of lysyl oxidase (see Siegel, 1979).] Lysyl oxidase activity was first identified and characterized by Pinnell and Martin (1968). Proof of its copper dependence then came from showing that purified lysyl oxidase contained copper (Harris *et al.*, 1974) and that the activity in copper-deficient tissue could be restored *in vivo* by copper injection and *in vitro* (in organ culture) by incubation with Cu(II), as could that of the purified enzyme dialyzed against chelating agents such as α,α'-dipyridyl or diethyldithiocarbamate (Siegel, 1979). [A few other ions, such Fe(II) or Co(II), were also partially effective.] Harris *et al.* (1974) reported that purified enzyme from chick aorta contained 0.14% Cu, or 1.4 atoms per 60,000 Da, while Shieh and Yasunobu (1976) found 1 Cu per 70,000 Da for lysyl oxidase from bovine lung. The more recent findings (for three forms of the enzyme of embryonic chick cartilage) are 0.20–0.32% Cu (Iguchi and Sano, 1985), averaging out to 1 Cu atom per subunit, which is probably the correct stoichiometry. (The individual 30,000–32,000-Da units are enzymatically active, and activity requires Cu.) Using copper-deficient chicks, Harris (1976) showed a rapid restoration of aortic lysyl oxidase activity after injection of 0.5–1.0 mg of $CuSO_4$/kg, amounting to 6–7- and 20-fold over that of untreated animals, at 2 and 4 h, respectively. Later work (Rayton and Harris, 1979) with aorta in organ culture indicated that restoration was much more efficient when nonalbumin serum proteins (versus ionic copper or Cu-albumin) were added to the medium. [Ceruloplasmin (and transcuprein) would be present in the nonalbumin fraction (see Chapters 3 and 4).] Homogenization of the tissue (or incubation in N_2) also inhibited the return of enzyme activity, in organ culture. Treatment with cycloheximide (to block 97% of protein synthesis) inhibited (but did not abolish) the ability of copper to restore enzyme activity. Overall, this implies that at least the major incorporation of copper is into new enzyme rather than into copper-deficient apoenzyme waiting to be activated directly and, also, that addition of copper (in deficiency) may trigger increased synthesis of the enzyme at the translational level. (Although effects of actinomycin D were also examined, and there was little effect, it was not clear that the doses used had inhibited RNA synthesis. Thus, nothing can be said about transcriptional control.) Interestingly, a histidine residue appears to be necessary for catalytic activity (Gacheru *et al.*, 1988). Whether it is involved in binding of the Cu is still unclear.

Lysyl oxidase also has another cofactor, long thought to be pryridoxal phosphate (see Siegel, 1979) but which more recent work suggests is either pyrroloquinoline quinone (PQQ) (Williamson *et al.*, 1986a; van der Meer and Duine, 1986) or 6-OH dopa (Fig. 6-17). Pyrroloquinoline quinone has been claimed as the cofactor of amine oxidase (Chapter 4), and dopamine β-hydroxylase (below), and its involvement with enzymes has recently been reviewed (Gallop *et al.*, 1989). However, Jane *et al.* (1990) provide compelling evidence that the actual cofactor for serum amine oxidase is 6-OH dopa ("topa") which can cyclase to a double ring structure resembling rings B and C of PQQ, so the involvement of PQQ in lysyl oxidase must also be re-evaluated. Clearly, a functional carbonyl is needed by lysyl oxidase to help oxidize the lysine amino group. Dietary pyridoxine deficiency had been reported to decrease or alter the cross-linking of elastin (Myers *et al.*, 1985). However, the cyanide spectra of the enzyme's prosthetic group and pyridoxal phosphate were not identical (see Williamson *et al.*, 1986b). Use of specific antibody against protein-linked pyridoxal phosphate in Western blots of the bovine aortic enzyme, as well as comparisons of phenylhydrazine spectra, confirmed that pyridoxal phosphate was not the source of the functional carbonyl in these studies. Nevertheless,

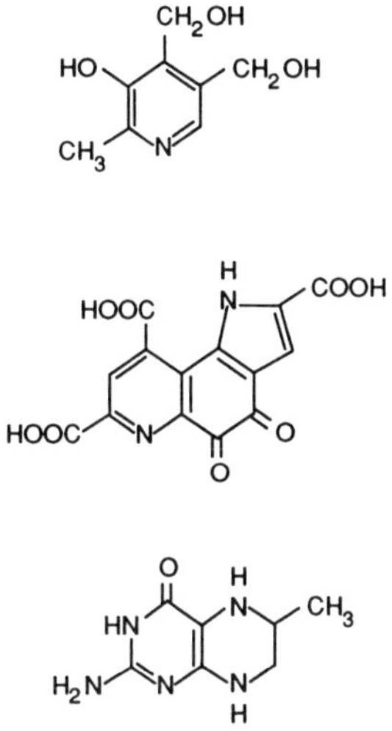

Pyridoxine

Pyrroloquinoline quinone

6 - Methyltetrahydrobiopterin

$NH_2-CH_2-C\equiv N$ β - Aminopropionitrile (BAPN)

Figure 6-17. Structures of potential cofactors and inhibitors of copper enzymes.

Bird and Levene (1982) reported that pyridoxal 5′-phosphate restored activity lost by purified chick aortal lysyl oxidase (during removal of urea) and that the fluorescence emission spectrum of this enzyme treated with semicarbazide resembled that of the pyridoxal phosphate derivative. (The spectra of the cyanide adducts did not, however, correspond.) PQQ deficiency of rats on special purified diets (PQQ is produced by bacteria and is normally widely prevalent in the diet) also appears to lower lysyl oxidase activity (Killgore *et al.*, 1989), supporting the concept that this is the true cofactor. Thus, the question of the nature of the lysyl oxidase cofactor remains open at this time.

Lysyl oxidase activity is found with connective tissue in all parts of the body, including the teeth (Table 6-12). It is especially prevalent in skin but also active in the vasculature and lung, while not so active in organs and softer tissue such as placenta (and the pulp of teeth). The enzyme is more concentrated is tissues during development, as indicated, for example, by the data on mouse bone (Sanada *et al.*, 1978) (not shown). This would be expected, as the laying down of collagen and elastin is necessary for tissue growth and bone mineralization. Lysyl oxidase is also found in the isthmus (membrame-forming region) of the hen oviduct next to the shell gland, where it is necessary for elastin cross-linking in eggshell formation (Harris *et al.*, 1980).

Apart from copper availability, little is known about what factors (if any) regulate gene expression and enzyme synthesis or activity. Long-term estrogen treatment (two injections per week, for 7 weeks) appears to enhance activity, as evidenced by a marked increase in the proportion of insoluble (mature) versus soluble collagen in mouse skin and increases in bone and skin enzyme activity (81 versus 63%) (Sanada *et al.*, 1978). Progesterone and testosterone had no such effects. Estrogen also increases plasma ceruloplasmin concentrations, and this may be connected to these observations (see Chapter 7). In contrast, malignant transformation (which reduces fibroblast tropocollagen release) severely decreases release of lysyl oxidase activity, as determined by screening the conditioned media of a variety of transformed and untransformed human cell lines in culture (Kuivaniemi *et al.*, 1986). After inactivation by BAPN (Fig. 6-17) or in copper deprivation, synthesis of lysyl oxidase increases markedly, reaching half of eventual maximal levels by 30 h in the aortas of 5–6-day-old chicks, after BAPN (Harris *et al.*, 1977), and by 10 h in chick aortas incubated with Cu(II) *in vitro* (Rayton and Harris, 1979). The half-life of the baby chick aorta enzyme *in vitro* may be 16–18 h (Harris *et al.*, 1977). The activity of the enzyme on collagen may also be modulated by heparin. Gavriel and Kagan (1988) found that high concentrations of heparin inhibited the tight binding of lysyl oxidase to developing collagen fibrils, which may explain the retarding effect of heparin on the healing of wounds.

6.4.2 The Brain and Dopamine β-Hydroxylase

The functions and distribution of copper (and other trace elements) in brain have recently been reviewed by Prohaska (1987). Table 6-14 shows that the human brain displays distinctive and sometimes dramatic variations in copper content from region to region, a phenomenon that we may assume is related to regional differences in brain function. Notable is the two-to three fold higher concentration in human gray matter compared to white matter [reflected in differences between cerebellar white matter and cortex? (also cerebral cortex)] and the very high concentrations in the substantia nigra and locus ceruleus. (One wonders if the "ceruleus" designation does not suggest the blue of some copper proteins.) In connection with this distribution pattern, one should remember that ceruloplasmin is synthesized (and presumably secreted into the interstitial and cerebrospinal

Table 6-14. Distribution of Copper among Regions of the Adult Human and Rat Brain[a]

Human[a]		Rat		
Region	Cu concentration (µg/g wet wt.)	Region	Cu concentration (µg/g wet wt.)	
			I[a]	II[b]
Cerebellar cortex	10.4	Cerebellum	4.5	5.6
Cerebellar white matter	4.3			
Cerebellar cortex (frontal)	7.8	Cortex	4.8	3.0
Corpus callosum	3.1	Midbrain	4.6	5.8
Substantia nigra	18.8	Medulla oblongata	3.5	
Pallidum	9.4	Pons and medulla		3.9
Putamen	10.3	Striatum	4.6	7.0
Locus ceruleus	62			
Hippocampus	6.6	Hippocampus	4.5	4.2
		Hypothalamus	5.8	12.8
Centrum semiovale	4.3			

[a] Rearranged and recalculated from J. R. Prohaska, *Physiol. Rev.* 67, 858–901 (1987), using a conversion factor of 5 for wet weight over dry weight.
[b] Data from Lai *et al.* (1985).

fluid) by the choroid plexus (Aldred *et al.*, 1987), which may encompass this high-copper region. Presumably, similar (though perhaps less dramatic) variations in copper content exist among comparable regions of rat brain, although the available data show only minor differences (Table 6-14). The apparent lack of differences may be due to the difficulty of separating (small) high-copper regions from surrounding low copper tissue. In the rat, the hypothalamus appears to be somewhat enriched in copper. This may reflect a need for copper in the release of neuropeptides that control neurohypophyseal hormone-releasing factors, such as, for example, gonadotropin-releasing hormone (Barnea *et al.*, 1988a), the hormone that controls luteinizing hormone (and ovulation and testosterone). (See more below and in Chapter 7, in relation also to prostaglandins.)

Focusing on the cell types in which copper may be most concentrated, histochemical staining with sulfide–silver (plus trichloroacetic acid) or rubeanic acid has shown copper localized in glial cells in rat cerebellar white matter and in the pyramidal cells of the cortex and also in the glial cells of the human locus ceruleus [see review by Szerdahelyi and Kasa (1984)]. In copper overload (Wilson's disease), copper was also detected in glial cells of the human putamen, in cells around the capillaries of the caudate nucleus, in astrocytes, and in basal ganglia. Recent studies with cultured cells have confirmed that glial cells are active in copper uptake, concentrating copper to levels of about 3 µg/g in the case of the chick (from medium containing 0.2 µg of Cu/ml) (Tholey *et al.*, 1988).

As already mentioned, subcellular fractionation of whole rat brain indicates that copper is concentrated in the mitochondrial and cytosol fractions (Matsuba and Takahashi, 1970). Data of Rajan *et al.* (1976) suggest there is considerable variation among subcellular fractions, in different brain regions, but the results may be unreliable. (They are much higher than what other investigators report.) As in other tissues, a number of copper-containing components in brain cytosol may be separated in gel permeation chromatography. These include components of the usual size (see Figs. 6-3 and 6-4), corresponding roughly in elution to metallothionein, superoxide dismutase, and something much larger (> 100,000 Da) (Terao and Owen, 1977; O'Dell and Prohaska, 1983; Hunt, 1980), at least in rodents. A small proportion of copper may also be with low-molecular-weight ligands (Hunt, 1980). Again, the nature of these components is still unclear and may be more complex than in the case of other tissues. Certainly, Cu/Zn superoxide dismutase is there and quite active (see later, Table 6-15), which may account for some or all of peak II. It is also fairly uniformly distributed throughout the brain (Prohaska, 1987), although the medulla (of the rat) may be enriched. As in other tissues, Cu/Zn SOD has been implicated in protecting the brain from peroxidative damage. Fraction III most probably contains some Cu-thionein (copper-bound metallothionein), at least in conditions where brain Cu is elevated (Waalkes *et al.*, 1982). In addition, there would appear to be an acidic protein, designated neurocuprein, with an M_r of 9000, containing one Cu atom per molecule (Sharoyan *et al.*, 1977; Nalbandyan, 1983). Although isolated from bovine brain, it is present also in brains of the human, pig, sheep, cat, and rabbit. The same protein may also be present in chromaffin granules of the adrenal medulla (involved in catecholamine metabolism), and the copper in the protein can be reduced by epinephrine and norepinephrine, although no enzyme activity of neurocuprein (conversion of substrate to product) has been demonstrated. (It is also not a superoxide dismutase or dopamine β-hydroxylase.)

The low-molecular-weight (10–15 kDa) fraction of the cytosol may also contain a protein described earlier as albocuprein II, of pale yellow color, with no EPR spectrum (Fushimi *et al.*, 1971). It is not the same as neurocuprein (Nalbandyan, 1983), but may be Cu-thionein, as it is rich in sulfhydryl groups and of the right size (Prohaska, 1987). This has not been pursued. "Albocuprein I" (Fushimi *et al.*, 1971), also yellow, was reported to have an M_r of 72,000 and some immunological resemblance to ceruloplasmin. This also has not been followed up. [As already indicated, ceruloplasmin is produced by certain parts of the brain, and its presence in brain tissue (above levels ascribable to blood contamination) has been reported (Linder and Moor, 1977), where it may be bound to cell membrane receptors for delivery of copper and/or protection against superoxide; see Chapters 3 and 4.]

In the brain mitochondria, as in other cells of the body, is cytochrome c oxidase. Prohaska (1987) has estimated that it accounts for about 20% of the cellular copper in brain tissue, but probably not for all of the copper in the mitochondria. Mitochondria and cytochrome oxidase are particularly rich in the synaptic terminal regions of major excitatory pathways leading to the hippocampus (Kageyama and Wong-Riley, 1982). Other (very active) axon terminals also have more mitochondria and cytochrome oxidase, while the trunks of axons tend to have very little. Another copper-dependent function of mitochondria may be the conversion of porphyrin a to heme a (which is part of subunit I or II of cytochrome oxidase; see earlier), as first demonstrated in yeast (Keyhani and Keyhani, 1980). [Heme a is decreased in copper-deficient brain (Prohaska, 1987).]

In the pituitary, another (new) copper-dependent enzyme has been reported, which amidates peptides containing a C-terminal glycine, by reacting the glycine with oxygen in the presence of ascorbate, to release glyoxylate, water, and oxidized ascorbate (Eipper et al., 1983), leaving behind the glycine α-NH_2 nitrogen as an amide group on the next-to-last (now "C"-terminal) amino acid. [See the recent review by Eipper and Mains (1988).] About half of bioactive neuropeptides have an α-amide group at the C-terminus that is essential for full biological activity, including calcitonin, oxytocin, and vasopressin. This α-amidating monooxygenase is therefore an important substituent of hypothalamic cells, in which it appears to be located mainly is secretory granules. The location of this enzyme and its apparent requirements for copper, O_2, and ascorbate mimic those of the better known enzyme dopamine β-hydroxylase (see below). The α-amidating enzyme has been isolated from bovine (Murthy et al., 1986) and porcine (Kizer et al., 1986) pituitary, and Stewart and Klinman (1988) have suggested that it cooperates with cytochrome b_{561} (in neurosecretory vesicles) to recycle ascorbate, in a manner that may be identical to cooperative actions of dopamine β-hydroxylase with the same kind of cytochrome in adrenal chromaffin granules. The same enzyme activity has been found in the various lobes of the pituitary, as well as in the submaxillary and parotid glands and in the heart atrium. Smaller concentrations are in the cerebral cortex, in the thyroid, and in endocrine tissue of the gastrointestinal tract (which latter may be quantitatively important). Activity is also measurable in blood plasma and in the cerebrospinal fluid. (The frog has a high activity in its skin and in skin secretions.) In the pituitary corticotrophic cells, α-amidation activity varies positively in relation to pro-adrenocorticotropin (pro-ACTH) (and endorphin) production, although ACTH is not a substrate for the enzyme.

Also of potential interest is the finding that serum enkephalin concentrations in man are sensitive to copper intake. Bhathena *et al.* (1986) reported that levels of leucine and methionine enkephalins dropped 70–90% during 11 weeks of low copper intake. (Copper deficiency was only marginal in these subjects, serum copper falling from 1.2 to 1.0 μg/ml.) Three weeks of copper supplementation completely reversed this condition. This suggests that enkephalin synthesis or release, or both, is copper dependent. It is noteworthy that enkephalins (like norepinephrine) are concentrated in chromaffin granules in the central nervous system, granules which also contain the copper-dependent dopamine β-hydroxylase (see below). These opiates are capable of binding copper with considerable affinity, and it has been postulated that they may be active and bind to their receptors in the form of Cu complex (Chapman and Way, 1980; Sadee *et al.*, 1982). Pettit and Formicka-Kozlowska (1984) have proposed that other neuropeptides (especially those with proline residues) may also be held in a biologically favorable conformation by Cu(II). Thus, copper may in several ways be necessary for normal central nervous system function—by acting as an enzyme cofactor for synthesis of certain neurotransmitters and agonists and by holding others in an active conformational state.

The enzyme of perhaps the greatest note in brain metabolism is *dopamine β-hydroxylase* (EC 1.14.17.1), responsible for the synthesis of the sympathetic neurotransmitter norepinephrine and necessary as well for epinephrine production in the adrenal. This enzyme is distributed in the brain in relation to the noradrenergic neurons, these being especially prevalent in the locus ceruleus, brain stem, and posterior hypothalamus (Prohaska, 1987) (see also Table 6-14). It may be present in the cytosol of the cell bodies, but also in the distal part of the axons, where it is in the norepinephrine storage granules. In the superior cervical ganglion, it appears to be concentrated much more in the "small intensely fluorescent" (SIF) cells than in the main neurons (Chang *et al.*, 1988). The enzyme is also present in the chromaffin granules of the *adrenal medulla*, where it functions (with ascorbic acid) in the pathway to epinephrine formation. The enzyme has four subunits (each about 75 Da) that appear to be arranged in disulfide-linked pairs (see Stewart and Klinman, 1988).

Dopamine β-hydroxylase catalyzes the reaction of hydroxyphenethylamine (dopamine) with O_2 to form norepinephrine + H_2O. The mechanism

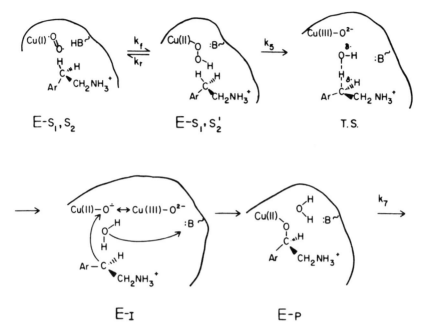

Figure 6-18. Mechanism of the reaction of dopamine β-hydroxylase with dopamine, proposed by Klinman and colleagues. Reproduced from Stewart and Klinman (1988), with permission.

of the reaction proposed by Klinman and co-workers is hown in Fig. 6-18. As with the α-amidation enzyme (see above), ascorbate is the source of electrons for the reaction (two electrons/dopamine molecule) that can be restored by NADH. Pyrroloquinoline quinone (PQQ) was again thought to be the cofactor, as first reported by van der Meer et al. (1988). The enzyme is a tetrameric glycoprotein, of about 290,000 Da (Wallace et al., 1973; Saxena and Fleming, 1983), that binds to concanavalin A (Colombo et al., 1987) and has not as yet been cloned (Stewart and Klinman, 1988). An extensive review of the adrenal enzyme has appeared (Stewart and Klinman, 1988), which, as far as structure and mechanisms are concerned, can be taken as applying also to the enzyme of the central nervous system. The question of the cofactor (PQQ versus 6-OH dopa), remains to be re-examined.

The exact copper content of the dopamine β-hydroxylase has been difficult to ascertain, suggesting that the metal ion is not so firmly bound. Values in earlier reports averaged to about two Cu atoms per tetrameric enzyme molecule. Convincing recent reports put the actual number at eight, or two per subunit (Klinman *et al.*, 1984; Ash *et al.*, 1984; Colombo *et al.*, 1984). This is more satisfying, as it fits with the concept that two electrons (as from two Cu atoms) should be available at each active site and that there is one active site per subunit. Some of these same authors still report values of only two Cu atoms per molecule in more recent studies in which enzyme was purified by otherwise superior methods that resulted in a high specific activity. However, Colombo *et al.* (1987) found that additional copper (up to about eight Cu atoms per molecule) would bind to purified dopamine β-hydroxylase *in vitro*, and this additional copper was then not readily removed by dialysis. As the maximum activity of the enzyme correlated with two Cu atoms per subunit (see Stewart and Klinman, 1988), addition of the extra copper also increased enzyme activity. The endogenous and extraneously added coppers give equivalent pulsed EPR spectra, spectra consistent with the binding of Cu(II) to imidazole side chains of histidine. As quantitated by spectral simulation (comparing spectra for two, four, and eight Cu atoms per molecule), it appears there are three or four ligands per Cu atom (McCracken *et al.*, 1988a). All this implies again that there are eight copper atoms per molecule (or two per subunit) and that the copper atoms are equivalent.

Figure 6-19 provides an overview of our limited knowledge of the involvement of Cu protein/enzymes in brain and central nervous system metabolism, and in catecholamine metabolism, in particular, based on what has already been presented. It is clear that Cu has several kinds of roles to play in the amidation, hydroxylation, and deamination of peptide and amino acid-derived hormones, involving production, release, and inactivation. Catecholamines are primarily inactivated by flavin-dependent monoamine oxidase (that contains no copper) in nervous tissue and elsewhere (see below). However, plasma (copper-containing) amine oxidase (EC 1.4.3.6) will deal with other amino acid-derived hormones such as histamine (and, theoretically at least, tryptamine, tyramine, dopamine, and serotonin) and, in some species, also with polyamines (see below).

Apart from exerting its effect via known (or partially known) copper-dependent enzymes (Fig. 6-19), copper has other roles in the brain that may not be mediated by the same enzymes and which, in any event, we do not as yet understand at the molecular level. These include a potential role in the mediation (with prostaglandins) of luteinizing hormone-releasing factor release, already cited (Barnea and Cho, 1987; Bhasker and Barnea, 1988; Barnea *et al.*, 1988a,b) and, more generally, a role in myelination, which is especially critical during growth and development. The role of copper in these and other processes becomes apparent during copper deficiency, especially when deficiency is instituted early (as in the gestational stage, before birth). Many effects of copper deficiency on brain

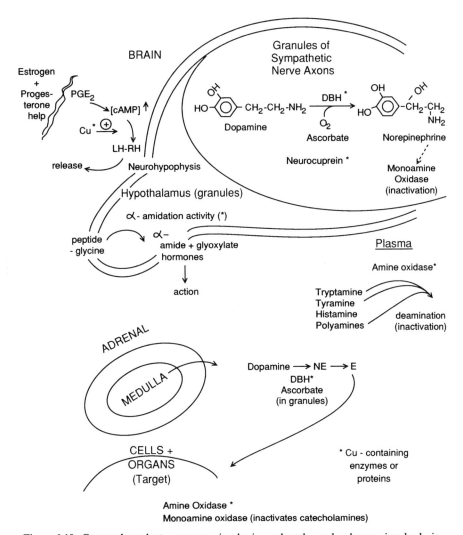

Figure 6-19. Copper-dependent enzymes in brain, adrenal, and plasma involved in catecholamine metabolism and in the metabolism of other hormones and amines. Starred (*) enzymes require copper for their activity. (Monoamine oxidase is not a copper-dependent enzyme.)

development and function have been studied, especially in normal and mutant mice that may be models for the inherited human disease of copper, Menkes' disease (Prohaska, 1987).

As Prohaska (1987) has described, the effects of copper deprivation are evident primarily before "neurons and glia have differentiated and the blood brain barrier has formed." Thereafter, the brain is fairly resistant to dietary deficiency, as if, once enough copper has been received and most brain growth has occurred, recycling of the copper present is efficient enough to support a fairly normal status quo. (This is interesting, in that the brain is its own producer of ceruloplasmin, for example, and this might continue normally in the brain as a self-contained system, even in the face of greatly reduced liver ceruloplasmin production, brought on by short-term dietary deficiency.)

A major feature of copper deficiency in the critical stages of brain development is reduced myelination, resulting in less total myelin and in myelin that is less mature, with less cholesterol, galactosides, and sulfatides, as well as a different phospholipid mix (with more phosphatidylcholine) (DiPaolo *et al.*, 1974; Prohaska, 1987). Underlying these changes may be reduced levels of UDP-galactose:hydroxy fatty acid ceramide galactosyl-transferase activity, as demonstrated in rats (DiPaolo *et al.*, 1974) and lambs (Patterson *et al.*, 1974). There is also a decrease in 2',3'-cyclic nucleotide 3'-phosphodiesterase (a myelin-associated marker enzyme). Both it and the galactosyltransferase are normally prevalent in the fully developed oligodendroglial cells that form the myelin. However, it is not known whether these enzymes (or others involved in myelin formation) are copper dependent (Prohaska, 1987). Are the oligodendroglial cells perhaps not fully developed in copper deficiency, and, if so, why not? Other direct possibilities for the involvement of copper in myelination have been elegantly described by Prohaska (1987) and include lack of ATP (to supply energy), lack of norepinephrine (to stimulate oligodendroglial cell activity), increased demyelination or myelin turnover (due to instability of immature myelin or less protection from Cu/Zn SOD), and vascular abnormalities (due to less lysyl oxidase and leading to some brain hypoxia). None of these possibilities appears to be very satisfactory; that is, there are facts that do not fit or exceptions to the general findings. [The defect of myelination found in *quaking* mice cannot be ascribed to copper deficiency (Prohaska, 1980; Cloez and Bourre (1987).]

Some of the effects of severe Cu deficiency on brain parameters of mice and rats are summarized in Table 6-15. Interestingly, there is not necessarily a reduction in brain weight (rats), although the cerebellum seems to be affected. Overall body growth is certainly less, and there tends to be anemia and a very low level of plasma ceruloplasmin (oxidase activity). Brain Cu levels and cytochrome oxidase activities are severely depressed, although ATP levels would appear not to be affected (Prohaska and Wells, 1975; Rusinko and Prohaska, 1985). The myelin-associated cyclic nucleotide phosphodiesterase activity is less, but Cu/Zn superoxide dismutase, glutathione peroxidase, tyrosine hydroxylase, and, notably,

Table 6-15. Brain Copper, Enzymes, and Catecholamines in Severe Copper Deficiency (Young Rats and Mice)

Parameter	Mean ± SD		
	Copper-adequate diet	Copper-deficient diet	Deficient plus Cu treatment[a,b]
Mice (C57), 11–12 days[b,c]			
Brain wt. (mg)	315 ± 15 (6)	234 ± 11 (7)	
Blood hemoglobin (g/dl)	9.8 ± 0.9 (11)	7.7 ± 0.8 (13)	
Plasma ceruloplasmin oxidase activity (IU/liter)	12–14 (24)	0.7–1.3 (24)	
Brain Cu (μg/g)	1.2 ± 0.2 (4)	0.3 ± 1 (4)	
Brain cytochrome oxidase activity (IU/g)	400 ± 36 (4)	75 ± 11 (4)	
Brain norepinephrine (ng/g)	256 ± 34 (5)	24 ± 7 (5)	263 ± 30 (7)
Brain dopamine (ng/g)[d]	445 ± 7 (4)	551 ± 99 (5)	451 ± 32 (7)
Dopamine β-hydroxylase activity (nmol/h per mg)	4.22 ± 0.40 (4)	5.26 ± 0.6 (4)	
Ascorbate (μg/g)	843 ± 86 (12)	595 ± 90 (5)[d]	802 ± 42 (7)
2′,3′-cAMP phosphohydrolase (IU/mg)	1.01 ± 0.02 (4)	0.58 ± 0.09 (4)	
Glutathione peroxidase (IU/g)	69 ± 4 (3)	71 ± 4 (4)	
Rats, 49 days[e]			
Cytochrome oxidase (U/g)	2.4 ± 0.5 (8)	0.4 ± 0.2 (7)	
Norepinephrine (ng/g)	426 ± 82 (15)	276 ± 80 (15)	
Dopamine (ng/g)	1050 ± 60 (10)	750 ± 45 (10)	
Cu/Zn superoxide dismutase (U/mg)	1.2 ± 0.4 (15)	0.8 ± 0.2 (15)	
Tyrosine hydroxylase (U/g)	23 ± 3 (10)	16 ± 5 (10)	
Rats, 28 days[f]			
Brain wt. (mg)	1540 ± 37 (4)	1510 ± 82 (4)	
Cerebellum wt. (mg)	210 ± 9 (4)	191 ± 18 (4)	
Cerebellum Cu (μg/g)	1.9	0.45	
Norepinephrine (nmol/g)	1.9	1.3	
2′,3′-cAMP phosphohydrolase (U/g)	282	224	

[a] Cu-treated mice were injected with 50 μg of Cu at 7 days of age.
[b] Prohaska and Cox (1983).
[c] Prohaska and Smith (1982).
[d] Prohaska and Cox (1983) *only*.
[e] Morgan and O'Dell (1977).
[f] Prohaska and Wells (1974).

dopamine β-hydroxylase (DBH) are not greatly changed. (Of the latter, only SOD and DBH are known copper enzymes.) Although DBH activity is not reduced, as measured *in vitro* (it is actually somewhat increased), there is a substantial decrease in total norepinephrine concentration, implying either decreased synthesis or enhanced degradation (or release) of this neurotransmitter. Prohaska and DeLuca (1988) showed that at least some of the decrease was in the hypothalamus (of mice), although Miller and O'Dell (1987) did not find this to be the case in rats. The reduction in brain ascorbate concentrations (Table 6-15), though not as dramatic on a total brain basis, nevertheless may be part of the story, as ascorbate is thought to be necessary for DBH action and tends to be concentrated (with the enzyme) in secretory/storage granules (Colombo *et al.*, 1987) (Fig. 6-19). (It is possible that ascorbate concentrations in these granules are more greatly reduced than in the brain as a whole.) Another possibility is that traces of copper in the assay mixture (*in vitro*) are sufficient to activate what was otherwise only a partially active enzyme, *in vivo*, leading to a false picture (overestimation) of the level of actual brain enzyme activity. This all needs to be studied further, but *in vivo* work supports the concept that decreased synthesis of norepinephrine is indeed occurring in severe copper deficiency (Hunt and Johnson, 1972) (also in the heart; Sourkes *et al.*, 1974; Gross and Prohaska, 1990). It is noteworthy that the levels of catecholamines (and ascorbate) respond quite readily to (and are restored to normal by) copper treatment (Table 6-15).

There have been conflicting reports that brain concentrations of *dopamine* are also reduced in copper deficiency (Morgan and O'Dell, 1977; Feller and O'Dell, 1982; Prohaska and Smith, 1982; Prohaska *et al.*, 1990). Miller and O'Dell (1987) found with various types of diets that the changes only occurred in a subpopulation of their rats and was evident in the corpus striatum, but not in the whole brain. This suggests that, at least in the corpus striatum, copper may in some way be necessary for the production of dopamine itself (as well as for its conversion to norepinephrine, via copper-dependent dopamine β-hydroxylase).

Changes in norepinephrine and dopamine pools may also secondarily affect levels of receptors for these (and perhaps other) hormones. This possibility is being studied, but the results are currently too confusing to interpret (see Prohaska, 1987). We are left with the knowledge that copper has profound and fundamental roles to play in the brain and central nervous systems, some of which are important to other tissues as well, but many of which may be unique. Our current, not inconsiderable, knowledge of this area provides a framework within which the missing pieces of major processes and molecular details must be placed as they become available from future work.

6.5 Other Intracellular Enzymes Requiring Copper

6.5.1 Tyrosinase and Pigmentation

Melanin pigments derived from tyrosine and other polyphenols (whose synthesis is initiated by the same enzyme) are widely distributed in nature. In mammals, melanin pigments are responsible for hair, skin, and eye color. In the eye, especially in the retina and uvea, they are also necessary for normal function (Silverstone et al., 1986). Formation of melanin is initiated by tyrosinase (monophenol, dihydroxyphenylalanine:oxygen oxidoreductase, EC 1.14.18.1), a copper-dependent enzyme located in melanocytes and particularly in the melanosomes (or vesicles), where the melanin is mainly formed and stored. Reviews on different aspects of tyrosinase structure and function include those of Lerch (1981) and Lerch and Germann (1988). The state of copper and its ligands in the protein, as well as its reaction mechanism and oxygen binding, have been of interest to bioinorganic chemists for some time. (Here, see also Solomon, 1988.) A more physiologically oriented review is that of Robb (1984).

Tyrosinase belongs to the hemocyanin family of copper monooxygenases. This conclusion is based on comparative spectroscopic properties of the proteins and on extensive sequence data, for a variety of tyrosinases in mammals, mushrooms, *Neurospora crassa*, and many other species (from plants to bacteria) and for molluscan and arthropod hemocyanins. [Hemocyanins function as circulating oxygen carriers in these species (see Chapter 10, Figs. 10-14 and 10-15)].

Tyrosinases appear to consist of one, two, or four subunits of 18–67 kDa (Table 6-16). The mammalian enzymes appear to be monomeric but have the larger units. Irrespective of species, the catalytic site appears to involve a binuclear Cu cluster. In most species, each Cu atom may be bonded to three histidine residues that are part of α-helices in two different regions of the polypeptide chain of each subunit. [In some cases, notably the mushroom and potato, binding sites for the two copper atoms may be on different subunits (see Chapter 10 and Kanagy et al., 1988).] The Cu binding site closer to the C-terminus in tyrosinases (and hemocyanins) (also known as the Cu_B site) is in a highly conserved sequence of 56 amino acids that may go back to a common ancestor with the hemocyanins (Fig. 10-14). The Cu_A site of the tyrosinases differs from that of the hemocyanins, although it may have arisen through duplication of the Cu_B region (Drexel et al., 1987; Lerch and Germann, 1988). A mouse gene (Shibahara et al., 1986; Ruppert et al., 1988) and a human gene (Known et al., 1988) have been cloned. The mouse gene consists of four sometimes widely separated exons. At least three alternatively spliced transcripts of the

Table 6-16. Molecular Properties of Tyrosinases[a]

Source	Molecular weight ($\times 10^{-3}$)	Subunit (MW $\times 10^{-3}$)	Cu/subunit	State of copper
Straptomyces glaucescens	29	29	2[b]	Binuclear
Streptomyces nigrifaciens	18	18	0.38[c]	—
Agaricus bispora (mushroom)	26–120	30	1	Binuclear
	120	13.4(L), 43(H)	—	—
Neurospora crassa	33–120	33	1	—
	46	46	2	Binuclear
Podospora anserina	100	42	0.21[c]	—
Potato	290	36	1	—
Beta vulgaris	40	40	2	Binuclear
Bombix mori	80	40	0.15[c]	—
Rana pipiens	200	54	0.15[c]	—
Human malignant elanoma	66.7	66.7	2	Binuclear

[a] Modified from Lerch (1981) for data from many reports.
[b] *Streptomyces* tyrosinase was originally reported to contain one copper atom per mole, but more recent data indicate a binuclear copper active site (Huber and Lerch, 1988).
[c] Percent copper in the enzyme.

gene are produced in melanoma cells (Ruppert *et al.*, 1988), which has potential implications for function and regulation. Also of great interest is the determination, by Ruppert *et al.* (1987; Kwon *et al.*, 1987), that the tyrosinase gene locates to the albino locus in the mouse. Thus, at least in this species, albinism is indeed synonymous with lack of tyrosinase. (It might have been due to defective expression of a trans-acting regulatory protein.) On the other hand, the agouti trait (also involving lighter hair color in the mouse) probably is due to the reduced availability of Cu to tyrosinase, perhaps because of sequestration by sulfhydryl compounds in cells of the hair bulb (Movaghar and Hunt, 1987). This results in a changeover from black to yellow pigmentation.

Tyrosinase synthesis is initiated on free polyribosomes, but at least a major portion of synthesis continues on polyribosomes attached to the endoplasmic reticulum. Within the reticuloendothelial or Golgi channels, carbohydrate (including sialic acid) is then added. From there, the enzyme is packaged into melanosomes (Mishima and Imokawa, 1983). Some of it is also free in the cytoplasm. Isoforms, as well as active and inactive forms of tyrosinase, have been isolated from different parts of various cells or tissues (Tomita *et al.*, 1983; Mishima and Imokawa, 1983). The presence of isoforms can be explained on the basis of mRNA variety (due to differential splicing) as well as by the presence or absence of added carbohydrate. Recent work of Martinez *et al.* (1987) indicates that the soluble and microsomal forms of the enzymes isolated from mouse melanomas were heterogeneous but contained very little sialic acid, as determined by migration is SDS electrophoresis, with and without prior neuraminidase digestion. In contrast, the enzyme in the melanosomes was larger, and considerable carbohydrate was released from it by neuraminidase, which appeared to convert it to the smaller form present in the soluble compartment. [Similar effects of neuraminidase treatment had been reported earlier by Hearing et al. (1978).] In agreement with work of Tomita et al. (1983) and Mishima and Imokawa (1983), Martinez et al. (1987) found that the soluble (cytosol) form of the enzyme had a low activity. Moreover, it could be activated by copper but had a lower affinity for Cu(II) than the form in melanosomes. Martinez et al. (1987) suggested that a lower affinity for copper, and thus a much lower enzyme activity (due to less sialic acid), would protect the soluble portions of cells from tyrosinase, which might otherwise cause toxic reactions. This would also agree with the finding that, in melanin-producing cells, copper concentrates in the melanosomes and is much less prevalent in the cytoplasm (Potts and Au, 1976). Also, when this goes awry, as in Parkinson's disease, accumulation of neuromelanin occurs in dopaminergic neurons, and this correlates with their degeneration (Hirsch et al., 1988). Neuromelanin accumulation can also be induced in astrocytes by norepinephrine, at least in cultured cells (Juurlink, 1988).

The growth and proliferation of melanocytes is dependent on copper availability and depends also on growth factors (basic fibroblast growth factor and insulin) and on a protein kinase C stimulator (such as phorbol ester), as shown for primary cultures of human foreskin melanocytes (Herlyn et al., 1988). Additional stimulation of cells (in serum-free medium) was obtained with α-MSH, FSH, and prostaglandin $F_{2\alpha}$. Further increases in cell proliferation occurred in the presence of added ceruloplasmin (at levels about one-tenth of normal serum concentrations) and $10^{-11} M$ tyrosinase. Ceruloplasmin also induced a change in cell shape. It is quite possible that ceruloplasmin can deliver copper to these cells (on specific cell surface receptors, etc.; see Chapter 4). Whether the tyrosinase that was added was also just a source of extra copper is unknown. In the case of the fungus *Neurospora crassa*, copper appears to stabilize (or activate) tyrosinase, so that there is more activity when extra copper is around (Huber and Lerch, 1987). However, copper does not regulate synthesis of the apoprotein.

Within the melanocyte, various "conversion" factors are active in regulating melanin production (Pawelek et al., 1980; Korner and Pawelek, 1980). The type of melanin produced within melanosomes also may be influenced directly by divalent metal ions, especially Cu(II) and Zn(II) (Palumbo et al., 1988). Figure 6-20 shows the series of reactions that are

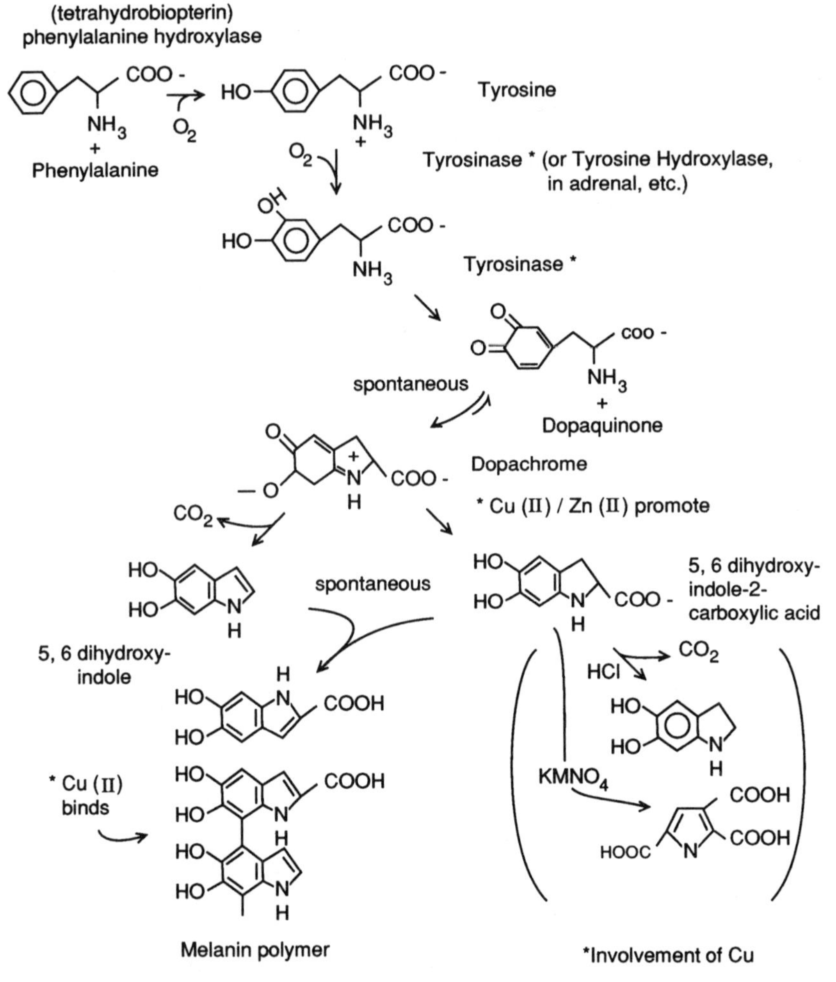

Figure 6-20. Conversion of tyrosine to melanins and the potential involvement of copper in different phases. Starred (*) enzymes or reactions require copper. Based on Palumbo et al. (1987, 1988), D'Amato et al. (1987), and Metzler (1977).

thought to underlie the formation of melanin pigments. Presumably, this normally occurs only in the melanosomes. Tyrosine is converted by tyrosinase to DOPA, by a two-step reaction that at each step involves binding of dioxygen to the binuclear copper cluster. [It may be noted that tyrosine *hydroxylase* (not known to be a copper enzyme) carries out the same conversion (to DOPA) in the first of the series of reactions catalyzed by tyrosinase to form melanins. The enzyme for conversion of phenylalanine to tyrosine (which is iron dependent in mammals and copper dependent in some bacteria) probably is not in melanocytes.] The reaction mechanism proposed for melanin formation is illustrated in Fig. 6-21. (Only the interactions of the phenol moiety are illustrated.)

Returning to Fig. 6-20, the DOPA formed is oxidized to dopaquinone (also with the help of tyrosinase). Dopaquinone then spontaneously cyclizes to form dopachrome, which then rearranges to form two products, 5,6-dihydroxyindole and 5,6-dihydroxyindole-2-carboxylic acid. These indoles are then further oxidized and polymerized to yield various melanins. The process can be carried out *in vitro*, with tyrosine or DOPA as substrates (or even either of the indoles alone) in the presence of tyrosinase (or in melanocytes given these substrates; Iyengar and Misra, 1987). The resulting "Gemish" will differ depending on the starting substrate. A number of intermediate polymerization products have been isolated, and much work suggests that authentic melanins are composed of the two indole moieties (with and without the 2-carboxylate group) (Palumbo *et al.*, 1987; Palumbo *et al.*, 1988). The presence of divalent metal ions may affect the degree to which the carboxylated indole is incorporated (Palumbo *et al.*, 1988). Cu(II) or Zn(II) ([as well as Ni(II) or Co(II)], at concentrations of 0.33 mM, all enhanced the process. Based on the availability of these metal ions *in vivo*, it seems most likely that copper (and perhaps zinc) would be the only ones involved, especially since copper is concentrated in the melanosomes. Whether there is copper in melanosomes other than that bound to tyrosinase is not clear. However, melanins themselves can also bind copper (D'Amato *et al.*, 1987; Potts and Au, 1976).

6.5.2 Phenylalanine Hydroxylase

Phenylalanine 4-monooxygenase (EC 1.14.16.1), already referred to in Fig. 6-20, catalyzes the conversion of phenylalanine to tyrosine. It is best known for its lack of expression in the rather common hereditary condition phenylketonuria (PKU). Lack of the enzyme blocks degradation of its substrate amino acid, resulting in increased blood phenylalanine concentrations that cause damage especially to brain cells during the early part of life. The enzyme is normally found primarily (perhaps even exclusively) in the liver, the organ particularly involved in supplying and degrading amino acids.

It has long been known that phenylalanine hydroxylase has a pterin cofactor, tetrahydrobiopterin (Fig. 6-17). The cosubstrate for the reaction, like that of most copper enzymes, is O_2 (Fig. 6-20). The mammalian

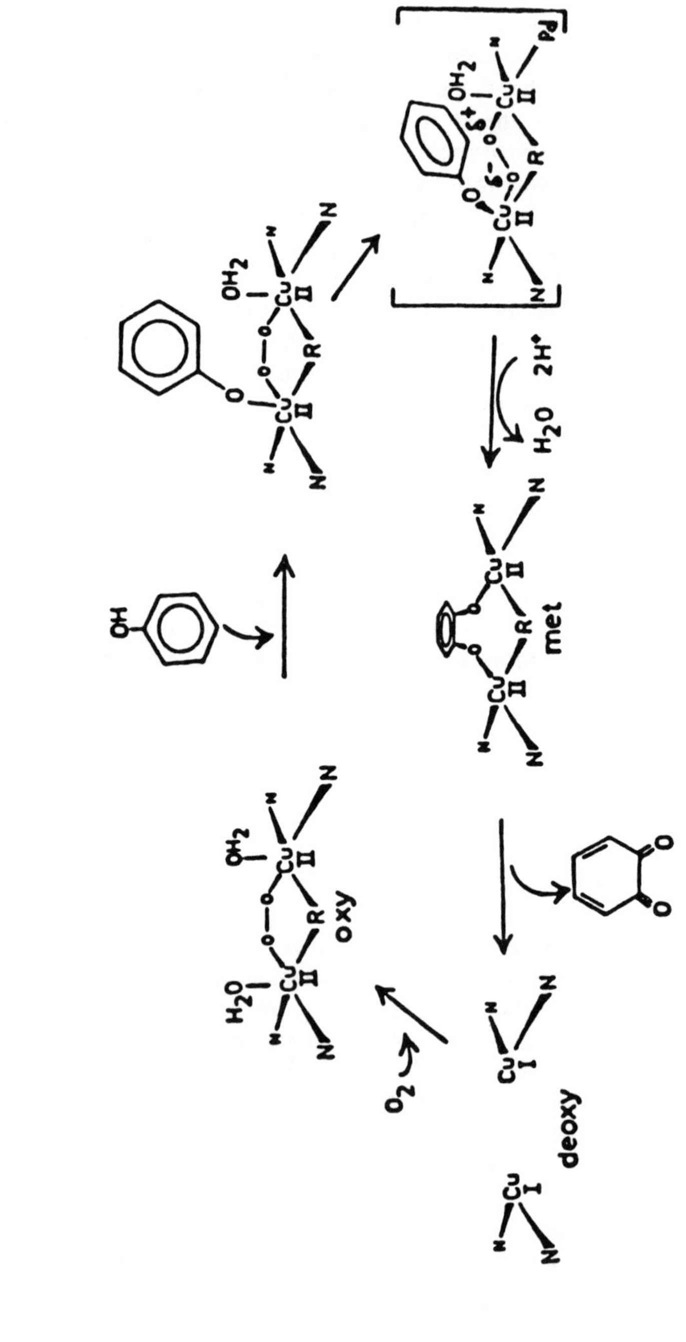

Figure 6-21. Reaction mechanism proposed for tyrosinase. Reprinted, with permission, from Solomon (1988).

enzyme also requires nonheme iron for activity (Fisher et al., 1972; Gottschall et al., 1982), which may funnel the electrons to oxygen. Now it has become apparent that the enzyme in *Chromobacterium violaceum* is copper rather than iron dependent (Pember et al., 1986). It contains stoichiometric amounts of type 2 Cu(II) (see Chapter 4, Section 4.1.1) that are directly linked to the N-5 of the cofactor, tetrahydropterin, next to C-4, where oxygen may attach (Pember et al., 1987). Similar cofactor reaction intermediates are observed for iron and copper versions of the enzyme (Pember et al., 1986). Both Cu(II) and Fe(III) must be reduced for the enzyme to be active, which indicates that the mechanisms by which oxygen is activated may be similar for the copper and iron forms.

6.5.3 Tryptophan Oxygenase

Tryptophan dioxygenase (EC 1.13.11.11), also known as tryptophan "oxygenase" or "pyrrolase" in earlier reports, is the key enzyme regulating bulk degradation of tryptophan to gluco- and ketogenic products and to niacin. Indeed, degradation of excess amino acids and gluconeogenesis are important functions of the liver, where this enzyme is localized. Tryptophan dioxygenase catalyzes the conversion of tryptophan to *N*-formylkynurenine, with the assistance of molecular oxygen. Its activity and concentration are regulated by glucocorticosteroids and by its substrate concentration, through increased transcription/translation and decreased degradation, respectively. As such, it has been one of the most studied enzymes in the elucidation of the mechanisms of mammalian enzyme regulation. Already early on, Tanaka and Knox (1959) reported that ferroheme (but not ferriheme) was a coenzyme, implicating it as the site for binding of the oxygen cosubstrate. This was later confirmed by Ishimura et al. (1970; Ishimura and Hayaishi, 1973), who showed that the oxyferroheme enzyme complex was an obligatory reaction intermediate. However, Brady et al. (1972) reported that pure preparations of the rat liver and *Pseudomonas (acidovorans)* enzymes also contained significant amounts of Cu. Moreover, they suggested that Cu might actually be the oxygen binding site. Copper was detected even after treatment with Chelex-100 (method not specified). Also, diethyldithiocarbamate and bathocuproine sulfonate (chelating agents with high affinities for copper) inhibited enzyme activity, and especially enzyme reactivation by ascorbate, after its oxidation. Reports of negligible copper in the *Pseudomanas* enzyme by Ishimura and Hayaishi (1974) and Poillon et al. (1969) were attributed to technical difficulties with the copper determination. However, some other facts clouded the picture for copper as an essential component. For one, different preparations of the liver and *Pseudomonas* enzymes had highly different contents of the metal (from 0.3 to 2.9 g-atom per mol for the bacterial protein), there being much less variation in heme content (1.2 to 1.8 mol per mol) (Brady *et al.*, 1972) for the same enzyme preparations. Moreover, the specific activity of the enzyme did not correlate with its copper content, and Ishimura and Hayaishi (1973) reported that addition of copper to the liver enzyme did not change its activity.

Later, Ishimura's group examined this question again for the pure rat liver enzyme (Makino and Ishimura, 1976) and showed that the turnover number of the enzyme (per enzyme-bound heme) was unchanged during the last stages of purification, while the protein-bound copper varied threefold, in seven separate purifications. Moreover, the final copper content of the enzyme was negligible, especially when determined by addition of pure tryptophan dioxygenase to known copper proteins (Cu/Zn SOD or laccase) prior to analysis. Again, the specific activity of the enzyme was not increased by adding copper during the purification, nor did dialysis against EDTA decrease it. From these various observations, it thus seems unlikely that a significant amount of copper is attached to this tryptophan-degrading enzyme. If it is, it is probably bound fairly loosely and is not necessary for enzymatic function.

6.5.4 Diamine Oxidases

In contrast to tryptophan oxygenase, most amine oxidases are truly copper dependent, as already described for serum amine oxidase (see Chapter 4) and lysyl oxidase (Section 6.4.1), which catalyzes the oxidative deamination of lysine to allow the cross-linking of elastin and collagen. There are, however, also flavin-dependent amine oxidases, such as monoamine oxidase, involved in the inactivation of catecholamines, and copper-dependent diamine oxidases have been isolated from (or detected in) intestinal (and gastric) mucosa, kidney, spleen, liver, placenta, granulocytes, and semen (as well as lymph, plasma, amniotic fluid, and urine). [For reviews, see Bombardieri *et al.* (1985) and Argentu-Ceru and Autuori (1985).] In general, diamine oxidases (DAO) are thought to function in the inactivation of histamine and polyamines. Polyamines are involved in cell proliferation, their concentrations being especially high in tissues such as regenerating liver or intestinal mucosa, where cells are dividing quite rapidly. It is thought that diamine oxidases may play a role inhibitory or limiting to excessive growth, by promoting polyamine catabolism (Maslinski et al., 1985). In this regard, it is of interest that normal (nonmalignant) hyperplastic tissues are rich both in diamine oxidase and in ornithine decarboxylase, the rate-limiting enzyme for polyamine production (Argentu-Ceru and Autuori, 1985). Parallel increases in both enzymes are observed during compensatory hypertrophy of the kidney (Perin et al., 1983) and heart (Desiderio et al., 1982) and in regeneration of the liver (Sessa et al., 1982) and skin. The other function of diamine oxidases is in inactivation of histamine, especially in the small intestine (where it enters from the stomach to stimulate acid secretion), and in allergic reactions all over the body, where histamine is released from mast cells (causing changes in capillary permeability, etc.).

The highest DAO activities are found in the small intestine, especially the ileum (Argentu-Ceru and Autuori, 1985). Activity is higher in omnivores and carnivores than in herbivores, perhaps to inactivate not just stomach histamine, but also dietary histamine and polyamines. Intestinal mucosa is also the major source of serum DAO, and (in the absence of pregnancy) the level of DAO in serum is considered indicative of the integrity of the intestinal mucous membrane (Bombardieri et al., 1985). Considerable DAO activity is also found in the kidney cortex of several species, including the human. Kidney DAO is especially rich in the proximal convoluted tubules, where diamines filtered by the blood would come in contact with it and be inactivated. Liver activity is generally low, and there is controversy about the exact levels in many other organs. However, the maternal portion of the placenta also has very high activity and is the main source of the increased plasma DAO in pregnancy. Placenta probably functions to inactivate the large quantities of amines produced by the developing fetus (Maskinski et al., 1985; Argentu-Ceru and Autuori, 1985).

The porcine kidney enzyme has been purified (see Mondovi and Riccio, 1985). It has a molecular weight of about 172,000, two Cu atoms (separately detected by EPR), and a pyridoxine-like cofactor that again may either be 6-OH dopa or pyrroloquinoline quinone (as in the case of serum amine oxidase and lysyl oxidase) (Fig. 6-17) (see earlier, under lysyl oxidase). It has a visible absorbance maximum at 480 nm, giving it a pinkish color. Human placental DAO is a glycoprotein that may be active as an 80-kDa monomer. Other forms have not as yet been purified.

6.6 Summary

Critical aspects of the structure and activity of the most established intracellular copper dependent enzymes are summarized in Table 6-17.

Table 6-17. Intracellular Copper Enzymes in Vertebrates, Their Cofactors, Substrates, and Functions

	Functions	Cosubstrate(s)	Copper basic unit	Cofactors
Cytochrome oxidase	Respiration, e-transport, energy metabolism	O_2, cytochrome c	Cu_A, Cu_B	Heme a Heme a_3
Superoxide dismutase	Superoxide dismutation (to H_2O_2)	O_2^-	Cu(II)/Zn(II)	—
Dopamine β-hydroxylase	Catecholamine biosynthesis	O_2, ascorbate	Cu (2)	Pyrroloquinoline quinone?[b]
Tyrosinase	Melanin biosynthesis	O_2	Cu_A, Cu_B	—
α-Amidating enzyme	Peptide hormone amidation	O_2	Cu	Pyrroloquinoline quinone?[b]
Lysyl oxidase[a]	Elastin and collagen cross-linking	O_2	Cu(II)	Pyrroloquinoline quinone?[b]
Diamine oxidase[a]	Histamine and polyamine inactivation	O_2	Cu (2)	Pyrroloquinoline quinone?[b]

[a] Mainly or partly extracellular.
[b] This will need to be re-evaluated, in light of recent findings that bovine serum amine oxidase actually has 6-OH dopa as a cofactor (Janes *et al.*, 1990).

Copper and Metabolic Regulation 7

7.1 Regulation of Copper Metabolism by Hormones

Copper is of fundamental importance to oxygen utilization and energy metabolism, to growth, in the defense of the organism against oxidation, and in other aspects of normal and defensive body function. The flow of this element and the concentration of specific copper substituents must therefore be regulated according to need. For this purpose, hormones, cytokines, and monokines are released, in response to normal and abnormal conditions, from pregnancy and stress to inflammation and cancer. In relation to these events, ceruloplasmin, the principal copper carrier in the blood plasma, and an acute phase reactant, has been the most studied parameter. Indeed, changes in ceruloplasmin are probably an excellent measure of changes in copper flow among body compartments, as ceruloplasmin is an important source of copper to cells and serves as a scavenger of superoxide for protection against oxidative processes (also involved in infection, inflammation, or tissue damage). These functions, of ceruloplasmin are discussed at length in Chapters 3 and 4. Ceruloplasmin is also very sensitive to dietary status (copper need), of which more will be said in the second half of this chapter, and it may play an important role in copper homeostasis and excretion (Chapter 5). Total serum copper concentrations change in the same direction as ceruloplasmin, because the latter normally accounts for 60% or more of plasma copper (Table 4-12). Other copper components of the plasma have not been as well studied, partly because their significance in terms of quantity and variety has only recently become apparent. Thus, much remains to be learned about changes that may occur in other copper-binding plasma fractions, their regulation, and their significance. With our current state of knowledge, only changes in ceruloplasmin may be adequately described.

7.1.1 Regulation of Plasma Cu and Ceruloplasmin

7.1.1.1 Production and Secretion

Ceruloplasmin in blood plasma is synthesized primarily by the liver (see Chapter 4), although it is formed (and presumably secreted) also in other parts of the body that have a "separate" circulation (like the brain). As demonstrated recently for cultured cells, especially by Parent and Olden and colleagues, ceruloplasmin is synthetized on endoplasmic reticulum (ER)-bound hepatocyte (or Hep G-2) polyribosomes (Newton et al., 1986; Aleinikova et al., 1987). From the channels of the rough ER, it travels into the Golgi apparatus and eventually out of the cells, via exocytosis (Parent et al., 1985). This is as would be expected for secreted proteins. The same investigators carefully documented for Hep G-2 cells (a human hepatoma cell line) that groups of plasma proteins were secreted at three discrete rates. Those secreted most rapidly appeared with a pulse-chase half-time of 30–40 min and included albumin, fibronectin, and α-fetoprotein. Those with an intermediate half-time (75–80 min) included ceruloplasmin and α_2-macroglobulin, and those secreted most slowly (half-time 110–120 min) included transferrin and fibrinogen (Parent et al., 1985; Yeo et al., 1985). They also found that it took 30 and 45 min, respectively, for ceruloplasmin and transferrin to leave the rough ER (Fig. 7-1A) and 30 and 65 min, respectively, for half of each to reach the medial Golgi. In terms of time, that puts ceruloplasmin right in the middle of normal plasma protein production and secretion, and ahead of transferrin, although we cannot be certain that this grouping holds exactly for hepatocytes, per se.

It appears that glycosylation may be a point of regulation for ceruloplasmin secretion and for most other plasma glycoproteins produced by the liver. Indeed, at least some glycosylation is necessary for ceruloplasmin secretion to occur [and the form of glycosylation may vary to some extent under the influence of monokines (see Section 7.2.1.2)]. Inhibitors of glycosylation, such as tunicamycin (Bauer et al., 1985) and 1-deoxynajirimycin (DNJ) inhibit transfer of ceruloplasmin from the RER to the Golgi (Parent et al., 1986) and thus inhibit its secretion from the cell (Fig. 7-1B). DNJ especially inhibits glucosidases I and II, which remove terminal glucose units from the carbohydrate moiety being built in ER channels. Swainsonine, an inhibitor of mannosidase II in the Golgi, did not have the same effect and, indeed, tended to stimulate secretion of newly synthetized [^{35}S] ceruloplasmin and other plasma proteins (except albumin) into the culture medium in these studies (Yeo et al., 1985). It is noteworthy that transferrin secretion was not affected by tunicamycin and DNJ, although it has somewhat similar carbohydrate units (see Chapter 4).

Parent et al. (1985) were also able to follow the conversion over time of ceruloplasmin precursor (with little or no carbohydrate) to the larger product (after carbohydrate addition) in SDS gel electrophoresis of cell lysates (Fig. 7-1C). Studies done with primary cultures of rat hepatocytes by Aleinikova et al. (1987) indicate that the appearance of radioactive ceruloplasmin in the medium (after exposure of cells to radioactive amino acid) takes about the same length of time as in the case of the Hep G-2 cells (30–40 min). Colchicine, another inhibitor of exocytosis (at low concentrations), inhibited the process. These researchers also estimated that the time required for translation of ceruloplasmin mRNA is about 3.5 min, in line with that for other plasma proteins.

7.1.1.2 Effect of Inflammation

Ceruloplasmin mRNA appears to be 20–28 S in different species (Gitlin, 1988a). It is 4.2 and 3.7 kb in length in humans and 3.7 kb in hamsters, rats, and mice. Ceruloplasmin mRNA concentrations have been shown to increase in inflammation, and regulation of message in other conditions is under current investigation. Gitlin (1988a) has shown in hamsters that transcription of ceruloplasmin mRNA increases within 3 h of induction of inflammation by turpentine injection, reaching a peak 2.5-fold above normal at 12–18 h (Fig. 7-2A). Concentrations of message (Fig. 7-2B) and serum ceruloplasmin (Fig. 7-2C) peak at successively later times. It is noteworthy that the increase in ceruloplasmin mRNA is almost 4-fold, which implies that increased transcription and decreased degradation of message must both be occurring. (The increase in transcription rate is only 2.5-fold.) These changes in ceruloplasmin mRNA in inflammation may be mediated by interleukin-1 (IL-1). Turpentine inflammation results in release of this (and other) monokines, and Mackiewicz et al. (1987 Mackiewicz, 1989) have shown that IL-1, IL-6, and tumor necrosis factor (TNF, or cachectin) cause increased accumulation of ceruloplasmin in the medium of cultured hepatic cells. They also cause increases in serum ceruloplasmin in rabbits. (Maximum levels were reached by 72 h under the conditions examined.) In humans, prostaglandin E_1 infusion over 72 h has been shown to cause a later rise in serum ceruloplasmin concentrations (Whicher et al., 1984), and IL-1 promotes synthesis and secretion of PGE. IL-1 is released from mononuclear phagocytes in response to a variety of stimuli related to infection and injury (Fig. 7-3). TNF is a different monokine released from the same kinds of cells and is named for its ability to induce hemorrhagic tumor necrosis (Beutler et al., 1985) and cachexia. It also shares many of the effects of IL-1 (Fig. 7-3), including stimulation of the synthesis of IL-2 and PGE_1 (Dayer et al., 1985; Perlmutter et al., 1986). IL-1 and TNF both

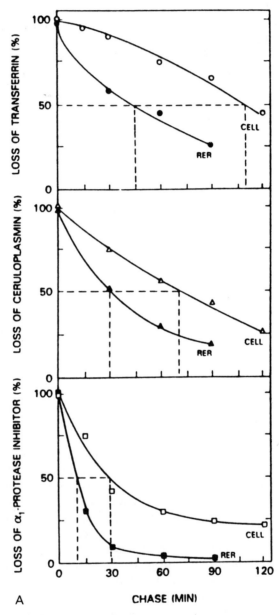

Figure 7-1. Rates of transfer of ceruloplasmin and transferrin from the rough endoplasmic reticulum (RER) to the point of secretion by Hep G-2 cells. (A) Transport kinetics, using

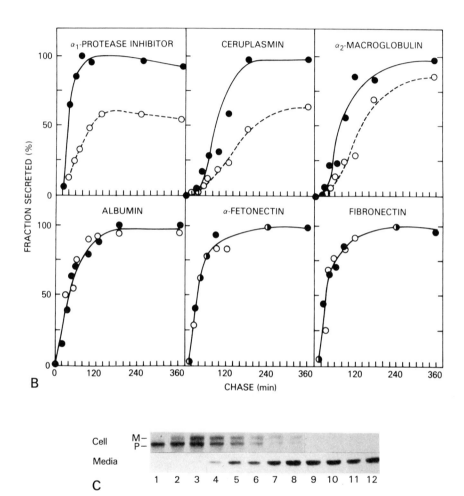

Figure 7-1. *(Continued)* a pulse/chase of [^{35}S]methionine, showing loss of the radioactive proteins from the RER and the whole cell; loss of the α_1-protease inhibitor is also shown. Reprinted, with permission, from Yeo *et al.* (1985). (B) Effect of inhibitor of RER glucosidase I and II on secretion of ceruloplasmin and other plasma proteins. Reprinted, with permission, from Yeo *et al.* (1985). (C) Precursor (P) and mature (M) forms of ceruloplasmin in whole cells separated by SDS electrophoresis at various times in the pulse/chase. Autoradiograph. Reprinted, with permission, from Parent *et al.* (1985).

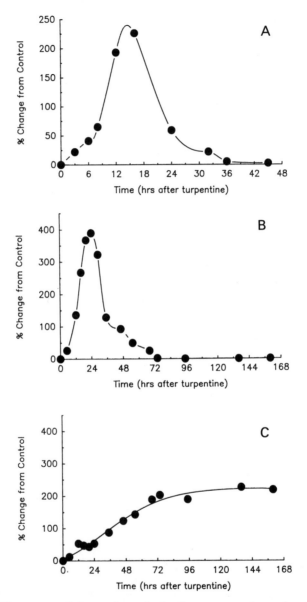

Figure 7-2. Time course of changes in ceruloplasmin gene expression and synthesis during turpentine inflammation, in the hamster. (A) Change in liver mRNA transcription, determined by *in vitro* run-on assays. (B) Change in concentration of liver total mRNA for ceruloplasmin, determined by densitometry of autoradiographs with hybridized [^{32}P]cDNA. (C) Change in ceruloplasmin oxidase activity of serum, determined with *o*-dianisidine. Reprinted, with permission, from Gitlin (1988a).

Copper and Metabolic Regulation

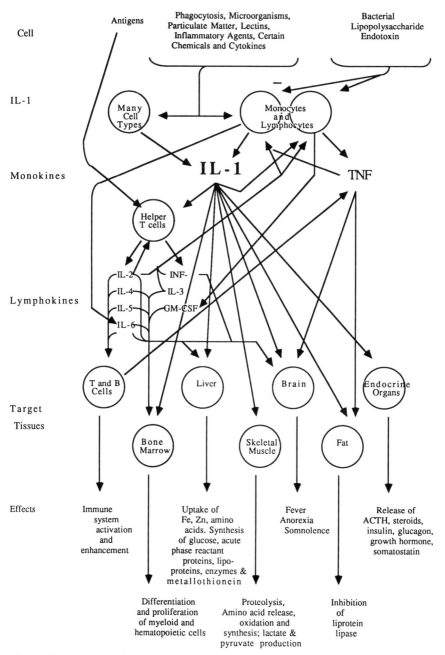

Figure 7-3. Overview of the regulation and effects of interleukin-1 (IL-1) and other lymphokines. Reprinted, with permission, from Beisel (1991).

stimulate the synthesis of many acute phase glycoproteins in hepatocytes, and this effect is greatly potentiated by corticosteroids (Baumann et al., 1987), the secretion of which is also enhanced by IL-1 via adrenocorticotropin (ACTH) (Berkenbosch et al., 1987; Bernton et al., 1987). [Pos et al. (1988) have shown that corticosteroids also enhance secretion of ceruloplasmin from hepatocytes of inflamed rats in culture.] Still another cytokine, interleukin-6 (IL-6), may also be involved, and its effect on ceruloplasmin is being examined.

Some acute phase monokines (but not IL-1 or TNF) also influence the glycosylation state of ceruloplasmin (Mackiewicz et al., 1987). This is now being studied by two-dimensional immunoelectrophoresis with (and without) lectins that bind and cause separation of different forms of ceruloplasmin in human serum. The results from this kind of a study are given in Table 7-1. Two forms of ceruloplasmin (Y and Z) were detected in human serum and appeared to differ in copper content based on the effects of dialysis in EDTA, followed by Cu(II), at pH 5. (The Y form increased upon EDTA dialysis and decreased upon Cu addition.) Further forms were separated in the presence of two kinds of lectins. It is noteworthy that inflammation decreased the proportion of "low-Cu" ceruloplasmin and increased one type of lectin binding fraction (peak 3) while decreasing another (peak 1). The presence of malignancy increased the proportion of "low-Cu" ceruloplasmin, while leaving the lectin binding proportional to that in healthy individuals. This suggests that the effects of tumors and inflammation on ceruloplasmin are not identical. Both increase synthesis of

Table 7-1. Heterogeneity of Ceruloplasmin in Normal and Disease States[a]

Parameter	Normal (30)	Patients with testicular cancer (24)	Patients with chronic colorectal inflammation (28)
Serum ceruloplasmin concentration (μg/liter)	338 ± 58	375 ± 97	398 ± 122
Percent low-Cu form (Y) (% of total)	23 ± 11	41 ± 9^b	7 ± 4^b
Percent binding LCA	33 ± 11	39 ± 10	36 ± 9
LCA electrophoresis forms (%)			
1	16 ± 9	24 ± 5^b	4 ± 2^b
2	58 ± 5	54 ± 4^b	63 ± 7^b
3	24 ± 8	21 ± 5	32 ± 6^b
4	2 ± 1	1 ± 1	1 ± 2

[a] Data of Hansen et al. (1988). Serum samples were subjected to cross affino-immunoelectrophoresis in the presence and absence of Lens culinaris agglutinin (LCA).
[b] Significant differences from the normal ($p < 0.01$).

ceruloplasmin polypeptide, but the inflammatory process also influences carbohydrate addition.

The observation that inflammation results in a higher proportion of copper-saturated ceruloplasmin would be consistent with the concept that the added ceruloplasmin is required for antioxidative protection in the inflamed area. The opposite being the case in cancer suggests that here ceruloplasmin is used more for copper delivery, perhaps to promote tumor growth. Whether the form of carbohydrate influences the capacity of ceruloplasmin to deliver copper to cells needs to be examined. (Altered glycosylation might, for example, allow ceruloplasmin to retain more copper and perhaps therefore be more effective in protection against oxygen radicals.) Inflammation (probably in most mammals) is accompanied by a precipitous drop in serum iron (on transferrin), an increased release of serum ferritin (which has a very low iron content; see Campbell et al., 1989a), and, as already mentioned, a fairly rapid increase in a number of serum proteins, including ceruloplasmin (Fig. 7-4; Table 7-2) (Linder, 1991a; Beaumier et al., 1984; Schreiber et al., 1978; Giclas et al., 1985; Pos et al., 1988). In rodents at least, but not in humans, transferrin concentrations also eventually rise, though more slowly than in the case of some other plasma proteins (but not albumin) (Beaumier et al., 1984; Mackiewicz et al., 1988; Thomas and Schreiber, 1989). Systemic infections with various bacteria produce the same syndrome of responses, again thought to be mediated by various monokines.

7.1.1.3 Steroid Hormones

Apart from monokines, several other hormones affect ceruloplasmin (and total copper) concentrations in the plasma (Table 7-2). The potential involvement of adrenocortical steroids has already been mentioned in connection with the response of hepatocytes to inflammation (which they potentiate). Administration of IL-1 has been shown to stimulate increased release of ACTH, which would increase release of cortisol or corticosterone from the adrenal. It would appear that this is not just a nonspecific reaction of experimental animals to the trauma of infection, in that IL-1 appears to directly enhance secretion of ACTH-releasing factor from cultured pituitary cells (Bernton et al., 1987). It is well know that low levels of corticosteroids are permissive agents in the actions of many hormones. This is not confined to a particular physiological state and may involve a general promotional effect on ribosomal RNA production. It is also well known that *high* concentrations of glucocorticoids (or potent man-made versions of these steroids, such as dexamethasone) have *anti*-inflammatory effects, involving decreased RNA and protein synthesis in lymphoid cells

Table 7-2. Effects of Hormones and Other Factors on Serum Copper and Ceruloplasmin

Factor or hormone	Species	Effect on:			References
		Serum Cu	Ceruloplasmin Activity	Ceruloplasmin Antigen	
Estrogens	Human (women or men)	↑	↑		Briggs et al. (1970)
		↑			Russ and Raymunt (1956)
		↑	↑	↑	von Ebeling (1975)
					Hambidge and Droegemueller (1974)
					Kaar et al. (1984)
				↑	Liukko et al. (1988)
	Rooster		↑		Planas and Frieden (1973)
	Rat	↑	↑		Sunderman et al. (1971)
		↑	↑		Sato and Henkin (1973)
			↑		Yunice and Lindeman (1975)
			↑		Karp et al. (1986)
Ovariectomy	Rat (adult)		No change (1 week)		Karp et al. (1986)
			No change (5 weeks)		Sato and Henkin (1973)
Progestogens	Human		↑		Ruokonen and Kaar (1985)
			↑		Kaar et al. (1984)
			↑		Sato and Henkin (1973)
	Rat		Little effect		Lei et al. (1976)

Copper and Metabolic Regulation

Androgens		No effect or slight	Barbosa et al. (1971)
			Sunderman et al. (1971)
Oophorectomy	Rat	No effect	Evans et al. (1970a)
Glucocorticoids	Rat	No effect	Evans et al. (1970a)
Betamethasone		←	Freeman et al. (1973)
Dexamethasone		←	Pos et al. (1988)
	Rat	Slight decrease	Weiner and Cousins (1983)
	Rabbit	←	Banaszkiewicz and Chodera (1962)
		→	Rowan and Meacham (1971)
Adrenalectomy	Rat	←	Evans and Wiederanders (1968)
		←	Evans et al. (1970a)
	Mice	No change	Schreiber et al. (1978)
		No change	Prohaska et al. (1988)
ACTH	Rabbit	←	Alias (1971)
	Chick	←	Rowan and Meacham (1971)
		←	Starcher and Hill (1965)
			Freeman et al. (1973)
	Rat	No effect	Evans and Wiederanders (1968)
Hypophysectomy	Rat	←	Evans et al. (1970a)
Human chorionogonadotropin	Rat	No effect	Sato and Henkin (1973)
Thyroidectomy + T$_4$	Rat	Returns to normal	Evans et al. (1970a)
Hyperparathyroidism	Human	Slight increase	Mallette and Henkin (1976)

(*continued*)

Table 7-2. *(Continued)*

			Effect on:		
				Ceruloplasmin	
Factor or hormone	Species	Serum Cu	Activity	Antigen	References
Inflammation (turpentine or other adjuvant)	Rat		↑		Schreiber et al. (1978)
		↑	↑		Cousins and Swerdel (1985)
					Pos et al. (1988)
					Giclas et al. (1985)
Infection	Rabbit		↑		Starcher and Hill (1965)
	Chick		↑		
	Mouse			↑	Beaumier et al. (1984)
IL-1	Human hepatoma cells			↑	Mackiewicz (1989)
	Rabbit, cultured cells			↑	Mackiewicz et al. (1988)
	Rabbit	↑			Morimoto et al. (1989)
IL-6	Human hepatoma cells			↑	Mackiewicz (1989)
Tumor necrosis factor (TNF)	Rabbit, cultured cells			↑	Mackiewicz et al. (1988)
	Human hepatoma cells			↑	Mackiewicz (1989)
	Rabbit	↑			Morimoto et al. (1989)
PGE_1	Human			↑ (Slight)	Whicher et al. (1984)
Epinephrine	Rat hepatocytes			↑ (Incorporation of ^{64}Cu)	Weiner and Cousins (1983)

Condition	Species	Effect	Reference
Exposure to 85% O_2	Rat	↑	Hart et al. (1989)
Exposure to O_3	Rat	↑	Ikemi and Ohmuri (1990)
High-thiamine diet (600 mg/kg)	Rat (on low-Cu diet)	↑	Ellerson and Hilker (1985)
Oxythiamine diet	Rat	↑	Schreiber et al. (1978)
6-Aminonicotinamide diet	Rat	Slight increase	Schreiber et al. (1978)
Xenobiotics in diet	Rat	↑	Ohchi et al. (1987)
Retinoic acid	Rat	↑	Cousins and Swerdel (1985)
High-ascorbate diet	Human	No change	Jacob et al. (1987)
Obesity	Mouse	↑	Prohaska et al. (1988)
	Rat	↑	Serfass et al. (1988)
Pregnancy	Many	↑	See Chapter 8
Copper deficiency	Rat	↓	Holtzman and Gaumnitz (1970a)
		Less	Linder et al. (1979b)
		↓	See Table 7-8
Round spermatid proteins	Sertoli cells	↑	Onoda and Djakie (1990)
$AgNO_3$ injection	Mouse	↓	Sugawara and Sugawara (1984)
Cu(II) injection	Various	±	See text
Smoking cigarettes	Human	↓	Pacht and Davis (1988)

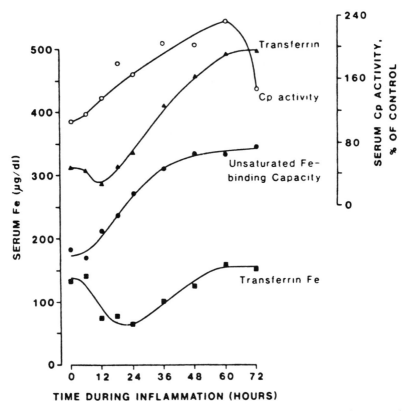

Figure 7-4. Changes in iron, transferrin, and ceruloplasmin in sera of mice during turpentine inflammation. Reprinted, with permission, from Beaumier *et al.* (1984).

and tissues. The potential direct involvement of glucocorticoids in enhanced production of ceruloplasmin is therefore a bit of a puzzle. The reported evidence on this is summarized in Tables 7-2 and 7-3 and is not without contradictions. Adrenalectomy, which would reduce circulating levels of corticosteroids and increase those of ACTH, appeared to raise plasma ceruloplasmin levels in rats but have no effect in mice. This may reflect a difference between these species; the mouse has much lover levels of ceruloplasmin oxidase activity and serum copper. (The alternative is that some of the data are wrong.) ACTH itself (which would increase release of glucocorticosteroids) was found to increase ceruloplasmin oxidase activity in rabbits and chicks but have no effect in rats. Glucocorticoids themselves have been reported by three different groups to increase cedruloplasmin activity and concentration (determined immunologically) in rats or rat

Table 7-3. Effects of Various Endocrine Functions on Liver Cu and Serum Ceruloplasmin in Rats[a]

Treatment	Liver Cu concentration (μg/g)	Plasma Cu[b] (μg/ml)	Plasma ceruloplasmin oxidase activity (relative units/100 ml)
None	4.1–4.3	1.25	100
Hypophysectomy	6.0[c]	1.64[c]	162[c]
Adrenalectomy	5.1[c]		185[c,d]
Adrenalectomy + corticosterone (daily)	3.7		100
Thyroidectomy + T$_4$ (daily)	4.4		118
T$_4$ treatment (no thydx)	—		65[c]
Oophorectomy	4.1		100
Estriadol (daily)	3.8		180[c]
Ovariectomy	—	0.88[c] (30% decrease)	55[b,c]

[a] Data are mean value values for groups of 6–12 3–4-month-old rats (sex not stated). From Evans et al. (1970a).
[b] Sato and Henkin (1973).
[c] Significant difference from controls ($p < 0.01$).
[d] Cpm of ^{64}Cu in bile 4 h after i.v. injection was 40% of that in control rats.

hepatocyte cultures. Here the data of Evans' group (while consistent in themselves; Table 7-3) are again at odds. [They suggest that in fact glucocorticoids depress plasma ceruloplasmin concentrations (if anything), perhaps by influencing the rate of bile production (see Chapter 5).] Whether the negative data of Evans et al. (1970a) were due to use of a less potent steroid or some other matter is not known. The same question applies to the two other negative reports (also from quite long ago) that concerned the rabbit (Table 7-2). The most recent studies (Pos et al., 1988) in rats would suggest that dexamethasone alone does indeed have a dramatic elevating effect on serum levels, not just of ceruloplasmin but also of haptoglobin and α_1-acid glycoprotein. Levels were 1.7–7.8-fold above normal by 24 h, but there was no effect on α_1-antitrypsin (a similar response to that seen 24 h after turpentine injection). Thus, the dexamethasone-induced elevation is not specific for ceruloplasmin, and it may only occur with high doses of such a potent (anti-inflammatory) corticosteroid. This will need to be examined further.

For steroids in general, the most studied response of serum copper and ceruloplasmin has been to estrogens and progestogens, the actions of which underlie what occurs in pregnancy and gestation as well as in women taking oral contraceptives. Estrogens, especially potent ones like diethylstilbesterol (DES), have dramatic effects on the levels of circulating ceruloplasmin, as measured by ceruloplasmin oxidase activity (Table 7-4) and also immunologically (Fig. 7-5). [Recent data suggest that the concentration of copper in some tissues is also changed (at least in rats) (Mehta and Fikum, 1989), notably a decrease in the liver and an increase in the brain.] The course of change that occurs in pregnancy is documented in Chapter 8 (Fig. 8-1). The same response to estrogen treatment occurs both in men and women and takes days and weeks of treatment to achieve, which suggests that it may not be a direct effect on the apparatus responsible for synthesis of ceruloplasmin. As already described in Chapter 5, the effect of estrogen is not mediated by the same mechanism as that occurring in cancer. The effects of both factors together are exactly additive (Table 7-4). Moreover, the turnover of whole body copper is decreased by estrogen treatment (Table 5-8), suggesting that ceruloplasmin turnover (degradation) may also be decreased. A reduction of the rate of turnover would, of course, explain a rise in plasma ceruloplasmin. These possibilities are being examined.

The rise in plasma copper and ceruloplasmin with estrogen treatment (as with most oral contraceptives) (Fig. 7-5) begins rapidly and then gradually achieves a new high plateau (Fig. 7-5A). It may, however, con-

Table 7-4. Effects of Estrogen Treatment and Tumor Implantation on Ceruloplasmin Oxidase Activity and Total Copper Concentration of Rat Plasma[a]

	Days after implant	Ceruloplasmin oxidase activity (10^5 IU/ml plasma) (mean ± SD)			
		Control	+ Tumor	+ Estrogen	+ Tumor + Estrogen
Normal female rats	7	25.4 ± 3.2	24.7 ± 2.2	49.2 ± 6.9[b]	46.9 ± 6.9[b]
	14		70.6 ± 1.1[b]	71.4 ± 14.2[b]	128 ± 23[b]
Ovariectomized rats	7	23.3 ± 6.3	18.2 ± 1.7	40.8 ± 5.5[b]	38.1 ± 6.0[b]
	14		48.3 ± 8.6[b]	92.1 ± 25.8[b]	109 ± 21[b]
		Total plasma copper concentration (µg/ml) (mean ± SD)			
Normal female rats	7	1.96 ± 0.25	1.73 ± 0.29	2.72 ± 0.40[b]	2.77 ± 0.24[b]
	14		3.79 ± 0.64[b]	3.40 ± 0.64[b]	4.79 ± 0.43[b]
Ovariectomized rats	7	1.57 ± 0.34[c]	1.55 ± 0.20	2.15 ± 0.31[b]	2.39 ± 0.19[b]
	14		3.19 ± 0.34[b]	4.17 ± 0.85[b]	5.40 ± 0.71[b]

[a] Data of Kamp et al. (1986). Tumor weights were 0.1–0.5 g 7 days post implantation and 6–17 g 14 days post implantation; no significant differences in tumor growth between groups were evident, based on t tests of values for tumor weight. Reprinted with permission of the publisher.
[b] $p < 0.001$ for difference from control; 5–6 rats per group.
[c] $p < 0.05$ for difference from nonovariectomized.

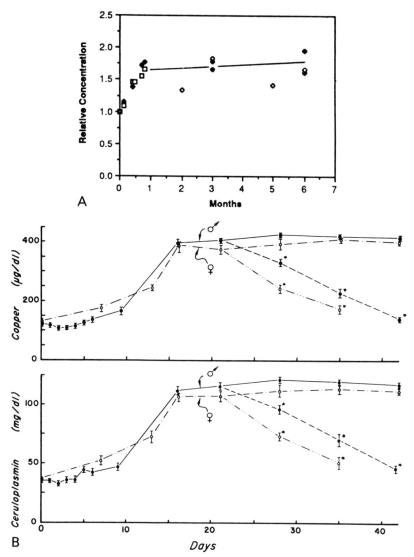

Figure 7-5. Changes in serum copper and ceruloplasmin during treatment with estrogens, including oral contraceptives. (A) Changes in serum copper ($\diamond$), ceruloplasmin oxidase activity ($\blacklozenge$), and ceruloplasmin antigen ($\square$) (measured by radial immunodiffusion), at various times after the start of oral contraceptive use by women. Data are relative to control (zero time) for various estrogen and progesterone combinations. Data from von Ebeling (1975) for ceruloplasmin antigen and oxidase activity are for orthinylestradiol; those of Rubinfeld et al. (1979) for total serum copper are for Orthonovin, Gynovlar, or Neogynon; those of Liukko et al. (1988) for ceruloplasmin oxidase activity are for orthinylestradiol plus levonorgestrel or desogestrel. (B) Effect of 17β-estradiol on total serum Cu and ceruloplasmin oxidase activities of serum of male and female rats. For two groups of rats (data points marked with asterisks), treatments ceased at day 21. Reprinted, with permission, from Sunderman et al. (1971).

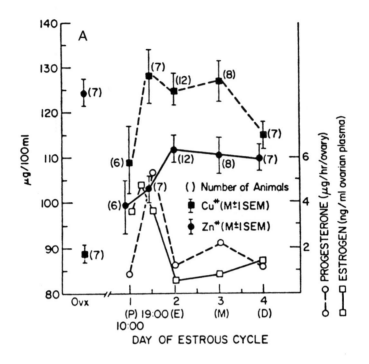
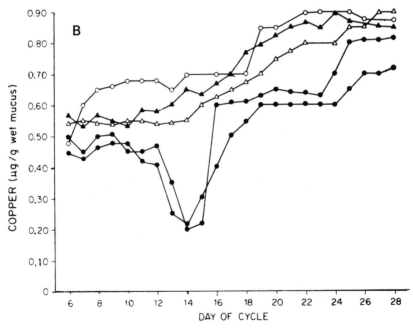

tinue to rise more slowly, perhaps indefinitely, even over years (Rubinfeld et al., 1979). This means that women on oral contraceptives may have vastly greater circulating levels of ceruloplasmin than is otherwise normal, for long periods of time. [The rise occurs equally in males and females (Fig. 7-5).] The increase achieved is sometimes beyond that occurring in the third trimester of pregnancy (Fig. 8-1), where concentrations rise about 100%, overall. So far, this has only been of theoretical concern, as no direct link has yet been established between this effect and any increased risk of malignancy (or stroke) that may sometimes accompany oral contraceptive use. (However, increased ceruloplasmin should serve as an increased, ready source of copper for new cell growth.)

It is noteworthy that effective contraception may, in fact, be mediated by the high concentrations of ceruloplasmin that are evoked by oral contraceptive steroids. It is well known that copper intrauterine devices (IUDs) are more projective against pregnancy than regular IUDs. Such IUDs release copper within the uterus (Salaverry et al., 1973; Singh, 1975), at a rate of about 50 μg a day. Copper can be spermicidal (Jecht and Bernstein, 1973; Zielski et al., 1974) and inhibit sperm motility (Kesseru and Leon, 1974), which may account for the special effectiveness of these IUDs (Hagenfeldt, 1972a,b). Ionic copper (at 10^{-6}–$10^{-5}\,M$) can also stimulate uterine contractions (Salgo and Oster, 1974). As a result, many fewer sperm arrive or stay in the uterus after vaginal insemination in women with copper IUDs (Tredway et al., 1975). [The motility of sperm cells, however, also positively correlates with the copper content of their midpiece, where mitochondria are concentrated (Battersby and Chandler, 1977).]

Copper may also alter the metabolism of the uterine endometrium to prevent implantation of the fertilized ovum. In this regard, investigators have shown that there is some variation in *plasma* copper concentrations in the menstrual/estrous cycle (at least in rats) (Fig. 7-6A) that can be attributed to changes in the secretion of estrogens and progesterone. Hints of this are also present in the early data of Studnitz and Berezin (1958) for women. The copper content of cervical mucus (in humans) also varies with the menstrual cycle (Fig. 7-6B). [This is not the case for the epithelium of

←

Figure 7-6. Changes in copper parameters associated with the menstrual/estrous cycle and the use of oral contraceptives. (A) Changes in serum copper (■) and zinc (●) in relation to estrogen (□) and progesterone (○) in the estrous cycle in rats. Reprinted, with permission, from Sato and Henkin (1973). (B) Changes in the copper content of cervical mucus during the menstrual cycle, in women taking (upper three curves) and not taking (lower two curves) oral contraceptive steroids (Ovral, Ovulin-21, Demulen). Reprinted, by permission, from Singh (1975).

the Fallopian tube (Patek and Hagenfeldt, 1974).] Of great potential interest is the observation that concentrations of copper in *cervical mucus* are dramatically lower just at the time when ovulation, fertilization, and implantation would be most likely. This suggests that concentrations fall, to better enable the process of conception to occur and, conversely, that conception is normally inhibited by the presence of copper in uterine and cervical secretions. (Presumably, cervical mucus would derive at least partly from the uterus.) In further support of this, Singh (1975) has shown that no such fall in secretory copper occurs in women using oral contraceptives (Fig. 7-6B) (who have high levels of plasma ceruloplasmin) and that this is true as well in users of copper IUDs. (It is not as true in users of noncopper IUDs.)

In connection with uterine metabolism, two other facts about copper are noteworthy. First, there is evidence that copper may be involved in modulating the affinity of uterine estrogen receptors, in conjunction with a low-molecular-weight, thermolabile factor that inhibits copper enhancement of estradiol binding and another (thermolabile) factor that promotes the copper effect (Fishman and Fishman, 1988). Although the optimal effect of copper was achieved with an unphysiologically high (30 μM) concentration of the metal ions added to cytoplasmic fractions (where it would be expected to bind to a variety of proteins), it is at least remotely possible that the observation has physiological relevance. The other fact is that mRNA for ceruloplasmin is expressed by the uterus itself, at least in the rat (Thomas and Schreiber, 1989), and this occurs not just if conception has occurred, but also in the virgin tissue. It could be rationalized that secretion of ceruloplasmin into the uterine cavity might protect the endometrium (and other uterine cells) during breakdown and re-formation in the menstrual cycle. If so, it should account for a proportion of the copper in uterine and cervical secretions. [Presumably (although this is not a certainty), ceruloplasmin made by the uterus is not secreted into the blood.] Either way, the forms of copper present in the uterine secretions need to be examined.

The potential effects of *progestogens* on ceruloplasmin and plasma copper have also been of interest to researchers, especially in connection with the use of oral contraceptives. A variety of different synthetic progestogens have been used in combination with estrogens, in different contraceptive mixtures. The overriding impression is that some of these in themselves can increase plasma ceruloplasmin concentrations or can potentiate the effects of estrogens (Table 7-2 and Fig. 7-5A). The effects of Neogynon and Gynovalar shown in Fig. 7-5A were ascribed by Rubinfeld *et al.* (1979) to differences in the type of progesterone analogue incorporated. Some forms of progestogens have been reported to have little or no effects on levels of ceruloplasmin (Table 7-2). Certainly, most have much less effect by themselves than do estogens alone. However, in pregnancy (at least in the rat), the time course of changes in total plasma copper follows that of progesterone secretion much more closely than that of secretion of estrogen (Sato and Henkin, 1973) (see Chapter 8; Fig. 8-1). Thus, some direct or indirect action of progestogens in conjunction with estrogens is certainly occurring in some physiological conditions. The changes of serum copper in the rat estrous cycle (Fig. 7-6A) and upon ovariectomy (Table 7-3) parallel changes in the secretion of both of these hormones (Sato and Henkin, 1973).

With such large increases in ceruloplasmin (and serum copper) concentrations, one wonders what the effects might be on tissue copper concentrations, especially in conditions (such as oral contraceptive use) imposed by human intervention. In normal pregnancy and with long-term estrogen treatment, copper does tend to accumulate within the body, as discussed in Chapters 8 and 5, respectively. [The total body content of Cu in the mother is increased (excluding the uterus, fetus, etc.) (Fig. 8-2), and turnover of whole body copper is slowed (Chapter 5; Table 5-8).] Certainly, there is more copper in the enlarged blood plasma compartment, which should also be reflected in increased copper (and ceruloplasmin) within the enlarged interstitial fluid compartment. The concentrations of copper in other tissues, however, may generally not change (Table 7-5), except in the case of kidney, and perhaps in other tissues that have not as yet been examined. [Onderka and Kirksey (1975) reported a *decrease* of about 25% in rat liver.] In the studies reported in Table 7-5, it is of special interest that the progestogen was more effective than the estrogen in increasing *kidney* copper concentrations and that both together had the greatest effect. (Their effects may have been more than additive.) Whether

Table 7-5. Effects of Estrogens and Progestogens on Copper Contents of Rat Tissues[a]

Tissue	Cu (mg/g ash)			
	Control	Estrogen	Progestogen	Both
Heart	0.39	0.38	0.40	0.36
Kidney	0.40	0.45[c]	0.56[c]	0.73[c]
Liver	0.27	0.26	0.25	0.24
Plasma	0.98[d]	—	—	1.18[c,d]

[a] Yunice and Lindeman (1975).
[b] Male rats i.m. injected daily for 12 weeks.
[c] $p < 0.001$ for difference from control.
[d] Units of $\mu g/ml$.

such changes in tissue copper concentrations also occur in women taking oral contraceptive steroids is not known. Also unknown is the intracellular binding and distribution of the extra copper, as well as its activities and functions. In contrast to estrogens and progestogens, androgens probably play no role in ceruloplasmin regulation (Tables 7-2 and 7-3), although they may influence lysyl oxidase (see later) (Bronson et al., 1987).

7.1.1.4 Other Hormones

Some other hormones have also been implicated in determining levels of plasma copper or ceruloplasmin, notably thyroxine (Table 7-3), and perhaps also parathyroid hormone (where serum copper may marginally increase). The effect of thyroxine would appear to be opposite to that of all of the other hormones mentioned so far, apparently decreasing ceruloplasmin oxidase activity. This has not been followed up. In other studies with rats (Oliver, 1975), it appeared that copper itself might be important for thyroid function, as its lack enhanced thyroid deficiency. Some effects of epinephrine (adrenaline) on the enhanced incorporation of ^{64}Cu(II) into ceruloplasmin have also been reported (Weiner and Cousins, 1983). Incorporation and turnover of ^{3}H-labeled amino acid in ceruloplasmin (12–24 h after treatment with radioisotope) were not altered. The potential significance of these observations (if any) is still obscure.

7.1.1.5 Nonhormonal Factors

A variety of nutrients, including Cu itself and Ag, have also been implicated in the modulation of plasma ceruloplasmin concentrations (Table 7-2). Increases in ceruloplasmin have been observed at high dietary concentrations of retinoic acid and analogues of thiamine and nicotinamide, but there is presently no clue as to what is going on to cause the effect and what its significance (if any) might be. The actions of high doses of ascorbic acid, on the other hand, may be understood to some degree, as ascorbate inhibits ceruloplasmin oxidase activity (which has often been used to assess levels of ceruloplasmin portein). Concentrations of ceruloplasmin protein (measured immunologicaly) did not decrease in the human studies of Jacob et al. (1987) in subjects with a high ascorbate intake. While there has been indirect evidence to suggest that high ascorbate intake might promote copper deficiency, this is still a matter of debate (see Chapter 2). [Copper deficiency would reduce ceruloplasmin concentrations, though not as much as ceruloplasmin oxidase activity (Holtzman and Gaumnitz, 1970a,b).] On the other side of the coin, high levels of ozone might also be expected to have an effect on ceruloplasmin oxidase activity. Indeed, Pierre et al. (1988)

have reported a significant but small (15%) decrease in ceruloplasmin concentrations measured immunologically in confined aluminum welders (with a high ozone exposure). Enzyme activity was not examined, and the drop in the mean value for total serum copper was not statistically significant.

The reason for the increase in o-dianisidine oxidase activity in genetically obese mice (Table 7-2; Prohaska *et al.*, 1988) is also obscure. Such mice would be expected to have increased insulin resistance, but a connection between insulin and copper has not been established (Nestler *et al.*, 1987; Davidson and Burt, 1973). There may be a connection through fat transport and metabolism (which is affected by copper status; see later), but that is also pure speculation.

There is no question but that copper deficiency has a dramatic and rapid effect on the oxidase activity of ceruloplasmin, its copper content, and turnover (see section 7.2, Actions of Copper, later in this chapter). Whether, on the other hand, copper regulates ceruloplasmin gene expression or synthesis under conditions in which dietary intake and copper status are adequate or normal is another matter, for which the answer appears to be negative. Data supporting this conclusion are also described in the later section 7.2 on Actions of Copper. The counteractive effects of silver (Sugawara and Sugawara, 1984) given in large doses ($AgNO_3$, 10 mg/kg) may be related to the direct competition of these two ions for similar intra- and extracellular protein binding sites, although this also is speculation. Ag(II) binds to metallothionein in the same manner as Cu(II) and has considerable affinity for this protein (see Chapter 6), although it also binds elsewhere (Sugawara and Sugawara, 1984). Silver injection caused a 47% drop in ceruloplasmin oxidase activity by 6 h, with a drop of only 30% in total serum copper. Oxidase activities and serum copper fell to a similar degree by 24 h, implying equilibration between ceruloplasmin and other serum compartments by that time.

7.1.2 Regulation of Other Aspects of Copper Metabolism by Hormones

Apart from whole-body turnover of copper, which is altered by estrogen treatment or by the presence of a malignant growing tumor (Chapter 5), and the changes in copper transport and tissue distribution associated with conception, gestation, lactation, and growth (Chapter 8), very little has been done to study how copper metabolism in general may be altered by hormones. The exceptions are some studies on lysyl oxidase and superoxide dismutase. Sanada *et al.* (1978) reported that estrogen increased lysyl oxidase activities in skin and bone and that there was a larger proportion of soluble collagen in animals in which estrogen levels had been reduced by ovariectomy. Since lysyl oxidase is a copper-dependent enzyme that may in fact derive its copper largely from ceruloplasmin (see Harris and DiSilvestro, 1981) and is responsible for cross-linking of collagen (leading to greater insolubility), it seems logical to suppose that elevated ceruloplasmin concentrations would provide more copper to lysyl oxidase for it to perform its cross-linking functions. Whether the opposite phenomenon (less estrogen and thus perhaps less ceruloplasmin) might be related to the increased likelihood of osteoporosis induced by menopause in women is not known. However, plasma ceruloplasmin concentrations do not drop in menopause or with increasing age thereafter (Linder et al., 1981). Might there be a drop in the infiltration of ceruloplasmin into these areas (skin and bone)?

Testosterone also may have effects on lysyl oxidase, in this case in vascular tissue. Bronson et al. (1987) have reported impressive stimulation of lysyl oxidase release by smooth muscle cells cultured from the aorta. We know of no other effects of testosterone on copper metabolism, although copper is necessary for sperm motility and effectiveness (see Sections 7.1.1.3 and 7.2.2.1).

Turning to Cu/Zn superoxide dismutase (SOD) and metallothionein (MT), DiSilvestro (1988) has reported that inflammation reduced the serum concentrations of Cu/Zn SOD activity, when rats were receiving diets adequate in copper. (This did not occur when rats were on copper-deficient diets.) The significance of the trace amounts of this enzyme in serum, and how its presence there comes about, remains to be defined. In inflammation, concentrations of liver and serum MT I can also increase (Bremner et al., 1987; Bremner and Beattie, 1990). In liver at least, this metal storage protein is induced by a variety of agents, and this occurs in states other than those involving increased dietary metal (especially zinc) intake (Table 7-6). Serum and liver MT I might thus be considered acute phase reactants, and they appear to be induced by infection, inflammation, endotoxin, interferon, and interleukin-1 (although IL-1 does not work directly on hepatocytes; Failla and Cousins, 1978). (Some stresses (like high O_2 exposure) also induce it in the lung (Hart et al., 1989).) Liver MT I accumulates also in response to physical and chemical stresses, the effects of which would be mediated by glucocorticosteroids and catecholamines. The effects of fasting (and X-irradiation, which results in fasting), as well as treatment with drugs like streptozotocin and isopropanol (see Bremner, 1987), would probably be mediated by glucagon. As already described in Chapter, 6, the promotor region of the MT I gene has a glucocorticoid regulatory element that allows increased transcription in stress (perhaps also in the case of stresses imposed by

Table 7-6. Factors and Physiological States That Induce Metallothionein in Liver or Cultured Cells[a]

State	Factor
Inflection	Endotoxin
Inflammation	Carrageenan
	Dextran
	Interleukin-1
	Interferon
Physical stress, laparotomy	Glucocorticosteroids
	Catecholamines
	Carbon tetrachloride
	Chloroform
	Ethionine?
	Ethanol?
	Alkylating agents?
Starvation, X-irradiation	Glucagon
	Isopropanol
	Streptozotocin
	Estrogen
	Progesterone

[a] Rearranged from Bremner (1987).

inflammation, etc.). The action of glucagon is by a nontranscriptional mechanism, synergistic with that of glucocorticoids (Etzel and Cousins, 1981; DiSilvestro and Cousins, 1984). The effect of endotoxin is to promote MT gene transcription and MT synthesis but also does not involve glucocorticoids, as it occurs in adrenalectomized animals (Etzel and Cousins, 1981; Quinones and Cousins, 1984).

How all of this may be related to copper metabolism is another question. Certainly, copper can (and will) displace zinc (and most other metal ions) from MT, an effect that may be mediated by glutathione (GSH) (see Chapter 6). However, of the various stressful situations mentioned, some result in the release of extra copper from the liver (on ceruloplasmin) rather than in liver copper accumulation; others appear to have little effect on tissue copper distribution (fasting, glucagon, even low levels of stress involving release of hydrocortisone or catecholamines). In contrast, there is a relationship between the accumulation of zinc in the liver (along with a shift of zinc from the plasma into liver cells) and the induction of liver metallothionein that occurs in inflammation and is accompanied by an increase in serum MT. These changes in MT probably have little impact on copper metabolism, although the metal (or metals) associated with the *serum* MT is (are) unknown (see Chapter 4). [In inflammation, concentrations of MT in serum only reach 80 ng/ml, as compared with 30 µg/g in liver (Morrison *et al.*, 1988), so this would be hard to determine.] A potential role of GSH in differentiating the zinc and copper responses to changes in MT, in these various conditions, may be worth examining.

Finally, increased urinary excretion of copper has been reported for humans with Cushing's disease (Henkin, 1974a,b), in which there is overproduction and release of glucocorticosteroids. How this may result in hyperexcretion of Cu(II) is unclear, although there is increased exchange of Na^+ and K^+ in the kidney tubules, leading to more excretion of K^+ as well. Whether there is any connection remains to be examined.

7.2 Copper Regulation of Metabolism and Hormones

7.2.1 Effects of Copper Deficiency

7.2.1.1 Changes in Copper and Copper Enzymes

As one would expect, the concentrations of copper in most tissues falls with time after the start of dietary copper deprivation. The same is the case for the activity of critical copper-dependent enzymes in these tissues. Some of these changes have been studied extensively, especially in rats. In general, gross dietary deficiency (rather than marginal deficiency) has been examined. From the observations, a general order-of-magnitude of the changes in individual tissues (and their enzymes) has emerged. The results of two of the more comprehensive studies are summarized in Figs. 7-7 and 7-8. [Reviews of the effects of copper deficiency are also contained in articles by Prohaska (1988 and 1990), Danks (1988), Paynter (1987), Fell (1987), and Davis and Mertz (1987).] From the data shown, it is evident that total concentrations of copper in plasma fall very rapidly, and more rapidly than those of other tissues, when animals are placed on a low-copper diet. Liver copper concentrations are the next most affected, while those of kidney (Table 7-7) and especially the bone (Fig. 7-7) are slower to respond. Changes in brain (and heart) copper concentrations also occur very slowly or less readily and are most effectively induced by instituting deprivation already in gestation (Prohaska, 1983a,b). Changes in other organs are shown in Fig. 7-8 for rats. Changes in the most abundant and key copper enzymes (Fig. 7-8) do not conform to a single pattern. Examining just the activities of ceruloplasmin, cytochrome *c* oxidase, and Cu/Zn

Copper and Metabolic Regulation

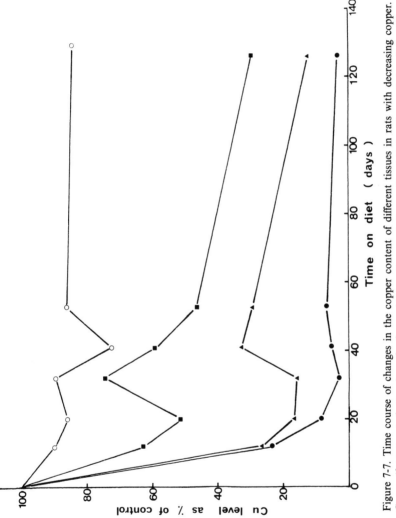

Figure 7-7. Time course of changes in the copper content of different tissues in rats with decreasing copper. Curves (from top to bottom) are for femur, kidney, liver, and plasma. Reprinted, with permission, from Alfaro and Heaton (1973).

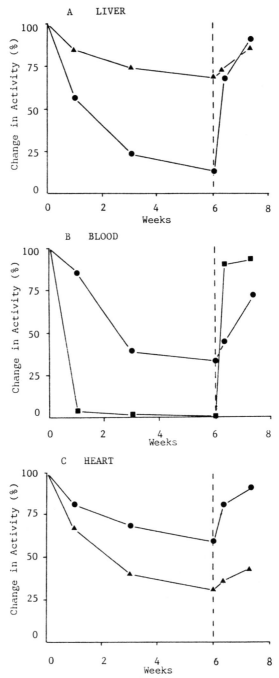

Figure 7-8. Changes in copper enzymes of different rat tissues during copper depletion. ●, Superoxide dismutase; ▲, cytochrome oxidase; ■, ceruloplasmin. Blood SOD was measured in erythrocytes, ceruloplasmin in plasma. Reprinted, with permission, from Paynter et al. (1979).

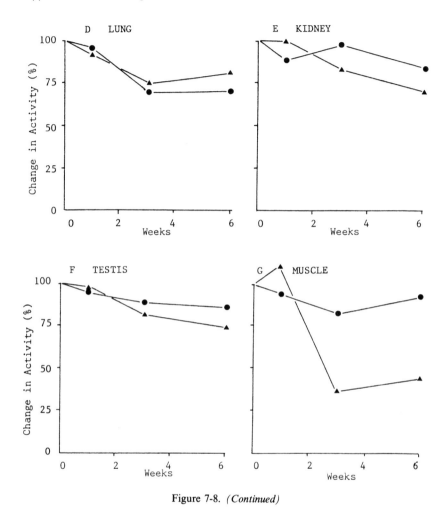

Figure 7-8. *(Continued)*

superoxide dismutase, the following sequence of change is observed. In terms of rate of decline, that of plasma ceruloplasmin and plasma copper is most rapid, liver SOD is next, followed sequentially by erythrocyte SOD, heart and skeletal muscle cytochrome oxidase, heart SOD, and liver cytochrome oxidase. Cytochrome oxidase and SOD in lung, kidney, and testis, and then SOD in skeletal muscle exhibit much slower rates of decline. This suggests a relative order of importance of these enzymes in various tissues or the existence of surplus enzyme in some tissues as compared with others. The level of copper and especially of ceruloplasmin

Table 7-7. Effects of Copper Deficiency on Copper Contents of Liver, Kidney, and Femurs in Rats[a]

Time on diet (days)	Copper content (μg)					
	Liver		Kidney		Femur	
	Normal	Deficient	Normal	Deficient	Normal	Deficient
20	40	11	4.3	2.5	2.3	2.1
41	38	14	5.4	3.2	4.1	3.0
126	26	5.3	12	4.5	6.5	5.2

[a] From Alfaro and Heaton (1973)

activity in the serum is therefore the most sensitive indicator of a copper deficiency in the diet, and the copper content of the liver is a close second. Ceruloplasmin is still synthetized, but at a much less rapid rate (Holtzman and Gaumnitz, 1970b; Linder et al., 1979b), and less copper may be incorporated into it during synthesis (or more removed by delivery to peripheral tissues requiring the element; see Chapters 3 and 4). [Copper-deficient rats have low levels of ceruloplasmin protein, which is mostly apoceruloplasmin (Holtzmann and Gaumnitz, 1970a).] In normal health, plasma ceruloplasmin concentrations and activities are held rather constant. However, rates of ceruloplasmin synthesis and turnover may be altered to increase concentrations of the protein in normal pregnancy, and in some other benign conditions (such as fibrocystic disease of the breast). (See Chapter 9.) Theoretically, at least, this could mask the presence of copper deficiency. For this reason, DiSilvestro (1988) has proposed the use of serum/plasma Cu/Zn SOD activity as a better measure of copper status, which may be the case in these circumstances. [The enzyme activity that he has followed is thought to be that of leaked or secreted *cellular* SOD, as opposed to "extracellular SOD" *per se*, which is a different, also copper-dependent, enzyme found in the plasma (see Chapter 4).] On the other hand, measurements of *erythrocyte* Cu/Zn SOD activity may well be of equal value (to measurement of plasma, cell-derived SOD), as changes in erythrocyte SOD in rats parallel those of liver SOD and are thus also in the group of factors that fall second most rapidly in deficiency. [A linear relationship has been observed between activities of liver and erythrocyte SOD (Paynter, 1987) in more than one species.] The finding that men placed on a high-fructose, marginal (1 ppm) copper diet showed a fall in erythrocyte SOD with no drop in serum ceruloplasmin (Reiser et al., 1985) suggests that this marker might also be better than ceruloplasmin.

However, a drop in erythrocyte SOD without an equal or more severe drop in ceruloplasmin oxidase activity has never been reported in the many other studies of copper deficiency where this has been examined, and it seems more likely that this an effect of the high fructose intake. Nevertheless, erythrocyte SOD should be considered a useful marker of copper status, as illustrated by the data of Uauy et al. (1985) for infants. Use of this marker may be better in situations (such as infection) in which ceruloplasmin levels are increased for other reasons and might mask a dietary deficiency. Blood macrophage SOD activities show a parallel dramatic decline (Carville and Strain, 1989); that of lymphocyte and granulocyte SOD activities is not nearly as great.

The fall in total concentrations of copper in plasma must reflect not just changes in ceruloplasmin copper but also in the copper content of its other binding proteins (transcuprein, albumin, ferroxidase II, etc.) (see Chapter 4). Figures 7-7 and 7-8 suggest that the fall in ceruloplasmin activity is greater than the fall in total plasma copper. It is presently unclear just how the other binding factors respond.

Concentrations of total copper in liver, and liver (or erythrocyte) Cu/Zn SOD activities, also are sensitive parameters for the assessment of copper status in most (or perhaps all vertebrate) species. During copper repletion, plasma and liver may be the last tissues to return to normal levels (Alfaro and Heaton, 1973; Paynter et al., 1979). This is consistent with the concept that copper in plasma (on its copper proteins) and in liver (which incorporates copper into plasma ceruloplasmin and also regulates excretion of copper via incorporation into the bile) together provide for and regulate the flow of copper to the other organs of the body. Copper itself may not be regulating synthesis of SOD, since Dameron and Harris (1987) showed normal levels of the enzyme protein to be present in copper-deficient aortas. Treatment with copper restored activity by 8 h in culture. Reduced levels of Cu/Zn SOD would be expected to result in enhanced oxidative damage to most cells of the body, as has been demonstrated directly for lungs of rats exposed to hyperoxia (Jenkinson et al., 1984) and indirectly in relation to the function of microsomal membranes of liver cells (Calviello et al., 1988).

Concentrations of lysyl oxidase in bone, lung, and other tissues are also sensitive to severe copper depletion, as examined by O'Dell, Harris, and Rucker and colleagues, mainly in chickens (Rucker et al., 1975; Harris, 1986). As in the heart and some other organs (Campbell et al., 1981), lysyl oxidase activity (and repletion) depends especially on ceruloplasmin copper (Dameron and Harris, 1987). Thus, when ceruloplasmin concentrations fall to low levels, this enzyme (and heart copper enzymes; Fig. 7-8) are especially vulnerable. The level of copper in the diet may also modulate parathyroid hormone action on bone, perhaps via the steroid hormone, vitamin D, as suggested by the work of Havivi and Guggenheim (1966) with mice. Deficiencies of liver Cu and connective tissue (especially vascular) lysyl oxidase activity appear to be enhanced by a high intake of ascorbate, at least in turkey poults (Simpson et al., 1971) and in chicks (Hunt et al., 1970). The mechanism of this effect is still unclear, but there is a remote possibility that it has something to do with sorbitol or fructose metabolism (see later) or with an enhanced loss or depressed intestinal absorption of copper (see Chapter 2), or both.

Amine oxidase activity in blood plasma also declines during dietary copper deprivation, as observed for cattle, sheep, and pigs (see Paynter, 1987). The rate of decline parallels changes in whole-blood Cu concentrations (Mills et al., 1976). This implies that the rate of decline of amine oxidase activity is not as rapid as that of ceruloplasmin but is more rapid than that of erythrocyte SOD activity. [Amine oxidase may thus also be a more sensitive measure of copper status than cellular SOD that enters the plasma (suggested by DiSilvestro; 1988).] Although this needs confirmation, the relatively high sensitivity of plasma amine oxidase may reflect an intestinal origin for this enzyme (see Chapters 4 and 6). If dietary copper intake is reduced, less copper would be available to the intestinal mucosa for incorporation into amine oxidase that is secreted into the blood.

The decline of tyrosinase activity in copper deficiency has been of some practical interest in relation to the pigmentation of the wool of sheep (Paynter, 1987). On test diets, it has been shown that the copper content of the hair and wool of sheep and cattle can indeed be indicative of chronic copper status and deficiency. Paynter suggested that the changes observed in hair and wool parallel those of erythrocyte SOD activity. The color of wool or hair will be affected by copper insufficiency when tyrosinase activity is decreased in the hair follicle (where it catalyzes melanin formation). (See Chapter 6.) Another effect on tyrosinase activity appears to be induced by thiomolybdates, which form when animals are grazed on forage high in molybdenum and sulfur. [The thiomolybdates form stable complexes with Cu(II) that render it unavailable for metabolic processes (Dick et al., 1975; Suttle, 1980; Mason, 1982).] Tyrosinase copper appears to be particularly susceptible to thiomolybdate inactivation (Paynter, 1984).

Changes in other copper enzymes, notably those in brain, also occur. Data for some of these changes are summarized in other chapters (Tables 6-4, 6-11, and 6-15). (See also Prohaska, 1987 and 1990). As in cases already cited, most of the changes demonstrated have occurred in severe copper deprivation. They are much less likely to occur in marginal deficiencies in which ceruloplasmin and erythrocyte liver SOD activities still decrease. Changes in the copper contents of metallothionein would also be expected to decline, although they may not be very high even normally. One circumstance in which such a decline is clearly observed is in the liver of the neonate, where the store of excess copper, stockpiled during gestation in the liver (mainly in cell nuclei), is gradually lost. Milk contains relatively little copper, and copper is released from the liver in response to the need for new tissue growth in the suckling period. [These changes are described at length in Chapter 8]. As in other conditions, especially those in which there is little copper storage, the concentration of metallothionein itself may not change appreciably in copper deprivation. Metallothionein concentrations are not regulated as much by copper as by zinc and hormones (see Chapter 4). In time, other copper-dependent enzymes may emerge that are sensitive to acute and chronic copper deficiency to the same degree as erythrocyte SOD, serum amine oxidase, or even perhaps ceruloplasmin. The depression in enkephalin release, for example, in copper deficiency is quite dramatic. (See Chapter 6, Section 6.4.2) This remains to be explored. Other less understood or less direct effects of copper deprivation are described in the next section.

7.2.1.2 Effects of Copper Deficiency on Other Areas of Metabolism

A number of effects of copper deprivation on other parameters, in all areas of metabolism and function, have been described. These are summarized in Table 7-8. For the most part, the effects observed fall into specific categories: those concerning (a) hematopoiesis and heme production; (b) cardiovascular integrity and connective tissue function; (c) glucose tolerance; (d) cholesterol and lipid formation, transport, or metabolism; (e) brain development; (f) fertility; and (g) immunity. The involvement of copper in brain myelination and neurohormone production is described in Chapters 6 and 8. Those concerning connective and elastic tissue function are described in Chapter 6, under lysyl oxidase. The other aspects are discussed below.

7.2.1.2a. Interactions of Copper with Iron Metabolism and Hematopoiesis. One of the "classic" hallmarks of copper deficiency is a hypochromic, microcytic anemia (Lahey *et al.*, 1952; Cartwright *et al.*, 1956). This has been demonstrated in a number of animal species, in human infants with malnutrition (Cordano *et al.*, 1964, 1966; Karpel and Peden, 1966), and also in adult women on total parenteral nutrition (Dunlap *et al.*, 1974; Zidar *et al.*, 1977). In the latter studies, defective granulocytopoiesis was also evident and, in fact, appeared to be more sensitive to copper deficiency (Dunlap *et al.*, 1974). [A case of copper-responsive anemia and neutropenia/granulocytopenia in a male subject on very high doses of oral zinc for acrodermatitis enteropathica has also been

Table 7-8. Effects of Severe Copper Deficiency on Noncopper Parameters in Mammals[a]

Parameter	Effect of copper deficiency	Reference(s)
Hematocrit/blood hemoglobin	Anemia (microcytic, hypochromic)	Lahey et al. (1952) O'Dell (1961a,b) Dunlap et al. (1974) Lee et al. (1976) Prohaska (1981a) Williams et al. (1983)
	Anemia especially in male rats	Kramer et al. (1988)
Serum iron	Decreased	Owen (1973) Williams et al. (1983)
Liver iron	Increased	Owen (1973) Alfaro and Heaton (1973, 1974) Williams et al. (1983)
Femur iron	Little change Slight decrease	Begona and Heaton (1973) Alfaro and Heaton (1973) Owen (1975)
Liver heme oxygenase	Increased	Williams et al. (1981, 1985)
Spleen heme oxygenase	No change	Williams et al. (1981, 1985)
Heme biosynthesis (reticulocytes)	Impaired	Williams et al. (1976)
Mitochondrial heme synthesis (hepatocytes)	Impaired	Williams et al. (1976)
Conversion of porphyrin a to heme a	Impaired (in yeast)	Keyhani and Keyhani (1980)
Brain heme a content	Decreased	Gallagher et al. (1956) Smith et al. (1976) Prohaska (1987)
Erythropoietin	Decreased (rats)	Zidar et al. (1977)
Blood neutrophils/granulocytes	Decreased (neutropenia)	Dunlap et al. (1974) Zidar et al. (1977)
Bone	Defective development Osteoporosis	O'Dell et al. (1961a,b) Danks (1988)
Heart weight/mitochondrial mass	Hypertrophy	Schultze (1939) Kelly et al. (1974) Prohaska and Heller (1982) Hassel et al. (1983)
Heart succinic dehydrogenase	Increased	Kelly et al. (1974)

Table 7-8. *(Continued)*

Parameter	Effect of copper deficiency	Reference(s)
Heart atrium	Thrombosis, abnormal EKG	Kllevay (1985)
Blood glucose	Glucose intolerance	Cohen et al. (1982) Hassel et al. (1983) Fields et al. (1983)
Liver glucose-6-phosphatase	Decreased No effect	Johnson et al. (1984) Moffitt and Murphy (1973)
Latent brain hexokinase	Activity decreased	Prohaska (1981)
Blood cholesterol	Hypercholesterolemia No hypercholesterolemia	Murthy and Petering (1976) Lei (1977) Allen and Klevay (1978) Klevay et al. (1984b) Lin and Lei (1981) Prohaska et al. (1985) Hunsaker et al. (1984)
Lipoprotein lipase (hepatic and endothelial)	Activity decreased	Koo et al. (1988)
Post-heparin lipoprotein lipase	Activity increased	Lau and Klevay (1981, 1982) Koo et al. (1988)
Lecithin–cholesterol acyltransferase	Activity decreased No consistent change	Harvey and Allen (1981) Lau and Klevay (1981) Lefevre et al. (1985)
Turnover of cholesterol in plasma	Slightly decreased (rate of fast turnover pool, in rats)	Lin and Lei (1981)
Synthesis of cholesterol	Unchanged	Lin and Lei (1981)
Triglyceride in plasma	Increased (rats) Increased (hyperinsulemic subjects)	Reiser et al. (1983) Bhathena et al. (1988)
Glycerolphosphate acyltransferase	Activity decreased	Gallagher et al. (1956)
Liver stearyl CoA saturase	Activity decreased	Wahle and Davies (1975)
Myelin formation	Decreased during growth	O'Dell and Prohaska (1983) (see Chapter 8)
Brain 2',3'-cyclic nucleotide 3'-phosphodiesterase	Decreased	Prohaska and Wells (1974)
Brain tyrosine hydroxylase	Decreased	Morgan and O'Dell (1977)

(continued)

Table 7-8. *(Continued)*

Parameter	Effect of copper deficiency	Reference(s)
UDP-gal hydroxy fatty acid ceramide galactosyltransferase	Decreased	DiPaolo et al. (1974)
Fertility	Decreased	O'Dell et al. (1961a,b) Hall and Howell (1969) Howell and Hall (1969)
Semen copper in infertility	No change	Stankovic and Davic (1976)
Sperm motility	Decreased	Battersby and Chandler (1977)
Thyroid deficiency	Exacerbated	Oliver et al. (1976)
Plasma glutamic–oxaloacetic transaminase	Increased	Fields et al. (1984)
Liver glutathione peroxidase	Decreased	Jenkinson et al. (1982)
Liver catalase	Decreased	Prohaska and Wells (1974)
Vasculature (especially arterial)	Lesions Low, abnormal elastin	O'Dell et al. (1961a,b) Hunt and Carlton (1965) Simpson et al. (1967) Waisman and Carnes (1967) Kitano (1980) Hunsaker et al. (1984)
Immunity	Decreased	See Table 7-12
Spleen weight	Increased (developing mice)	Prohaska et al. (1983)
Thymus weight	Decreased	Mulhern and Koller (1988)

a Taken, in part, from Prohaska (1988).

reported (Hoogenraad et al., 1985).] With regard to anemia, the findings are summarized in Table 7-8 and detailed in some typical recent studies (Table 7-9). Copper deficiency results in a reduction in red cell mass and hemoglobin in the blood, a decreased concentration of plasma iron (bound to transferrin), and an increase in liver iron concentrations, as if the "flow" of iron from turnover of aged erythrocytes were slowed. However, the concentration of iron in spleen is not increased (Owen, 1973), nor is that of the marrow markedly decreased (Owen, 1973), which suggests that the slowdown in iron recycling is specific to the liver, and not a general finding for the reticuloendothelial system. There also are no consistent changes in the concentrations of iron in other tissues. Although Cartwright and

Table 7-9. Effects of Copper Deficiency on Blood (and Other Tissue) Parameters[a]

	Cu in diet			
	Male rats		Female rats	
	0.6 ppm	5.6 ppm	0.6 ppm	5.6 ppm
Body weight gain (g)	213–239	277–280	132–136	135–140
Food intake (av. g/day)	15	20–21	14	15–17
Hematocrit (%)	27–28[b]	45	41–44	44–45
Hemoglobin (mg/dl)	7.9–8.0[b]	14.8–14.9	12.8–14.4	15.0–15.3
Serum Cu (μg/dl)	3–5[b]	114–120	2–30	122–124
Serum Fe (μg/dl)	63–76[b]	309–346	223–373[b]	475–532
Ceruloplasmin oxidase (ΔOD)	>1[b]	130–148	123	161–180
Liver Fe (μg/g)				
Low-Fe diet	498	222	1025	772
High-Fe diet	399	427	794	1777
Kidney Fe (μg/g)				
Low-Fe diet	158	212	194	316
High-Fe diet	59	264	50	402

[a] Data of Kramer et al. (1988) for Lewis rats placed on low- or adequate-copper diets for 42 days, starting at 28 days of age. Except for liver and kidney, data for low (50 ppm) or adequate (300 ppm) Fe intake are combined.
[b] Significant difference ($p < 0.01$) between values for low (0.6 ppm) and adequate (5.6 ppm) Cu intake.

Wintrobe and their colleagues (Gubler et al., 1952b; Lee et al., 1968b) reported that release of iron from the intestine to the blood was impaired in copper-deficient pigs, several other researchers have found no accumulation of iron in the intestine in copper-deficient rats (Owen, 1973; Williams et al., 1983). The increase in liver iron is accompanied by an increased activity of heme oxygenase (which degrades heme) in that tissue (Table 7-8), while that in spleen is unchanged.

In support of the concept that the recycling of iron is impaired in copper deficiency are the observations that serum ion concentrations are much lower than normal. Typical data are given in Table 7-9 for rats. In the severe deficiencies usually induced in these studies, ceruloplasmin oxidase activity is very low or undetectable in the plasma. In the studies shown, serum Fe concentrations were still significant, although they were four- to five fold below the normal. (Presumably, the iron is attached almost exclusively to transferrin.) Of further interest are findings that the release of (extra) iron from the liver to the plasma occurs quite rapidly when copper is refed to deficient experimental animals (Evans and Abraham, 1973). The exciting observations of Ragan et al. (1969) with pigs and those of Osaki and Johnson (1969) with perfused dog livers that intravenous infusion of *ceruloplasmin* resulted in release of iron from the liver (and perhaps also from other cells in the body) to the blood plasma offered an attractive means of reconciling many of the interactions of copper with iron metabolism and anemia (Frieden, 1970, 1971). In both systems, release of iron (or increased plasma iron) occurred within minutes after infusion of ceruloplasmin (from the same species). $CuSO_4$, HCO_3^-, citrate, apotransferrin, and albumin (and nutrients like glucose and fructose) did not have this effect (Osaki et al., 1971). Concentrations for the half-maximal effect of human ceruloplasmin in dog liver perfusates were about 0.1 μM (about 5% of normal human plasma concentrations). Ragan et al. (1969) and Roeser et al. (1970) also found that ionic copper had little or no effect compared to ceruloplasmin, and that desialylated ceruloplasmin did not work. The latter is of special interest, in that asialoceruloplasmin would be better at entering hepatocytes than the holoenzyme (see Chapters 3 and 4). In pigs, Roeser et al. (1970) found that the hypoferremia of copper deficiency was not apparent, however, until active ceruloplasmin fell below 1% of normal concentrations. This implies that ceruloplasmin is normally present in many times the concentrations needed for its effect on iron mobilization, which would also explain why there are no problems with anemia and iron metabolism in Wilson's disease (Johnson et al., 1967; Frieden, 1970).

Ceruloplasmin's capacity to oxidize iron (its ferroxidase activity) was discovered by Curzon and O'Reilly (1960; Curzon, 1961) and confirmed by Osaki and Frieden and collaborators in the mid-1960s (Osaki et al., 1966). [A second ferroxidase, named ferroxidase II, also copper dependent, was discovered in plasma later on (see Chapter 4).] Ceruloplasmin (ferroxidase I) and ferroxidase II both catalyze the oxidation of Fe(II) to Fe(III) in the presence of molecular oxygen (see Chapter 4, Fig. 4-8). The assay is normally performed in the presence of apotransferrin, which forms a complex with the Fe(III) that is produced, with an absorbance at 460 nm (Johnson et al., 1967). Frieden and colleagues have proposed that the direct link between ceruloplasmin and iron metabolism is precisely at this point, namely, the oxidation of iron to Fe(III) which would enable it to attach to, and be transported by, transferrin (thus increasing transferrin iron concentrations). The proposed scheme is depicted in Fig. 7-9. At the present time, it is still impossible to prove or disprove this hypothesis. It is noteworthy, however, that intravenous infusion of ferroxidase II into copper-deficient animals had the same effect as ceruloplasmin, stimulating release of iron into the serum (Topham et al., 1980). Thus, two different (though both copper-containing) proteins with the same enzymatic capacity had the same effect.

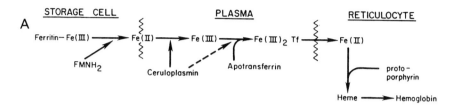

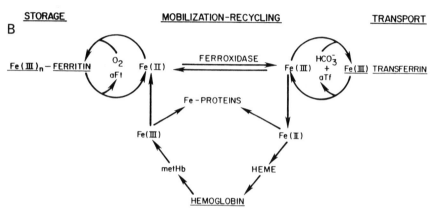

Figure 7-9. Proposed function of ceruloplasmin ferroxidase in the flow of iron from storage sites in cell ferritin to transferrin in the plasma. (A) Iron may be released as Fe(II) from ferritin, oxidized by ceruloplasmin (in the plasma), and bound to transferrin as Fe(III), for transport to reticulocytes and other cells. (B) A different depiction of the Fe(II) and Fe(III) "cycles" that may be involved. Reprinted, with permission, from Frieden (1983).

From the studies described, it would seem that the action of ceruloplasmin is in the plasma or interstitial fluid and is not due to its uptake (or uptake of its copper) by hepatocytes or other cells. The effect on the liver is much greater with sialo- then with asialoceruloplasmin (Roeser et al., 1970). The effect is also very rapid, leaving no time for conversion of the copper or ceruloplasmin degradation. Ionic copper only works after ceruloplasmin appears in the plasma (Roeser et al., 1970). If so, then one might expect that release of iron would be impaired not just in the liver but in other parts of the organism (such as the spleen) that are actively involved in red cell degradation and turnover. As already explained, this does not appear to be the case.

We know little about the form in which iron becomes available for release from storage sites or hemoglobin during degradation. Iron is released from porphyrin upon action of heme oxygenase. Presumably, iron is still in the ferrous state in the porphyrin before cleavage of the tetrapyrrole ring. In the process, Fe(II) is thought to be oxidized to

Fe(III), with the release of CO (Metzler, 1977). What then happens to the Fe(III) is not known exactly. Certainly, one of the immediate consequences in the case of normal reticuloendothelial (RE) cells is the release of iron to transferrin, the sequestration of iron in RE cell ferritin, and the release of small amounts of iron also on serum ferritin, as shown for example by Siimes and Dallman (1974). If iron is released during degradation as Fe(III), it would be ready to bind to transferrin right away. The iron entering ferritin would probably undergo reduction and reoxidation to do so, as suggested by numerous *in vitro* studies with apoferritin and iron salts (see Munro and Linder, 1978). In both spleen and liver, under normal as well as Cu-deficient conditions, most of the iron would be found in cellular ferritin. In the liver, this would be mainly in the hepatocytes (Van Wyk *et al.*, 1971), and, indeed, there is evidence that iron originally released by RE cells during degradation of erythrocytes is later transferred to hepatocytes. Transfer may even be in the form of ferritin, as there are receptors for ferritin, particularly on hepatocytes (Mack *et al.*, 1981).

In order for iron to be released from ferritin, however, it is thought that reduction and chelation are required (as they are for *in vitro* iron removal). Nevertheless, it is not absolutely certain that this is the normal mechanism of iron release *in vivo*. [Alternatively, degradation of ferritin protein may be required, which would make the inherent Fe(III)–OOH available to metabolites and other factors involved in a hypothetical "iron transit pool."] Thus, iron may indeed emerge from storage in ferritin in the reduced (ferrous) state chelated to citrate, ascorbate, flavin mononucleotide, or other available agents. This might also be more common in liver hepatocytes than in other cells involved in iron turnover and flow. However, it is also possible that this not the form in which iron is released by cells for eventual attachment to transferrin. To bioinorganic chemists, it seems strange that only ceruloplasmin would be capable of catalyzing the oxidation of the iron, rather than some of the many other intra- and extracellular components with a higher redox potential. In any event, the fact that ferroxidase I activity must fall so very low before a reduction in serum iron is observed supports the concept that ceruloplasmin is not a regulatory factor in normal iron metabolism, although it might be involved in the oxidation step.

An additional point worth mentioning is that the phenomenon of reduced iron "flow" from the liver (and perhaps other RE cells) to the marrow occurs not only in copper deficiency, but also in inflammation (see Section 7.1.1.2). However, in inflammation, this is accompanied by a *rise* in serum ceruloplasmin, which again would suggest that it does not regulate the "flow" of iron. Letendre and Holbein (1984) reported that infusion of additional (human) ceruloplasmin intravenously into copper deficient,

Neisseria meningitidis infected mice that had almost no ceruloplasmin activity increased serum iron concentrations (and reduced resistance against the infective organism). This suggests that the impairment of iron "flow" is important for resistance against infection (for which there is plenty of other evidence) and that the presence of ceruloplasmin will enhance iron flow and make the infection worse. If so, the increase of ceruloplasmin that occurs in infection should be harmful, which it normally seems not to be. Alternatively, the mouse is unusual in its response or the use of a heterologous form of ceruloplasmin has an anomalous effect.

A reduction in iron flow is probably not sufficient to explain all of the anemia of copper deficiency, and moreover, the development of anemia, *per se*, does not always occur. A number of researchers have shown that oral or intramuscular iron treatments reversed some or all of the effects of copper deficiency relating to iron metabolism or, more importantly perhaps, that not all experimental animals with undetectable (or very low levels) of serum ceruloplasmin (ferroxidase activity) develop anemia. Williams *et al.* (1983) showed that iron treatment increased hematocrits of copper-deficient rats by normalizing red cell volume and increasing hemoglobin concentrations. It also increased serum iron but did not restore it (or the hematocrits) to normal values. Kramer *et al.* (1988) (Table 7-9) found that a 300-ppm iron diet maintained normal hematocrits in female rats during copper deficiency but did not have this effect in male rats. Cohen *et al.* (1985), also using female rats, were unable to demonstrate any anemia (nor a fall in serum iron or increased liver iron) in copper deficiency in which ceruloplasmin activity was very low. Working with adult male mice, Prohaska *et al.* (1984) found that copper deficiency induced a variable degree of anemia and that, even with no detectable ceruloplasmin activity, hematocrits were close to normal in about half the animals (Fig. 7-10). Brindled, 11-day-old mice, with a "natural deficiency" in liver copper and ceruloplasmin activity also had normal hematocrits (Prohaska, 1981a). [However, they did not grow nearly as rapidly as their heterozygous litter mates without the expressed brindled trait, so their prebirth copper stores might have "gone further" to prevent anemia.] Weisenberg *et al.* (1980) found that a reduction in liver ATP in copper deficiency also could be reversed by iron treatment. In line with other observations, female rats were found to be less sensitive to copper deficiency enhanced by sucrose or fructose diets than male rats (Fields *et al.*, 1983a, 1984, 1986b,c). Ceruloplasmin activity was not detectable in either group, but only the male rats were anemic. All of these studies suggest that mediation of the "flow" of iron through ceruloplasmin oxidation is not critical to the development of the anemia of copper deficiency, if it occurs at all. The apparent differential effectiveness for copper

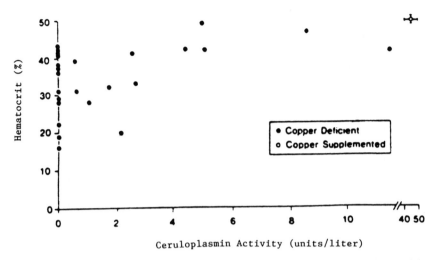

Figure 7-10. Relationship between hematocrit and ceruloplasmin oxidase activity (*o*-dianisidine) in mice with different degrees of copper deficiency. Reprinted, with permission, from Prohaska *et al.*, 1985.

deficiency in male versus female rats also remains to be sorted out, although in the human species, women can develop an anemia due to copper deficiency (Dunlap *et al.*, 1974; Zidar *et al.*, 1977). This still does not preclude ceruloplasmin from functioning in iron oxidation for transferrin binding, when it is present.

Another point where copper may be functioning in iron metabolism, and particularly in hemoglobin and red cell formation, is in the biosynthesis of heme. Early reports in fact suggested that δ-aminolevulinate synthetase, the rate-limiting enzyme in porphyrin (heme *a*) biosynthesis, might be a cuproenzyme (Iodice *et al.*, 1958), an observation not supported by most later work (Wilson *et al.*, 1959; Cheh and Nielands, 1973). [Work of Orten *et al.* (1975) suggested that in certain species the otherwise zinc-dependent enzyme may require copper.] No matter what the mechanism may be, the synthesis of heme appears to be reduced in copper deficiency, at least in copper-deficient pigs and male rats with anemia (Table 7-10). In support of this are findings that iron incorporation into heme by pig copper-deficient reticulocytes was reduced as was iron incorporation into heme in liver mitochondria (Williams *et al.*, 1976) (Table 7-10); that the brain has lower concentrations of heme *a* (Prohaska, 1987); that yeast cells are dependent on copper for insertion of iron into porphyrin to form heme *a* (Keyhani and Keyhani, 1980); and that Pb treatment, which normally causes iron accumulation in mitochondria (where it is waiting for insertion

Table 7-10. Effects of Copper Deficiency on Reticulocyte Function and Liver Mitochondria of Pigs[a]

Parameter	Normal	Copper deficient	
		Acute	Chronic
Hematocrit (%)	42	35	25
Serum Cu (μg/ml)	1.8	0.09	0.17
Serum Fe (μg/ml)	1.6	0.82	1.9
Reticulocytes			
No. ($\times 10^6$/ml)	245	200	90
Iron uptake (nmol/10^9 per 2 h)	6.7	6.5	3.5
Heme synthesis (nmol/10^9 per 2 h)	4.2	2.3	1.4
Liver mitochondria			
Cytochrome c oxidase			
activity (nmol/min per mg protein)	244	72	
Heme a (nmol/mg protein)	0.28	0.07	
Copper (ng/mg protein)	61	4	
Iron (μg/mg protein)	1.2	4.9	
Heme synthesis			
(nmol/mg protein per h)	2.6	1.3	

[a] Data from Williams *et al.* (1976).

into porphyrin), instead results in accumulation of iron by cytoplasmic vesicles developing in copper-deficient erythrocytes (Goodman and Dallman, 1969). In more recent studies, Williams *et al.* (1985) reported that *uptake* of iron by copper-deficient liver mitochondria was *not* impaired in copper deficiency, although incorporation into heme was reduced. In this *in vitro* work, iron was provided either as ferritin (with or without flavin monucleotide, to effect reduction and chelation) or $FeCl_2$. Others have shown that incorporation of iron into mitochondria and utilization by ferrochelatase (the enzyme that catalyzes insertion of iron into porphyrin) requires its reduction to Fe(II) (Barnes *et al.*, 1972; Flatmark and Romslo, 1974). In other words, energy is also required, in the form of reducing equivalents, if not also for other reasons. (Ferrochelatase is not thought to be a copper-dependent enzyme). Taken together, these data suggest that copper deficiency impairs heme biosynthesis (in erythropoietic cells and also others) and that it does this by inhibiting the availability or uptake of Fe(II) that would be incorporated into heme by ferrochelatase in the mitochondria. The question is then how a lack of copper brings this about. Certainly, cytochrome oxidase activity is markedly reduced in many cells, including those of the liver and reticulocytes (Table 7-10). This would be expected to reduce the flow of electrons available for reductive processes.

However, Gallagher and Reeve (1971) showed already some years ago that ATP synthesis by liver mitochondria was not impaired even in a fairly severe copper deficiency, although they suggested that the transport of ATP/ADP into and out of mitochondria was reduced. (The binding capacity of mitochondria for ADP was much less.) Whether this is relevant to the provision of electrons for reduction of incoming iron (to allow heme insertion) is another matter. We are left with the possibility that the availability of electrons (and electron flow) within the mitochondrial membranes may be limiting for heme formation and, if so, that a lack of copper resulting in much lower than normal levels of active cytochrome oxidase might indirectly be responsible. Whether copper might be involved in some other way is still an open question.

As regards cytochrome oxidase, the thought lingers that reductions in its activity (perhaps not in liver as much as in reticulocytes) might indirectly play a role by reducing the energy available not just for heme, but also for hemoglobin formation. While Weisenberg *et al.* (1980) found that iron treatment restored levels of ATP and protein synthesis to normal in Cu-deficient liver cells of rats, we do not know that this would also be the case in hematopoietic cells in the bone marrow. Presumably, an effect of energy deprivation would be more apparent in rapidly proliferating cells (such as those of blood-forming tissues) where new copper must constantly arrive to allow for new synthesis of cytochrome oxidase.

A final point about interactions of copper with iron metabolism is that the fragility and life span of erythrocytes may also be compromised when little copper is available. Already in 1956, Bush *et al.* showed that turnover of erythrocytes was enhanced in the deficient pig, and Jain and Williams (1988) later found that red cell membranes of rats had malonyldialdehyde adducts (as well as more cholesterol and phospholipid) in copper deficiency. This may be explained both by a lack of erythrocyte Cu/Zn superoxide dismutase and by a lack of ceruloplasmin extracellularly, both of which cuproproteins protect red cell membranes against oxidative damage (see Chapter 4). [In contrast, findings that red cells of copper-deficient human subjects also have more arachidonate and docosahexanoate and less oleate (Cunnane, 1988) seem unlikely to be connected to a lack of ceruloplasmin and SOD.]

7.2.1.2b. Copper and Cardiovascular Function. As already described in relation to lysyl oxidase function (Chapter 6), copper deficiency has serious consequences for connective and vascular tissue function, which is dependent upon collagen and elastic fibers. Lysyl oxidase, a copper-dependent extracellular enzyme, catalyzes the oxidation of lysine amino groups, allowing proper cross-linking of the fibers. Therefore, severe copper deficiency will result in a weakening of the blood vessel wall, leading to hemorrhage and a greater likelihood of aneurysms (Shields *et al.*, 1962; Carlton and Henderson, 1963; Hunsaker *et al.*, 1984) and even aortic rupture (O'Dell *et al.*, 1961b).

Another consequence of copper deficiency only partly explained by failure to produce normal elastic fibers is cardiac hypertrophy (Table 7-8). This is accompanied by an abnormal electrocardiographic pattern and a greater likelihood of atrial thrombosis (Klevay, 1985) or cardiac failure (Bennetts *et al.*, 1948; Abraham and Evans, 1971; Gubler *et al.*, 1957; Waisman and Cames, 1967; Coulson and Cames, 1967; Waisman *et al.*, 1969a,b; Kitano, 1980). Also well documented is a swelling and increased mass of the mitochondria of the myocardial cells (the same occurs in hepatocytes), an increase in succinic denydrogenase activity (Table 7-8), and, of course, a marked drop in cytochrome oxidase (Fig. 7-8). Prohaska and Heller (1982) have documented that perfused hearts from deficient rats had slower rates of spontaneous contraction, developed less systolic pressure, and took on more fluid. Indeed, at least a major portion of the increased heart weight can be attributed to the change in mitochrondrial mass. These and other observations related to copper and cholesterol metabolism led Klevay to hypothesize that copper deficiency is connected to the prevalence of cardiovascular disease, including ischemic heart disease and atherosclerosis, so prevalent in Western society (Klevay, 1981, 1983, 1985) and underlying myocardial infarction (heart attack).

Part of the reason for the hypertrophy of cardiac mitochondria may well be related to a dysfunction of electron transport. (It is well known, for example, that isolated mitochondria swell when they are uncoupled or lack substrate.) The fall in cytochrome oxidase activity of heart tissue is especially pronounced in, and a relatively early consequence of, copper deprivation in rats (Fig. 7-8). Kopp *et al.* (1983) have reported that this is accompanied by decreased levels of ATP and phosphocreatine. The same defect may underlie some of the electrocardiographic changes that parallel the changes in mitochondria, as observed in rats (Kopp *et al.*, 1983). Prohaska and Heller's (1982) observation that norepinephrine concentrations in the left heart ventricle are reduced and that turnover is enhanced (Gross and Prohaska, 1990) may also have a bearing on these matters. [The production of catecholamines is also copper dependent through dopamine-β-hydroxylase (see Chapter 6).] Electrocardiographic changes have been recorded for healthy human volunteers undergoing copper deprivation: in one of three young men examined by Klevay *et al.* (1984b; Davis and Mertz, 1986) and in four subjects in the study of Reiser *et al.* (1985), a study that had to be terminated prematurely. In none of these cases (and especially in the latter study) was the deficiency very severe, as measured by serum ceruloplasmin or erythrocyte superoxide dismutase activities.

However, the study of Reiser *et al.* (1985) was complicated by the use of a high-fructose diet, which may in itself have had adverse consequences for heart muscle function relating to excessive sugar alcohol accumulation (Fields *et al.*, 1989). Since, apart from the study of Reiser *et al.* (1985), myocardial irregularities were only noted in one of a total of more than ten cases of copper deficiency in the human adult (Dunlap *et al.*, 1974; Vilter *et al.*, 1974; Palmissano, 1974; Zidar *et al.*, 1977), and this case was one with a relatively low serum copper concentration (the ones in the Reiser study were not), it is far from certain that the human differs from other species in having a heart that is *supersensitive* to copper deficiency (as suggested by Davis and Mertz, 1986). Rather (and certainly in the absence of confounding factors), the effects on myocardial rhythm would be more likely to occur only in severe copper deprivation. In humans, as in animals, sucrose (or fructose) may enhance the effectiveness of copper deprivation (Petering *et al.*, 1986; Cornatzer and Klevay, 1986) and enhance a marginal deficiency.

Another reason for the cardiac hypertrophy observed may have to do with the alterations in iron flow that accompany copper deficiency. Prohaska (1990) (and others before him, less definitively) have shown in mice that anemia *precedes* the development of hypertrophy. Fields and her colleagues (Fields *et al.*, 1990 and 1991a) have shown in rats that red cell transfusion prevents or reverses most of the cardiac hypertrophy of copper deficiency (as well as the pancreatic damage). Together these studies suggest that a lack of oxygen transport may be involved. However, other studies suggest instead that there is a lack of protection against oxidation; and/or that oxidative damage increases, with copper deprivation: Johnson and Saari (1989) report that both the cardiac hypertrophy and anemia are reduced by dietary antioxidants; while Fields *et al.* (1990b) found that desferrioxamine, the Fe (III) chelator that reduces tissue iron contents (and thus might reduce Fe-catalyzed inappropriate oxidations) reduced the tissue damage and anemia of fructose-enhanced copper deficiency. It thus seems likely that we will soon better understand the origins of these deficiency symptoms.

More "beneficial" effects of copper deficiency would be a reduction in systolic blood pressure, as reported by several research groups for rats (Fields *et al.*, 1984; Wu *et al.*, 1984; Medeiros *et al.*, 1984). Data of Moore and Klevay (1988) suggest that one of the intermediates here may be angiotensin I-converting enzyme, plasma levels of which are somewhat lower in copper deficiency.

It is well known that hypercholesterolemia is a risk factor in atherosclerosis and heart disease. In experimental animals, hypercholesterolemia has been produced by feeding diets low in copper by a number of investigators (Table 7-8). The results of others have been negative. In humans (with a less severe deficiency), a statistically significant but slight hypercholesterolemic effect has also been observed, as first reported by Klevay *et al.* (1986) (Fig. 7-11) and later by Reiser *et al.* (1987) for a group of 24 men. In the earlier human studies, mostly on severely ill patients given total parenteral nutrition (Dunlap *et al.*, 1974; Vilter *et al.*, 1974; Palmissano, 1974; Zidar *et al.*, 1977), nothing about cholesterol was noted, and it may not have been examined. Menkes' disease patients, who also suffer from a form of copper deficiency, did not demonstrate hypercholesterolemia (Blackett *et al.*, 1984), not did brindled mice (Prohaska *et al.*, 1985) that have a somewhat similar defect. The variability in response suggests that other factors (not yet identified or controlled) may be important for the development of hypercholesterolemia in copper deficiency. This is borne out by studies of Stemmer *et al.* (1985) with rats given purified diets based on egg white, casein, or milk powder as a protein source (plus lactose and cornstarch) and 20 or 140 ppm zinc (Table 7-11). Cardiac hypertrophy and lesions appeared to be worse for the rats on the milk powder diet than for those on the egg diet. Elevated cholesterol and triglycerides in the serum were only consistently obtained with the casein-based diet. (The cow's milk-based diet would also be high in casein, but the

Table 7-11. Effects of Different Diets on the Occurrence of Symptoms of Copper Deficiency[a]

Parameter	Egg	Casein	Milk powder	Control purified diet (20 ppm of Zn and 8 ppm of Cu)	Purina Lab Chow
Heart/body weight (g/100 g)	0.38, 0.47	0.36, 0.48	0.52	0.30–0.32	0.33
Heart lesions (score)	1.7, 1.8	1.5, 2.0	2.8, 2.5	0.2–0.5	0
Blood vessel lesions (score)	0.2, 1.3	0.7, 1.3	0.3, 1.2	1.0–1.5	0
Serum cholesterol (mg/dl)	104, 135	156, 154	142	112–120	89
Serum triglyceride (mg/dl)	105, 82	150, 162	117	83–105	76
Liver cholesterol	5.4, 5.5	5.5, 5.2	4.9	5.6–5.7	4.7

[a] From Stemmer *et al.* (1985) for studies on male rats on purified diets with different protein sources (20%), with 0.5 or 8 ppm of Cu and 20 or 140 ppm of Zn. Data for low and high zinc are shown sequentially for the egg and casein groups. (In the case of milk powder diets, only the lower zinc intake was studied.) Purina Chow with 18 ppm of Cu and 58 ppm of Zn, was studied in parallel. (It had 5% fat versus 10% corn oil in the other diets.)

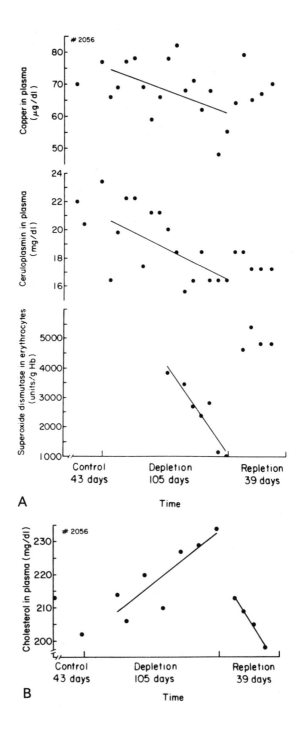

effect in rats on the milk powder diet was less.) The egg white protein diet deficient in copper produced no more triglyceridemia or cholesterolemia than the same diet with adequate copper. There were no effects on liver cholesterol concentrations, and no induction of vascular lesions attributable to copper deficiency. In further support of missing "factors" modulating these symptoms of copper deficiency, it is noteworthy that the purified diets with *adequate* copper and zinc produced some of the same pathology as that ascribed to copper deficiency, when compared to the results for rodent chow (Table 7-11). This was especially true for the vascular lesions. Another potential variable is sucrose (or fructose) intake (Petering *et al.*, 1986) (see Section 7.2.1.2c).

Other variables that may be relevant to possible effects on serum lipids include age and sex. Prohaska *et al.* (1985) demonstrated that hypercholesterolemia was not induced in male mice and rats by dietary deficiency instituted during gestation and continued in lactation. Other studies (Kramer *et al.*, 1988; Table 7-9) referred to in relation to hematopoiesis indicated that female rodents were often refractory to the apparent symptoms of copper deficiency, as has also been observed by Fields *et al.* (1989) in studies of rats on high-sucrose and high-fructose diets. Still another study has come up with a statistically significant positive correlation between serum copper and cholesterol in women, but not in men (Abu-Farsakh *et al.*, 1989), although, in general, women are less likely to have hypercholesterolemia than men and may have slightly higher ceruloplasmin concentrations (Linder *et al.*, 1981). A potential connection between genetic obesity and an increase in serum copper and ceruloplasmin has been reported for the Zucker rat (Serfass *et al.*, 1988). The reasons for these various findings are still a mystery.

Given the development of hypercholesterolemia and triglyceridemia in many instances during copper deficiency and its at least partial prevention by added copper, the potential mechanism of copper involvement in this occurrence must be examined. The serum cholesterol concentrations of fasting subjects (or animals) are the combined result of rates of secretion of very low density lipoprotein (VLDL) into the blood, mainly from the liver but also from the small intestine (see Linder, 1991b), and its rates of removal from the serum as high-density lipoprotein (HDL) and low-den-

←

Figure 7-11. Effects of copper depletion and repletion on serum copper parameters and cholesterol in a human subject. A healthy young man was placed on a diet with 0.8 mg of Cu per day, down from 1.3 mg in the control phase, and later repleted with 4 mg of extra copper per day. (A) Total serum copper, serum ceruloplasmin (determined by RID), and erythrocyte superoxide dismutase activity. (B) Total serum cholesterol concentrations. Reprinted, with permission, from Klevay *et al.* (1984b).

sity lipoprotein (LDL). Dietary factors (not directly related to copper) that influence these processes include some forms of saturated fats, n-3 fatty acids (as in fish oils), and fiber. In this connection, the effects of copper deficiency on the activities of enzymes involved in serum lipid disposal have been examined by a number of researchers, sometimes with contradictory results (Table 7-8). Lin and Lei (1981) reported that liver synthesis of cholesterol was unchanged in copper-deficient rats but that turnover of the rapid turnover plasma pool was a bit slower. Unfortunately, in these experiments the control rats (on adequate copper) used to determine the first two critical points in the curve were hypercholesterolemic, so the reliability of the findings is unclear. However, Valsala and Kurup (1987) have reported an increased activity of hydroxymethylglutaryl-CoA reductase and increased incorporation of [^{14}C] acetate into free liver cholesterol, implying that hypercholesterolemia might be the result of increased release of VLDL into the blood. This will need to be confirmed. Two research groups have reported that the activity of lecithin–cholesterol acyltransferase (L-CAT), important for the conversion of VLDL to LDL (and the esterification of cholesterol), is reduced in copper deficiency in rats (Table 7-8). The data of Harvey and Allen (1981) indicate that fractional and molar activities decreased about two-thirds in severe copper deficiency and resulted in about 50% more nonesterified cholesterol in the plasma. The turnover of cholesterol-containing lipoproteins might therefore also be slower, which would also promote plasma accumulation. However, Lefevre *et al.* (1985) found no consistent effect of deficiency on L-CAT, although cholesterol esterification was reduced, but HDL cholesterol and its lipoproteins were increased. Post-heparin lipoprotein lipase activity, which degrades chylomicrons and VLDL triglyceride in plasma for localized fatty acid release and fat storage, appears to be lower in copper deficiency (Table 7-8), though the changes reported are not very large (20–40%). This might in turn adversely affect the removal of lipid from the blood. In this connection, Lau and Klevay (1982) cited the finding of Huang and Lee (1979) that apolipoprotein B (an important component of LDL and VLDL) can be made soluble in aqueous solution by the addition of ionic Cu(II). However, as free ionic or chelated copper is not present in plasma in concentrations even approaching those used here (see Chapter 4), this seems an unlikely linkage between copper and LDL (or VLDL). [Moreover, the effect is due to extensive oxidation (Herzyl *et al.*, 1987).] In addition, Koo *et al.* (1988) observed an initial enhacement of chylomicron removal in copper-deficient animals, and Hassel *et al.* (1988) reported enhanced binding of apoE-rich HDL to hepatic membranes in copper deficiency, all suggesting an enhanced capacity for removal of cholesterol from the blood in copper deficiency.

It is clear from these various studies that the significance of copper availability to normal cholesterol transport and metabolism (and to normal triglyceride distribution) needs further documentation and that the mechanisms of direct copper action, if any, remain to be unraveled. To add to the confusion, there are several reports linking copper *supplementation* (or injection) with the opposite effect, namely, enhanced hypercholesterolemia. Klevay (1982) found that hypercholesterolemia induced in rats by feeding cholesterol and cholic acid was exacerbated by giving extra dietary copper (60 ppm above base levels in chow), at least for the first two months of these studies. (The difference was abolished at later times.) Since then, two groups of investigators have reported that treatment (by injection) increased serum cholesterol concentrations of rats (Tanaka *et al.*, 1987) and hamsters (Ohguchi *et al.*, 1988), perhaps by increasing the activity of hydroxymethylglutaryl-CoA reductase (Tanaka *et al.*, 1987). [Cholesterol ester hydrolase activity (of liver) was also decreased.] Conversely, using guinea pigs (which do not synthesize ascorbic acid), Veen-Baigent *et al.* (1975) found that diets very high is ascorbate (600 mg/day) lowered total serum copper concentrations but that this was accompanied by a *reduction* in levels of total and nonesterified cholesterol. It has been assumed that ascorbate can lower serum copper and ceruloplasmin concentrations by compromising copper absorption or excretion, or both, although this remains far from proven (see Chapters 3 and 4). With regard to the guinea pig studies cited, the critical question is whether there really was a copper deficiency. A decrease in serum copper and ceruloplasmin activity could be due to enhanced release (and cellular uptake) of copper from ceruloplasmin.

Relationships between other aspects of lipid metabolism, especially those involving myelination, have been reviewed in connection with brain function (Chapter 6) and brain development (Chapter 8).

7.2.1.2c. Copper and Carbohydrate Metabolism. Another area that remains poorly understood is that involving connections between copper and the metabolism of carbohydrates. There are several reports of glucose intolerance induced by copper deficiency in rats and humans. In the rat studies of Hassel *et al.* (1983), release of insulin appeared to be delayed but prolonged, suggesting that insulin action was less efficient. This was not evident in the studies of Klevay *et al.* (1986) on two human subjects given intravenous glucose. Cohen *et al.* (1982) reported a diminished insulin release in rats over the 2-h period after oral glucose and an enhanced release of insulin (with or without copper depletion) when rats were on diets high in sucrose as compared to starch. The total content of insulin in the pancreas in the fructose- or sucrose-fed rats was also greater (as if less had been released) (Recant *et al.*, 1986).

It appears unlikely that the effects of copper deficiency on glucose tolerance involve an abnormality in insulin production or secretion, as only the exocrine part of the pancreas appear to be affected in copper deficiency (Muller, 1970; Smith et al., 1982; Fell et al., 1982, 1985). [This also results in a decreased release of digestive enzymes (Lewis et al., 1987).] On the basis of in vitro studies, Fields et al. (1983b, 1984) suggested that the glucose intolerance may somehow be connected to impaired glucose oxidation and lipogenesis in fat cells, perhaps even involving a direct need for copper to enhance insulin action at the cell surface. However, no data were offered to support this concept (Fields et al., 1983b), and it was explained that addition of copper to adipose tissue in vitro (form and concentration unspecified) only had an ameliorative effect when copper was given 60 min after insulin (at which point most insulin would have been internalized and degraded). Insulin-like effects of copper on adipocytes were also observed by Saggerson et al. (1976), but copper was given as CuCl or $CuCl_2$ at concentrations far exceeding those ever encountered in vivo (6 mM) and in the absence proteins, which makes the physiological relevance of the observation questionable.

As concerns "the other side of the coin," Failla and colleagues have reported that steptozotocin diabetes in the rat leads to an accumulation of copper (on metallothionein) in both kidney and liver (Failla and Kiser, 1981, 1983; Chen and Failla, 1988). The underlying basis for this phenomenon is not clear, although the extra glucagon and corticosteroids circulating in these animals (Chen and Failla, 1988) would induce liver (but not kidney) metallothionein production, which then might trap extra copper (but this would not explain the extra kidney accumulation). Nevertheless, they found that insulin treatment reversed the copper accumulation (Failla and Kiser, 1981). However, in normal humans, two studies showed no effects of a glucose load or insulin infusion on serum copper concentrations over several hours (Davidson and Burt, 1973; Nestler et al., 1987). In genetically obese mice (that would be expected to have some glucose intolerance), ceruloplasmin activity was markedly increased, and the degree of increase was positively related to the amount of food consumed (mainly carbohydrate) (Prohaska et al., 1988), suggesting some other relationship to carbohydrate intake and adipose tissue function. A similar relation between genetic obesity and ceruloplasmin has been observed in the Zucker rat (Serfass et al., 1988). The inhibition of glucose-6-phosphatase activity, in the direction of glucose-6-phosphate synthesis from carbamoylphosphate and glucose, by large (10 μM) concentrations of Cu(II) has also been reported in vitro (Johnson and Nordlie, 1977), although, again, this seems physiologically unlikely except as a toxic effect of excess copper.

One intriguing new area where copper and carbohydrate metabolism may merge is in relation to sorbitol production from fructose. Sorbitol, the sugar alcohol of glucose or fructose, is implicated in many of the long-term pathological effects of diabetes (cataracts, angiopathy, etc.) (Linder, 1991). Fields *et al.* (1989; Lewis *et al.*, 1990) have shown with copper deprivation that a higher fructose (or sucrose) diet enhances tissue sorbitol accumulation, at least in male rats. This is likely to have long-term diabetic-like consequences and might help explain some of the eventual insulin resistance, hyperlipemia, and other symptoms of copper deficiency that may arise. The exact connection between sorbitol metabolism and copper, however, is currently still unknown. Fields *et al.* (1989) reported that both sorbitol dehydrogenase and aldose reductase activities of liver were reduced in fructose-fed animals on copper-deficient (versus copper-adequate) diets, but this was not the case in the kidney nor when starch was the carbohydrate source, suggesting that these enzymes do not depend on copper for their activities. Sorbitol itself has a high affinity for Cu(II) (Mitera *et al.*, 1989), with a stability constant that may be as high as 45—! It might thus keep copper from binding to important enzymes. However, one might also then expect the copper to accumulate within cells, along with the sorbitol, but this does not happen. Instead, sorbitol accumulation in copper deficiency is paralleled by accumulation of glyceraldehyde. This may have other damaging effects, especially when superoxide dismutase is low (Fields *et al.*, 1989). All of this needs to be further explored.

7.2.1.2d. Copper and Immunity. An association of genetic or dietary copper deficiency with increased susceptibility to infection has been evident for some time (Danks *et al.*, 1962b; Pedroni *et al.*, 1975; Mulhern *et al.*, 1987; *Nutr. Rev.* 1982, p. 107). Experimental evidence for a copper requirement of the immune system was first provided in 1981, by Prohaska and Lukasewycz, who examined deficient mice, and by Boyne and Arthur, working with deficient steers. Since then, there have been many reports from these and other laboratories indicating that a number of aspects of immune function may be dependent directly on copper availability. The main findings are summarized in Table 7-12.

Depending upon the severity of the deficiency, here induced by dietary means, there are decreases in the development (growth and weight) of the thymus gland and of T-cell subpopulations of the spleen, as well as splenomegaly and an increase in the percentage of spontaneously cycling spleen cells (cells in $S + G_2$ phases), which may be B cells or hematopoetic stem cells (Mulhern and Koller, 1988). Spleen cells are more resistant to mitogenic stimuli. Phagocytes, including natural killer cells (NKC), are less active. There is less thymic hormone secretion and sometimes a lower antibody titer. The acute (and perhaps the delayed) inflammatory response

Table 7-12. Changes in Immune Function in Rodents on Copper-Deficient Diets

Change	Reference
Hyporesponsiveness of lymphocytes (or lymphoid cells) to sheep RBC and T- and B-cell mitogens	Prohaska and Lukasewycz (1981) Vyas and Chandra (1983) Lukasewycz and Prohaska (1983) Lukasewycz et al. (1985) Blakely and Hamilton (1987) Mulhern and Koller (1988) Kramer et al. (1988)
Alterations in lymphoid tissues	Prohaska et al. (1983)
Decreased thymus weight, increased spleen weight	Koller et al. (1987) Mulhern and Koller (1988) Lukasewycz and Prohaska (1990)
Less thymic hormone	Vyas and Chandra (1983)
Decreased splenic T-cell subpopulations	Lukasewycz et al. (1985) Mulhern and Koller (1988)
Increased susceptibility of host to pathogens	Newberne et al. (1968) Jones and Suttle (1983)
Decreased antimicrobial activity of phagocytes	Boyne and Arther (1981) Jones and Suttle (1981)
Increased acute (and delayed) inflammatory response No change in delayed inflammatory response	Jones (1984) Koller et al. (1987)
Less IL-2 production (in inflammation)	Flynn et al. (1984)
PGE_2 production unchanged PGE_2 production decreased (in liver)	Koller et al. (1987) Lampi et al. (1988)
Decreased antibody titer (males more susceptible)	Blakely and Hamilton (1987) Failla et al. (1988) Koller et al. (1987) Prohaska and Lukasewycz (1989a) Lukasewycz and Prohaska (1990)
Decreased NKC activity	Koller et al. (1987)
Decreased resistance against transplanted leukemia	Lukasewycz and Prohaska (1982)
Less effective Con A stimulation of spleen lymphoid cells Not much change in Con A stimulation of spleen lymphoid cells	Kramer et al. (1988) Prohaska and Lukasewycz (1989a)
Decreased response to, and effect on, mixed lymphocytes	Lukasewycz et al. (1987)
Changes in lipid composition of lymphocyte plasma membranes	Korte and Prohaska (1987)
PGF_2 production decreased (in liver)	Lampi et al. (1988)

may be enhanced, along with less release of IL-2 (but not IL-1) and PGE_2. All of this points to a suppression primarily of T-cell rather than B-cell function. As in the case of hematopoiesis and the response to fructose intake, male rodents appear to be more susceptible than females to these symptoms. Age of onset of copper deprivation may also make a difference. Prohaska and Lukasewycz (1989a) reported that mice in whom copper deprivation was instituted at weaning rather than birth had lymphocytes that responded normally to concanavalin A, although B-cell numbers and antibody titers were reduced, and there was general evidence of severe deficiency. Deprivation during gestation or lactation alone had little effect (Prohaska and Lukasewycz, 1989b).

The actual molecular basis of the changes observed, related to copper function, is not yet understood. As in the case of hematopoiesis, one may speculate that a contributing factor could be a reduced availability of energy, due to less cytochrome c oxidase. Certainly, cell proliferation (so important in T-cell function) would require copper for cytochrome oxidase and Cu/Zn superoxide dismutase (SOD), if not also for other as yet unknown reactions or processes. Data from Prohaska and Lukasewycz (1989b) on cytochrome oxidase activities of spleen and thymus show marked (50%) reductions already after one week of copper deprivation. Much less copper, especially in the form of ceruloplasmin, would be available in the circulation in copper deficiency. [In the case cited, ceruloplasmin oxidase fell to 5% of normal in weanling mice after 3 days of deprivation (Prohaska and Lukasewycz, 1989a).] In this connection, the protective actions of ceruloplasmin against oxidative damage (in regions of the body where enhanced phagocytic activity is occurring) would also be diminished (Chapter 4) (as might the intracellular protective action of Cu/Zn SOD), although this may have no relevance to the immune response, *per se*. This may enhance the damage and swelling that occur in the inflamed area, as suggested by data of Kishare *et al.* (1990) on carrageenan-induced edema. As already detailed earlier in this chapter, the inflammatory response itself enhances production of ceruloplasmin (and other acute phase reactants) by the liver, which means that it alters the organ distribution and circulation of the element and could make ceruloplasmin copper more available. It is possible that part of this response (involving the actions of monokines) is to provide added copper for formation of new T cells.

Korte and Prohaska (1987) have also noted a change in the fatty acid composition of splenic lymphocyte and erythrocyte plasma membranes (but not those of thymic lymphocytes). Differences in fatty acid composition were evident, with more unsaturation. This has implications for membrane fluidity that might affect membrane function. A differences in membrane protein composition was also observed. The potential connections between these changes and copper functions are obscure at this time.

7.2.2 Other Potential Actions of Copper

7.2.2.1 Copper and Fertility

Already in the early 1960s, O'Dell et al. (1961a) discovered that rats on copper-deficient diets produced markedly fewer litters and offspring and that the offspring were born with more serious problems involving hematopoiesis and bone and vascular development than others on control diets or diets deficient in manganese or iron. Howell and Hall (1969) found that mating appeared to be normal but that litters were not born, mainly because fetal tissue was resorbed after the second week of pregnancy. Although there appeared to be no correlation between the total copper content of human sperm and infertility (Table 7-8), sperm motility was found to be dependent upon the copper content of the midpiece of the sperm (Battersby and Chandler, 1977; Morisawa and Mohri, 1972), where the mitochondria and cytochrome oxidase are located. Morisawa and Mohri (1972) suggested that most of the copper in the sperm is accounted for by this enzyme, at least in the case of the sea urchin. A similar lack of copper for new cell production and respiration may account for the resorption of fetal tissue during gestation. Lack of antioxidant protection, as provided by superoxide dismutase and ceruloplasmin, may also lead to damage and removal of the embryo. Whether other aspects of copper function are involved is unknown, although angiogenesis may be another critical feature (see below).

Another connection between copper and gonadal function that has been examined over several decades has derived from the observation that copper can induce or enhance ovulation in experimental animals (Suzuki et al., 1972). Cu(II) added to crude hypothalamic extracts injected into rabbits was much more effective in stimulating ovulation than the crude extract alone. Recent studies in this area have focused on the effect of 150 μM Cu(II)-histidine on the secretion of gonadotropin-releasing factor from explants of the median eminence in culture (Barnea et al., 1988a) and the synergism between copper and prostaglandin E_2 that may be involved (Barnea et al., 1988b). Although this may mean that copper is necessary for the release of gonadotropin (luteinizing hormone) or involved in regulating the release, these experimental phenomena must be viewed in the context of what we know about the availability of Cu(II) or Cu-amino acid chelates under normal conditions, as opposed to the experimental conditions employed. Rabbits were injected with intravenous doses of 400 μg of Cu/100 g of body weight, a very large dose. [Doses of 100 μg/100 g of body weight will cause hemolysis in some strains of rats (Linder *et al.*, 1979b).] However, in other studies (Sharma *et al.*, 1976), a much lower dose (300 μg per kg) given intravenously was also effective. In the *in vitro* studies with explants, Cu was added at concentrations greatly exceeding those likely to be found in the form of free ions or amino acid chelates in the blood plasma or interstitial fluid (see Chapter 4; also Tables 4-6 and 4-12).

A different kind of connection between copper and ovulation is that relating to egg formation in the chicken oviduct (Schraer and Schraer, 1965). In the immature oviduct, concentrations of Cu are about twice as high in the isthmus as in the magnum and shell glands. During ovulation and egg formation, isthmus concentrations rise even further. These changes in copper concentration may reflect a special need for transport of the element to the developing egg yolk or its involvement with the copper enzyme, lysyl oxidase, that will help lay down connective tissue in preparation for egg-shell mineralization.

7.2.2.2 Copper and Angiogenesis

As already mentioned, there are suggestions that copper has interactions with prostaglandin metabolism. The effect of copper salts on ovulation is synergistically enhanced by PGE_2 (Barnea *et al.*, 1988b); liver PGE_2 production and/or release is decreased in copper deficiency (Lampi *et al.*, 1988); also, Cu(II) has been shown to suppress formation of PGE_2 and enhance $PGF_{2\alpha}$ production in kidney slices (Fujita *et al.*, 1987). Another area of interaction may be in angiogenesis. Angiogenesis is the still rather mysterious process of developing and laying down a network of capillaries and blood vessels, to allow delivery of oxygen and nutrients to cells in new tissue and cast away their metabolites. Angiogenesis is important in normal growth, beginning with embryonic development. It may occur abnormally in connection with the development of malignant, as well as benign, tumors and has therefore been of interest to cancer research. [An excellent review is provided by Gullino (1978).]

The stimulus for angiogenesis is thought to be transferred by one or more cytokines or growth factors released locally to stimulate the proliferation of endothelial cells (or other cells) to form the blood vessels (D'Amore and Thomson, 1987; Zetter, 1988). Enhancement of collagenase activity or other proteases (Gullino, 1978) may also be necessary for this to go forward. Candidate peptides for stimulation include angiogenin, first isolated by Vallee and colleagues (Fett *et al.*, 1985; Shapiro *et al.*, 1987);

acidic and basic fibroblast growth factors (Fett et al., 1987; Schweigerer, 1988; Mignatti et al., 1989); transforming growth factors a and b (see Boboru, 1987); and tumor necrosis factor (Frater-Schroder et al., 1987; Schweigerer et al., 1987). All of these factors can be expressed by tumor cells and are thought to contribute to the development of tumor blood vessels. Angiogenin, however, is produced particularly by the liver, especially in the adult and least during fetal development, leading Weiner et al. (1987) to suggest that this peptide may not be critical to angiogenesis. Vallee's group found that angionenin is present in quite high concentrations in blood plasma (Shapiro et al., 1987) and induces angiogenesis *in vivo* (Fett et al., 1985). However, its action may be indirect, in that it elicits synthesis and secretion by endothelial cells of prostacyclin (Bicknell and Vallee, 1989), a prostaglandin not implicated in angiogenesis. Angiogenin is also a ribonuclease (Strydom et al., 1985; Lewis et al., 1989), which may yet prove of some significance. It does not appear to be a copper protein.

Information on the potential connection of angiogenesis stimulation to copper originally came from studies done in two different laboratories. McAuslan et al. (1980) reported that angiogenic factors were produced not just by tumors, but by normal tissues such as salivary gland and liver. Using *in vivo* assays involving vascularization of the kidney and cornea and *in vitro* assays using migration of endothelial cells in culture, McAuslan and colleagues found that $CuSO_4$ (at concentrations of 1 μM *in vitro*) induced vascularization or endothelial cell migration. Commercial ceruloplasmin preparations and extracts of bovine parotid gland, but not hemocyanin (another copper protein), were also active, and their interest focused on the possible involvement of a basic peptide of low molecular weight that might be a fragment of ceruloplasmin. In the laboratory of Gullino, it was observed that there was accumulation of copper in corneal tissue, prior to its vascularization by penetration of capillaries from the limbus (Ziche et al., 1982). Moreover, ceruloplasmin, some of its copper-containing fragments, Cu-heparin, and Cu-glycylhistidyllysine all stimulated angiogenesis (Raju et al., 1982) in the rabbit cornea assay (Table 7-13). These effects were enhanced by PGE_1 (Ziche et al., 1982). Moreover, indomethacin inhibited the angiogenic response to ceruloplasmin, supporting the concept that prostaglandin formation underlies (and is permissive for) the copper effect. In later work with an *in vitro* system, in which the mobilization of capillary endothelial cells was followed in the absence of the cornea (Table 7-13), only the Cu-heparin complex emerged as directly active (Alessandri et al., 1983). Ceruloplasmin and PGE_1 were ineffective, as were Cu-glycylhistidyllysine and an 11-kDa fragment of ceruloplasmin. A chemotactic factor (or factors) released from corneal tissue was implicated in mediating the *in vitro* effect of cerulo-

Table 7-13. Effectors of Angiogenesis in Vivo and in Vitro[a]

	Mobilization of capillary endothelium		
	Gelatin–agarose assay	Boyden chamber assay	Gelatin–agarose assay
Control	427[b]	21	461
PGE$_1$	2520[b]	441[b]	398–496
PGE$_2$	478	36	378–495
Ceruloplasmin fragment	1679[b]	241[b]	407–452
Apoceruloplasmin fragment	415	46	414–475
Heparin + Cu	2007[b]	267[b]	600–1003[b]
Heparin alone	453	25	375–423
Ceruloplasmin	—	—	398
Apoceruloplasmin	—	—	412
Glycylhistidyllysine + Cu	—	—	350–486
GHL alone	—	—	358–463
CuCl$_2$ alone	—	—	455–572

[a] Data of Alessandri et al. (1983).
[b] Significant effect.

plasmin and the other agents (Raju et al., 1984). Most recently, Ziche et al. (1987) have reported that stimulation of angiogenesis in corneal tissues begins with remodeling events that may involve copper-dependent amine oxidases, the activities of which are altered. (Benzylamine oxidase activity was increased.) This remains to be further explored.

We are left with strong feelings that copper, perhaps as a heparin complex, is somehow involved in angiogenesis. Apart from heparin, growth factors may also be active as copper complexes, particularly fibroblast growth factors (FGFs), where special affinity for copper has been recognized [and even used as an affinity tool for purification (Shing, 1988)]. Some of the copper-binding components of lower molecular weight fractionated by size exclusion chromatography (see Chapter 4 and Fig. 4-16B) may, in fact, be these same factors. Finally, still another monokine, named angiotropin, has been implicated, this one containing not only copper but also RNA (rich in guanine) (Wissler et al., 1986). This 25-kDa RNA (and 1 Cu), purified and studied extensively by Wissler's group, may be a critical link.

The tripeptide glycyl-histidyl-lysine (GHL) as a copper complex may be another example of a Cu-dependent cofactor, although questions linger about its actual occurrence in vitro (see Chapter 6). Recent data from several laboratories have confirmed earlier work indicating growth regulatory effects of the complex. These include stimulation of hepatocyte and hepatoma cell growth (Morris hepatoma 7777) (Barra, 1987); enhancement of the settlement and metamorphosis of barnacle larvae, at about $10^{-9}\,M$ (Tegtmeyer and Rittschof, 1988); and stimulation of collagen synthesis in human fibroblasts, beginning at concentrations as low as $10^{-12}\,M$ (maximum at $10^{-9}\,M$) (Marquart et al., 1988). (The presence of the GHL triplet in the α-chain of type I collagen was also noted.) It thus seems likely that this or similar peptides (complexed with copper) might have such actual regulatory effects in vivo. A review of potential actions has been provided by Pickart and Lovejoy (1987).

Copper in Growth and Development

8.1 Gestation

8.1.1 Maternal Responses

Gestation involves changes in copper metabolism and balance that allow transfer of significant amounts of copper from mother to fetus. Presumably, this occurs already very soon after implantation, as the new cells in the growing embryo will at minimum each need their complement of cytochrome oxidase for energy generation through oxidative phosphorylation (see Chapter 6). Copper may also be needed for Cu/Zn superoxide dismutase and ceruloplasmin to inhibit damage by free radicals (see Chapters 4 and 6) and for specialized enzymes in brain, connective tissue, and other areas. The maternal organism must therefore rise to the occasion, and it does so in several ways that we can currently understand may serve fetal needs.

One of the principal changes accompanying pregnancy is a rise in the concentration of circulatory estrogens and progesterone. This may be responsible for the substantial rise in serum ceruloplasmin (Fig. 8-1), since ceruloplasmin is responsive to estrogen (and progestogen) administration (see Chapter 7). Ceruloplasmin, in turn, is likely to be the major source of copper for transfer to the fetus (see Chapter 3 and Lee *at al.*, 1991). Concomitant with the rise in ceruloplasmin, the total serum copper concentration increases. Indeed, total Cu and ceruloplasmin increase more in pregnancy than in almost any other normal (or pathological) state. Already in 1944, Nielson reported that total serum copper rose from 1.2 µg/ml to 2.7 µg/ml in the first three months of human pregnancy (Nielson, 1944a, b). Later findings suggest that the initial increase may be more modest, from about 1 µg/ml to 1.5 or 1.6 µg/ml in the first trimester (Fig. 8-1), with further increases to 1.9 µg/ml in the second trimester, and

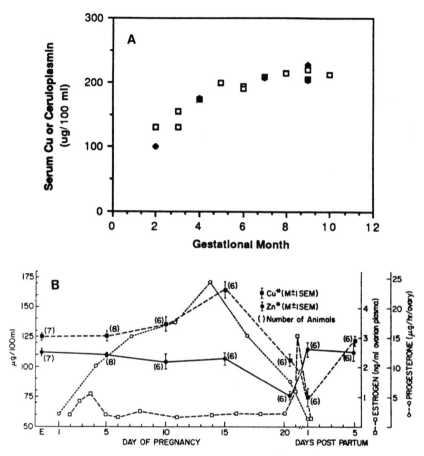

Figure 8-1. Changes in the serum copper and ceruloplasmin during pregnancy. (A) Data for the human, compiled from Studnitz and Berezin (1958) and Hambidge and Droegemueller (1974) for ceruloplasmin (□) and from Denko (1979), Kiiholma et al. (1984a), and Yamashita et al. (1987) for serum copper (◆). (B) Changes in plasma Cu and ceruloplasmin oxidase activity in the rat. Data of Sato and Henkin (1973).

to about 2.2 μg/ml at term, returning to normal by about the fifth week postpartum (Kiiholma et al., 1984b). [Others report 1.9–2.0 μg/ml in the first trimester (Ozgunes et al., 1987) and similar or lower values at term: 1.7 μg/ml (Bogden et al., 1978); 1.50 μg/ml (Hillman, 1981); 1.9–2.1 μg/ml (Yamashita et al., 1987).] Ceruloplasmin concentrations (determined by immunodiffusion) also increase in pregnancy (Fig. 8-1) The additional ceruloplasmin probably more than accounts for all of the increase in total serum copper observed, although it only accounts for 60% of the copper in *normal* adult serum (see Chapter 4). Recent data (Kiiholma et al., 1984a; Salmenpera et al., 1986) indicate that maternal Cu and ceruloplasmin concentrations at term are 2.0–2.4 µg/ml and 0.60–0.75 mg/ml, respectively, as compared with 1.2 µg of Cu and 0.32 mg of ceruloplasmin per ml in nonpregnant women. Assuming that the increased ceruloplasmin contains an average of six Cu atoms per molecule (which may be high), it would entirely account for the additional 0.8 to 1.2 µg of Cu/ml in maternal serum.

The changes in serum copper seen in the human are seen to a lesser extent in other species that have been examined. In the horse (with a somewhat longer gestation), serum copper increases significantly from 0.96 to 1.18 µg of Cu/ml in the first three months, reaching the highest values (of 1.35 µg/ml) at six months, and gradually tapering down (to about 1.10 µg/ml) at term (Auer et al., 1988). In the rat, serum copper and ceruloplasmin activity (Sato and Henkin, 1973) have their major increase in the second week of pregnancy (Fig. 8-1B). This is followed by a sharp fall during the third week, with values returning to normal in the first week postpartum. Serum copper increases a total of 40–50%, and ceruloplasmin activity almost doubles. The lesser rise in serum copper and ceruloplasmin in the rat may reflect the lesser maturation of the fetuses during gestation in this species and (directly after birth) the greater transfer of copper from the mother through the milk (see later in this chapter). However, it also suggests that not just estrogen but other hormones may be involved, since the time courses of changes in estrogen secretion, serum copper, and ceruloplasmin are not strictly parallel (Sato and Henkin, 1973).

Another effect of pregnancy appears to be a greater retention of dietary copper. King and Wright (1985), studying copper balance in pregnant women, found that more copper was retained (and/or absorbed) and less excreted during gestation. Consistent with this, Terao and Owen (1977) reported that for rats the concentration of copper in bile declined dramatically during gestation and only returned to normal after the suckling perod. [The bile is probably the major route for copper excretion (see Chapter 5).] This effect is probably due to estrogen, since Linder et al. (1986) found that estrogen treatment of female rats reduced the excretion rate of whole body copper, measured with ^{67}Cu. The effect of pregnancy on copper retention by rats was also monitored by Williams et al. (1977), who found that the total copper content of the maternal body increased about 10% by the middle of pregnancy (Fig. 8-2A) and much more rapidly thereafter, attaining a maximum about 50% above normal by the end of term. These values do not include uterine, placental, and fetal copper, which also accumulate, especially in the last trimester (third week of pregnancy in rats) (Fig. 8-2B). Placental superoxide dismutase activity also increases

304   Chapter 8

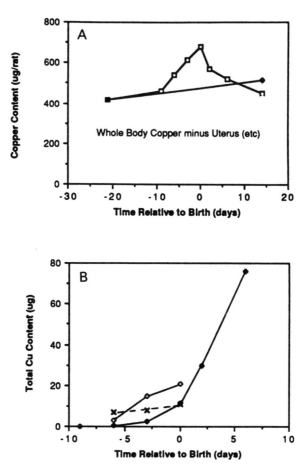

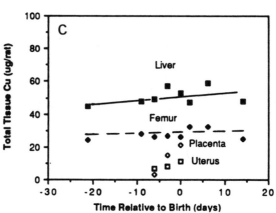

throughout gestation (Van Hien *et al.*, 1974), as measured in humans. In rats, the liver and bones of the mother do not account for the accumulation of copper in the maternal body, as their copper content remained very constant (Fig. 8-2C). The doubling of total serum copper and the enlarged blood volume of the pregnant rat would only account for perhaps 7% of the additional copper accumulated (assuming an increase of 10% in body weight, not including the uterus, etc.). One wonders therefore where the rest of the accumulation has occurred. [Some of it may be in the mammary gland (see below).]

8.1.2 The Fetus

In the rat, the accumulation of copper in the fetus itself appears to accelerate in the very last period before birth and continues at a fast rate in the first weeks of life (Fig. 8-2B). The copper contents of the uterus and placenta are greater than that of the fetus until birth (Fig. 8-2B), at least in the rat (and the mouse; Masters *et al.*, 1983). As the rat is relatively immature at birth, the rapid rise in fetal copper accumulation probably occurs earlier (well before birth) in the human and other species. Changes in liver copper probably reflect changes in total body copper, and these reach a maximum about two weeks after birth in the rat (corresponding to about six months of age in the human) but are maximal well before birth in the case of the human (see below). Nevertheless, it appears that probably in all species the most rapid period of copper accumulation is in the latter part of gestation. In general (copper being adequately available in the diet), the accumulation of fetal copper does not reduce the copper stores of the mother, although such a reduction may occur in cows and sheep, where liver (Gooneratne and Christensen, 1985) and blood (Butler, 1963) copper concentrations have been noted to decline during gestation.

As regards copper transport to the fetus, the very earliest uptake may be from the uterus itself, after implantation. The bases for this conjecture are recent findings by Thomas and Schreiber (1989) that mRNA for ceruloplasmin is expressed by cells of the normal (nonpregnant) or pregnant uterus—in fact, in concentrations half as great as those of normal liver (where most ceruloplasmin is produced). This suggests that cerulo-

---

Figure 8-2. Changes in the copper content of the whole body and tissues of the mother during pregnancy in the rat. (A) Total copper content of two groups of the same rats with (□) and without (■) pregnancy. (B) Total copper content of uterus (x), placenta (◇), and fetus (◆) before and after birth. (C) Copper contents of maternal tissues: liver (■), femur (◆; values have been multiplied by 10), uterus (□), and placenta (◇). Data of Williams *et al.* (1977).

plasmin secreted by the uterus, perhaps to the mucous surface, is available there (a) to provide copper for the implanted developing embryonic cells (see Chapter 3), and (b) to offer antioxidant protection to them.

In later stages of gestation, the placenta (and yolk sac plus uterus in the rat) probably plays the major role in copper delivery. The mechanisms of delivery have not yet been elucidated. However, ceruloplasmin (which is elevated in the maternal blood) may deliver copper to placental cells via specific cell surface receptors. This copper may, in turn, be incorporated into new ceruloplasmin and released into the fetal bood. In support of this concept, Mas and Sarkar (1988) have shown that $100,000 \times g$ supernatants of placental yolk sac homogenates of rats near term fractionated on columns of Sephadex G-150 contain a peak labeled with ^{67}Cu eluting in the position of ceruloplasmin, 2 and 24 h after intravenous administration of ionic radiocopper to the mother. (See also Figs. 4-19, 6-3, and 6-4.) (Peaks corresponding to metallothionein and perhaps even transcuprein are also present.) Moreover, the specific activity of copper in placenta (and fetal liver) was much lower than in maternal liver and kidney at 2 h and increased with time, implying that incorporation of copper into maternal ceruloplasmin precedes its transfer to the fetus. Recent studies of Linder *et al.* (1990a; Lee *et al.*, 1991) indicate that, at least in the rat, ceruloplasmin is the most important and possibly even the only source of copper for the fetus. This conclusion is based on a comparison of the rapidity and efficiency of ^{67}Cu-uptake by placenta and fetus when infused intravenously as ^{67}Cu-ceruloplasmin or ^{67}Cu(II) attached to transcuprein and albumin. Thus the ceruloplasmin found in placental yolk sac extracts by Mas and Sarkar (1988) certainly comes at least partly from residual blood in the maternal circulation. However, placental (or yolk sac) expression of the ceruloplasmin gene also occurs near term, as demonstrated in the rat by Thomas (1989), and more recently by Linder *et al.* (1990a; Lee *et al.*, 1991). Consequently, some of the copper delivered by maternal ceruloplasmin may also be released into the fetal circulation on new ceruloplasmin. However, at least in the rat, most of the copper in the fetal circulation is initially found not on ceruloplasmin but in association with transcuprein and α-fetoprotein (and albumin) (Lee *et al.*, 1991). In the rat, the yolk sac may also be important, and Thomas and Schreiber (1989b) have proposed that the yolk sac epithelium may take up copper from ceruloplasmin secreted by the uterine epithelium, in the last five to six days of gestation, after the decidua capsularis and Reichart's membrane break down. (This would not occur in humans.) In the rat, yolk sac cells (and uterus) express much more ceruloplasmin mRNA than the placenta. The timing of decidua capsularis/Reichart's membrane breakdown does somewhat coincide with that for accelerated accumulation of copper in the rat fetus, near term (Fig. 8-2B). In this period, expression of mRNA for ceruloplasmin by the fetal liver has also begun (see below). (It begins much earlier in humans.) In addition, copper appears to be released into the fetal circulation from the placenta (or elsewhere) in ionic form, attaching to transcuprein, α-fetoprotein, or albumin for further transport to the liver and other tissues (Lee *et al.*, 1991) [Genes for albumin and α-fetoprotein are expressed by fetal liver well before birth (Tilghman and Belayew, 1982).]

Very recent studies of McArdle and Erlich (1990) indicate that, in the mouse, uptake of copper by placenta and fetus occurs rapidly when ionic copper is used as a source. This copper would bind to albumin and transcuprein (and maybe even other proteins) in the maternal circulation, which might then deliver copper to the placenta for transfer to the fetal circulation. Evidence that this might be the case comes from the observation that small but significant amounts of copper did appear to enter the placenta almost immediately after intravenous infusion of radiocopper into the mother. [The uncertainty here is only that the radioactivity recorded might have been due (at least partly) to residual blood.] Another important point, however, is that less of the copper infused appeared to enter the maternal liver than is the case for nonpregnant rodents, suggesting a shunting of incoming copper away from the liver and to the fetus. Nevertheless, only about 12% of the injected copper had entered the 14-day fetuses two days after administration of the copper radioisotope (assuming 12 pups per litter), with another 3–4% in the placentas. (About half of the radiocopper in the fetuses was in their livers.) At this stage, about 10% of the original dose remained in the maternal circulation, and only about 3% in the maternal liver. One wonders where the rest of the copper went.

Transport of copper entering the fetal circulation and amniotic fluid from the mother probably involves the usual transport proteins found also in adult blood plasma and other fluids, although this has not been well studied. Gitlin and Biasucci (1969) reported the presence of ceruloplasmin already at 6.5 weeks postconception in the blood of the human fetus, rising until 27 weeks and then staying about the same until term. (Ceruloplasmin was measured as oxidase activity.) In the rat, ceruloplasmin mRNA is expressed by the liver by the end of the first gestational week. Not all of the circulating ceruloplasmin may be derived from the liver, however, as the yolk sac and uterus express the ceruloplasmin gene (at least in the rat) (Thomas, 1989; Thomas and Schreiber, 1989). [In fact, there is a report that maternal ceruloplasmin may cross into the fetal circulation, at least in pigs (Malinowska, 1986).] Fetal lung also appears to be a major producer of ceruloplasmin (Fleming and Gitlin, 1990), and this continues through the first weeks of life in the rat (but not in the adult). [The guinea pig may be exceptional, as Bingle *et al.* (1990a, b) have reported a lack of ceruloplasmin oxidase activity (in the serum) or mRNA (in the liver) before birth, but its rapid induction within hours after birth.] The amniotic fluid also appears to have ceruloplasmin, as Chan *et al.* (1980) have reported the presence of oxidase activity, measured with *o*-dianisidine (one of the common substrates used to monitor ceruloplasmin concentrations in adults). The data are difficult to interpret, as the units are arbitrary and values for normal adult serum were not given for comparison. Two to three "units" of activity per liter were present from 20 to 38 weeks of gestation, with little variation. Amniotic fluid has also been shown to have some copper (Table 4-13; Nusbaum and Zettner, 1973; Chez *et al.*, 1978; Shearer *et al.*, 1979; Chan *et al.*, 1980); and data obtained by Gardiner *et al.* (1982) with gel filtration and immunoprecipitation suggest that ceruloplasmin accounts for much of the copper in human amniotic fluid. However, Cu/Zn superoxide dismutase is also present in significant quantities (at levels of about 140 ng/l) (Holm *et al.*, 1980) and some of the copper is also with components of low molecular weight. Total Cu concentrations are roughly in the range of those for fetal serum and may decrease with gestational age, as the volume of amniotic fluid expands. Chan *et al.* (1980) reported values of 0.17 μg of Cu/ml at 20–30 weeks, 0.13 μg/ml at 30–35 weeks, then falling to 0.08 μg/ml and 0.07 at 35–38 and >38 weeks, respectively. Ceruloplasmin may thus play an important role in copper delivery to fetal tissues, after its release from fetal liver or from other tissues associated with the gestational apparatus.

Albumin and α-fetoprotein are also ready to aid in copper delivery well before birth, and the same may be true for transcuprein and low-molecular weight ligands, although this has not been well studied. [Data of Lau and Sarkar (1984) for human cord blood at term suggest that transcuprein and low-molecular-weight ligands (Fig. 4-11) are important copper-binding components in fetal blood (see Chapter 4); this has been confirmed by Lee *et al.* (1991).] With regard to albumin and α-fetoprotein, both genes are expressed in mouse liver very early in gestation (Tilghman and Belayew, 1982) and are also expressed in yolk sac, fetal gut, and some other tissues. It is noteworthy that the initial expression of both genes follows a parallel pattern, expression rising rapidly in the third (last) week of the pregnancy. Albumin gene expression then rises to a maximum by the end of the first week postpartum, while α-fetoprotein gene expression falls to almost nothing. This implies that α-fetoprotein is not just substituting for albumin in gestation. Certainly, in most species, albumin is capable of (and active in) copper transport, especially for ionic copper released into the serum from the diet on its way to the liver and kidney (see Chapters 3 and 4). α-Fetoprotein, which has a similar critical amino acid sequence at the N-terminus, involving a histidine (Fig. 4-9), would, theoretically at least, be capable of participating as well. [Mouse α-fetoprotein, on the other hand, has the N-terminal sequence Leu-His-Glu-Asn (Gorin et al., 1981) and is missing the terminal aspartate (or threonine in human α-fetoprotein), which may also be important.] Based on studies of infants born at various stages of immaturity (Zlotkin and Casselman, 1987), serum albumin concentrations were reported to rise from a median of 24 g/liter at 24 weeks to about 28–30 g/l by 32 weeks until term. Together with low-molecular-weight ligands, these various copper transport proteins offer the fetus plenty of options for distributing copper to its different body parts.

8.1.3 Fetal Liver

In the fetus, the liver is a major site for copper accumulation in the last part of gestation. Indeed, in the rat (with three weeks of gestation), the percentage of total fetal copper found in the liver rises from about 15% at six days to perhaps 30% at three days before birth and to a much higher percentage at birth itself (Fig. 8-3A). In the rat, this accumulation then continues postpartum, while in the human (and perhaps in most species) it reaches a maximum before birth (Fig. 8-3B). [In the mouse, there may also be a slight rise in the first five days after birth (Keen and Hurley, 1980).]

The precipitous increase in liver copper in the latter part of gestation (as exemplified by that in the rat from three or four days to birth; Fig. 8-4A, B) is a much studied phenomenon and has been found to involve deposition of copper bound to metallothionein. [The concentration of metallothionein in the liver of the rat is, however, not regulated by the copper, but by the zinc entering this organ (Paynter et al., 1990).] Cu-metallothionein accumulates mostly in the nucleus (see Table 6-1 and Chapter 6). Evans et al. (1970c) and later Riordan and Richards (1980) and Kern et al. (1981) demonstration *indirectly* that accumulation was in the nucleus and/or mitochondria. Cherian and co-workers (Panemangalore et al., 1983; Nartey et al., 1987b) then definitely demonstrated nuclear deposition. In the rat, liver metallothionein (in the nuclei) continues to accumulate after birth (Fig. 8-4B). It is lost preferentially from the nuclei toward the end of the suckling period (see also Table 6-1). Metallothionein mRNA (Fig. 8-4B) becomes detectable at about the same time before birth as metallothionein protein. Presumably, these events also occur in other species.

The deposition of copper in the liver attached to metallothionein means that a store of excess copper accumulates there, in preparation for

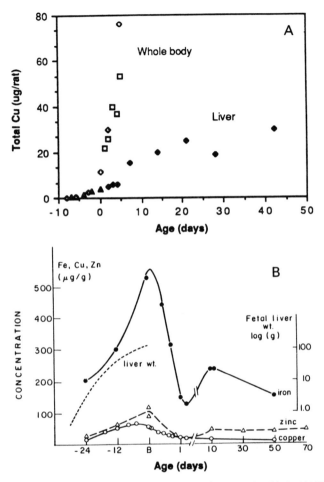

Figure 8-3. Copper content of fetal tissues in the rat, before and after birth. (A) Total copper content (μg) of the fetus (□) and liver (♦) of rat pups before and after birth. Data from Kern *et al.* (1981), Williams *et al.* (1977), and Terao and Owen (1977) have been combined. (B) Concentrations of copper, iron, and zinc in human liver before and after birth. Reprinted, with permission, from Linder (1978a). Changes in liver weight (— — —) are also shown. Dates on x-axis are in weeks (before birth: B) or years (after birth).

the demands of rapid growth during the suckling period. The accumulation of this store is dependent upon the copper status of the mother, as demonstrated in the mouse by Masters *et al.* (1983). Maternal mice with severely reduced tissue copper concentrations (including one-sixth the serum copper and one-third the kidney copper) had fetuses close to term with liver copper concentrations one-seventh of the normal (1.9 μg of Cu/g

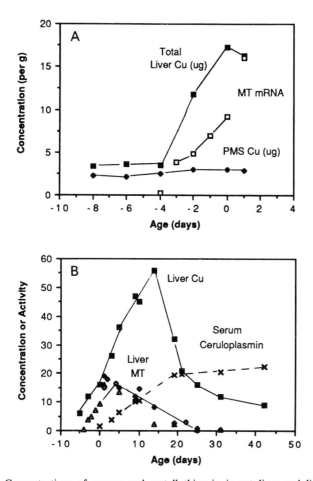

Figure 8-4. Concentrations of copper and metallothionein in rat liver and liver fractions before and after birth. (A) Concentration of copper in total liver (■) and postmitochondrial supernatants (♦), as well as of metallothionein mRNA (□) from Andrews et al. (1987). Data for liver copper are from Kern et al. (1981), and those for metallothionein are from Cherian et al. (1987). (B) Concentrations of total copper in the liver (■) from Evans et al. (1970c) and metallothionein protein ($\triangle$, ♦) ($nmol/g \times 10^{-1}$) from Cherian et al. (1987), Gallant and Cherian (1987), and Panemangalore et al. (1983) are shown in relation to age for rats. Activities of ceruloplasmin (X) (Evans et al., 1970c) is also indicated.

wet weight). It is noteworthy that, in the deficient state, a larger percentage of the copper ingested by the mother went to the products of conception (including the pups). Indeed, it is a general finding that newborn mammals have a store of copper in the liver and, indeed, have liver copper concentrations far exceeding those of adults (Table 8-1). (Concentrations of copper

Table 8-1. Concentrations of Copper in Livers of Fetal, Newborn, Suckling, and Adult Mammals[a]

Species	Copper concentration (µg/g dry wt.)			
	Fetus (3rd trimester)	Newborn	Suckling	Adult
Deer[b]	1400	—	—	(40)[c]
Horse[d]	320	219	—	31
Cow	170	550	—	70
Human	160	300	89[e]	25
Sheep[f]	(190)	(260)	—	25
Pig[g]	—	(230)	(180)	35
Dog[h]	—	200	300	300
Guinea pig	122	203	—	17
Mouse[i]	64	119	169	20
Rat[j]	45	80	88	18

[a] Modified from Linder and Munro (1973).
[b] Reid et al. (1980).
[c] Values in parentheses are wet-weight values multiplied by 4 to make them equivalent to those for dry weight.
[d] Egan and Murrin (1973) and O'Cuill et al. (1970).
[e] Lapin et al. (1976).
[f] Moss et al. (1974), Gurtler et al. (1973), and Owen (1982a).
[g] Linder (1978a), Owen (1982a), and Chang et al. (1976).
[h] Keen et al. (1981).
[i] Camakaris et al. (1979) and Keen and Hurley (1979).
[j] Kern et al. (1981) and Terao and Owen (1977).

in newborn deer livers are particularly high.) The dog (and sometimes the sheep) appears to be different in that the high liver levels at birth are maintained or exceeded in later life, due to a decreased capacity for biliary copper excretion (see Chapter 5).

Since storage of copper in fetal liver is such a general phenomenon and begins well before birth, copper may be obtained more easily from the maternal circulation than via the milk. Milk (the principal food of the newborn and infant) is not rich in copper. (The same is true for iron and zinc.) Exceptions may be the rat, pig, and mouse (Table 8-2), and, in fact, the liver of the rat pup continues to accumulate copper after birth. The same thing may happen in the mouse: mouse liver has more copper during the suckling period than at birth (Table 8-1). In the pig, on the other hand, as in human liver, copper concentrations fall already during the suckling period (see Table 8-1 and below). There may also be a connection between the rate of growth of the newborn and the copper content of the milk in different species. Certainly, the rate of growth of mouse and rat pups is much greater than that of pigs, and the rate for the latter will again be

Table 8-2. Copper Concentrations in the Milk of Different Mammals[a]

Species	Milk Cu (μg/ml)	Cu secreted at peak yield (μg/day) (A)	Maternal body wt. (kg) (B)	Ratio A/B (Cu/kg)	Newborn liver Cu[b] (μg/g dry wt.)
Human	0.25 (1.2)[c]	330	60	5.5	300
Cow	0.15	4500	500	9.0	550
Goat	0.15	750	50	15	—
Sheep	0.25	830	50	17	260
Pig	0.75	5250	150	35	230
Rat	1.5 (1.2)[c]	150	0.3	450	80
Mouse	3.2[d]				119

[a] Modified from Suttle (1987).
[b] From Table 8-1.
[c] Serum Cu, normal adults.
[d] Prohaska (1989).

greater that for the human infant. [It is noteworthy that the concentrations of copper in rat (and pig) milk are unusually high and do not reflect species differences in *serum* copper concentrations (Table 8-2).]

8.2 After Birth

8.2.1 Serum Copper and Ceruloplasmin

Just as newborn mammals are born with stores of extra copper, so serum copper and ceruloplasmin concentrations are low at birth (Fig. 8-5A, B). In the human, serum copper and ceruloplasmin concentrations (the latter determined by radial immunodiffusion) at parturition are about one-third those of the adult (Tyrala et al., 1985; Salmenpera et al., 1986; Fay et al., 1949; Henkin et al., 1973; Scheinberg et al., 1954). Both rise rapidly during the first 10–12 weeks of life. Ceruloplasmin reaches adult levels by about five months and remains at or above that for the first year. Total serum copper achieves adult levels more slowly and does not rise above them (Fig. 8-5A). In concert with its immaturity, the rat has an even lower concentration of serum ceruloplasmin (and copper) at birth (Fig. 8-5B), as determined by measuring p-phenylenediamine oxidase activity (see Chapter 4). However, this also rises rapidly after birth, reaching levels half of those in the adult by the middle of suckling (Fig. 8-5B) and climbing more slowly thereafter. Ceruloplasmin mRNA is expressed by the

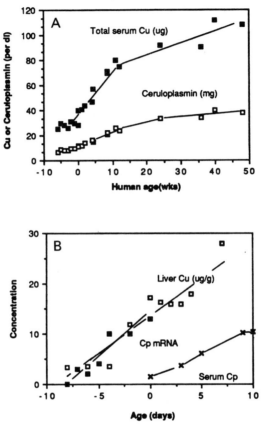

Figure 8-5. Serum copper and ceruloplasmin after birth. (A) Concentrations of total serum copper (■) and ceruloplasmin protein (determined immunologically) (□) in human infants. Data for premature infants were combined with those for full-term infants, with and without breast feeding, because the patterns of accumulation after term were identical. Combined data are from Tyrala *et al.* (1985) and Salmenpera *et al.* (1986). (B) Serum ceruloplasmin, liver copper, and ceruloplasmin mRNA, in the rat. Data for ceruloplasmin oxidase activity (x) (in relative units per ml) are from Evans *et al.* (1970d); those for liver copper (□) are from Terao and Owen (1977) and Kern *et al.* (1981); and those for ceruloplasmin mRNA (■) are from Thomas (1989).

liver of the fetal rat already at the start of the third trimester and rises very much in parallel with the accumulation of liver copper (Fig. 8-5B). Pigs would appear to be an exception, as Chang *et al.* (1976) reported that they are born without plasma oxidase activity against *p*-phenylene diamine but that it develops very rapidly and is detectable already 15 h after birth. This is not just a reflection of the presence of an inactive ceruloplasmin, as no precipitin band was detected in Ouchterlony double immunodiffusion of newborn serum against anti-adult pig ceruloplasmin. In contrast, human infants born an average of six weeks before term have considerable circulating ceruloplasmin (with concentrations about 20% of those in adults) (Fig. 8-5A); they have concentrations of total serum copper greater than 20% of those of adults (Fig. 8-5A). The pattern of increase seen in the premature infants, after birth, with no addition of copper to parenteral feeding mixtures (Tyrala *et al*, 1985) (so they are drawing on internal stores) appears to follow what is normal for age irrespective of maturity at birth: The curves for premature and full-term infants merge after birth. Breast feeding does not change the kinetics of these parameters, when prematurely born and normal-term infants (with and without breast feeding) are compared. However, the values for ceruloplasmin among groups match more closely than those for total serum copper. Serum copper appears higher in the premature infant, which may reflect differences in other copper components.

8.2.2 Lactation

Maternal responses to birth and lactation have already been touched upon (see captions to Figs. 8-1 and 8-2). In the rat, the total copper content of the mother's body returns to normal very rapidly (Fig. 8-2A), within a week after birth (corresponding to about three months in the human). That of liver and bones remains constant throughout (Fig. 8-2C). However, concentrations of serum Cu (and ceruloplasmin) appear to be almost "normal" at parturition (Fig. 8-1B) (Spoerl and Kirchgessner, 1975; Terao and Owen, 1977), and ceruloplasmin oxidase activities may even fall initially (Terao and Owen, 1977). In humans, serum Cu and ceruloplasmin concentrations are high at parturition (Fig. 8-1A) and return to normal within a month or two. Recent data of DiSilvestro (1986) show that, in lactating women, these serum parameters are still well above normal one month postpartum, with plasma copper values 65% and ceruloplasmin oxidase activities 90% above those of nonlactating women.

Copper concentrations of the milk also fall during lactation, as illustrated by data in Fig. 8-6 for humans. The high concentrations of Cu in early milk are attributable to the release of colostrum (the rich milk secreted during the first days after birth). The rest of the fall in milk copper concentrations may parallel the fall in maternal body copper content (Williams *et al.*, 1977), observed in rats (Fig. 8-2A). Apart from possible accumulation in mammary tissue, it is still unclear just where the extra copper in the mother is located (see Section 8.1.1), copper that would be

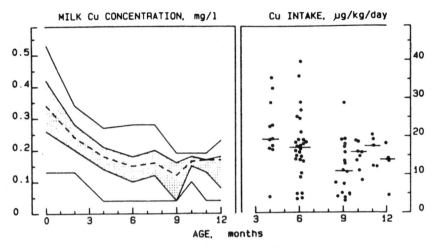

Figure 8-6. *Left*: Concentration of copper in human milk during lactation, for mothers feeding exclusively by breast, as a function of age of the infant. The dashed line shows the median values; the areas on either side of this line indicate the 50% and 90% confidence limits. *Right*: Daily copper intake by the infants in the same study. Reprinted, with permission, from Salmenpera *et al.* (1986).

used to help supply the milk. However, dietary copper intake and absorption are also a factor that can influence the copper content of the milk, especially in deficiency (Suttle, 1987). While there is a definite trend to decrease with time during the lactation period, the concentration of copper in milk is also extremely variable among individuals (Fig. 8-6). As shown for the rat (Keen *et al.*, 1980a), supplementation of the dam's diet with Cu(II) (in this case, given as the nitrilotriacetate complex) increased the copper concentration of the milk in the second and third weeks of suckling but not in the first week postpartum. In the rat, excess maternal stores of copper are released in the first week and would no longer be available in weeks 2 and 3, which would account for the results. Supplementing diets already adequate in copper does not increase the copper content of the milk, as indicated by several studies on women (see Salmenpera *et al.*, 1986; also Munch-Peterson, 1950) as well as goats an cows (Elvehjem *et al.*, 1929).

8.2.3 Copper in Milk

As already indicated, the content of copper in the milk of different species varies not just with time and diet, but also with the species (Table 8-2). The milk of the rat is especially rich in copper. The concentration in pig's milk is also high, whereas that of bovine, sheep, and cow's milk is much lower and much like that of human milk.

The substances to which copper attaches in the milk have not been well studied. Martin et al. (1981, 1984), using human, bovine, and goat defatted and lyophilized whole milk or ultrafiltered fractions, found that a portion (less than 30%) was attached to macromolecules (presumably proteins) that precipitated with acid and that the remainder was associated with one or more very small ligands ($M_r > 500$), based on chromatography on Sephadex G-15 in Cu(II) buffer (1 μg/ml) and ultrafiltration with Amicon UM-5 membranes. It is noteworthy that human milk contained three or four distinct copper-binding peaks, whereas cow's milk had only one. (It should be acknowledged that this technique would measure additional components to which copper would bind, as well as those on which it would normally be found.) The dominant peak in both milks eluted in the position of Cu(II)-citrate; others (in human milk) eluted in the positions for Cu-glutamate and glutamine. Using this information, Nelson et al. (1986) did a computer simulation to verify the nature of the complexes involved in the case of the milk *per se* and during its transit through the digestive tract (where it encounters a lower pH). For this, values for known concentrations of citrate and amino acids and the dissociation constants of their copper complexes were employed. The findings are summarized in

Table 8-3. Low-Molecular-Weight Ligands for Copper in Milk[a,b]

Ligand	Assumed milk concentration (μM)	
	Human	Bovine
Histidine	21	9
Glutamate	1470	128
Glutamine	589	12

pH	Predominant copper complex	
	Human	Bovine
3	Free Cu(II)	Free Cu(II)
4	[Cu-citrate]⁻	[Cu-citrate]⁻
5	[Cu-citrate]⁻	[Cu-citrate]⁻
6	[Cu-citrate]⁻	[Cu-citrate]⁻
6.5	Cu-His$_2$	[Cu-citrate]⁻

[a] Data of Nelson et al. (1986).
[b] Other metal ions assumed to be present and considered in the calculations were inorganic phosphate, Ca(II), Mg(II), and Zn(II).

Table 8-3 and suggest that at neutral pH (or above 6.5), copper would mainly be in the form of a dihistidine complex in human milk, but it would be present as the citrate complex in cow's milk, while at lower pHs (such as encountered in the stomach) binding to citrate (1:1) would predominate in both cases. (For zinc, citrate binding predominates even at neutral pH.)

The question of whether the copper in these various milks is equally available (for example, to the human infant) has not adequately been addressed. Widdowson *et al.* (1974) reported that copper in human breast milk appeared to be much more available to small-birth-weight infants than that in formula. However, these (balance) studies were only carried out on six babies. More recently, Johnson and Canfield (1989) found with stable isotope that normal infants absorbed/retained 90% of Cu from breast milk and 81% from milk formula of the type in current use. Nevertheless, even in breast feeding, intake and absorption vary greatly among individuals (see below).

8.2.4 Copper Intake in Infancy

Substantial amounts of copper are secreted in the milk. Suttle (1987) has calculated that, in various species, the ratio of copper secreted (per day) to body weight of the mother varies inversely with body weight (Table 8-2). It may also vary to some extent with basal metabolic rate and life span. An inverse relationship may exist between this ratio and the concentration of copper in newborn liver, reflecting the size of the copper stores with which a creature is born. Part of the difference may also be that the species with the higher ratios are multiparous (Suttle, 1987). This is especially so when it comes to rats and pigs. In the case of the rat, neonates may receive a maximum of 150 μg of Cu per day in the milk (Suttle, 1987), which may then be divided among up to 15 or so pups (or only a few, in some instances), giving variable amounts to the pups, but an average of perhaps 15 μg per pup, on a daily basis. In the newborn human infant, the mother may supply up to 300 μg of Cu per day, usually only to one infant. (This is only about double the total copper secreted by the maternal rat in her daily milk!) Other reasons for these species differences may also exist that need to be uncovered.

Measurements of actual copper intakes of newborn human infants were originally made by Widdowson (1969), who also measured fecal and urinary losses. In her studies, intakes of 68-day-old infants in Great Britain averaged 350 μg per day, with losses of about 40 μg. Estimates of the intake of exclusively breast-fed infants in Finland made more recently (Solmenpera *et al.*, 1986) are much lower but were measured between 4 and

12 months postpartum, when intakes would be more modest. Median values for total intakes were quite constant (at about 130 µg/day), and median intakes per infant (divided by body weight) fell about 40 % in the period from 4 to 9 months (from 19 to 11 µg/kg per day). Variations in individual intakes, however, were enormous (ranging from 30 to 292 µg of Cu/day). In any event, much more copper would appear to be retained than excreted by the infant. Thus, most of the copper ingested is also absorbed, which fits with the concept (and fact) that copper is needed for the formation of new tissue during growth.

8.2.5 Copper Absorption

Intestinal absorption of copper during suckling seems to have been studied mainly in rodents (using ^{64}Cu). Table 8-4 shows data for mice, which indicate that net absorption is much higher in neonates than adults and that the muscles and blood of the neonate retain proportionately more of the newly absorbed copper (at 5 h) than is the case for mature mice. Otherwise, the observed tissue distribution of copper (that had entered from the intestine) was similar for newborns and adults, with about 60 % going to the liver over the time span examined.

Interpretation of the data is more difficult, because of the very large doses of copper administered, and also because the data for the adult mice

Table 8-4. Absorption and Distribution of Radiocopper in Suckling and Adult Mice[a,b]

	Net absorption (% of dose)		Distribution[c]	
	Suckling (10–12 days)	Adult	Suckling	Adult
A. Total	16	9.3		
Urine	0.45	0.20	2.8	2.2
Liver	9.3	5.5	58	59
Kidney	1.1	0.6	6.9	6.5
B. Lung			1.4	0.5
Gut			16.4	15.4
Muscle (per g)			3.4	0.2
Blood (per g)			7.4	0.9

[a] Data of Mann et al. (1979).
[b] Radioactivity recovered in tissues 5 h after intubation (into small intestine) of ^{64}Cu(II): 5.4 µg of Cu for (A); 0.9 and 10 µg Cu for suckling and adult rats in (B), respectively. Data for adult rats in (B) are from 3.5 h.
[c] Percentage of the absorbed dose.

were (partly) for 3.5 h as compared with 5 h for the 10–12-day-old mice. (Also, doses were not standardized per gram of body weight.) It is likely that, as with other metals, absorption of copper by the newborn is, in general, more efficient than that by the adult, especially before the intestinal tract becomes fully mature and uptake by pinocytosis is reduced or lost as a mechanism (Linder, 1978a).

8.3 Changes in Tissue Copper during Growth

8.3.1 Liver and Blood

As already described at some length, the concentration of copper in the *liver* of the newborn and neonate is much higher in almost all species than it will ever again be under normal circumstances. Moreover, in most species (and even in rats) (Fig. 8-3), liver concentrations of copper fall quite rapidly in "infancy" and continue to fall in "childhood". Attainment of the low adult levels appears to occur after sexual maturation and cessation (or stabilization) of growth, as seen in the human and the rat (Fig. 8-3). Most of the excess copper is probably in the nuclei (and partly in the cytosol) attached to and "stored" by metallothionein. This liver store is used up in connection with growth of tissues and organs, presumably to supply critical, fundamental enzymes and other factors with copper in all parts of the developing organism.

Although it has not been studied, the mechanism(s) by which stored copper is taken from the liver and delivered to other tissues may very well involve its incorporation into ceruloplasmin within hepatocytes; secretion of ceruloplasmin into the blood and plasma; specific binding of ceruloplasmin to receptors on cell surfaces in peripheral tissues; and finally, transfer of copper to the cell membrane, releasing apo- (or partially copper desaturated) ceruloplasmin back into the plasma. (These potential steps are discussed at great length in Chapters 3 and 4; see Fig. 4-20.) Clearly, there is synthesis and secretion of ceruloplasmin by the liver already in the fetus well before term in most species (see earlier); so this aspect of the mechanism is in place. Moreover, this mechanism would be a logical way of moving copper out of the liver to other cells. Whether ceruloplasmin receptors capable of receiving its *copper* are in place as well has not as yet been examined.

Although most milk is not particularly rich in copper (Table 8-2), a significant amount of copper that may be used for growth does enter the infant's blood also from the diet. Probably, this copper arrives at the surface of the intestinal mucosa (and ultimately enters the blood) in the form of ionic Cu(II). In the blood plasma, it probably binds mainly to the same two plasma proteins to which it binds in adults: albumin and transcuprein. (Evidence for the involvement of these proteins in the initial phase of copper transport is detailed in Chapter 3.) In support of this, Lau and Sarkar (1984) have observed that ionic radioactive copper added to human cord blood (= infant blood) *in vitro* binds to a component (or components) eluting in the void volume (transcuprein) and a component (or components) coinciding with albumin in gel permeation chromatography. It also binds to some components of low molecular weight and does so more avidly than is the case for adult serum (see Fig. 4-11A). In cord blood plasma, more of the radioactivity is associated with the void volume component (and components of low molecular weight) than is the case for adult blood. This suggests that transcuprein concentrations may be higher in the infant. As regards albumin, the gene for this abundant plasma protein is expressed already well before birth (see earlier). Thus, albumin is also available at birth to take on the task (with transcuprein) of transporting dietary copper from the intestine to the liver and kidney. (α-Fetoprotein may also help in some species.)

Of the copper that enters from the diet and is used for growth, the question is whether this goes directly to most tissues or whether it is first taken up by the liver and kidney (as in adult rats), then resecreted from the liver on ceruloplasmin and distributed from there. The available data from the mouse 5 h after intestinal administration suggest, first, that deposition in the liver and the kidney is still the major initial event for copper entering the newborn circulation from the diet (Table 8-4). [The same was observed by Mistilis and Mearrick (1969) for the suckling rat.] However, the concentration of radioisotope in muscles and blood of rat pups (at 5 h) was considerably higher than in the case of adults. Whether the high "concentration" of isotope in the muscles simply reflected the higher blood concentration of ^{67}Cu is not clear, although it would only partly explain the difference. The reason for the higher blood radioactivity is also uncertain. While there was a greater rate of influx from the diet, was there also a much more rapid (or much less rapid) incorporation into ceruloplasmin (secreted back into the blood)? Was much more copper (and radioactive copper) retained in the form of low-molecular-weight components in the serum or on transcuprein? (The finding in question may also reflect use of unphysiologically large doses of ^{64}Cu.)

In this connection, the data of Henkin *et al.* (1973) on 130 human infants suggest that the copper associated with ultrafiltrable serum components is not present in greater *concentration* than in the adult but that it is a larger *proportion* of the total serum copper in this period when ceruloplasmin concentrations have not as yet reached those of adults. (Infants ranged in age from a few hours to 21 months, and a filter with a cutoff of about 50 kDa was employed.) Newborn infants had a total serum Cu of 0.29 µg/ml, with 0.04 µg of Cu/ml ascribable to components of lower molecular weight (Table 8-5). Already after a day or two, total Cu concentrations rose considerably (about 0.11 µg/ml), with little rise in ultrafiltrable Cu. However, the latter doubled by the end of four weeks and did not change thereafter. [The value of 0.11 µg/ml of Cu in components of low molecular weight of adult serum is entirely consistent with the recent data from Wirth and Linder (1985). The more recent data of Barrow and Tanner (1988), however, for neonatal serum fractionated on Sephadex G-150 are much lower.]

It is still unclear just how much of the copper in infant serum is ascribable to ceruloplasmin. Current estimates for the human adult are almost 65% (Table 4-12), based on furnace atomic absorption analysis of serum fractions from gel permeation chromatography (Table 4-10). In the adult, concentrations of ceruloplasmin protein are about 0.32 µg/ml, which suggests a mean Cu content of about 2 µg/mg for ceruloplasmin (or 0.20%). [Fully copper-loaded ceruloplasmin has about 0.30% Cu, but delivery of copper by ceruloplasmin would reduce its copper content (see Chapters 3 and 4).] If the same Cu:protein ratio applies to the newborn infant, then ceruloplasmin in its blood (Salmenpera et al., 1986) would account for about 0.20 µg of Cu/ml (Table 8-5). This would be about two-

Table 8-5. Estimates of Proportions of Serum Copper That May Be Associated with Various Components in Infancy

Age	Total serum Cu^a (µg/ml)	Total ceruloplasmin^b (mg/ml)	Estimated ceruloplasmin Cu^c (µg/ml)	Ultrafiltrable Cu^d (µg/ml)	Copper remaining^e (µg/ml)
Newborn	0.29 (0.35)^f	0.10 (0)	0.20 (0)	0.04 (?)	0.05
1–2 days	0.40	0.13	0.26	0.05	0.09
12–15 days	(2.12)		(1.0)		
1 month	0.48 (1.64)	0.15	0.30 (0.9)	0.11	0.07
2 months	0.75	0.22	0.44	0.11	0.20
6 months	0.95	0.33	0.66	0.11	0.18
1 year	1.10	0.39	0.78	0.11	0.21
Adult	1.10	0.32	0.64	0.11	0.35

^{a} Based on data of Tyrala et al. (1985), Salmenpera et al. (1986), and Henkin et al. (1973).
^{b} Data of Tyrala et al. (1985) and Salmenpera et al. (1986), determined by radial immunodiffusion.
^{c} Assuming a Cu:protein ratio of 0.2%, as determined for adults.
^{d} Determined by Henkin et al. (1973).
^{e} Difference between values in column 1 and those in columns 3 and 4.
^{f} Values in parentheses are from Chang et al. (1976) for pigs.

thirds of the total serum copper measured by Henkin et al. (1973), as well as by the Salmenpera and Tyrala groups (Fig. 8-5A). The remaining serum copper (about 0.10 µg/ml) (or a greater amount if the average saturation of ceruloplasmin was less) would be associated with transcuprein, albumin (α-fetoprotein), and lower molecular weight ligands. As already discussed, the latter would comprise about half or less of the nonceruloplasmin copper, leaving the rest for larger proteins (Table 8-5). Following through on these estimates, the ultrafiltrable serum Cu fraction achieves adult levels soon after birth. Ceruloplasmin *copper* may reach adult levels by six months and may rise above adult levels for a period at the end of the first year of life (as does the ceruloplasmin protein concentration). On the other hand, the nonceruloplasmin, nondialyzable portion of serum copper (last column, Table 8-5) may stabilize at about 0.20 µg Cu/ml by two months of age and only achieve adult levels much later. As the pig may be born without ceruloplasmin in its serum (see earlier; Chang et al., 1976), it would have to rely on nonceruloplasmin proteins (perhaps transcuprein and albumin) or other blood ligands as sources of copper during prenatal and early postnatal growth. While the no-ceruloplasmin condition is very transient (ceruloplasmin oxidase activity being detectable by 15 h and rising rapidly to high levels by 8–10 days of age), the total Cu content of pig plasma at birth is the same as that in the human infant (Table 8-5) but rises much more rapidly. Estimates by Chang et al. (1976) of the copper in pig serum attributable to ceruloplasmin (based on *p*-phenylenediamine oxidase activities) suggest that, even well into the suckling period, half or more of the plasma copper is not on ceruloplasmin in this species. Therefore, nonceruloplasmin copper may be a more important copper source for growing tissues in the pig. All of these estimates are, of course, based on an assumption about (a) the average copper saturation of ceruloplasmin (in infancy and childhood), or (b) the ratio of ceruloplasmin copper to ceruloplasmin oxidase activity, an assumption that may be wrong. The true situation will only emerge from further work.

In the pig studies, it was also noteworthy that a low-Cu diet had no initial effect on the appearance of (and rise in) plasma ceruloplasmin oxidase activity, implying that the copper from ceruloplasmin was initially coming from liver stores. However, in these pigs, liver copper stores were largely used up in the first two weeks, and the rise in ceruloplasmin could not be sustained. Thus, the pig (and the rat) may rely more heavily on the copper in milk to supply the element for growth. [Indeed, the milk of pigs (and rats) is especially rich in copper (Table 8-2)].

There have been a few studies in which the nature of neonatal human ceruloplasmin has been compared with that of the adult. Young and Curzon (1974) found that neonatal ceruloplasmin behaved identically to adult ceruloplasmin during purification by DEAE-Sephadex and hydroxyapatite chromatography. The ratio of absorbance at 610 nm to Cu content was also identical, as was the UV–visible absorption spectrum. The kinetics of ceruloplasmin oxidase activity against N,N'-dimethyl-p-phenylenediamine were also the same. Although this does not answer the question about the average copper saturation of mixed forms of ceruloplasmin in neonatal human plasma, it serves to suggest that the form of the protein expressed and secreted by neonatal liver is the same as that of the adult. In contrast, Milne and Matrone (1970) reported that the adult pig had two distinct forms of ceruloplasmin separable by DEAE and hydroxyapatite chromatography and in nondenaturing electrophoresis, with different copper contents and specific activities (in the p-phenylenediamine oxidase assay). Neonates expressed only one of these forms. This needs to be examined further. [There still is considerable confusion about how many forms of ceruloplasmin occur in any species (see Chapter 4).] In any event, the pig may differ in terms of its developmental elaboration of ceruloplasmin, as already indicated.

8.3.2 Intestine

Changes in the intestinal copper of rats after birth have been studied by Mason *et al.* (1981). Already before birth, concentrations of Cu in the whole intestine were quite high, at 17 μg/g wet weight (Fig. 8-7). However, they rose enormously (more than eight fold) by two days of age, fell more gradually (until about 12 days of age) and then more precipitously (at two weeks of age), to mature levels at the end of the three-week suckling period. [Maturation of the intestinal tract of the rat is stimulated by maturation of the adrenal-pituitary axis at two weeks of age (Greengard, 1970).] In intestinal tissue, two-thirds or more of the copper was recoverable in the postribosomal/microsomal supernatant throughout, leaving a significant one-third (as much as 50 μg/g) in undefined larger organelles. (Could it be in the nuclei, with metallothionein?) Of the Cu in the cytosolic/extracellular fluid portion of the tissue, almost all was associated with a sulfhydryl-labile component of relatively low molecular weight (Fig. 8-7B) that, in hindsight, probably is metallothionein. (A portion was also with components of even lower molecular weight, and a much smaller fraction with the void volume.) In support of the identification of this component as metallothionein, the main component formed a distinctive peak eluting in the appropriate position from columns of Sephadex G-75, especially when tissue was homogenized (and fractions purified) in buffers containing 2 mM dithiothreitol, and exposure to air (oxygen) was mini-

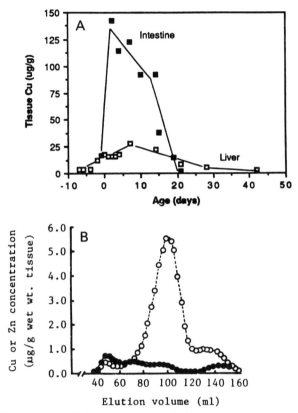

Figure 8-7. Changes in intestinal copper and metallothionein in rats in the perinatal period. (A) Total copper concentrations of the whole small intestine (■), from Mason *et al.* (1981). Data for changes in liver copper concentrations [from Terao and Owen (1977) and Kern *et al.* (1981)] (□) are included. (B) Chromatography of copper components in the 100,000 × g supernatant of intestinal tissue, separated on Sephadex G-75. Open circles indicate the elution of copper, determined by atomic absorption spectroscopy. Closed circles show the elution of zinc. Metallothionein probably elutes at about 100 ml. Reprinted, with permission, from Mason *et al.* (1981).

mized. Also, this copper component displayed the anomalous and difficult behavior typical of Cu-thionein (which tends to aggregate and/or polymerize during purification; see Chapter 6, Section 6.3.1). On the other hand, [^{35}S]-cystine administered 4 h before death was incorporated into a component eluting somewhat later than "authentic" (low-copper) metallothionein from adult rat liver. (Interpretation is difficult because the exact elution position of monomeric metallothionein seemed to vary in these different experiments.) One possible explanation for any discrepancy that may exist is that the Cu-metallothionein of interest ("stored" in the intestine) was already polymerized, or at least dimerized, when [^{35}S]-cystine was injected. Also, the material may not even have been synthesized by the intestine, since it is possible for the neonatal rat to absorb milk proteins by pinocytosis, and these proteins may or may not then be released into the cell without degradation. The potential maternal origin of the protein was suggested by Mason et al. (1981) and is certainly something that should be examined. (The protein-binding components to which copper attaches in milk have not been identified.) Copper is lost from this intestinal "metallothionein" during the suckling period, only traces remaining by 21 days of age in the rat.

The rapid accumulation and storage of copper by the intestine in the case of the neonatal rat is consistent with a relatively high concentration of copper available to it from the milk of the dam (Table 8-2). Indeed, as already indicated, an average of 15 µg of Cu may enter the suckling pup's digestive tract from its mother every day. More specifically, Mason et al. (1981) estimated that 42 µg of Cu might be ingested per pup (from colostrum) in the first day or two (and half as much later on), of which two-thirds would appear to be retained by the intestine. That a substantial portion of copper initially remains in the intestine suggests that the mechanisms for transferring copper to the blood, across the serosal surface of mucosal cells (or by other means that may exist in the immature digestive tract of neonates), or for transporting it from there to other parts of the body (on transport proteins) are limiting (in the face of such large intakes) or less developed than in later life. This is borne out by data from Mistilis and Mearrick (1969), who observed with ^{64}Cu that rates of transfer from the intestine to the liver and carcass were independent of uptake by the ileum and indeed that a large portion of the administered radioactivity remained in the intestine. (See also Table 8-4 for the mouse.) Whether this sequence of events observed in rodents also holds for other species is unknown. However, it seems likely that the rodent pattern reflects a greater immaturity of its digestive tract at birth, as well as an unusually high concentration of copper in the milk. Thus, human infants and calves, for example, may be much less likely to accumulate high concentrations of copper in their intestines in the early suckling period. Nevertheless, it is possible that most species are born with higher concentrations in the intestine than are present in the adult intestine (which are quite low) and that, overall, concentrations fall during suckling.

8.3.3 Brain and Kidney

In contrast to liver and intestine, the copper concentrations of brain and kidney (and blood) rise considerably in early life, at least in the rodent

(Fig. 8-8). In the kidney, the concentration rises gradually, reaching adult concentrations at sexual maturation. In the case of the brain, the copper concentration stabilizes at three to four weeks of age (Fig. 8-8), when brain weight has stabilized as well (Terao and Owen, 1977). Similar results have been obtained by Camakaris *et al.* (1979) and Keen and Hurley (1979) for normal and crinkled mice, where "stabilization" (at about 3 μg/g) occurs at the end of three weeks. [See also Cloez and Bourre (1987) for quaking mice.] Lai *et al.* (1985) examined changes in the Cu contents of different parts of the rat brain, reporting concentrations that were five- to tenfold higher and *may thus* be wrong. However, if they are correct at least in a relative way, they indicate that different parts of the brain accumulate copper at different rates (which would not be surprising). Specifically, the pons and medulla, as well as the hypothalamus, seemed to accumulate copper more rapidly between days 5 and 10 than between days 10 and 21, while the cortex, midbrain, and cerebellum had a greater rate of accumulation between 10 and 21 days of age (Fig. 8-9A, B). The striatum seems to have a more constant accumulation rate. These changes do not strictly parallel differential rates of tissue growth, as the cerebellum and hypothalamus more than tripled in weight during the period from 5 to 10 days, while other brain parts only doubled. Between 10 and 21 days, most of the parts almost doubled again, but there was more growth of the cerebellum, pons, and medulla (Lai *et al.*, 1985). The importance of copper to known brain enzymes, including cytochrome c oxidase, dopamine β-hydroxylase, and α-amidating enzymes (in the hypothalamus), has already been discussed (see Chapter 6, Section 6.4.2). Myelination is also a copper-dependent

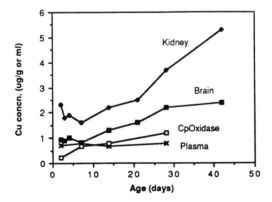

Figure 8-8. Copper concentrations of kidney and brain of rats after birth. Data for kidney and brain are from Terao and Owen (1977); those for plasma copper are from Evans and Wiederanders (1967b). Ceruloplasmin oxidase (CpOxidase) activity is from Evans *et al.* (1970c).

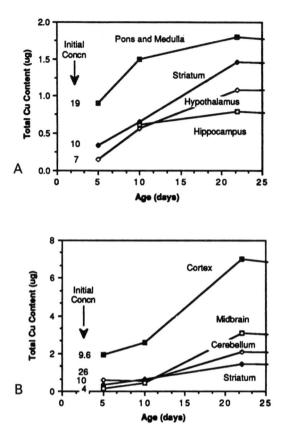

Figure 8-9. Changes in the copper contents of different brain fractions of the rat, after birth. Data are from Lai *et al.* (1985). Total copper content in different parts of the brain is plotted against age, in days. The initial *concentration* of copper in each brain area (at five days of age) is also indicated. (A) and (B) show changes occurring in different parts of the brain. Striatum is included in both. Values are probably five to tenfold too high (see text).

process, and the brain (of the developing rodent, at least), is most sensitive to copper deprivation in early life when the blood–brain barrier has not yet been formed and neural and glial cells have not yet differentiated. In other mammals (that are born more mature), this development may occur largely before birth.

Copper in the brain overall appears to be almost completely localized in the cytosol and extracellular fluid, as shown by Terao and Owen (1977) for two-day-old rat pups. (Copper in $100,000 \times g$ supernatants accounted for only 60% of the total in adult rat brain.) Of the small portions of radioactive $^{67}Cu(II)$ administered intraperitoneally to the pups that entered the brain (0.03–0.04% of dose), most was first associated with a component eluting roughly in the position of metallothionein (MT) on columns of Sephadex G-100 (see Figs. 6-3 and 6-4 and discussion in Chapter 6). [Whether glutathione mediates the initial binding to MT is unknown, but this is certainly possible (see Chapter 6).] A small portion of the ^{67}Cu tracer was with an intermediate peak roughly in the elution position of superoxide dismutase, and a larger portion was in the void volume. With time, more of the radioactivity appeared in the intermediate component. A similar pattern was obtained for adult rat brain, although the rise in the radioactivity of the intermediate component corresponded with a fall in the proportion of radioactivity in the "MT" fractions. The overall uptake of radioactivity by the brains of the pups was three- to fourfold greater than in the adults. This does not, however, take into consideration any differences in doses that may have been present. (Although carrier-free ^{67}Cu was employed, it is not certain that only trace amounts of copper were actually injected.) Nevertheless, the early growth of the brain would require the uptake of more copper, which might then be quite efficiently recycled within the central nervous system. Another marked difference between neonates and adults was in the distribution of radioactive copper to intracellular organelles. In the pups, this was only about one-third initially, but rose to 70% by 48 h. In the adult, it remained at 30 to 50% throughout. Both of these differences emphasize the importance of copper to the developing brain that may be critical to the whole organism (Prohaska, 1983a, b, 1987; also see Chapter 6).

8.3.4 Bone and Skin

Much less is known about the copper contents of bone and skin during development and aging. Nevertheless, Sanada *et al.* (1978) studied lysyl oxidase activity and collagen solubility, in male and female mice from 2 to 99 weeks of age. As shown in Fig. 8-10, this copper-dependent enzyme (involved in the cross-linking of collagen and elastin fibers) had its highest activity during early growth, reaching a peak a week or so before sexual maturation and then declining first more rapidly and then more slowly, during aging. Female rats appeared to have higher levels of enzyme activity during what would correspond to the period between puberty and young adulthood, although they also had a somewhat larger proportion of insoluble collagen in their bones than the male rats (which seems contradictory). In both bone and skin, the ability to form insoluble collagen in females was dependent upon estrogen and could be explained by the finding that lysyl oxidase activity is at least partly dependent on this

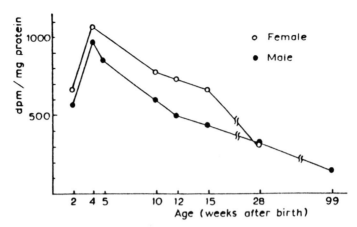

Figure 8-10. Lysyl oxidase activity of bone in mice during growth and aging. Reprinted, with permission, from Sanada *et al.* (1978).

hormone. (Estrogen treatment increased lysyl oxidase activity in both skin and bone.) In contrast, orchidectomy or testosterone treatment of male rats had little effect on the enzyme, but estrogen treatment enhanced activity. If the same conditions prevail in humans, the reduction in estrogen production that occurs in women after menstruation ceases, and the consequent diminution of lysyl oxidase activity, may contribute to the enhanced loss of bone mineral that is observed at menopause.

Copper and Disease*

9.1 Inherited Diseases of Copper Metabolism

9.1.1 Wilson's Disease

9.1.1.1 Introduction

Wilson's disease (hepatolenticular degeneration) was first characterized by S. A. Kinnear Wilson in 1912 (Wilson, 1912). It is an inherited, progressive, and, if untreated, ultimately fatal disease of copper accumulation in the human body, particularly in the liver, brain, and kidney (see Table 9-1). Indeed, the demonstration of excess copper in the liver is a requisite for the diagnosis of Wilson's disease. The disease is usually seen as hepatic dysfunction in early adolescence, although it has been found in patients as young as four years. Symptoms are nonspecific, commonly with degenerative changes in the brain and cirrhosis of the liver. The incidence of the disease is high among Arabs, Japanese, Chinese, Indians, and Jews (Bearn, 1960). The gene for Wilson's disease appears to be located on chromosome 13 (Frydman et al., 1985).

In Wilson's disease, serum copper is usually decreased about four- to five fold from normal (Mallet and Aquaron, 1983; Graul et al., 1982) although there have been several reports of no change or even a slight increase in serum copper levels (Barrow and Tanner, 1988). Non-ceruloplasmin-bound copper (i.e., that associated with albumin, transcuprein, and low-molecular-weight complexes) is elevated in Wilson's disease, as demonstrated by fractionation of serum (Table 9-2) (Barrow and Tanner, 1988) and also using an indirect approach (Sass-Kortsak

---

*Chapter 9 contributed by Dr. Christina A. Goode, Department of Chemistry and Biochemistry, California State University, Fullerton, California 92634.

Table 9-1. Copper in Wilson's Disease

Serum ceruloplasmin	
Oxidase activity	Decreased in 95% of cases (to various degrees—typically 25% of normal)
Antigen	Usually decreased to the same extent as oxidase activity, but marked discrepancies can occur (see text and Gibbs and Walshe, 1979)
Tissue copper levels	Increased in brain, liver, and kidney
	Increased in cornea
	Increased in spleen[a]
	Decreased in bile
	No change in saliva and gastric juices
Serum copper[b]	Low to normal
Nonceruloplasmin-bound Cu	Increased during later stages of the disease
Urine copper	Increased

[a] Patient had high muscle copper concentration and was likely supersaturated with Cu (Leu *et al.*, 1971).
[b] See Table 9-2 for distribution of copper in serum.

et al., 1959; Walshe, 1989). Urinary excretion of copper is elevated (Walshe, 1989), particularly during the later stages of the disease, and this may be due to increased levels of the low-molecular-weight copper complexes in serum (see Fig. 4-1) (Lau and Sarkar, 1984). The most obvious and usually confirmatory sign of Wilson's disease is the appearance of brownish Kayser–Fleischer rings in the cornea of the eyes due to copper deposits at the periphery of the cornea in Descemet's membrane.

For a complete review of the clinical manifestations of Wilson's disease, see, for example, Sass-Kortsak (1975), Seymour (1987), and Walshe (1989). A more comprehensive treatment of all aspects of Wilson's disease can be found in Owen (1981).

9.1.1.2 Possible Defects of Copper Metabolism in Wilson's Disease

Several possible defects in Wilson's disease that would lead to the observed abnormalities in copper metabolism have been proposed (Table 9-3). One of the most consistent diagnostic hallmarks is a decrease in the serum levels of the copper-containing protein ceruloplasmin, which is synthesized by the liver (Scheinberg and Sternlieb, 1970). An extensive survey by Gibbs and Walshe (1979) found that the levels of ceruloplasmin (as measured by oxidase activity) are decreased in about 95% of Wilson's disease patients, with an average decrease to about 20–25% of normal values (Table 9-4). Early on, it was suggested that the reduction in plasma ceruloplasmin was due to a decrease in its hepatic synthesis. This hypothesis is supported by recent work of Czaja *et al.* (1987) on liver

Table 9-2. Serum Copper Distribution in the Normal Adult and in Cases of Wilson's Disease and Indian Childhood Cirrhosis (Hepatic Copper Overload)[a]

Serum protein	Copper content	
	ng/ml	% of total
Normal adult ($N=8$)		
Transcuprein	86	8
Ceruloplasmin	703	63
Albumin	298	27
Low molecular weight	23	2
Total	1110	
Wilson's disease ($N=4$)		
Transcuprein	159	12
Ceruloplasmin	242	19
Albumin	844	66
Low molecular weight	37	3
Total	1282	
Indian childhood cirrhosis ($N=6$)		
Transcuprein	240	14
Ceruloplasmin	690	40
Albumin	744	43
Low molecular weight	54	3
Total	1728	

[a] Data of Barrow and Tanner (1989), for analysis of column chromatographic fractions (see Chapter 4, Table 4-10) and ceruloplasmin determinations by RID using an assumption of 6 atoms of copper per molecule of ceruloplasmin.

Table 9-3. Proposed Defects in Wilson's Disease

Defect	Reference(s)
Increased intestinal absorption of copper	Bush et al. (1955)
	Bearn and Kunkel (1955)
Decreased excretion of copper via bile	Bush et al. (1955)
	Walshe and Potter (1977)
	Iyengar et al. (1988)
Decreased lysosomal excretion of copper into bile	Sternlieb et al. (1973)
Decreased ceruloplasmin gene expression	Czaja et al. (1977)
Reduced incorporation of copper into ceruloplasmin	Matsuda et al. (1974)
Failure to switch from fetal mode of copper metabolism	Epstein and Sherlock (1981)

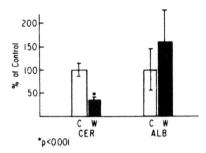

Figure 9-1. Steady-state mRNA content for ceruloplasmin (CER) and albumin (ALB) in control (C) and Wilson's disease (WD) patients. Each bar represents the mean ± SE for five patients. mRNA was quantitated by densitometric analysis of autoradiograms of dot blots hybridized with a ^{32}P-labeled probe. Reproduced, with permission, from Czaja et al. (1987).

ceruloplasmin mRNA, where variations in serum ceruloplasmin among five Wilson's disease patients could, at least in part, be related to variations in concentrations of ceruloplasmin mRNA. This may in turn reflect differences in the rate of transcription of the ceruloplasmin gene. Ceruloplasmin mRNA was detectable in samples from all the patients including one with no detectable *serum* ceruloplasmin (Fig. 9-1). (The authors, however, did not report how the serum ceruloplasmin was measured.) Levels of ceruloplasmin mRNA were, however, considerably lower than in control patients with other types of liver dysfunction. Furthermore, the rate of ceruloplasmin mRNA transcription was reduced, as determined by nuclear run-on assays (Fig. 9-2). [In contrast, levels of albumin mRNA in Wilson's disease patients appeared to be elevated relative to those of the controls (Figs. 9-1 and 9-2).] Thus, at least in these cases, Wilson's disease appears

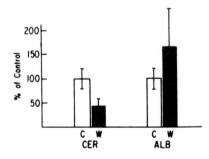

Figure 9-2. Ceruloplasmin (CER) and albumin (ALB) gene transcription in Wilson's disease (WD) and control (C) patients, as determined by nuclear run-on assays. Control values are mean ± SD for three patients; WD values are mean ± SD for four patients. Reproduced, with permission, from Czaja et al. (1987).

Table 9-4. Serum Ceruloplasmin Concentrations

	Ceruloplasmin concentration[a] (mg/dl)	
	Male	Female
Normal adults	33.3 ± 6.1 (83)	36.6 ± 9.3 (82)[b]
Liver disease controls[c]	38.0 ± 17.5 (36)	37.3 ± 23.6 (42)
Heterozygotes for Wilson's disease[d]		
Fathers/mothers	25.4 ± 8.6 (60)	29.3 ± 10.7 (73)
Sons/daughters	27.2 ± 3.2 (12)	25.2 ± 3.2 (15)
Wilson's disease patients		
Untreated	6.3 ± 8.8 (75)[e]	
Treated[f]	4.5 ± 8.0 (64)[e]	

[a] Measured as enzyme activity; mean ± SD (N). Data of Gibbs and Walshe (1979).
[b] If a group of 30 females taking oral contraceptives is included in the population, the mean ceruloplasmin level rises to 41.0 ± 13.0 (112) (see Chapter 7).
[c] These "hepatic controls" had either cirrhosis or chronic aggressive hepatitis.
[d] Parents or children of patients with Wilson's disease.
[e] Male and female.
[f] Patients were studies at intervals after the start of treatment (to remove copper), varying from 6 months to 20 years.

to be associated with decreased ceruloplasmin gene expression leading to reduced ceruloplasmin synthesis and reduced secretion of copper from the liver into the blood.

It seems unlikely, however, that this is the primary defect in Wilson's disease, as there are patients with normal or near-normal levels of ceruloplasmin (measured as enzyme activity and, in some cases, antigen) (Gibbs and Walshe, 1979; Shokeir, 1971; Cartwright et al., 1960). Also, there are heterozygotes with low plasma ceruloplasmin activity but no clinical manifestations of copper accumulation (Cartwright et al., 1960). Indeed, levels of ceruloplasmin (enzyme activity) among untreated Wilson's disease patients are highly variable and cover a very broad range from 0 to 53.8 mg/dl (normal range 25.0–45.0 mg/dl). In other studies, Gibbs and Walshe (1979) surveyed 21 patients for plasma ceruloplasmin, using both oxidase activity and an immunological technique. In six of the patients, there was a marked disagreement between the methods, with the more frequently used oxidase assay giving the relatively lower values. It is thus clear that the finding of a low ceruloplasmin activity in the serum is not evidence of an absence of this protein, unless confirmed by immunological techniques.

An accumulation of liver copper might result from a decreased incorporation of copper into ceruloplasmin during its synthesis in the liver. Upon administration of $^{64}Cu(II)$ or $^{67}Cu(II)$ to normal humans or animals, radiolabel is first associated with albumin and transcuprein and rapidly disappears from the circulation (see data in Chapters 3 and 4). It then appears in the liver, reaching a maximum in the rat approximately 6 h after isotope administration (Fig. 4-18). Following this, there is a steady decline in specific activity of copper in the liver paralleled by an increase in specific activity of circulating plasma ceruloplasmin, reflecting the incorporation of label into newly synthesized ceruloplasmin and its secretion into the blood. However, in Wilson's disease patients there is no second rise in the radioactive copper content of the blood (Fig. 9-3) (Bush *et al.*, 1955; Bearn and Kunkel, 1955), or the rise is dramatically reduced (Walshe and Potter, 1977). This is so even in those patients who have normal or near-normal levels of circulating ceruloplasmin activity (Sass-Kortsak *et al.*, 1959). This reduced incorporation of radiolabel into ceruloplasmin is sometimes used to distinguish Wilson's disease from other types of hepatic dysfunction (Vierling *et al.*, 1978). Whole-body monitoring (Walshe and Potter, 1977) and liver biopsy indicate that the bulk of the radiolabeled copper remains in the liver, although a small amount does appear in the plasma, reaching in one case approximately 15% of the total administered radioactivity by 24 h (Biesold and Gunther, 1972). This may be due to the release of other forms of copper, such as that bound to amino acids (Neumann and Sass-Kortsak, 1967), or to the (albeit reduced) synthesis and release of ceruloplamin, or both. Failure of ceruloplasmin to incorporate adequate

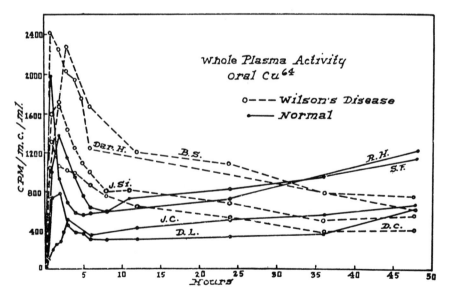

Figure 9-3. Plasma radioactivity following oral administration of ^{64}Cu to various individuals with and without Wilson's disease. Reproduced, with permission, from Bush *et al.* (1955).

amounts of copper during its synthesis could lead to a buildup of copper in the liver, in the absence of normal mechanisms for funneling the excess directly into the bile. The release of low-copper or apoceruloplasmin from the liver might have serious consequences for those organs and enzymes that may depend on ceruloplasmin as a source of copper, such as the brain (Orena et al., 1986). (See Chapters 3 and 4, concerning the role of ceruloplasmin in copper delivery.) As evidence for this, Shokeir and Shreffler (1969) reported that Wilson's disease patients had markedly reduced levels of cytochrome c oxidase activity in their leukocytes. [Cytochrome c oxidase is a copper-dependent enzyme (see Chapter 6).]

Epstein and Sherlock (1981) have suggested that Wilson's disease results from a failure to incorporate copper into apoceruloplasmin due to a mutation in a controller gene, leading to the continued repression of the adult mode of copper metabolism. This hypothesis stems from data of Shokeir (1971) that indicated that neonates had adult levels of ceruloplasmin, measured immunologically, but only about 20% of adult serum copper, suggesting that most of the ceruloplasmin was in apo form. [The normal infant is born with all the biochemical features of Wilson's disease, namely, a high liver copper content and low serum copper (Chapter 8).] However, data from Matsuda et al. (1974), using antibodies specific for holoceruloplasmin and apoceruloplasmin, demonstrated no difference in apoceruloplasmin levels in serum from Wilson's disease patients, compared to normal controls, although there was a significant reduction in holoceruloplasmin levels in the Wilson's patients' serum (Table 9-5). Furthermore, in umbilical cord blood of normal newborns, there is significantly less of *both* apo- and holoceruloplasmin (see also Chapter 8), suggesting a reduced total synthesis of the protein. Thus, the observed reduction in the amount of holoceruloplasmin released by the liver in Wilson's disease may simply reflect the decrease in ceruloplasmin synthesis and gene expression already mentioned.

Table 9-5. Serum Levels of Holo- and Apoceruloplasmin[a]

	Holoceruloplasmin (mg/dl)	Apoceruloplasmin (mg/dl)
Control (normal adult)	35.7 ± 10.0 (21)	3.3 ± 3.1 (21)
Wilson's disease patients	4.8 ± 4.0 (10)[b]	2.7 ± 2.0 (10)
Maternal blood	89.0 ± 19.5 (11)[b]	9.9 ± 4.1 (11)[b]
Newborn infants	10.9 ± 6.7 (11)[b]	0.7 ± 0.5 (11)[b]

[a] Data of Matsuda et al. (1974). Levels were measured using antibodies specific for holo- and apoceruloplasmins. Values are means $\pm$ SD (N).
[b] Differs significantly from control, by the Student t-test, $p < 0.01$.

An alternative (or additional) explanation is that reduced plasma ceruloplasmin might result not from decreased synthesis, but from a defect in secretion of ceruloplasmin by the liver into the plasma, as suggested by Graul *et al.* (1982) and Danks (1983). Indeed, ceruloplasmin-reactive material was localized immunocytochemically in the hepatocytes of liver biopsy samples from patients with Wilson's disease (Graul *et al.*, 1982). However, the staining pattern was the same as for normal liver. (The antibody used also did not differentiate between holo- and apoceruloplasmin.)

Copper is also lost from the liver via the bile (see Chapter 5), and there is evidence that this mechanism may be defective in Wilson's disease (Sass-Kortsak, 1959). Whole-body counting of normal individuals following an intravenous dose of radiocopper shows the progression of the label first to the liver, followed by its appearance in the bile (Fig. 9-4A) and subsequent fecal excretion (Walshe, 1983). This normal pattern is not seen in Wilson's disease patients, who show radioactivity distributed in all tissues (Bush *et al.*, 1955; Walshe and Potter, 1977) (Fig. 9-4B). This appears to be true even for patients who have been "decoppered" (Gibbs and Walshe, 1980). Gallstones from Wilson's disease patients characteristically also are low in copper (Walshe, 1983). Thus, there has been considerable interest in looking at copper-binding components in the bile of Wilson's disease patients. As discussed in Chapter 5, plasma ceruloplasmin appears to be a significant source of copper destined for excretion via the bile. Thus, if ceruloplasmin synthesis, incorporation of copper, or secretion from the liver is reduced in Wilson's disease, this would lead to a decrease in biliary copper excretion and exacerbate the copper loading of the liver. Iyengar and his co-workers (1988) have provided evidence that a large 40-kDa copper-binding component found in bile is a protease-resistant fragment of ceruloplasmin destined for excretion. [Alternatively, it may be an aggregate of a 5-kDa copper-containing component (see Chapter 5).] Interestingly, this component is absent from bile of Wilson's disease patients (Iyengar *et al.*, 1988; Sternliev *et al.*, 1973). However, calculations show (Iyengar *et al.*, 1988) that, even normally, this component would only account for about 360 μg (or 15%) of the approximately 2500 μg of copper excreted by adult humans in the bile every day. An absence of this component (presumably from lack of ceruloplasmin) would therefore only partly explain the markedly reduced excretion of copper in the bile of patients with Wilson's disease. It is obvious that other aspects of biliary copper excretion also must be defective.

An earlier theory advanced by Sternlieb *et al.* (1973) and others, based on histological studies, is that the primary defect in Wilson's disease is in the release of *lysosomal* copper into the bile. As copper accumulation

Copper and Disease

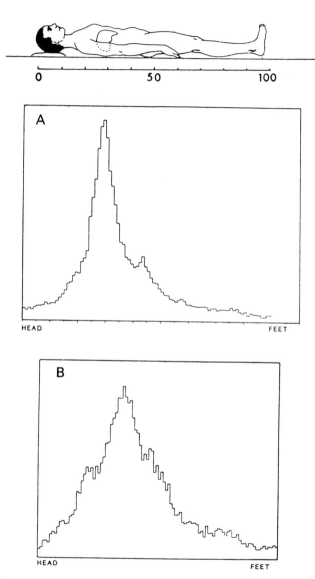

Figure 9-4. Time sequence of radiocopper distribution in normal subjects and those with Wilson's disease. Profile scan in a whole-body monitor 24 h following intravenous radiocopper. (A) Normal individual, showing excretion of copper in bile, as evidenced by development of a secondary peak over lower abdomen; (B) untreated Wilson's disease patient, showing radioactivity in all tissues and no clear secondary (bile) peak. Reproduced, with permission, from Walshe (1983).

increases during the early stages of the disease, there are dramatic changes in the morphology of the mitochondria of the hepatocyte, including increases in the electron density of the matrix and folding of the inner membrane (Goldfischer et al., 1980) and a large increase in mitochondrial copper content (see Table 6-1). With increasing cirrhosis, these morphological changes in the mitochondria regress (Goldfischer et al., 1980), and the copper becomes diffusely scattered in the cytoplasm (bound to metallothionein). As the disease progresses still further, the copper appears in increasing amounts in lysosomes until, in the advanced stages, it is almost exclusively associated with these organelles, as shown by subcellular fractionation (Sternlieb et al., 1973) and histochemical staining techniques (Goldfischer et al., 1980). However, in any circumstance of hepatic copper accumulation, the excess copper binds to metallothioneins in lysosomes (Haywood et al., 1985; Lesna. 1987), protecting the cytosol from the toxic effects of copper. Thus, what occurs in Wilson's disease may just be a "normal" response to the excess copper accumulation. It has been further suggested (Sternlieb, 1973) that the lysosomes in Wilson's disease patients are somehow especially damaged by the included copper (perhaps when metallothionien synthesis cannot keep pace with the incoming copper) and are therefore unable to have their contents excreted via the bile. It is also possible that excess copper in the lysosomes of livers from Wilson's disease patients somehow inhibits lysosomal enzymes and thus prevents the partial digestion of the copper-containing proteins (including ceruloplasmin) destined for excretion , all of which might inhibit progression of this copper into the bile. However, Seymour (1987) demonstrated by release of marker enzymes that the liver lysosomes isolated from Wilson's disease patients were not significantly more fragile (easily damaged) *in vitro* than those of controls. *Mitochondria* from copper-loaded Wilson's disease livers, however, *were* more fragile and showed a reduced capacity for oxidative phosphorylation (Seymour, 1987). The increased copper found in lysosomes may thus partly derive from the damaged mitochondria.

Excess accumulation of total body copper could also result from an increase in the intestinal *uptake* of dietary copper. The idea of increased absorption of copper has been extensively studied over the years with contradictory results. Tissue uptake of radiolabeled copper given orally has been measured using several techniques, including whole-body counting (Walshe and Potter, 1977; Walshe, 1983; Strickland et al., 1972) and appearance of radiolabel in plasma (Strickland et al., 1972). These studies, however, demonstrated no significant difference between the intestinal absorption of copper by Wilson's disease patients relative to controls ($62 \pm 17\%$ for control vs. $57 \pm 14\%$ for Wilson's disease patients; Fig. 9-5).

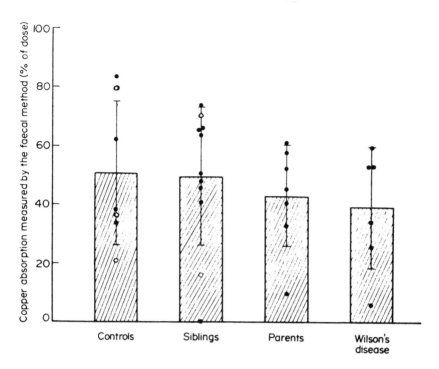

Figure 9-5. Copper absorption by patients with Wilson's disease (mean ± SD). Absorption was determined by simultaneous administration of ^{64}Cu orally and ^{67}Cu intravenously and calculated as the mean ratio of ^{64}Cu body retention to ^{67}Cu body retention at 3 and 4 days. Reproduced, with permission, from Strickland *et al.* (1972b).

Earlier reports (Bush *et al.*, 1955; Bearn and Kunkel, 1955) based on studies using oral and intravenous administration of ^{64}Cu concluded that there *was* increased intestinal absorption of copper in Wilson's disease. However, these were balance studies and so did not distinguish between changes in copper absorption and excretion. (They did show increased *retention* of copper in Wilson's disease.)

Although it appears that disturbances in copper metabolism involving ceruloplasmin are important in Wilson's disease, the finding (Frydman *et al.*, 1985) that the putative gene for Wilson's disease is on chromosome 13 and that for ceruloplasmin is on chromosome 3 (Chapter 4) indicates that the basic genetic defect of Wilson's disease is not in the ceruloplasmin gene *per se*. However, it might reside in other genes related to copper metabolism or in a controller gene for ceruloplasmin synthesis.

9.1.1.3 Treatment of Wilson's Disease

Treatment of Wilson's disease is normally aimed at "decoppering" the patient or removing the excess metal from the body before it causes damage. One of the earliest and best copper chelators used in the treatment of Wilson's disease was British anti-lewisite (BAL), or 2,3-dimercaptopropanol, discovered in 1945 by Peters and his group (Peters *et al.*, 1945). BAL was first used to treat Wilson's disease patients in 1951 (Cummings, 1951). It led to an increase in urinary copper excretion and a marked degree of clinical improvement. However, the intramuscular (i.m.) administration of BAL proved to be extremely painful, and the large doses needed were toxic to many patients (Hollister *et al.*, 1960). The search for other copper chelators led to the introduction of penicillamine in 1956, when Walshe first reported the successful treatment of two patients (Walshe, 1956). Penicillamine has proven extremely effective in "decoppering" most patients, and, with early presymptomatic diagnosis, many of the debilitating effects of excess copper now can be avoided. The toxic effects of the drug appear to be far fewer than for BAL and are usually of an immunological nature (Walshe, 1975), although renal and hematological complications have also been observed (Owen, 1981). Walshe (1981) reported that some 7% of patients with Wilson's disease develop some form of penicillamine reaction, but this can often be managed by a reduction of the dose or a temporary cessation of treatment. Penicillamine appears to lead to an increased excretion of copper by the kidney (Walshe, 1964) and a restoration of near-normal handling of copper by the liver (Walshe and Potter, 1977). Interruption of penicillamine therapy leads to a return to the abnormal copper-loaded pattern seen in the untreated disease, and treatment must therefore be continued throughout the life of the patient, with typical doses of 500–2000 mg/day (i.m.). One potential drawback is the concomitant chelation and loss of zinc, but this is easily overcome by dietary zinc supplementation. The current prognosis for Wilson's disease patients diagnosed early in life is therefore good.

9.1.2 Menkes' Disease

9.1.2.1 Introduction

Menkes' disease was first described by Menkes and his co-workers in 1962 (Menkes *et al.*, 1962). However, it was not until 1972 that it was recognized as a disorder of copper metabolism, when Danks *et al.*

(1972a,b) demonstrated that the disease is due to a severe and long-term copper deficiency. Menkes' disease is an X-linked genetic defect, with a incidence of 1 in 50,000 to 1 in 100,000 live births (Danks and Camakaris, 1983). It is characterized by progressive neurological deterioration and is usually fatal by the age of 3 years, often as a result of a ruptured or occluded artery. [Lysyl oxidase cross-linking of elastin and collagen is impaired (see Chapters 6 and 7).] Physically, patients present with poor growth and failure to develop, with light pigmentation, and with abnormal hair. The latter has been referred to as kinky hair or steely hair syndrome (due to its similarity to steel wool pot scrubbers) and is caused by spiral twisting of the hair shaft or pili torti (Danks et al., 1971). In 1974, it was reported (Hunt, 1974) that a series of mouse mutants, the mottled mutants, exhibited a very similar disturbance of copper transport to that seen in Menkes' disease. The mottled genes (Mo) have also been mapped to the short arm of the X chromosome in the mouse (Cox and Palmiter, 1982), and there are five known alleles at this locus. The heterozygous female shows a patchy distribution of pigmented and unpigmented fur. Two of the allelic homozygous males are extensively used as models of Menkes' disease, the blotchy mutant (Mo^{blo}) and the brindled (Mo^{br}) mutant, the latter of which usually dies at around 14 days, a life span comparable to that of Menkes' patients.

All of the characteristic defects of Menkes' disease and the mouse mutants can be explained by a deficiency in copper-containing enzymes important in various aspects of copper metabolism (Prohaska, 1989). The steely hair is due to a lack of inter- and intramolecular disulfide bonding in the hair keratin, a reaction in which copper has been implicated (Gillespie, 1973). The weakened arteries of Menkes' patients are due to the decreased cross-linking of elastic fibers, leading to elongation and narrowing (Oakes et al., 1976). Cross-linking of these fibers is a process normally catalyzed by the copper enzyme lysyl oxidase (O'Dell et al., 1961b; Chapter 6), the activity of which has been shown to be significantly lowered in the skin and cultured fibroblast cells of Menkes' patients (3–30% of normal; Royce et al., 1980), as well as of brindled mice (50–60% of normal; Royce et al., 1982) and of blotchy mice (30% of normal; Rowe et al., 1977). Other copper-dependent enzymes that are suggested to be decreased in Menkes' disease include tyrosinase (involved in pigmentation); cytochrome c oxidase [its decrease leading to an inability to control body temperature (Danks et al., 1972a,b) and enlarged mitochondria in cerebral cortex and thalamic nuclei (Yajima and Suzuki, 1979)]; and also dopamine β-hydroxylase (implicated in the neurological disturbances characteristic of Menkes, disease) (Grover et al., 1982). (For details on functions of these enzymes, see Chapter 6).

Table 9-6. Copper Levels in Menkes Disease[a]

Tissue	Copper level (μg/g dry weight)			
	Menkes' disease baby[a]	Normal baby (6–12 mo)[a]	Menkes' disease fetus[b]	Normal fetus[b]
Brain	1–7	20–30	5	2
Liver	10–20	50–120	47	142
Kidney	240	10–20	69	3
Duodenum	50–90	7–29	50–90	ND[c]

[a] From Danks (1987).
[b] From Heydorn et al. (1975). (Converted to μg/g dry weight from μg/g wet weight using a conversion factor of 5 for brain and 4 for other tissues.) Values for normal fetus are means of 4 determinations; value for Menke's fetus is for a single case.
[c] ND: Not determined.

9.1.2.2 Copper Metabolism in Menkes' Disease

Menkes' disease and the mouse mutations are characterized by dramatic changes in copper metabolism and distribution in tissues. Danks and his colleagues (Danks et al., 1972a, 1973) demonstrated that Menkes' disease patients have very low levels of copper in liver and brain, in contrast to high levels in the kidney, intestine, and fibroblasts (Table 9-6). This same pattern is seen in the mottled mouse mutant (Hunt, 1974; Mann et al., 1981; Camakaris et al., 1979; Prohaska, 1981a), with lower copper concentrations in all tissues (except the kidney and intestinal mucosa) (Table 9-7). In the fetus of the mottled mouse (Horn, 1981; Mann et al., 1980) (and in the single aborted human fetus that has been examined; Heydorn et al., 1975) the pattern is different, in that all tissues show increased levels of copper, with the notable exception of the liver, where again a decrease is observed. Prohaska (1989) has recently shown that a

Table 9-7. Copper Levels in Tissues of Brindled Mice[a]

Tissue	Copper level (μg/g dry wt.)		
	Normal Mo[+]	Mutant Mo[br/y]	Heterozygote Mo[br/+]
Brain	12 ± 3 (6)	3 ± 0.4 (6)	8 ± 1 (6)
Liver	197 ± 93 (6)	13 ± 1 (5)	25 ± 3 (5)
Kidney	6 ± 1 (6)	50 ± 10 (6)	112 ± 28 (6)
Duodenum	16 ± 3 (7)	41 ± 10 (8)	44 ± 17 (7)
Stomach	19 ± 3 (3)	13 ± 1 (4)	25 ± 9 (4)

[a] Adapted, with permission, from Camakaris et al. (1979). Values are for 11-day-old mice and are expressed as means $\pm$ SD. p values for all differences were >0.05 by the student's t test.

deficiency of dietary copper during gestation and lactation leads to a more severe copper deficiency in the offspring of brindled mutant mouse dams ($Mo^{br/+}$), as evidenced by lower organ copper levels and decreased activities of Cu/Zn superoxide dismutase (SOD) and ceruloplasmin.

Studies of the intestinal absorption of copper in Menkes' disease patients following oral administration demonstrate that there is very little uptake across the intestinal mucosa (Danks et al., 1972b; Lucky and Hsia, 1979). In contrast, intravenous administration of radiolabeled copper leads to the normal incorporation of label into plasma ceruloplasmin (Danks et al., 1972a; Lucky and Hsia, 1979), indicating a normal uptake and utilization of copper by the liver.

Studies with the mottled mouse mutants have further shown that the initial rate of uptake of radiolabeled copper by the liver (Mann et al., 1980) (following duodenal injection), or by hepatocytes in vitro (see Fig. 3-6), is similar to that of normal (control) animals (Table 9-8 and Fig. 9-6). However, within 2 h most (50–95%) of the absorbed copper in the livers of mutant mice had been lost to other tissues, particularly the kidney, in contrast to events in the normal mouse (or rat; Fig. 4-18). This same pattern was seen when distribution was investigated three days after a subcutaneous injection of radiolabeled copper (Prohaska, 1983a). The further observation (Hunt, 1974) that there is no difference in absorption of copper across the brush border of the intestinal mucosa in the brindled mouse compared to that of controls suggested that a defect in Menkes' disease involves the transfer of copper from the intestinal cells to the blood. It is now widely accepted that the defect in mottled mice and in Menkes' disease is actually a defect in copper release from *all* cells, with the notable exception of those in the liver. Indeed, in a situation in which there is sufficient copper, as seen in fetal gestation, most cells (except in the liver) will

Table 9-8. Absorption of Copper by Normal and Brindled Mice[a]

	Normal $Mo^{+/y}$	Mutant $Mo^{br/y}$
Total absorption (%)	16	7
Percent of absobed dose in:		
Liver	57 ± 1	26 ± 11
Kidney	5 ± 2	16 ± 8
Remaining carcass	38 ± 4	55 ± 7

[a] Adapted from Mann et al. (1979). Mice were injected with $^{64}CuCl_2$ (5.4 mg of Cu) into the duodenum and killed at 2 h. Results are means ± SD for three observations. Absorbed dose is total radioactivity recovered in carcass minus radioactivity in gut wall and intestinal contents.

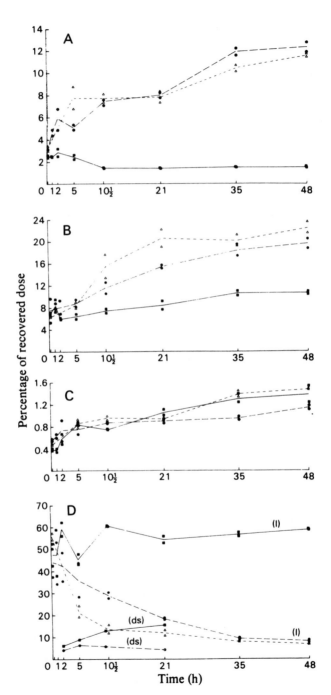

accumulate large concentrations of copper (Table 9-6). This contrasts with the case for normal mice, which (like other species) accumulate copper mainly in the liver (Chapter 8). Postnatally, the observed decrease of copper in internal organs is due to copper being retained by the intestinal mucosa and not transferred to the blood; the copper is then lost when the cells are sloughed off. Much of the evidence for this mechanism has come from studies with a variety of cultured cells of different types, from Menkes' disease patients and mottled mice. These cells accumulate copper increasingly with time (Horn, 1976; Goka *et al.*, 1976; Chan *et al.*, 1978b; Camakaris *et al.*, 1980). The degree of accumulation of intracellular copper appears to be dependent upon several factors, including the length of the incubation and the copper concentration of the medium. (It should be pointed out that copper was delivered to these cells mainly in the form of histidine complexes, and at unphysiologically high concentrations.) Copper accumulation in culture is also observed in normal cells (see Chapter 3), but Menkes' cells appear to accumulate more copper than normal cells when grown in media with very low levels of copper (Fig. 9-7; Camakaris *et al.*, 1979; Herd *et al.*, 1987). When cultured with copper at a level of 0.01 µg/ml, the Menkes' cells had approximately nine times the intracellular copper content of normal cells. As the concentration of copper in the medium increased, however, the normal cells accumulated proportionately more copper than before, now reaching a maximum level half that of the Menkes' cells (Herd *et al.*, 1987). These observations can be explained by proposing that, at low copper concentrations, Menkes' cells have most of their copper sequestered in a nonexchangeable pool (Camakaris *et al.*, 1979b; Danks, 1987; Herd *et al.*, 1987). Normal cells invoke such sequestration only at much higher copper concentrations. This would again be compatible with the hypothesis that Menkes' disease is due not to a defect in copper delivery to, or absorption by, the cells but rather to a defect in the release of copper from the cells.

In identifying the nature of this putative nonexchangeable pool (reducing copper release), much attention has focused on metallothionein (MT), since excess copper in tissues and cultured cells from Menkes' disease patients is associated with this protein (Riordan and Jolicoeur-Paquet,

---

Figure 9-6. Distribution of ^{64}Cu in suckling normal and brindled mice. For each of the time intervals shown, two normal (■), two mutant (●), and two heterozygote (▲) mice received an intracardiac injection of ^{64}CuCl$_2$ (2.7–4.6 µCi of ^{64}Cu; 0.5–0.9 µg of Cu) and were killed at the specified time interval thereafter. Results for mice killed at 35 h and 48 h are expressed as percent of administered dose, because a total recovery of the dose was not possible owing to urinary excretion. (A) Kidney; (B) skin; (C) brain; (D) liver (l); and ligated duodenum and stomach (ds). Reproduced, with permission, from Mann *et al.* (1979).

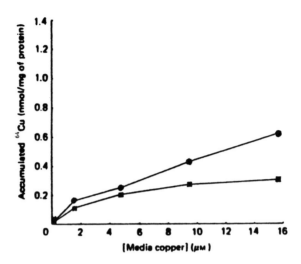

Figure 9-7. Effect of copper in medium on accumulation of ^{64}Cu in cells. Normal (■) and Menkes' (●) lymphoid cells were incubated in RPMI 1640 + ^{64}Cu (0.16, 1.57, 4.73, 9.44, 15.7 μM Cu) for 2 h at 37°C. The cells were harvested and washed twice with chilled 10 mM EDTA in PBS, pH 7.4. The levels of ^{64}Cu and protein in each pellet were determined. Reproduced, with permission, from Herd et al. (1987).

1982; Crane and Hunt, 1983; Chan et al., 1978a; Prins and van den Hamer, 1980; Bonewitz and Howell, 1981). An increase in the nonexchangeable copper pool would be due to an increase in MT production, which, since MT has a relatively long half-life and binds copper tightly (see Chapter 6), theoretically might lead to a decreased availability of copper for copper-dependent enzymes. Alternatively, Menkes' disease might be due to a defective metallothionein protein, which binds copper more avidly. These possibilities have not been ruled out. However, the discovery that the human metallothionein gene is located on chromosome 16 (Schmidt et al., 1984) and is absent from the X chromosome indicates that a structural mutation of the MT gene is not the primary defect in Menkes' disease.

Little is presently known about the regulation of metallothionein production in Menkes' disease. Leone et al. (1985), using cultured Menkes' fibroblasts, have measured metallothionein mRNA expression under various conditions. Interestingly, they found that lower levels of copper (although still unphysiologically high, at 50–400 μM CuSO$_4$) dramatically stimulated metallothionein synthesis in Menkes' cells. Normal cells were only stimulated at levels of Cu greater than 700 μM, and to a much lesser extent. (These authors did not, however, show that this was a specific effect of copper.) They further showed that the increased metallothionein protein synthesis could be ascribed to increased levels of metallothionein mRNA. Riordan and Jolicoeur-Paquet (1982) also demonstrated increased intracellular levels of metallothionein in Menkes' lymphoblasts, as compared with normal lymphoblasts, cultured with high levels of copper. However, they did not see any difference at low copper concentrations.

Leone *et al.* (1985) further showed (by transfection experiments with a cloned metallothionein fusion gene) that the increased synthesis of metallothionein mRNA was due not to a structural change in the metallothionein gene, but rather to a defect in a regulatory protein involved in either metallothionein transcription or in another aspect of copper metabolism. The involvement of an as yet unknown protein in the metal regulatory pathway also had been suggested by Danks and his co-workers (Danks, 1987) and Horn (1984). They proposed that, normally, intracellular copper is bound to a protein that provides an exchangeable copper pool for delivery to copper enzymes. In Menkes' disease, this protein might be deficient in cells other than those of the liver, so that "excess" copper (not bound) is able to stimulate production of metallothionein, which then binds the copper and does not release it to copper enzymes. In normal cells, as the concentration of extracellular copper increases, the capacity of this protein is exceeded and metallothionein synthesis is stimulated. A lack of this protein in Menkes' cells would explain their increased sensitivity to copper (and enhanced metallothionein synthesis) (Camakaris *et al.*, 1980), because the *only* means of protecting the cell against excess copper would be metallothionein. Leone *et al.* (1985) reported that copper induced the synthesis of two other proteins in Menkes' cells, tentatively identified as heat-shock proteins, which might contribute to the toxic effects of copper in these cells. Whether these might be proteins of the postulated exchangeable pool is not known. The very recent discovery of Freedman *et al.* (1989) that glutathione mediates uptake of Cu by metallothionein suggests that it (rather than a protein) might in fact be the factor that is reduced in Menkes' cells. This has not as yet been further examined.

An alternative explanation is that the Menkes' defect arises from a mutation in a protein that regulates the transcription of the metallothionein gene, leading to overproduction of metallothionein. A problem with this hypothesis is that it cannot explain the toxic effects of copper on Menkes' cells, since increased production of metallothionein would actually *decrease* "free" intracellular copper concentrations.

9.2 Other Disorders of Copper Metabolism

9.2.1 Copper and Inflammation

Inflammation is a normal response to cell tissue injury or damage and functions to minimize infection and to initiate wound healing. The course of inflammation begins when local blood vessel dilation allow access of mediators and leukocytes to the site of the injury. Increased vascular permeability induced by histamine causes exudation of plasma proteins and neutrophils and results in swelling. The production of reactive oxygen species by the neutrophils and macrophages, as well as the development of hypoferremia, are necessary to reduce the proliferation of invading microorganisms. The production of free radicals also leads to the release of tumor necrosis factor (TNF) from monocytes and macrophages (Clark et al., 1988). TNF is a 1700-Da polypeptide hormone implicated as a mediator of the inflammatory response (Beutler and Cerami, 1987). It has been shown to increase the expression of the histocompatibility antigens necessary to initiate an immune response (Collins et al., 1986; Pfizenmaier et al., 1987) and to induce the release of the cytokine interleukin-1 (IL-1) (Dinarello, 1986), which (among other things) mediates the release of the acute phase reactant proteins, including ceruloplasmin, by the liver (Ramadori et al., 1986). Many of these proteins are proteinase inhibitors and act to prevent inappropriate breakdown of healthy tissue by proteinases released from lysosomes and phagocytic cells in the area of the injury. Chronic inflammation, as is seen in rheumatoid arthritis, results when these various chemotactic signals persist and cannot be eliminated by phagocytic leukocytes.

9.2.1.1 Copper Metabolism during Inflammation

Total serum copper and ceruloplasmin concentrations increase significantly in many inflammatory and infectious diseases including hepatitis (Ritland et al., 1977), periodontal disease (Sweeney, 1967), and tuberculosis (Bogden et al., 1977). The observed rise in ceruloplasmin levels in inflammation appears to correlate well with the increase in serum copper (Scudder et al., 1978a; Conforti et al., 1982). In the chronic inflammatory condition of rheumatoid arthritis, there is a significant increase in serum copper levels during the active phase of the disease and a return to normal (Sorenson, 1978; Conforti et al., 1983b) or below normal levels (Milanino et al., 1985) during periods of remission. [In contrast, degenerative joint disease, a noninflammatory condition, is not accompanied by changes in serum copper levels (Conforti et al., 1983b).] Ceruloplasmin concentrations (as measured by oxidase activity and immunologically) in synovial fluid from patients with rheumatoid arthritis are elevated relative to those of normal or osteoarthritic patients (Bajpayee, 1975; Scudder et al., 1978a,b; Lunec et al., 1982; Dixon et al., 1988). Of considerable interest are cases of spontaneous remission of adult/juvenile rheumatoid arthritis (Whitehouse, 1976) in conditions, such as pregnancy, in which serum copper levels are significantly elevated. These observations have led to the suggestion that the development of hypercupremia is a natural *anti*inflammatory response of the organism (Sorenson, 1977; Milanino et al., 1979).

Experimentally induced inflammation is characterized by elevations in the levels of serum copper and ceruloplasmin that are independent of the agent used to induce the condition. For example, increases are seen in *Neisseria meningitidis* infection in the mouse (Letendre and Holbein, 1984); with turpentine injection in the mouse (Beaumier et al., 1984), rat (Aldred et al., 1987), and hamster (Gitlin, 1988a); with carrageenan foot edema in the rat (Conforti et al., 1982); and in the chronic inflammation model, the adjuvant rat (Milanino and Velo, 1981) (see also Table 7.2).

Within 12 h after infection in mice, the level of circulating plasma ceruloplasmin begins to increase significantly, reaching a maximum level after about 60 h (Beaumier et al., 1984) (Fig. 7-4). However, as already mentioned in Chapter 7, changes in the transcription rate for ceruloplasmin mRNA appear even earlier (in hamsters), with an increase in transcription by 3 h of induction by turpentine (Gitlin, 1988a) and a peak level at 12–18 h (see Fig. 7-2A). The concentration of ceruloplasmin mRNA also increases, but at a slower rate, with a significant increase after 12 h and a maximum (of 3.5–5.0-fold above control) reached within 36 h in hamsters (Gitlin, 1988a) and rats (Aldred et al., 1987). The regulation of ceruloplasmin synthesis in inflammation, as with that of several other acute phase plasma proteins studied (Schreiber, 1987), appears to be at the level of the mRNA and probably involves both increased transcription and a decrease in degradation of the message (Gitlin, 1988a). IL-1 released from leukocytes upon an injury or infection has been shown to increase ceruloplasmin secretion by cultured hepatocytes (Mackiewitz et al., 1987) and to increase oxidase activity in the serum of rabbits and rats (Pekarek et al., 1972; Cousins and Barber, 1987). Interestingly, Cousins and Barber (1987) showed that copper status will modulate the outcome of the effect of IL-1 on ceruloplasmin, as measured by immunoprecipitation of [^{3}H]ceruloplasmin in serum. Rats fed a copper-deficient diet, which reduced serum copper levels to 4% of normal, did not show any increase in ceruloplasmin oxidase activity following stimulation with IL-1, in contrast to an almost two fold increase for the copper-sufficient group. However the synthesis of the ceruloplasmin protein (as measured by a 2-h pulse label of [^{3}H]leucine) *was* stimulated, although not to quite the same extent (maximal stimulation was less at 18 h, as compared to that at 12 h in the copper-sufficient animals). When copper (as copper acetate) was given by injection to the copper-deficient animals together with IL-1, there was a dramatic (30-fold, by 24 h) increase in oxidase activity of the serum, but copper alone did not significantly increase the rate of ceruloplasmin synthesis.

Copper distribution in other tissues during inflammation is not as well understood. In *acute* inflammation, the copper contents of liver and kidneys do not appear to change (decrease) significantly (Conforti *et al.*, 1983a; Milanino *et al.*, 1984) in parallel with the increases in plasma copper. Using ^{64}Cu, Whitehouse (1984) has made reference to *changes* in the copper turnover of various tissues during inflammation without specifying a direction. The increase in total *serum* copper may be a result of change(s) in the dynamics of copper distribution. Copper metabolism in *chronic* inflammation has been less well studied, and there have been several conflicting reports regarding the copper concentration of various organs. Kishore *et al.* (1984) reported a 30% increase in liver copper 21 days after adjuvant injection in the rat, while Karabelas (1972), using similar experimental conditions, found a dramatic 180% increase. In contrast, Feldman *et al.* (1981) reported a 50% decrease in liver copper following injection of Freund's complete adjuvant into dogs. Whitehouse (1984), using radiolabled copper, has investigated a wider range of rat tissues and found that, with the exception of the liver, in which there was an increase, there was no significant change in the copper concentrations of tissues in response to chronic inflammation.

9.2.1.2 Role of Copper in Inflammation

Copper may be implicated at several stages in the inflammatory process (see Tables 7-2, 7-12, and 9-9). It is a potent activator of histamine [for example, it has been shown to enhance histamine action *in vitro* (Walker *et al.*, 1975)], which, upon release from the mast cells by IL-1, induces the initial vascular permeability. Histamine release also plays an important role in hypersensitivity reactions, and copper depletion leads to an enhancement of delayed immune inflammatory responses (Jones, 1984). Furthermore, copper may be involved (perhaps via IL-1) in the modulation of prostaglandin synthesis, in that it decreases the synthesis of PGE_2 and increases the synthesis of PGF_2 (Maddox, 1973; Lee and Lands, 1972; Boyle *et al.*, 1976). Copper may also participate in the stabilization of

Table 9-9. Proposed Roles of Copper in Inflammation

As component in copper-dependent enzymes involved in inflammatory response
Modulation of lymphocyte response
Regulation of histamine
Modulation of prostaglandin synthesis
Ceruloplasmin involvement in iron metabolism
Stabilization of lysosomal membranes
Regulation of free-radical metabolism

---

lysosomal membranes and decrease lysosomal permeability (Chayen et al., 1969).

As discussed in Chapter 7, copper is involved in modulation of the lymphocyte response (Table 7-12). Copper deficiency in mice and rats leads to immunosuppression, as demonstrated by impaired responsiveness to contact sensitizing agents (Jones, 1984; Kishore et al., 1984) and reduced numbers of antibody-producing cells following immunization with sheep red blood cells (Prohaska and Lukasewycz, 1981; Prohaska et al., 1983). Copper deficiency also leads to an exacerbation of the inflammatory response, as shown in the carrageenan model of acute inflammation (Milanino et al., 1978a). These authors showed that copper deficiency in rats, induced by administration of a low-copper (0.4-ppm) diet for three months, led to a significant enhancement of foot swelling. Furthermore, the degree of copper deficiency (as determined by copper levels not only in plasma but also in the tissues) appeared to play a major role in the development of the immune response to challenge by an inflammatory agent. Thus, animals fed the same diet for only one month had similar plasma copper levels to (14.4 μg/100 ml) as after three months (6 μg/100 ml) but did not exhibit the marked pro-inflammatory response.

Copper appears to be essential for immune function, especially cellular immunity, there being less natural killer (NK) cell activity and decreased phagocytosis of macrophages in copper deficiency (Table 7-12). Lipsky (1981, 1982, 1984) proposed that copper (and ceruloplasmin) potentiate the antiarthritic effect of D-penicillamine by inhibiting T-lymphocyte proliferation. He suggested that D-penicillamine, and other sulfhydryl-containing compounds such as captopril, act [together with copper(II) ions] to oxidize the free thiol and produce hydrogen peroxide locally in the synovial tissue. This might then inhibit T-cell (Lipsky and Ziff, 1980) and NK-cell (Seaman et al., 1982; El-Hag et al., 1986) function and thus the immune response. Earlier reports had suggested that, in direct contrast to the situation with acute inflammation, the development of adjuvant arthritis was impaired in young copper-deficient rats (Milanino et al.,

1978a; West, 1980). Subsequently, Kishore *et al.* (1984) were able to show that such young copper-deficient rats were in a state of immunosuppression, presumably, in large part, due to their undeveloped immune system, but they were unable to confirm that copper depletion inhibited development of adjuvant arthritis.

Copper is also an important component of at least three enzymes involved in inflammation: lysyl oxidase, involved in the repair of damaged tissues (Chapter 6); Cu/Zn superoxide dismutase (SOD), which acts intracellularly to dismutate superoxide radicals released in inflammation (Halliwell and Gutteridge, 1984); and ceruloplasmin, which functions as an extracellular scavenger of free radicals (Goldstein and Charo, 1982; Gutteridge, 1985) (see below). Ceruloplasmin in the plasma [and also traces of an unrelated extracellular Cu SOD (Chapter 4)] may thus protect cells against inappropriate lipid peroxidation by superoxide and other radicals released from neutrophils and macrophages during the inflammatory reaction (Ward *et al.*, 1983). It should have a similar effect in the interstitial fluid of damaged tissue.

It has been suggested that the damage of rheumatoid arthritis is due to destruction and depolymeriztion of hyaluronic acid by oxygen radicals, resulting in decreased viscosity of synovial fluid (McCord, 1974). In support of this are findings of no significant SOD activity in synovial fluid from patients with rheumatoid arthritis (Blake *et al.*, 1981) and significantly decreased levels of erythrocyte SOD activity in similar patients (Banford *et al.*, 1982). It also has been shown that intraarticular injections of SOD or Orgotein (bovine liver SOD) are effective in reducing the joint swelling occurring in rheumatoid arthritis (Goebel *et al.*, 1981; Wolf, 1982). More direct evidence for the role of ceruloplasmin (and SOD) comes from the observation (Denko, 1979) that ceruloplasmin, but not albumin, will reduce the acute inflammatory response in rats when coinjected with the irritant. This has led to one current hypothesis that the anti-inflammatory action of ceruloplasmin (and copper) is related to its role in disposing of free radicals (Milanino and Velo, 1981; Scudder *et al.*, 1978a,b). The free-radical-scavening activity of ceruloplasmin may also result in the inhibition of release of TNF. Kobayashi *et al.* (1984) have demonstrated that hydroxyl radical scavengers inhibit TNF-B production by as much as 77%.

Ceruloplasmin may also be involved in another way in bacterial infection, as a function of its role in iron metabolism. Bacteria require iron for growth, and, during the course of an infection, this is usually provided by circulating host transferrin. Severe bacterial infection is often characterized by the development of hypoferremia (Weinberg, 1974; Holbein, 1980) and an increased sequestration of iron by liver and spleen. This reduction in circulating iron available to the invading microorganisms may result from reduced turnover of heme by the reticuloendothelial system in inflammation, and a resultant decrease in delivery of iron to circulating transferrin. The latter process it thought to normally involve the oxidation of Fe(II) to Fe(III) and has been proposed to be catalyzed by ceruloplasmin in its role as a ferroxidase (Frieden, 1979; Chapter 7). Letendre and Holbein (1984) observed that mice with no detectable plasma ceruloplasmin activity (because of a dietary deficiency) were more resistant to infection with *Neisseria meningitidis* than those with normal amounts of ceruloplasmin. They explained this on the basis of a reduction (to 45% of normal) of circulating transferrin iron levels in the ceruloplasmin-deficient animals. Supplementation with human ceruloplasmin (by injection) markedly exacerbated the infection (Fig. 9-8), presumably due to the increase in serum iron levels that would result (Osaki *et al.*, 1971; Williams *et al.*, 1974) (although this was not checked). The importance of this aspect of ceruloplasmin function to inflammation is, however, questionable since, as already mentioned, a deficiency of copper (and hence of ceruloplasmin activity) causes a significant *enhancement* of the acute and chronic inflammatory response. Furthermore, the relatively rapid rise in serum ceruloplasmin that normally occurs in infection would theoretically lead to an *increase* in serum iron available to invading microorganisms, and, as already mentioned, the opposite occurs.

Ceruloplasmin may also be needed in inflammation in its role as a copper transport protein (Chapters 3 and 4) to deliver copper to copper-dependent enzymes, including lysyl oxidase and Cu/Zn SOD. Finally, in

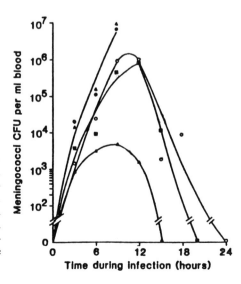

Figure 9-8. *N. meningitidis* infection in ceruloplasmin activity-deficient mice. Groups of control (○, ●) and hypoferremic, ceruloplasmin activity-deficient (△, ▲) mice were injected with saline (○, △) or 5 mg of iron dextran (●, ▲) immediately before injection of 10^4 CFU of *N. meningitidis*. Ceruloplasmin activity-deficient mice were also supplemented with exogenous ceruloplasmin (■) immediately before infection with meningococci. Two mice from each group were sacrificed every 3 h, and bacterial counts in blood were determined by plate counts. Reproduced, with permission, from Letendre and Holbein (1984).

some species such as the rhesus monkey (Hampton et al., 1972) and domestic fowl (Butler and Curtis, 1977), ceruloplasmin itself may also have some histaminase activity. This might enhance the removal of histamine and perhaps reduce the extent of the immune response, although serum amine oxidase (also a copper enzyme; see Chapter 4) is probably the main enzyme that performs this role.

9.2.1.3 Copper Therapy

The use of copper therapy in inflammation is not a new idea. Simple copper salts were reportedly used to treat eye infections already during the reign of the pharaohs 3500 years ago (Whitehouse and Walker, 1978). Copper bracelets as a remedy for arthritis have been part of folklore medicine for many hundreds of years, although their efficacy has only recently been scientifically examined. Walker and Keats (1976; Walker, 1982) have shown that these bracelets are capable of delivering significant amounts (13 mg/month) of copper, as a neutral complex with glycine, formed when the metal dissolves in sweat. In recent years, there has been considerable interest in the action of preformed copper complexes as anti-inflammatory agents. In 1976, Sorenson demonstrated that copper chelates of inactive ligands, such as acetate and 3,5-diisopropylsalicylate (DIPS), were effective in reducing experimentally induced inflammation (carrageenan-induced paw swelling). The work was then extended to include copper chelates of anti-inflammatory drugs, such as acetylsalicylate (aspirin) and D-penicillamine (Table 9-10). [Penicillamine readily forms

Table 9-10. Anti-inflammatory Activity of Some Copper Complexes[a]

Compound	Lowest active dose given (mg/kg of body wt.) against:		Cu (%)
	Carrageenan paw edema	Adjuvant arthritis	
$Cu(I)_2(acetate)_4$	8	Inactive	31.8
Aspirin	64	200 i.g.	—
$Cu(II)(aspirinate)_4$	8	10	15.0
3,5-DIPS	Inactive	Inactive	—
$Cu(II)$-$DIPS_2$	8	1.2	12.5
D-Penicillamine	Inactive	Inactive	—
$Cu(I)$D-pen$(H_2O)_{1.5}$	8	10	26.7
$Cu(II)$(D-pen disulfide)$(H_2O)_2$	8	25	15.4

[a] Modified, with permission, from Sorenson (1987). All compounds were given by subcutaneous injection unless indicated as intragastric (i.g.).

stable copper(II) complexes and is used extensively in the treatment of Wilson's disease to accelerate the urinary excretion of copper by forming *in vivo* copper chelates (see Section 9.1.1.3).] For all the chelates tested, it was shown that the copper complex of the drug was more effective in reducing the inflammation than the drug alone. In addition, the copper complex of penicillamine was less ulcerogenic than the drug alone, as demonstrated by a decreased mortality and reduced severity of ulcers in the Shay rat (Sorenson *et al.*, 1982b). However, oral administration of some of the complexes, such as copper salicylate, were found to cause gastric irritability, presumably due to liberated copper ions resulting from the breakdown of the complex. To avoid this, these complexes are often coadministered with sunflower oil (Rainsford, 1982) (to provide a source of arachidonic acid for the synthesis of prostaglandins) or delivered as a topical ointment, such as Alcusal (Walker *et al.*, 1980). To date, over 70 copper complexes of various drugs and ligands have been tested and have been shown to be effective in a multitude of inflammatory states associated with increased plasma copper concentrations ([reviewed by Sorenson (1982)]. The mode(s) of action of these chelates is still uncertain. They may simply be a vehicle for delivery of copper to intra- and extracellular copper-dependent enzymes that work against inflammatory damage, and, in this regard, it may be noted that many of these complexes are highly lipid soluble. The question naturally arises as to the rationale for giving exogenous copper in a condition in which there is already an elevated serum copper. This can be explained by an alternative, or additional, role of the chelate in the direct scavenging of the reactive oxygen species (by the complexes themselves) implicated in the inflammatory process. CuDIPS, in particular, has been shown to be a potent scavenger of superoxide radicals and hydrogen peroxide *in vitro* (Reed and Madhu, 1987; Oberley *et al.*, 1983). These possibilities appear to be the most reasonable. Sorenson (1982) has proposed that the role of copper chelates may also be to facilitate the uptake of copper by cells or to cause *in vivo* chelation of copper, resulting in a decrease of the exchangeable serum copper pool (see Chapter 4). However, with the exception of penicillamine, it remains doubtful that anti-inflammatory drugs, including corticosteroids and $Cu_2(aspirin)_4$, have the ability to significantly reduce plasma copper or (ceruloplasmin activity) (Brown *et al.*, 1980).

Thus, it appears that copper and perturbations in the metabolism of copper and copper-containing proteins may be involved at several stages in the inflammatory response. The role of ceruloplasmin and Cu/Zn SOD, as well as some therapeutic copper chelates, in free-radical scavenging would appear to be of great importance in reducing the damage and autopromotion of inflammation. Copper may also serve to modulate several aspects of the inflammatory process and as a component of other mediators.

9.2.2 Alterations of Copper Metabolism in Cancer

It has been well established that a number of different kinds of cancer are associated with elevations in the concentrations of serum copper and ceruloplasmin (Linder et al., 1981). The degree of increase in both parameters is related to the severity of the disease (Turecky et al., 1984), and remission of the disease (upon surgical removal of the malignant tumor) correlates with a return to normal levels (Linder et al., 1981). Individuals in high-risk groups, such as heavy smokers at risk for lung cancer, who have elevated serum copper and ceruloplasmin concentrations are more likely to develop malignant tumors (Linder et al., 1981). Persistent or recurrent elevation of ceruloplasmin levels in treated cancer patients is associated with a poor prognosis. Thus, there has been much interest in using serum copper and ceruloplasmin assays in evaluating the degree of malignancy of tumors, in following individuals at high risk for cancer, and in assessing the effects of therapy (see Linder et al., 1981; Linder and Moor, 1978; Diez et al., 1989). The biochemical basis for (and effect of) the changes in copper metabolism that appear to occur in various disease conditions is still, however, unclear.

An increase in the circulating copper concentration may be required for tumor growth. Neovascularization is essential for the growth of some tumors (Folkman, 1987; Brem, 1976) and may be important for malignant

Table 9-11. Tissue Uptake of Copper from ^{67}Cu-Labeled Ceruloplasmin Following Intravenous Administration to Normal and Tumor-Bearing Rats[a]

	Normal rats	Tumor-bearing rats[b]
Total uptake (1 h)[c]	82	96
Tissue ^{67}Cu concentration (% of dose/g)		
Liver	6.6	4.7
Kidney	4.1	3.6
Heart	5.6	4.9
Spleen	8.6	ND[d]
Brain	0.5	1.2
Tumor	ND	1.2
Total Cu absorbed (μg)[e]	5.5	11.7

[a] Modified with permission from Campbell et al. (1981). Data are mean values 1 h after i.v. injection into Fisher rats.
[b] Dunning mammary tumor DMBA-5A: tumors weighed 37 g.
[c] For tissues shown (% of total dose).
[d] ND: Not determined.
[e] Calculated on the basis that 70% of plasma Cu is ceruloplasmin.

transformation in some cell lines (Ziche et al., 1982); the importance of copper in the process of angiogenesis has been documented (Raju et al., 1982; Chapter 7). More recently it has been shown that copper depletion (to 50% of normal serum levels) inhibits the growth of central nervous system tumors (Alpern-Elran and Brem, 1985; Zagzag and Brem, 1986). Copper is also required for tumor cell function in the form of copper-containing enzymes such as cytochrome c oxidase. In support of this, Chakravarty et al. (1984a) showed that chemically induced tumors had twice as much copper in their mitochondrial and microsomal fractions as normal tissues of similar origin (muscle). Linder and her colleagues (Cohen et al., 1979; Campbell et al., 1981) showed with radioisotopes that the tumor itself had a substantial capacity for copper uptake and became a repository for considerable copper. Evidence for this is provided in Table 9-11 by *in vivo* data from rats given ^{67}Cu as ceruloplasmin as well as from studies in which uptake of ^{67}Cu tracer was examined in tissue culture (Table 9-12). Measurements of the copper content of a variety of tumors showed them to have copper concentrations close to those of liver and higher than in many other organs (Table 9-13). In tumors, some of the excess copper is associated with metallothioneins, as, for example, in thyroid tumors (Nartey et al., 1987a), hepatomas, and differentiated teratocarcinoma cells (Andrews et al., 1984). The expression of metallothionein mRNA in these tumors lends further support to the idea that the tumor is able to retain excess copper. It has been previously suggested (Linder, 1983) that ceruloplasmin, in its role as a copper transport protein, may be the source of the extra copper needed by tumors. Campbell et al.

Table 9-12. Copper Uptake by Tumor Cells in Tissue Culture[a]

Cells	Incubation with:		
	Transcuprein	Ceruloplasmin	Albumin
Ehrlich ascites tumor cells			
Cu added (ng)	1.8	16.1	9.4
Cu bound (pg)	25	74	94
Cu absorbed (pg)	31	52	47
BALB/c fibroblasts (transformed)			
Cu added (ng)	1.8	16.1	9.4
Cu bound (pg)	14	56	66
Cu absorbed (pg)	36	27	38

[a] From Weiss et al. (1985). Plasma containing ^{67}Cu tracer was fractionated on borohydride-treated columns of Sephadex G-150. Aliquots (10 μl) from each fraction were incubated for 1 h at 0° or 37°C with 10^8 cells, in serum-free medium. Cu bound reflects total cpm associated with washed cells. Cu absorbed is the difference between cpm for cells incubated at 37° versus 0°C.

Table 9-13. Copper Content of Transplanted Hepatic and Mammary Tumors in the Rat[a]

Tissue	Total copper concentration (μg)	
	Per gram tissue[b]	Per 10^8 cells
Morris hepatoma		
7777	5.0 ± 0.9 (4)	1.02
5123tc	4.4 ± 0.5 (3)	0.83
7800	4.8 ± 0.7 (5)	2.09
9618A	3.9 ± 0.3 (3)	1.95
Mammary tumor 5A	3.6 ± 0.9 (4)	2.24
Normal liver, adult female	4.9 ± 0.2 (5)	2.22

[a] Reprinted, with permission, from Linder and Moor (1978). Samples of apparently viable tumor tissue and livers from mature, female Fischer rats were assayed for their total content of copper. Homogenates made with distilled-deionized water were wet washed with a mixture of perchloric, sulfuric, and nitric acids, prior to the copper determination.
[b] Mean ± SD; values in parentheses are numbers of samples assayed.

(1981) found that, at least in the rat, circulating ceruloplasmin is the predominant copper source for malignant as well as normal cells (see Chapter 3). This would suggest that the increases in plasma ceruloplasmin that occur in cancer could be beneficial to the tumor and might even promote its growth. The unexpected observations of Chakravarty *et al.* (1984b), who reported that ceruloplasmin oxidase activity was *reduced* in serum from tumor-bearing mice compared to that for the control mice, might be due to a loss of copper from ceruloplasmin as a result of delivery of its copper to the tumor. However, Murillo and Linder have shown that ceruloplasmin activity is elevated in mice with L1012 tumors (Murillo, 1985).

In contrast to ceruloplasmin, copper in the form of chelates has been shown to inhibit the growth of tumors *in vitro* (Howell, 1958; Petering, 1980; Willingham and Sorenson, 1986) and *in vivo* (Linder *et al.*, 1981). It is now generally believed that the production of reactive oxygen species, including superoxide and oxyradicals, can lead to mutagenesis and to cell death (Cerutti, 1985). Normally, cells are protected from the harmful effects of these free radicals by enzymes such as Cu/Zn SOD and Mn SOD, as well as catalase. However, tumor cells, in general, appear to have lower than normal levels of SOD activity (Yamanaka and Deamer, 1974; Sahu *et al.*, 1977), and this is especially the case for Mn SOD activity, which is greatly reduced in all tumors so far tested [reviewed by Oberley and Buettner (1979)]. A reduced level of SOD activity in the tumor cell would be expected to lead to an increase in the levels of superoxide and superoxide-derived radicals. These observations have led to the development of two seemingly opposite therapeutic strategies. The first uses the higher than normal levels of radicals to selectively kill the tumor cell. It is believed that the mechanism of action of a number of antitumor agents, including mitomycin C and adriamycin, is mediated by species such as O_2^- and HO $\cdot$. One example of these is the antibiotic bleomycin, which has been shown to be effective against ascites and solid forms of Ehrlich carcinoma and sarcoma-180 (Rao et al., 1980; Takita et al., 1978). It is believed that bleomycin acts to cause cell death by participating in strand cleavage of DNA in a reaction that requires Fe(II) and oxygen. Metal ions, including Cu^{2+}, inhibit the production of hydroxyl radicals, while reducing agents such as thiols and O_2^- greatly enhance the reaction (Oberley and Buettner, 1979; Lown and Sim, 1977; Sausville et al., 1978). Since the tumor cell has lower than normal SOD activity, it will have an increased level of O_2^- and will thus be more sensitive to cytotoxic effects of bleomycin. The normal cells with Mn SOD will presumably have some additional protection against the harmful free radicals produced (Lin et al., 1978). Attempts have also been made to modulate the toxic effects of these antitumor drugs on normal cells. One approach by Lin et al. (1980) used a copper chelator, diethyldithiocarbamate (DDC), in conjunction with bleomycin. The effect of DDC is twofold in that it will reduce the concentration of Cu^{2+} in the tumor and thus enhance the bleomycin effect and it will inhibit SOD activity in the cells, leading to an increase in O_2^- (Frank et al., 1978; Heikila et al., 1976). It is found that, in the cell line tested (Chinese hamster V79), the DDC-treated cells were more susceptible to the cytotoxic effects of the drug, and the cellular toxicity became dose dependent. Interestingly, copper complexes of other antitumor drugs have been widely tested and found in many cases to be extremely effective (Petering, 1980), although copper-free complexes were in some cases less toxic to cells (Umezawa et al., 1968) and may act to chelate copper in the organism, to produce the active form of the drug (Rao et al., 1980). Another potential additional mechanism of damage to DNA may be through action of the copper chelate itself (Sigman et al., 1979; Marshall et al., 1981; Reich et al., 1981).

The alternative approach to drug therapy involves increasing or replacing the SOD activity in the tumor cell. As proposed by Oberley and his colleagues (Oberley et al., 1982), Mn SOD is a protective enzyme, and the loss of this activity in cancer cells leads to changes in key subcellular structures brought about by increased levels of oxygen-derived radicals. Thus, addition of SOD activity to these cells might lead to cessation of cell division and tumor growth. This hypothesis was initially tested by adding natural Cu/Zn SOD to a tumor and measuring tumor growth after implantation into mice (Oberley et al., 1982). The treatment was found to have a small effect on tumor growth and increased animal survival by approximately 20%. Several additional copper complexes with SOD activity and high liposolubility are now being tested (Oberley et al., 1982; reviewed in Sorenson, 1989) and have been reported to produce rapid and marked reductions of tumor growth, accompanied by an increase in host survival.

Changes in copper metabolism in cancer have been most studied using the rat as a model system. Rats with implanted tumors undergo changes in copper metabolism that are very similar to those seen in humans with various forms of cancer (Linder, 1983). Among these are fundamental changes in the compartmentation of copper within the organism. Of particular note is that liver copper concentrations (normally very carefully maintained) are decreased, as are those of the kidney, whereas concentrations in blood plasma, intestine, and heart increase substantially (Table 9-14).

The biochemical mechanisms that underlie the changes in serum copper and ceruloplasmin levels in cancer are only partly understood. The available evidence suggests that the synthesis of plasma ceruloplasmin is increased. Linder et al. (1979a) measured the incorporation of [^{14}C]leucine into plasma ceruloplasmin in rats with and without implanted tumors and

Table 9-14. Ceruloplasmin Oxidase Activity and Tissue Copper in Normal Rats and Those with Implanted Tumors[a]

	Normal rats	Tumor-bearing rats[b]	Tumor weight (g)
Ceruloplasmin oxidase activity (10^5 IU/g)	24.3[c]	23.7	<0.1
		34.6[d]	1.3
		46.1	14.5
Plasma Cu (μg/g)	1.62[c]	1.50	<0.1
		2.43	1.3
		2.52	14.5
Liver Cu (μg/g)	3.9	3.0	~15
Kidney Cu (μg/g)	9.3	4.7	~15
Heart Cu (μg/g)	3.1	6.2	~15
Spleen Cu (μg/g)	2.1	2.7	~15
Intestine Cu (μg/g)			
Whole	0.7	4.6	~15
Mucosa	1.7	2.8	~15

[a] Modified from Linder (1983). Data for ceruloplasmin and plasma Cu are mean values from Linder et al. (1979a). Tissue data are from Murillo (1985).
[b] Rats implanted subcutaneously with Dunning mammary tumor DMBA-5A.
[c] Similar values were obtained for sham-operated and liver-implanted rats.
[d] For all rats with tumors >1 g, values for ceruloplasmin and tissue copper concentrations were significantly altered. ($p < 0.01$ for difference from normal controls, 4–19 rats per group.)

showed that there was a threefold greater incorporation on the part of the rats with tumors, presumably due to increased ceruloplasmin synthesis by the liver (Table 9-15). It is possible that mediators such as prostaglandins and interleukins may be involved in these changes. IL-1, for example, has been reported to increase ceruloplasmin synthesis in cultured hepatocytes (Mackiewicz et al., 1987, 1988). In some cases at least, the tumor itself may also be secreting ceruloplasmin (Chu and Olden, 1985; Raju et al., 1982; Saito et al., 1985), although this not appear to be the case for all tumor cell lines (DiSilvestro and David, 1986; M. C. Linder and J. R. Moor, unpublished observations). Chu and Olden (1985) also demonstrated that some untransformed nonhepatic cells were able to produce ceruloplasmin-immunoreactive material, but the significance of this finding is uncertain, since any cultured cell line is, almost by definition, transformed to some extent.

While it is well established that both total serum copper and ceruloplasmin concentrations generally increase in the cancer patient, the degree of increase in total serum copper is not as great as that of ceruloplasmin, if measured as oxidase activity (Linder and Moor, 1978; Linder et al., 1981). Also, the increase in oxidase activity is greater than the increase in ceruloplasmin antigen, measured by radial immunodiffusion. This may indicate that the form of ceruloplasmin circulating in the cancer patient has more enzyme activity per molecule, and perhaps less copper per molecule, than ceruloplasmin circulating in the normal person. However, the increase in serum copper may also reflect increases of other copper-containing serum components such as transcuprein.

Table 9-15. Effects of Tumors and Estrogen on Ceruloplasmin and on Intestinal Copper Absorption

	Normal rats	+Tumor	+Estrogen	+Tumor +estrogen
Ceruloplasmin oxidase activity[a] (10^5 IU/ml)				
Intact rats	25.4	70.6[b]	71.4[b]	128[c]
Ovariectomized rats	23.3	48.3	92.1	109
Plasma Cu[a] (μg/g)	1.57	3.19	4.17	5.60
Copper absorption[b] (% dose vs. control)	100	182	17	—
Radioactivity retained in mucosa[b] (vs. control)	100	40	174	—

[a] Data from Karp et al. (1986).
[b] All values for estrogen-treated and tumor-bearing rats are significantly different from those for controls ($p < 0.001$).
[c] Data from Cohen et al. (1979).

An alternative or additional explanation for the changes in serum copper and ceruloplasmin seen in cancer is that there is *decreased* degradation of ceruloplasmin, leading to serum accumulation and perhaps also reduced excretion of copper. Upon cleavage of two or more terminal sialic acid residues, ceruloplasmin is normally removed from the circulation via asialoreceptors on the liver (Van den Hamer *et al.*, 1970; Chapter 5). Fisher and Shifrine (1978) have suggested that the increased hepatic sialyl transferase activity reported in cancer patients (Kessel and Allen, 1975) results in resialylation of asialoceruloplasmin and reduced ceruloplasmin turnover. This would probably reduce the biliary excretion of copper. Indeed, Linder *et al.* (1981) have shown that turnover and excretion of whole-body copper diminished in rats implanted with Dunning mammary tumors (DMBA-5A) and injected with $^{67}Cu(II)$. They found that the half-life of at least the fast-turnover whole-body pool (Chapter 5) was significantly increased in comparison to that in sham-implanted controls. (Estrogen also diminished whole-body copper excretion.)

The observed increased levels of serum copper and ceruloplasmin may also be due to an increased intestinal absorption of copper. Using $^{64}CuCl_2$, Cohen *et al.* (1979) found that the transfer of copper from the lumen of the gut to the blood was doubled over a range of copper doses in tumor-bearing rats. This occurred even with tumors of a small size, and at the same time or earlier than the increases in ceruloplasmin during tumor growth. The level of ceruloplasmin in the blood did not appear to be regulating intestinal absorption since, in estrogen-treated rats (which also have elevated serum ceruloplasmin), copper absorption was markedly *decreased* (17% of normal). The increased copper absorption of tumor-bearing animals appeared to be inversely related to a decreased binding of copper to metallothionein in the intestinal mucosa as compared to that in normal or estrogen-treated animals (see Chapter 2). Since intestinal absorption appears to be enhanced and excretion of copper diminished, the implication is that the tumor-bearing host must be accumulating copper. Certainly, that appears to be the case for the tumor itself (which can grow very large in a tumor-implanted rat), and for some of the other tissues that have been examined (Table 9-14). Thus, it seems that in cancer there are a variety of fundamental changes in copper metabolism leading to changes in tissue distribution. Most of the mechanisms and reasons for these changes, and their significance, remain to be elucidated.

9.2.3 Coronary Artery Disease

A direct connection between copper and heart disease was first suggested by Klevay in 1973 in his co-called "Cu–Zn hypothesis." Klevay proposed that a relative or absolute copper deficiency caused by an increase in the dietary Zn:Cu ratio and a resultant decrease in copper absorption across the intestinal mucosa "is a major factor in the etiology of ischemic heart disease" (Klevay, 1977).

This suggestion came from observations [reviewed by Klevay (1977)] that rats fed diets deficient in copper developed hypercholesterolemia, a risk factor for coronary artery disease. However, as discussed in Chapter 7, several other groups (Fisher et al., 1980; Caster and Doster, 1979; Geiger et al., 1984; Hunsaker et al., 1984) have failed to observe hypercholesterolemia in rats even with diets that have Zn:Cu ratios as high as 30:1 (the normal/recommended ratio is 5:1). A major problem with these studies is the great variation in parameters such as basal diet composition, length of study, strain of rat, and exact trace element composition [reviewed and summarized by Samman and Roberts (1985)]. In addition, there are many other dietary interactions to be considered, such as the source of protein (see Chapter 7). However, it does appear that the absolute level of dietary copper itself (rather than the Zn:Cu ratio) is important and, if low, usually induces hypercholesterolemia (see Chapter 7).

Klevay and others have looked at enzymes involved in cholesterol metabolism to attempt to determine the factor(s) responsible. Several reports have suggested that the activity of lecithin–cholesterol acyltransferase (L-CAT) is decreased in copper-deficient rats (Lau and Klevay, 1981, 1982) although this has not been confirmed by others (Lefevre et al., 1985; Suzue et al., 1980; Prohaska et al., 1985). Recently, it was reported that human low-density lipoprotein (LDL) was cleared at a *faster* rate in copper-deficient rats, suggesting an alteration in the apo B, E receptors responsible for LDL uptake (Koo and Lee, 1989). (Other reports are reviewed in Chapter 7.) It seems likely that if copper does play a role in lipid metabolism, it is in the uptake/disposal of LDL versus very low density lipoprotein (VLDL).

A drawback to the Cu/Zn hypothesis as applied to humans is that the data supporting it are from studies with only a very few individuals (Fig. 7-11; Klevay et al., 1984b). Also, individuals with a genetic copper deficiency (Menkes' disease) do not appear to develop elevated plasma cholesterol (Blackett et al., 1984). It is thus difficult at the present time to assess the relative importance of copper and copper metabolism to hypercholesterolemia.

Another putative role for copper in coronary artery disease is in the development of hypertension. This effect seems, at least in animals, to depend upon the stage of growth. Dietary copper restriction of rats at (or prior to) weaning leads to a decrease in systolic blood pressure (Wu et al., 1984; Medeiros et al., 1984; Fields et al., 1984). This may be similar to the effects seen in Menkes' disease and in the blotchy mouse mutant (see Section 9.1.2), in which a deficiency of lysyl oxidase has been implicated, since this leads to defective cross-linking of collagen (and elastin) in arteries and cardiac tissue. In contrast, copper deficiency instituted after weaning or with adult rats leads to *hyper*tension (Klevay, 1986; Medeiros, 1987). This picture is further confused by the observations of Liu and Medeiros (1986) that *excess* copper in the diet of eight-week-old Wistar and spontaneously hypertensive rats can also lead to the development of some hypertension. Wistar rats fed excess copper showed a very small, but statistically significant, increase in blood pressure (approximately 17% compared to 12–14% for the controls) above the normal observed for the growing animal (Box and Mogenson, 1982). The spontaneously hypertensive rat also had a slight (but not significant) elevation in blood pressure when fed excess copper. It remains to be seen therefore whether a direct link between dietary copper and hypertension exists or whether it is a nonspecific effect.

Copper in Nonvertebrate Organisms 10

Probably all living cells in organisms of almost all phyla contain the fundamental enzyme of respiration, cytochrome c oxidase. On this basis alone, copper is probably essential for all living creatures. Cytochrome oxidase has been discussed in detail in Chapter 6, including the one- to three-subunit forms found in bacteria and the somewhat larger forms found in plants and other creatures. It will not be discussed again in any detail in this chapter. However, the other important copper enzymes and redox proteins and the distribution and functions of copper in nonvertebrate organisms will be described and summarized herein.

10.1 Copper in Bacteria

The transport, metabolism, and function of copper in bacteria have not been well studied, with one exception, the small, blue, single-copper redox protein, azurin. The latter has been of special interest to bioinorganic chemists in relation to the evolution of blue copper proteins in general, which include the plastocyanins of plants, laccase and ascorbate oxidases (also of plants), and ceruloplasmin (in mammals). [See, for example, the review on evolution of blue copper proteins by Ryden (1988).] On the other hand, copper has been recognized as an antibacterial agent (Barber *et al.*, 1955; Erardi *et al.*, 1987), contributing to the benefits of copper piping in maintaining the purity of drinking water. Some bacteria have developed interesting copper resistance mechanisms. Otherwise, there are isolated reports of copper-dependent proteins reminiscent of those found in mammalian cells.

10.1.1 Copper Transport and Resistance

Bacterial copper transport is being studied in various copper-sensitive and resistant mutants of *Escherichia coli* by Rouch *et al.* (1989a,b), who have identified two membrane transport systems for Cu(II). Both have apparent K_ms of about 10 μM (which is similar to that of the mammalian cell transport system identified by Ettinger and colleagues; see Chapter 3). One of the bacterial systems appears also to transport Ni(II), the other Zn(II). (Each can selectively be knocked out by mutations.) This supports the concept that at least a limited amount of copper is useful or important to bacteria, as it would be for cytochrome oxidase but perhaps also for some other purposes. Certainly, copper is needed for azurin in one class of eubacteria for electron transport (see below). However, whether all bacteria really need any copper is not known. At least in some bacteria, efflux systems for copper are also well developed. Rouch et al. (1989a,b) found that copper-resistant *E. coli* strains do not differ from sensitive strains in terms of copper uptake, but they accumulate less copper. This implies that a special copper efflux mechanism can imbue the bacteria with resistance to toxicity. Rouch et al. (1989a) also found that this efflux capacity involved four genes, two coded by a plasmid, one of which (gene C) is responsible for a 25-kDa copper-binding protein found in the cell fluid (which is probably a dimer). This protein is probably not a metallothionein (J. Camakaris, personal communication). In some way, then, it would appear that this protein is responsible for the enhanced (and energy-dependent) release of copper that had first entered the cells via the transport systems. Two additional genes may code for the membrane export system; and the efflux mechanism (and synthesis of the 25-kDa protein) is regulated via a repressor protein that is also expressed by the copper-resistant plasmid (Rouch et al., 1989b).

Existence of efflux-dependent copper resistance in bacteria was initially reported for *E. coli* by Tetaz and Luke (1983), for *Pseudomonas syringae* pv. *tomato* by Bender and Cooksley (1986), and for *Mycobacterium scrofulaceum* by Erardi et al. (1987). The latter group has also uncovered another resistance mechanism, involving inactivation of copper by its precipitation as insoluble CuS (Erardi et al., 1987). This mechanism requires a source of sulfur in the medium (for example, sulfate) and results in intracellular accumulation of the black sulfide material. Sulfide production is also associated with copper resistance in yeast (*Saccharomyces ellipsoideus*; Ashida and Nakamura, 1959). One wonders whether this may have a connection to H_2S production that occurs in *E. coli* (Orskov and Orskov, 1973) and some other microorganisms. The innocuous intracellular accumulation of this form of copper would serve not only to prevent its toxic effects on the cells it is in, but would allow sequestration of the copper to prevent it from having toxic effects on other cells and neighboring organisms. [In these studies, copper was concentrated more than 100-fold with respect to its concentration in the original medium, which was 1 mM (Erardi et al., 1987).] Similar mechanisms may pertain to Hg detoxification (see Erardi et al., 1987).

Apart from the accumulation as CuS, and attachment to transport proteins for entry and efflux, copper also can associate with bacterial cell walls and membranes. Gram-*positive* bacteria have outer walls of peptidoglycan, coupled with teichoic or teichuronic acids (Beveridge and Fyfe, 1985). Gram-*negative* bacteria have an additional outer lipid bilayer, mainly containing phosphatidylethanolamine, but with some polysaccharide. Sandwiched between it and the inner plasma membrane is a thinner peptidoglycan layer. All of these outer layers or walls are negatively charged at pHs in the neutral or slightly acidic range, and they bind many kinds of metal ions. Beveridge and colleagues have compared the abilities of various bacterial cell walls and membranes to bind Cu(II) and other metal ions (as reviewed by Beveridge and Fyfe, 1985). In the presence of 5 mM metal ion, the walls of gram-positive *Bacillus subtilis* bound 3 μg of Cu per mg dry weight, an amount only exceeded by that of Fe(II) (3.6 μg) (but not Na^+, K^+, Co^{2+}, Mn^{2+}, Ni^{2+}, and Au^{2+}). This membrane is composed of peptidoglycan-containing linear strands of 10–50 N-acetylmuramyl-(β-1 $\rightarrow$ 4)-N-acetylglucosamine units with covalently linked teichoic acid. Removal of the latter had only a small effect on copper binding, but neutralization of the carboxylate residues (on the peptidoglycan) cut copper binding in half. The cell walls of another gram-positive organism, *B. licheniformis*, which has less peptidoglycan and has teichuronic as well as teichoic acid, bound only one-sixth as much copper, and there was much less bound to the inner envelope (or outer membrane) of gram-negative *E. coli*. Thus, at least in the gram-positive bacteria, a first step in copper acquisition might involve binding to carboxylate and nitrogen groups in the outer cell wall, which might concentrate copper for binding to (and transfer across) the cell membrane. The actual affinity of these binding sites for copper is unknown but is probably not all that high.

10.1.2 Bacterial Copper-Binding Proteins

As already indicated, apart from cytochrome oxidase, the copper proteins of greatest note in bacteria are the azurins, a series of small, blue, single-Cu proteins found in one of four subgroups of bacteria. First isolated from *Pseudomonas aeruginosa* by Horio (1958), the blue pigment was also found in the closely related *Bordetella* and *Alcaligenes* species. Related but different, small, blue copper proteins have been found in other bacterial strains (Table 10-1). Blue copper proteins, in general, are characterized by absorption of visible light in the range of 600 nm, with a molar absorption

Table 10-1. Types and Properties of Smaller, Blue, Single-Copper Proteins in Bacteria and Plants[a]

Protein	Source	Molecular weight	Peptide length	Carbohydrate (%)	pI	Reputed function
Azurin[b]	Pseudomonas	4600	128–129	Absent	5.4	(Electron transfer)
	Alcaligenes		128–129			
	Bordetella		129			
Pseudoazurin	Pseudomonas[c]		123			(Electron transfer)
	Actinomobacter[c]					
	Alcaligenes[d]					
Amicyanin	Pseudomonas[c]		99			Oxidation of methylamine
	Paracoccus[e]		129			
	Methylmonas[e]		121			
Rusticyanin[f]	Thiobacillus	16,500	(159)			Iron oxidation
Plastocyanin	Green plants	11,000	99	Absent	4.1	Photosynthesis
	Green algae	11,500	98	Absent		
	Cyanobacteria		105	Absent		
Phytocyanin[g]	Lacquer tree (stellacyanin)	19,000	107	40	9.9	(Electron transfer)
	Horseradish (umecyanin)	14,000	125	3.7	5.9	
	Cucumber seedlings	10,100	96		10.6	

[a] Modified from Ryden (1988, 1984). Values in parentheses are best approximations or guesses.
[b] Ambler (1971).
[c] Ambler and Tobari (1985).
[d] Husain and Davidson (1986).
[e] Hormel et al. (1986).
[f] Cox and Boxer (1978).
[g] Katoh and Takamiya (1961).

coefficient of 3000–5000 (Ryden, 1984), and a narrow hyperfine splitting in the EPR spectrum (Farver and Pecht, 1981). (See further below and Table 10-2.) In the case of the small, blue copper proteins mentioned, which have 1 mol of Cu/mol, the Cu serves as a single-electron redox center and, at least in some cases, serves to shuttle electrons between protein subunits embedded in lipid membranes, for overall respiration (as in the case of plastocyanin found in plants). Azurins (and perhaps even pseudoazurins) may have such electron transport functions in bacteria (Farver and Pecht, 1981), although their exact functions are still uncertain (Ryden, 1988). The functions of a few other small, blue copper proteins detected in bacteria and plants have been more clearly defined (Table 10-1).

As will be described in detail in Section 10.4, plastocyanins are at the end of photosystem II of plants and green algae and, as such, function to transfer electrons from cytochrome f to P-700 (in photosystem I) for photosynthesis, where oxygen is regenerated from H_2O. (See Fig. 10-5.) The structures of both an azurin and a plastocyanin have been determined by X-ray crystallography (Fig. 10-1A), and the data show that the two proteins are remarkably similar in structure. This strongly suggests a similar electron transport role for the azurins in bacteria. Although not that homologous in terms of amino acid sequence (Fig. 10-2), both are composed of two sets of four (mostly β) strands, forming two elongated halves of the molecule that face each other with predominantly hydrophobic residues. Both bind copper in the same way and at the same structural position, involving two histidines, one cysteine, and one methionine (Fig. 10-1B) (see Ryden, 1988, or Chothia and Lesk, 1982). (The environment of the copper is otherwise rather hydrophobic). The main differences are that azurins have additional amino acids, mostly added as a loop on the fifth stand (Fig. 10-1A), and a greater angle between the two halves of the molecule. Pseudoazurins, amicyanins, and phytocyanins probably fit into the same kind of configuration, with short extensions on the ends of the first loop, a few less residues in the fifth loop, and 31 more amino acids at the C-terminus in the case of pseudoazurin and some phytocyanins (Ryden, 1988). The potential evolutionary relationship between these proteins is pictured in Fig. 10-3. (A related protein, not thought to bind copper, is the major allergen of ragweed). All of this is based upon the extensive amino acid sequence data available for these proteins from different bacterial and plant species (see also Fig. 10-2).

The spectral properties and redox potentials of many of these proteins are given in Table 10-2. In general, binding of copper is probably to two histidines and two sulfurs, forming a flattened tetrahedral structure (Solomon et al., 1980). However, azurins and plastocyanins have an axial EPR spectrum, while those of stellacyanin (a phytocyanin) are rhombic (Ryden, 1984). The redox potentials of all the copper sites are higher than that of aqueous tetragonal Cu(II) (153 mV). The highest values are those of the plastocyanins, which are clearly involved in photosynthetic electron transport, and rusticyanin (involved in ferrous ion oxidation), while those of nonphotosynthetic plant proteins and azurins are lower. This would tend to play down the concept of azurins having a role in normal bacterial electron transport.

Farver and Pecht (1988) have nevertheless suggested that cytochrome c-551 and bacterial cytochrome c oxidase are the "natural partners" of azurin—that is, that azurin is part of the normal electron transport chain in the outer membranes of some bacteria. Indeed, they have demonstrated rapid electron exchange/transfer between azurin and either protein *in vitro*.

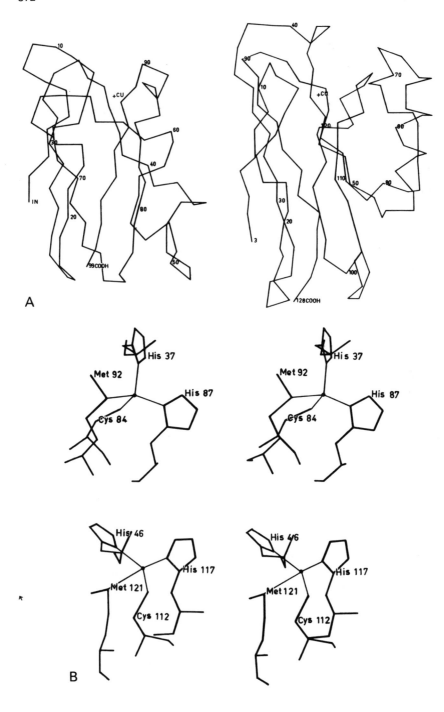

A

B

They identified two transfer sites, one (probably involving histidine-35) for uptake of electrons from cytochrome c-551 [which is perturbed and inhibited by Cr(III) bound nearby], the other, probably involving histidine-117 (exposed to solvent but surrounded by hydrophobic residues), transferring electrons to cytochrome oxidase. [This in not perturbed by Cr(III) binding.] (For more information on bacterial cytochrome c oxidase, see Chapter 6.)

A few other bacterial copper proteins are recognized. The 25-kDa protein involved in cytoplasmic transport and efflux of copper in some bacteria has already been mentioned (Rouch *et al.*, 1989a,b). "Bacteriocuprein" or bacterial copper/zinc superoxide dismutase has also been isolated, beginning with Puget and Michelson in 1974. This was despite the prediction that bacteria (like mitochondria) would, in general, contain only a Mn-dependent SOD. Further information on Cu/Zn SODs may be found in Chapter 6. No doubt, Cu/Zn SOD functions in these bacteria, as it does in mammalian cells, to shield internal proteins and lipids (including cell membranes) from peroxidative damage. Overall, it is still true that bacteria usually contain the Mn form of the enzyme (which has no homology with the Cu/Zn protein).

Some metallothionein-like proteins have also been isolated from bacteria. (See, for example, Fig. 6-6 and Table 6-6.) Presumably, these function as in other cells to sequester excess divalent copper and other metal ions. Whether metal ions (and copper in particular) are capable of inducing the thioneins, as they are in eucaryotic cells, has not been adequately explored. So far, expression of such proteins has not been implicated in resistance against copper toxicity. Indeed, the relative rarity of these proteins in bacteria may explain why copper tends to be toxic to them.

In *Chromobacterium violaceum*, a form of phenylalanine hydroxylase has been detected that appears to have substituted Cu(II) for Fe(III), the usual metal ion used by other organisms (including mammals) (Pember *et al.*, 1986, 1987). Phenylalanine hydroxylase (EC 1.14.16.1) catalyzes the synthesis of tyrosine from phenylalanine. Both the copper (and nonheme) iron forms of the enzyme also require a pterin cofactor (6-methyl- or 6,7-dimethyltetrahydropterin) (Fig. 6-7). The Cu(II) comes into direct associa-

---

←

Figure 10-1. Comparison of the X-ray structure of plastocyanin and azurin and of the ligands associated with the copper in these proteins. (A) The carbon framework of the structure of poplar plastocyanin (left) and *Pseudomonas* azurin (right), reprinted, with permission, from Farver and Pecht (1988) and Ryden (1988), respectively. (B) Copper sites of copper plastocyanin (top) and *Pseudomonas* azurin (bottom). Reprinted, with permission, from Ryden (1984).

Azurins	1	10	20	30	40	50	60	70	80	90	100	110	120
		AECSVDIQGNDQMQFNTNAITVDKSCKQFTVNLSHPGNLPKNVMGHNWVLSTAADMQGVVTDGMASGLDKDYLKPDDSRVIAHTKIIGSGEKDSVTFDVSKLKEGEQYMFFCTFPGHSALMKGTLTLK-											
Ps. aeruginosa													
Ps. denitrificans	E	SVDIQGN	Q	QFSTNAITVD	S	T	VN	S P	SLPKNV	MAA	IDKN	V DG T VI	KII S K V F S KAGDA AF S SAM K TLT K-
Ps. fluorescens B-93	E	KTTIDST	Q	SFNTKAIEID	S	T	VE	S	SLPKNV	LSA	IDKN	L EG T VI	KVI A K L I S NAAEK GF S ISM K TVT K-
Ps. fluorescens C-18	E	KVTVDST	Q	SFDTKAIEID	S	T	VD	K S	NLPKNV	MAA	IDKN	L EG T II	KII S K V F S KADGK MF S IAM K TVT K-
Ps. fluorescens D-35	E	KVDVDST	Q	SFNTKEITID	S	T	VN	T S	SLPKNV	MAA	IDKD	L PG S VI	KII S K V F S TAGES EF S NSM K AVV K-
Bordetella bronchiseptica	E	SVDIAGT	Q	QFDKAIEVS	S	Q	VN	K T	KLPRNV	IAA	IDKN	L AG T VL	KVL G S V F A AAGDD TF S GAL K TLK VD
Alcaligenes denitrificans	Q	EATIESN	A	QYDLKEMVVD	S		VH	K K	KMAKAV	MAA	LAQD	V AG T VI	KVI G S V F S TPGEA AY S WAM K TLK SN
Alcaligenes faecalis	-	DVSIEGN	S	QFNTKSIVVD	T	E	IN	K T	KLPKAA	MKA	LNND	V AG E VI	WSI T EIK GS
Alcaligenes spp	E	SVDIAGN	G	QFDKKEITVS	S	Q	VN	K P	KLAKNV	MAA	LDNN	V KD A VI	KVI G T V F S AAGED AY S FAL K VLK VD

Plastocyanins	1	10	20	30	40	50	60	70	80	90	100		
		ETYTVKLGSDKGLLVFEPAKLTIKPGDTVEFLNNKVPPHNVVFDAALNPAKSADLAKSLSHKQLLMSPGQSTSTTFPADAPAGEYTFYCEPHRGAGMVGKITVAG											
Anabaena (cyanobacterium)	-DVTVK	ADS	A V E	SSVTIKA	ETVTWV	AGF	I	EDEV	SAGNAEAL	--HEDY	NAP	ESYSAKF	--DT-A T GYF E Q K TI Q-
Chlorella (green algae)	--IEIK	GDD	A A V	GSFTVAA	EKIVFK	AGF	I	EDEV	AGVDASKI	MSEEDL	NAP	ETYAVTL	--SE-K T SFY S Q V KV Q-
Rumex (dock)	VEIL	GE	S A	SVPS	EK T	V		SA	S DD	P	TYS T	TE K T K S	N
Sambucus (elder)	VEIL	GG	S A L	GD	SVAS	EE V		AS	S AA	S	ED	P	TYK T TE K T K S N
Spinachia (spinach)	VEVL	GD	S A I	ND	SIAK	EK V		EI	S AG	N	ED	A	TYK A TE A T S A K
Cucurbita (marrow)	IEVL	GD	S A I	ND	SIAK	EK T		EI	A	S D	N D	A	TYE A TE A T T A K
Capsella (shepherd's purse)	LDVL	GD	S A I	GN	SVSA	EK T		EI	A	S	A ED	A	TYS T SE K T T A N
Solanum (potatoe)	IEVL	SD	G A V	GN	SVSA	EK T		EI	A	S	P ED	P	TYS T SE K T T S Q
Solanum crispum	VEVL	AS	S A V	NS	EVSA	DT V		EI	A	S	P ED	P	TYS K DA K T K S N
Vicia (broad bean)	LEVL	SG	S V V	SE	SVPS	EK V		EI	A	AV	P EE	P	TYV T DT K T S S K
Phaseolus (French bean)	LDVL	SD	E A V	NN	SVPS	EK T		EI	A	S	D AD	P	TYA T TE K T S A K
Mercurialis (dog's mercury)	AEVL	SS	G V E	ST	SVAS	EK V		EI	A	AS	S	ED	P TYA T TE K T S S N
Lactuca (lettuce)	IDVL	AD	S A V	SE	SISP	EK V		SI	S	AS	S	ED	K TFE A SN K E S N
Populus (poplar)													

Plant glycoproteins	1	10	20	30	40	50	60	70	80	90	100		
Lacquer tree (stellacyanin)	TVYTVGDSAGWKVPFFGDVDYDKKWASNKTFHIGDVLVFKYDRRFHNVDKVTQKNYQSCNDTTPIASYNTGNDRINLKTVGQKYYICGVPKHCDLGQKVHINTVRS												
Horseradish (umecyanin)	ED D GDME R --S PKFYIT TG RV E E DFAAGM D AV KDAFDN KKEN SHMT PPVK M X T PQ T G........												

Figure 10.2. Amino acid sequences of azurins, plastocyanins, and other small, blue copper proteins. Reprinted, with permission, from Ryden (1984). (Single amino acid code: A = ala; C = cys; D = asp; E = glu; F = phe; G = gly; H = his; I = ile; K = lys; L = leu; M = met; N = asn; P = pro; Q = gln; R = arg; S = ser; T = thr; V = val.)

Table 10-2. Spectroscopic and Redox Properties of Small, Blue Copper Proteins[a]

Protein source[b]	Redox potential (mV)(pH)	λ_{max} (nm)	ε_M ($M^{-1}cm^{-1}$)	g_Z	$g_{\parallel} \, g_{\perp}$ g_Y	g_X	A_Z	$A_{\parallel} \, A_{\perp}$ A_Y	A_X
Azurins (bacterial proteins)									
Pseudomonas aeruginosa	300 (7.0)	631	3800		2.260	2.052		0.006	0
Pseudomonas denitrificans	230 (6.8)	620			2.26	2.055		0.006	
Pseudomonas fluorescens		625	3500		2.261	2.052		0.0058	0
Paracoccus denitrificans	230 (7.0)	595, 448[c]	1530		2.290	2.052		0.0077	
Thiobacillus ferrooxidans (rusticyanin)	680 (2.0)	597	2200	2.229	2.064	2.019	0.0045	0.0020	0.0065
Phytocyanins (nonphotosynthetic plant glycoproteins)									
Lacquer tree latex (stellacyanin)	184 (7.1)	617	3550	2.287	2.077	2.025	0.0035	0.0029	0.0057
Horseradish roots (umecyanin)	283 (7.0)	610	3500		2.317 2.05			0.0035	
Rice bran	275 (7.4)	600	4300					<0.002	
Green squash fruits (mavicyanin)	285 (7.0)	600	5000	2.287	2.077	2.025	0.0035	0.0029	0.0057
Plantacyanins (nonphotosynthetic plant small proteins)									
Cucumber seedlings	317[d]	597, 443[c]	3500, 900[c]	2.207	2.08	2.02	0.0055	0.001	0.006
Spinach leaves		593	800						
Plastocyanins (photosynthetic proteins)									
French bean	347	606	3260	2.226	2.053		0.0063		<0.0017
Chlorella	390	597	4700						

[a] Modified from Ryden (1984).
[b] Cf. Table 10-1.
[c] A secondary maximum.
[d] Guss et al. (1988).

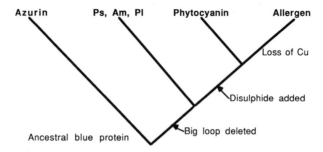

Figure 10-3. Evolutionary relationships postulated between small, blue copper proteins. Ps, pseudoazurin; Am, amicyanin; Pl, plastocyanin. Reprinted, with permission, from Ryden (1984).

tion with the cofactor during binding, and the EPR spectra are consistent with the pterin serving as a direct electron donor ligand to the copper through N-5 (Pember *et al.*, 1987). Recent data from McCracken *et al.* (1988b) indicate that two histidine imidazoles are also involved in the copper binding. Some further information on this enzyme is given in Chapter 6.

10.2 Copper and Fungi

Fungi (and plants), but apparently not bacteria, contain Cu-dependent amine oxidases. Bacteria contain the FAD-dependent enzymes, which also occur in fungi (and plants) but have a preferential specificity for polyamines (Kunasai and Yamada, 1985). Two forms of fungal Cu-dependent amine oxidases have been isolated. The plant and fungal enzymes appear to be similar and indeed appear to resemble the mammalian enzymes from serum or kidney. (Considerable information on the latter is given in Chapters 4 and 6.)

Table 10-3 summarizes many of the properties of the fungal and plant enzymes that have been isolated. Like some of the mammalian enzymes, they are composed of two apparently identical subunits, each with one atom of copper. Probably they also all carry the cofactor 6-OH dopa (or pyrroloquinoline quinone) (Fig. 6-17), which is also responsible for most of the visible absorption, as first deduced by Ameyama *et al.* (1984). The enzymes are pink, with an absorption maximum at about 500 nm (primarily due to the carbonyl cofactor, the spectrum of which is perturbed by the copper) (Table 10-3). The functions of amine oxidase in bacteria are still obscure. In general, the enzyme tends to appear in connection with

Table 10-3. Properties of Copper-Dependent Amine Oxidases from Fungi and Plants[a]

Name	Source	Molecular weight	Subunits	Carbohydrate (%)	Copper content (g-atom/mol)	Cofactor	Substrates	Absorption spectrum (λ_{max}) UV/vis (nm) Native	Absorption spectrum (λ_{max}) UV/vis (nm) Copper-free	EPR (g_m)
Fungal										
Amine oxidase	*Aspergillus niger*	252,000	2	?	2	Pyrroloquinoline quinone (pqq)[b]	Aliphatic amines, diamines, benzylamine, phenylethylamine	480–500, 410, 280		
Agmatine oxidase	*Penicillium chrysogenum*	160,000	2	?	2	?	Agmatine ≫ histamine, putrescine			
Plant										
Amine oxidase	Latex: *Euphorbia characias*	140,000	2	12	2	Probably pqq[b]	Putrescine, cadaverine, aliphatic diamines	490, 278		2.07
Amine oxidase	Epicotyls: *Lens esculenta*	(176,000)	(2)	14	(2)	?	Putrescine, cadaverine, aliphatic diamines, spermidine	495, 278	480, 278	2.07
Amine oxidase	Epicotyls: *Pisum sativum*	180,000	2	13	2	?	Putrescine, cadaverine, aliphatic diamines, spermidine	500, 280	480, 280	

[a] Modified from Kumagai and Yamada (1985) and Rinaldi *et al.* (1985). Values in parentheses are best approximations or guesses.
[b] More recent findings suggest that the cofactor is 6-OH dopa (topa) rather than PQQ (Janes *et al.*, 1990) (see Chapter 6 for further details).

growth and cell proliferation. In the fungi, it is present in the mycelia, and activity is highest during the early stationary phase of growth (Kumasai and Yamada, 1985). In the pea seedling, activity appears during early germination, is highest in the growing regions, and decreases when the seedling reaches maturity (Rinaldi et al., 1985). The enzyme appears to be feedback regulated through auxins and their metabolites. Inhibitors of the enzyme also may be present, even in germinated seeds. From the substrate specificities of the copper enzymes in comparison with those of FAD-dependent enzymes (especially the polyamine oxidases) in the same organisms, the main function is probably not to inactivate polyamines that may be regulating (and accelerating) growth and cell proliferation. There are separate polyamine oxidases that have this action and indeed provide the only pathway for spermine and spermidine degradation. However, diamines and histamine are reported to be stimulators of growth and development in plants (Mazlinski et al., 1985; Galston, 1983), and the copper amine oxidases may be responsible for their inactivation. The fact that putrescine and cadaverine are also good substrates, at least for the plant copper enzymes, suggests a regulatory role in polyamine biosynthesis as well or a role in inactiviating these diamines which may in themselves be involved in growth regulation.

Another enzyme apparently unique to certain fungi (*Dactylium dendroides*, formerly misclassified as *Polyporus circinatus*) is galactose oxidase (EC 1.1.3.9). This enzyme has been of interest to bioinogranic chemists, as it is the only known single-copper enzyme that catalyzes a two-electron oxidation without the help of another cofactor or metal ion (Kosman and Driscoll, 1988). It is also an example of an enzyme with only "type 2" copper. (See 10.5.2 below.) It has been useful as a specific tool for galactose assays, as it has no activity against D-glucose. It reacts with raffinose and galactoside as well as substituted galactose units, and it also oxidizes some unsaturated alcohols and pyridinium carbinols (Ettinger and Kosman, 1981), as well as simple molecules like dihydroxyacetone, hydroxypyruvate, glycerol, and propanediols (Hamilton et al., 1978). The reaction results in the oxidation of alcohols to aldehydes, with molecular oxygen being transformed into peroxide.

Galactose oxidase consists of a single, apparently nonglycosylated polypeptide chain of 68,000, with a high content of basic amino acids and a pI of 12. It has not as yet been cloned or crystallized. (In solution, it often aggregates, and it sticks tenaciously to glass.) The isolated enzyme is green, with two visible absorption peaks at 630 and 445 nm. The copper probably has two histidine ligands, with a pseudo-square planar geometry. The third ligand is probably water and the fourth is still uncertain (Ettinger and Kosman, 1981). Whittaker and Whittaker (1988) have provided evidence that the protein itself may participate as the second redox center in the enzyme reaction. This suggestion should spark further work.

The protein is copiously secreted by fungi and has been estimated to account for 11% of the extracellular protein accumulated at the end of the growth phase (Ettinger and Kosman, 1981). In copper deficiency, secretion of the apoprotein is not hindered, and addition of Cu(I) or Cu(II) will bring it to activity. Despite this copious output, the function of galactose oxidase is unknown.

Fungi also contain or secrete three other enzymes that are found in plants: a form of tyrosinase (also known as catechol oxidase or phenolase) that appears to be involved in the production of the melanin of fruiting bodies in certain species; laccase, which is mainly secreted from cells (like galactose oxidase) and is capable of many of the same kinds of reactions as tyrosinase but more likely is involved in humic acid production and in lignin decomposition by soil fungi; and ascorbate oxidase, again with substrates somewhat similar to those of laccase and tyrosinase, also found in some bacteria (as well as in plants), the function of which is even more obscure. These enzymes are discussed in detail in Section 10.4. Finally, fungi also contain the "standard" copper enzymes, cytochrome oxidase and superoxide dismutase (SOD), as well as two small forms of metallothionein. Both of these forms bind six Cu atoms and are homologous with the β domain of the mammalian protein (Winge et al., 1988). (Further aspects of metallothionein structure and function are described in Chapter 6.) Shatzman and Kosman (1978) have shown with *Dactylium dendroides* that copper is utilized preferentially by cytochrome oxidase when there is a scarcity, and that deficiency enhances Mn SOD production while reducing synthesis of Cu/Zn SOD (Shatzman and Kosman, 1979a). In deficiency, galactose oxidase is also secreted in the apo form, but production of the protein is not impaired. In contrast, when copper is sufficient (at concentrations of Cu above 5 μM), Cu/Zn SOD synthesis is high and *holo*galactose oxidase is secreted. Schatzman and Kosman also reported that copper is rapidly concentrated when fungal cells are grown in a low-copper medium (<30 nM), and by a non-energy-dependent process. In the presence of high copper concentrations, some of the copper in the cytosol associates with a small peptide (M_r 1350) containing glycine and lysine (and no cysteine or histidine). The nature of this complex is still unclear.

10.3 Copper in Yeast

Yeasts contain the usual vital copper enzymes, cytochrome c oxidase and Cu/Zn superoxide dismutase, both of which have been studied extensively and strongly resemble those of other phyla in terms of their most basic structural units (see Chapter 6). Yeasts have been of interest in the study of cytochrome c oxidase regulation and assembly, in which, however, the role of copper (if any) is still unclear. (Certainly, copper is needed for enzyme activity, but the full apoprotein may be present in deficiency.) As described in Chapter 6, the "core" three subunits of cytochrome c oxidase are encoded by mitochondrial DNA, and the others derive from nuclear expression. Pioneering work of Poyton and Schatz and their colleagues, in the mid to late 1970s, first showed that mitochondrial protein synthesis and gene expression were dependent upon cytoplasmic proteins (Poyton and Kavanagh, 1976), indeed even (as we now know) for nuclei-derived RNA (and DNA) polymerases (see Gloekener-Gruissem *et al.*, 1988). Moreover, the formation and assembly of cytochrome oxidase (at least normally) is dependent upon the availability of heme (Saltzgaber-Muller and Schatz, 1978). Heme regulates expression of nuclei-derived subunits V and VII (Gollub and Dayan, 1985; Trueblood *et al.*, 1988), which are necessary for assembly of the active enzyme. (In the absence of these subunits, other subunits are present but not assembled.) Most recently, Trueblood *et al.* (1988) have reported that heme regulation of nuclear subunit V is more complex. When heme is present, it binds to a protein (product of the HAP2 gene) that activates expression of the subunit Va isoform, while expression of the Vb isoform is suppressed (through the action of another nuclear gene product, RZEO1). In the absence of heme, the alternative (Vb) isoform is expressed. These investigators postulated that subunit Vb forms a cytochrome oxidase that (along with another form of cytochrome c) is better adapted to anaerobic conditions.

Relatively little is known about unique aspects of copper metabolism ascribable to yeasts, except perhaps in the regulation of yeast metallothionein (MT), under study particularly by Hamer and colleagues (Hamer, 1986; Furst *et al.*, 1988). The yeast form of MT, like that of vertebrates, is initially expressed as a 61-amino acid polypeptide (Table 6-6 and Fig. 6-6) from which the eight N-terminal amino acids are then cleaved (Wright *et al.*, 1987). Coordination and geometry of the 12 Cu atoms bound are also similar to those of mammalian MTs, despite extensive differences in amino acid sequence (see discussion in Chapter 6). Considerable work has been done on regulation of the yeast copper thionein, which depends upon the "*CUP1*" locus upstream to the initiation site for transcription and is activated by high (stressful) copper concentrations (Furst *et al.*, 1988). Activation occurs by binding of a trans-acting gene product (a protein expressed by the ACE-1 locus; Culotta *et al.*, 1990; also designated "*CUP2*"; Buchman *et al.*, 1990) directly to the sensitive region of the MT promoter (*CUP1*) region (Furst *et al.*, 1988). [An *in vitro* transcription system with these components has even been described (Culotta et al., 1990).] Binding of the ACE1 product requires the presence of Cu and can also be induced by Ag(I), but not by Zn(II) or Cd(II), and occurs via a conformational change in the N-terminal part of the activator protein. Furst et al. (1988) have proposed that the resulting S-coordinated Cu cluster may resemble that of copper thionein itself (George et al., 1988).

Apart from MT, polymers of α-glutamylcysteine (see Fig. 10-4), called phytochelatins because of their presence in plants, are used for storage of excess Cu by certain yeasts (see under Plants, low molecular weight ligands). Indeed, recent EXAFS studies in which a copper resistant yeast and MT were directly compared (Desideri et al., 1990) suggested that phytochelatins rather than MT can be the main ligands for excess copper. The phytochelatins are induced by heavy metals (including copper), and

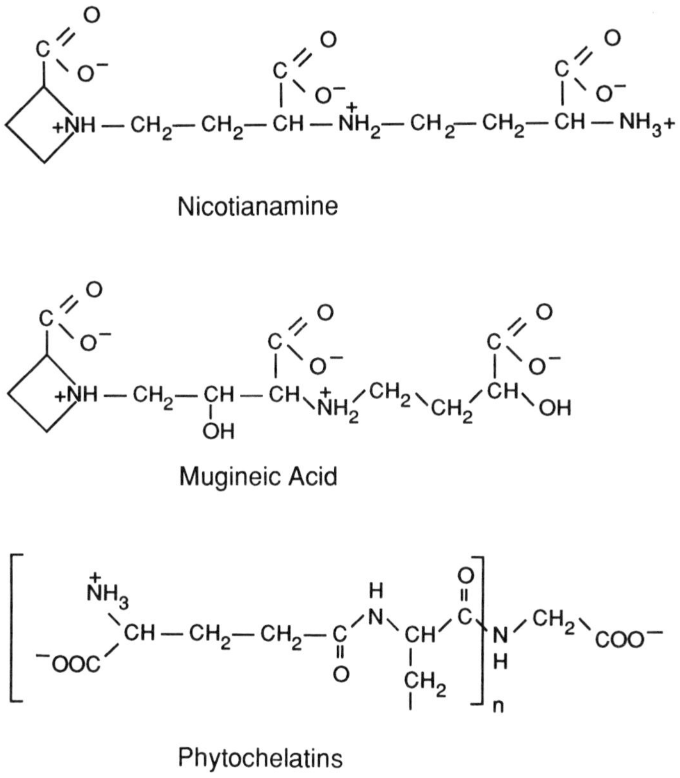

Nicotianamine

Mugineic Acid

Phytochelatins

Figure 10-4. Structures of plant substances that may chelate copper. (Nicotianamine from Budesinsky et al., 1980; mugineic acid from Sugiura et al., 1981.)

are formed from glutathione. [See more under Plants, and Fig. 10-13.] It is interesting that glutathione appears to be important also for copper binding of MT (see Chapter 6).

Resistance of yeasts to copper depends (at least partly) on the number of copies of the CUP1 gene for copper thionein (MT) (Welch et al., 1983; Khan et al., 1988). Copper toxicity, in turns, is dependent on the availability of O_2 and is exacerbated by a relative lack of Cu/Zn or Mn SOD (Greco et al., 1990). This is not surprising, in view of the capacity of copper complexes to form oxygen radicals, and SOD to dispose of superoxide.

A few studies have also been done on uptake of copper by yeasts. In general, yeasts are relatively insensitive to copper toxicity in comparison with other microorganisms, because of their capacity to induce MT and phytochelatins to sequester the free metal ions. [Growth is only inhibited 50% by 1–4 mM copper (Imahara et al., 1978).] Most of the copper accumulates in the cytosol (Imahara et al., 1978), where MT is found. Using *Debaromyces hansenii* in the stationary phase of their growth, Wakatsuki et al. (1979) found that there was an initial rapid uptake followed by a slower phase with possible Michaelis–Menten kinetics. At low concentrations, Cu uptake occurs by a saturable process with a K_m of about 10 μM and a capacity of about 12 nmol of copper per mg of cell dry weight per min (Wakatsuki et al., 1979). Uptake involving additional lower affinity sites occurred at higher copper concentrations. The high-affinity uptake mechanism was inhibited by a variety of divalent metal ions, but especially by Zn(II), suggesting that the carrier was not specific for Cu(II). However, mole per mole, the affinity for copper seemed to be higher than that for zinc, manganese, calcium, nickel, and cobalt (in that order) (Wakatsuki et al., 1979). A different energy-dependent uptake mechanism, not inhibited by Cu(II), appears to be used by zinc (Failla et al., 1976). Most recently, Lin and Kosman (1190) have found, with *Saccharomyces cerevisiae* in log phase growth, that uptake was by an energy-dependent, saturable process, with an apparent K_m of 4.4 μM. Uptake velocity (V_{max}) was enhanced by a protein synthesis-dependent mechanism upon pre-incubating with copper, implying that copper induced a protein that aided uptake. The transport system had a high specificity for Cu(II), being only mildly inhibited by Ni(II) or Zn(II). A strain of yeast with no CUP1 metallothionein regulatory locus took up similar amounts of copper, and in the same time frame, but its transport system appeared to have a four- to five fold higher affinity for copper (Km 0.7 μM). This needs to be confirmed. Studies by Funk and Schneider (1989) have identified an additional, non-energy dependent transport system in the same organism which can be induced by an analog of vitamin K (menadione).

10.4 Copper in Plants (and Algae)

10.4.1 Copper Absorption and Distribution

Aspects of the general subject of copper in plants have recently been reviewed by Delhaize *et al.* (1987), and reviews on the distribution of copper in plant tissues have also appeared (Loneragan, 1981; Graham, 1981). Some of the same information has already been covered in Chapter 1 of this book, and extensive listings of the copper contents of different food plants may be found in Appendix A.

10.4.2 Copper and Roots

Copper is obtained by plants primarily through the roots. The mechanisms by which copper is absorbed from the soil (or soil water), and indeed the forms of copper that are available or involved in absorption, are still largely unknown (Delhaize *et al.*, 1987). The nature of the soil will, of course, make a great deal of difference with regard to the forms of copper that may be available, including their inorganic versus organic nature, soil microbiology, and pH (especially in more lifeless soils). (Some aspects of this are discussed in Chapter 1.) Based on studies with chelated copper ions in nutrient solutions, concentrations of 10^{-8} to 10^{-9} M available (soluble) copper are adequate, at least for some plants. [This contrasts with recommendations that overall concentrations of copper in inorganic soils (not soil water) should exceed 6 $\mu g/g$ (or 10^{-4} "M") and those in muck soils (high in some kinds of organic matter) 30 $\mu g/g$ (or 0.5 "mM").] Apparently, the endogenous nitrogen content of the roots is also an important variable. Addition of nitrate to the growth medium, or the presence of nitrogen-fixing bacteria in the root nodules of legumes, enhances copper uptake (Jarvis, 1984). However, under field conditions, addition of nitrogen fertilizers often enhances copper deficiency (Gartell, 1981), perhaps by overstimulating tissue growth. As in the case of intestinal copper absorption, a large excess of zinc will depress root copper uptake, presumably due to competition for carrier systems in the membranes of root cells. Whether a similar antagonism is exerted by iron, manganese, or other divalent metal ions that may be present has evidently not been studied. In the roots, much of the copper is associated with the polysaccharide cell wall, from which it can be displaced by Pb(II), but not by other divalent cations. The copper appears to be bound to nitrogen, sulfur, and oxygen ligands (Graham, 1981).

The roots are also usually the main site for deposition of excess copper that may enter from the soil, and they have also been the sources from which copper and sulfur-containing polypeptides (initially thought to be metallothioneins) have been isolated. Apparently, the expression of such proteins is associated with a tolerance for excess copper in soils or media. Plant "MT" was first identified as a small, high-cysteine, copper-containing protein in *Agrostis* (Rauser and Curvetto, 1980) and later also in cucumber roots (Lolkema *et al.*, 1984), both of which species have a high copper tolerance. (Many plants do not have a good tolerance for high concentrations of copper.) However, the *Agrostis* "Cu thionein" was found to have a molecular weight of only 1700 (Rauser, 1984), which is close to that of MT from the fungus *Neurospora crassa* (see Table 6-6 and Fig. 6-6). In high-tolerance plants, leaves may also store excess copper (Woolhouse and Walker, 1981). It now appears that the copper-binding peptides are not metallothioneins, but a rather unique series of polymers that have been renamed phytochelatins. (They are discussed at the end of this section.)

Apart from "phytochelatins," other low-molecular-weight binding proteins may be involved in copper storage and detoxification in the root. Tukendorf *et al.* (1984) have reported the presence of two nonthionein copper proteins in copper-tolerant spinach roots. These proteins (M_r 2500 and 9500) accounted for most of the excess copper in the root but were also present under conditions of low availability of the metal. Delhaize *et al.* (1987) suggested that some of the nonplastocyanin small, blue copper proteins found in plants (Tables 10-1 and 10-2) (and discussed in relation to azurin and bacteria) may also play a role in detoxification and storage of copper. Indeed, some of these have been found in roots and in other nonphotosynthetic tissues. However, the fact that they have a strict 1 g-atom/mol copper stoichiometry, a potential for involvement in redox reactions, and structures that may be very similar to those of plastocyanins (and bacterial azurins) (Fig. 10-1) (Ryden, 1988) suggests that they are more likely to function as enzymes. They would also be rather inefficient as sequestrants of excess copper ions.

10.4.3 Copper Transport in the Plant

From the roots, copper is transported to the rest of the plant mainly through the sap. A wide variety of potential ligands for copper are present in the cell sap, including most amino acids, as well as small peptides (Yoshida *et al.*, 1982). These are likely to be important for copper transport, as there are probably no carrier macromolecules, and amino acids or peptides (especially those with histidine) have a very high affinity for copper (see Chapter 4 and Table 4-5). (The K_D for the dihistidine complex is about $10^{-17}\ M$.) Thus, as in vertebrate organisms, the concentrations of free Cu(II) are likely to be extremely low. Whether other small molecules more unique to plants are also involved is an open question. Delhaize et al. (1987) pointed out that nicotianamine (an amino acid that chelates iron; Budesinsky et al., 1980) and mugineic acid (a phytosiderophore secreted by roots to chelate iron; Sugiura et al., 1981) form complexes with several other metal ions, including copper. Both substances are widely distributed among plants, and their structures are shown in Fig. 10-4. However, iron and copper do not usually associate with the same complexing agents present in living organisms, and the affinities of both of these agents for copper is likely to be less than that of histidine or some other amino acids [The copper complex of mugineic acid was prepared in the presence of mM Cu(II) (Nomoto et al., 1981).]

10.4.4 Copper in Other Plant Parts

Significant amounts of copper are present in shoots and leaves, where they play an important role in respiration and photosynthesis. A large proportion of the copper in both tissues is bound to plastocyanins (and perhaps also other protein factors) associated with the chloroplasts. The copper content of shoots and leaves tends to be highest in young plants and declines with age. This is the case in conditions of copper sufficiency, but it will be reversed if a progressive deficiency arises, as then, the new leaves will have less copper available from the soil and roots, while the older leaves already have copper and retain it. Thus, once it has entered the cells of the leaf, the copper absorbed seems not to flow easily to other parts of the plant, a situation very different from that of the mammal. Why the copper concentrations of leaves should otherwise fall with age is still unclear, although one can speculate that the plant may need less energy (per weight) as it matures, and therefore its photosynthetic capacity in general may be less. [The initial growth of plants (in terms of rates of increase in mass) is probably greater early after germination than later on.] It has also been suggested that the available copper in the soil local to the roots may decline with time and that this may explain part of this trend in the leaves (Delhaize et al., 1987), but one could argue as well that this would have the opposite effect, less being available for new shoots.

The young parts of flowers, seeds, and fruits are also those where the highest concentrations of copper are found in these organs: the anthers of flowers; the germ (or embryonic plant) in seeds; and the aleurone and pericarp (of fruit). In general, legumes and herbs have somewhat higher copper concentrations in their shoots and seeds than those of cereals and grasses. A limitation of the nitrogen supply and differences in tendencies to accumulate nitrogen will also affect the copper accumulated by plant seeds and shoots. Legumes tend to accumulate more copper and protein than cereals, and a deficiency in the availability of soil copper will depress the copper content of grains (Delhaize *et al.*, 1987). There are also substantial differences in copper accumulation ascribable to genetic differences among strains or subspecies, as, for example, rye versus wheat, where a "copper efficiency factor" has been associated with the rye chromosomes (see Graham, 1981). (Rye tends to accumulate twice as much copper as wheat.)

10.5 Plant Copper Proteins and Their Functions

10.5.1 Plastocyanins (and Cytochrome Oxidase)

Apart from the ubiquitous and essential function of copper in respiratory electron transport as part of mitochondrial cytochrome c oxidase, it has the equally (or even more) important function in plants of being involved in photosynthesis (and O_2 production) via plastocyanins in chloroplasts. Plants require cytochrome oxidase-based electron transport in nonphotosynthetic tissues (such as the roots) to supply ATP and otherwise (in all cells) during periods when light is not available as an energy source. As mentioned in Chapter 6, plant cytochrome oxidase is intermediate in size between that of bacteria and mammals, with about 5 subunits (versus 1–3 in bacteria and 12 or 13 in mammals). As in other phyla, it requires at least 2 g-atoms of copper per mol. The enzyme from sweet potato roots has been the most studied in this regard (Maeshima and Asahi, 1978) and has a total molecular weight of about 130,000.

Plastocyanin is probably on the inside of the membrane of the chloroplasts, where it is part of the end of the electron transport chain of photosystem II, receiving electrons from cytochrome f and transferring them to the P-700 unit of photosystem I (Fig. 10-5). Plastocyanin (PC) was first isolated from a green alga, *Chlorella*, and later from spinach leaves, by Katoh (Katoh, 1960; Katoh and Takamiya, 1961). In most leaves, it probably accounts for one-third or more of the total copper present. Klyavinya *et al.* (1982) have reported that it accounts for as much as 50–70% of the copper in spinach, barley, or cabbage leaves, while earlier reports estimated it as one-third of the copper in chloroplasts (Plesnicar and Bendall, 1970). The redox potentials for cytochrome f and PC are so close (365 versus 347–390 mV, respectively) that the exact position of PC

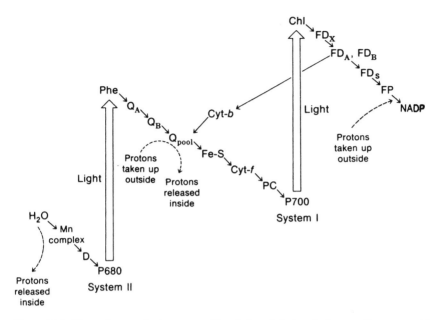

Figure 10-5. Plant photosynthetic systems II and I and their substituents. Component D (=Z) is unknown; Phe; pheophytin a; Q_A and Q_B, plastoquinones; Q_{pool}, a large pool of plastoquinones; cyt-b, cytochrome b-563; cyt-f, cytochrome f; PC, plastocyanin; Chl, chlorophyll a; FD_X (sometimes called X or A_2), FD_A, and FD_B, Fe-S proteins (bound ferredoxins); FD_s, soluble ferredoxin; FD, flavoprotein (ferredoxin:NADP oxidoreductase). Dashed arrows refer to events across the thylakoid membrane of the chloroplast. Reprinted, with permission, from Zubay (1983).

in the electron transport chain may be questioned (Ryden, 1984). However, in analogy with their work on azurins (see Section 10.1.2), Farver and Pecht (1988) have found that P-700 rapidly accepts electrons from reduced PC and have concluded that it probably associates with this photo unit via its negatively charged hydrophobic patch (with four carboxylate residues) near Tyr-83 (Fig. 10-1A). By analogy with azurin, they suggested that histidine-87 (one of the copper ligands exposed to the solvent but surrounded by hydrophobic amino acids) is the site for transfer of electrons from the cytochrome b_6/f complex in the membrane. (It would be attracted to hydrophobic parts of the complex in the membrane.) In the case of azurin, the analogous site binds (and accepts electrons from) cytochrome b-551 (see Section 10.1.2). In contrast, Beoku-Betts et al. (1984) have reported that the distinctive negatively charged patch around Tyr-83 (seen in poplar plastocyanin) is the site for cytochrome f binding. This region is altered and less negative in PC from the green alga *Scenedesmus obliquus* (Moore *et al.*, 1988) and even more so in the case of the blue-green alga *Anabaena variabilis*, where it is replaced by neutral and even positively charged amino acids (Jackman *et al.*, 1987). (This PC binds no cationic electron transfer units.) How electron transfer occurs in this organism in conjunction with photosynthesis is therefore an open question.

Transfer of electrons through PC occurs with minimal movement of the copper or its ligands. Plastocyanins are about 100 amino acids long (Table 10-1), and (as in the case of azurin) two histidines (Fig. 10-1B), a cysteine sulfur, and a methionine sulfur bind the copper in one corner of the region between the two sets of β-like polypeptide strands (Fig. 10-1A). (See Section 10.1.2) The combination of "hard" (imidazole N) and "soft" (S) ligands makes transitions easier between the Cu(I) and Cu(II) states (Delhaize *et al.*, 1987), as is ideal for involvement in an electron transfer system. The amino acid sequences of PCs from different plant sources are very similar (Fig. 10-2) and also close to those of plastocyanins from photosynthetic cyanobacteria and blue-green (or green) algae. A functionally equivalent but heme-containing cytochrome (c-553) can substitute for PC in procaryotic blue-green algae and in most eucaryotic algae (where PC is not expressed except in the green forms) (Sandmann, 1986).

The average photosynthetic unit of a plant chloroplast probably contains one unit each of plastocyanin and the cytochromes per four or more units of plastoquinone and multiple units of chlorophyll (Table 10-4). As already indicated, PC accounts for much, but not all, of the Cu in the chloroplast. It has been suggested that the fraction I protein of chloroplasts may be where the rest of the copper is bound, though this is still controversial (see Ryden, 1984). Fraction I consists of the ribulose diphosphate–carboxylase–oxygenase complex (of 550 kDa), also photoactivated, that is the most abundant protein in leaves and most important for formation of the carbon skeletons of the sugars that are used to form the basic carbohydrates, fats, amino acids, and vitamins of plants (and other living organisms). It has been reported that much of chloroplast copper associates with the high-molecular-weight soluble component fraction after homogenization of the leaves. This component could be the fraction I complex, but there are contradictory findings about the copper content (if any) of the isolated complex itself that still need to be resolved (Ryden, 1984).

As already mentioned, a cytochrome (c-553) replaces PC in some eucaryotic and procaryotic photosynthetic algae. In some algae subspecies and strains, either can be expressed, depending upon copper availability. In the eucaryotic green algae, it was shown by two groups of investigators that PC is expressed when sufficient copper is available but that c-553 is substituted (and PC disappears) when the medium is low in copper (Bohner and Boger, 1978; Wood, 1978). More recently, Sandmann (1986)

Table 10-4. Approximate Composition of an Average Photosynthetic Unit in a Spinach Chloroplast[a]

Component	Number of molecules or atoms
Chlorophyll a	160
Chlorophyll b	70
Carotenoids	48
Plastoquinone A	16
Plastoquinone B	8
Plastoquinone C	4
α-Tocopherol	10
α-Tocopherylquinone	4
Vitamin K_2	4
Phospholipids	116
Sulfolipids	48
Galactosylglycerides	490
Iron	12
Ferredoxin	5
Cytochrome b-563	1
Cytochrome b-559	1
Cytochrome f	1
Copper	6
Plastocyanin	1
Manganese	2
Protein	928 kDa

[a] Reprinted by permission from Metzler (1977).

found that procaryotic blue-green algae are of two types, some that substitute PC for c-553 when copper is available and some that do not (Fig. 10-6). Most of them do substitute it, and maximal production of PC seems to occur at about 0.8 μM Cu. [Half-maximal expression is at about 0.4 μM Cu, although there is residual expression at much lower concentrations in some strains (such as *Pseudoanabaena*; Fig. 10-6).] Thus, copper serves to induce PC in at least some algae. One wonders whether that is also the case in some higher plants and by what mechanisms it occurs.

10.5.2 Other Small, Blue Copper Proteins in Plants

The existence of a variety of other small, blue copper proteins in nonvertebrate organisms has already been mentioned and partly described. A number of them have been isolated from different higher plants, and the properties of some of these are listed in Tables 10-1 and 10-2. Most of these

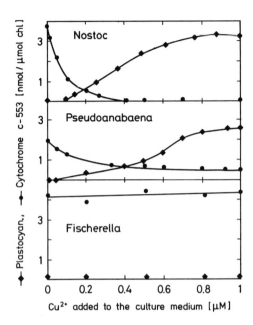

Figure 10-6. Induction of plastocyanin by copper added to various species of blue-green algae in the culture medium. Reprinted, with permission, from Sandmann (1986).

have been found in nonphotosynthetic tissues, like roots or latex (Table 10-2), which indicates that they clearly have a nonphotosynthetic function. Most of them also are glycoproteins, and some are extracellular (such as the stellacyanin of lacquer tree latex). They are nevertheless all closely related proteins (Ryden, 1984, 1988) (Fig. 10-3). In stellacyanin, which has almost twice the molecular weight of plastocyanin because of added carbohydrate, the overall size (Fig. 10-2 and Table 10-2) and configuration of the polypeptide portion may still be quite similar, at least as regards the copper environment. The visible and circular dichroism (CD) spectra of stellocyanin and plastocyanin are not as close as those of azurin and plastocyanin, but there is still a marked similarity. Copper is coordinated with two histidines and two sulfur ligands, in a flattened tetrahedral structure (Solomon et al., 1980). However, in stellacyanin, the second sulfur is not a methionine but may involve a disulfide bond (Ryden, 1984). The redox potentials of stellacyanin and others of these nonphotosynthetic proteins are also less than those of plastocyanins, but close to those of the azurins and still higher than that of aqueous Cu(II). In the case of stellacyanin, carbohydrate is attached to asparagines in three parts of the structure. Umecyanin of horseradish roots is very similar in amino acid sequence (Fig. 10-2) but has a different carbohydrate content and attachment, as well as a quite different pI (Table 10-1). Again, the copper configuration has been conserved. No X-ray structure for these two proteins is yet available.

The X-ray structure of a quite different "phytocyanin," however, has recently been resolved. This small, blue copper protein comes from cucumber seedlings (Tables 10-1 and 10-2), has no carbohydrate, but, like stellacyanin, has a basic pI. It turns out that the structure of this protein is closer to that of azurin and plastocyanin than the structures of stellacyanin and umecyanin are likely to be (Guss *et al.*, 1988). [The redox potential is also closer to that of PC (Table 10-2).] The copper configuration involves histidine-39, histidine-84, cysteine-79, and methionine-89 (Fig. 10-7), which even numerically are close to the residues involved in the case of azurin and PC (Fig. 10-1B). There is also a great deal of similarity

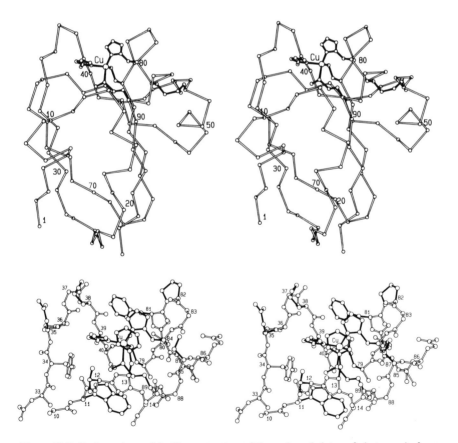

Figure 10-7. Various views of the X-ray structure of the carbon skeleton of phytocyanin from the cucumber. Reprinted, with permission, from Guss *et al.* (1988).

in overall structure, although the conformation is broader. Guss *et al.* (1988) pointed out that the main difference between azurin and plastocyanin is in strand 5, which for azurin is part of the β barrel (but with a side flap) (Fig. 10-1A), whereas for plastocyanin it is irregular and not part of the β structure. In the cucumber protein, involvement of strands 4 and 5 in the β sandwich is even less, and they twist out, away from the main structure (Fig. 10-7). The rest of the strands (six in all) still form the basic sandwich, with three strands in each half.

As already discussed in the case of the small, blue bacterial proteins, the function of these proteins (probably as redox enzymes) is unknown. Certainly, they are found in diverse parts of the plant. The latex (where stellacyanin is found) may be considered a secretion induced by injury and therefore should contain enzymes that have a protective function. Another of these, laccase, is also copper dependent.

10.5.3 Laccase

The structure and function of this enzyme have been most recently reviewed by Reinhammar (1984). Laccase (benzenediol:oxygen oxidoreductase EC 1.10.3.2) belongs to the more complex group of blue copper oxidases that may also have evolved from the same small, blue copper protein ancestor as azurin, plastocyanin, etc. (Fig. 10-8) (Ryden, 1984,

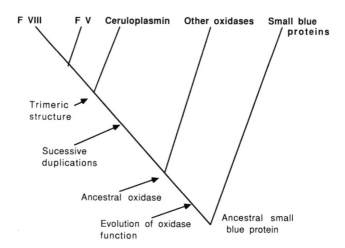

Figure 10-8. Evolution of blue copper oxidases and other copper proteins, based on amino acid sequence. Reprinted, with permission, from Ryden (1984).

1988). The blue copper oxidases include plant ascorbate oxidase and mammalian ceruloplasmin. They have multiple copper atoms of at least three types: type 1 (the blue), with visible absorbance peaking at about 600 nm and an EPR signal; type 2, which is also EPR detectable; and type 3, which is EPR silent, diamagnetic, has 330-nm absorbance, and is thought to represent pairs of spin-coupled Cu(I) atoms. (See also Chapter 4, Section 4.1.1 and Tables 4-2 and 4-3.) Laccase can also be classified as a polyphenol oxidase (Mayer and Harel, 1979) and, as such, would group with another copper-dependent plant (and animal) protein, tyrosinase (or catechol oxidase). Laccase, ceruloplasmin, and ascorbate oxidase have a great deal of similarity in terms of their spectra and the primary states of their multiple copper ions (Fig. 10-9). Ceruloplasmin has even been thought of as a mammalian laccase. All three proteins have a very broad substrate specificity, although there are also some differences. Laccase, but not ceruloplasmin (Mondovi and Avigliano, 1984), has ascorbate oxidase activity, and ascorbate oxidase does not have as broad a substrate roster as the others, although it does oxidize a number of similar phenols (Reinhammar and Malmstrom, 1981). Laccase and ceruloplasmin oxidize p-diphenols as well as *ortho*- and *meta*-substituted phenols (such as catechol and resorcinol). (Like tyrosinase, they can oxidize monophenols.) However, the specific activities of laccase are about tenfold higher than those for ceruloplasmin, with the same substrates. The blue (type 1) copper sites in these enzymes, like those of the small, blue copper proteins, involve two histidine nitrogens and a sulfur, which in the case of laccase is methionine (Fig. 10-10A). (A generalized model for the site is shown in Fig. 10-10B.) Here, there even appear to be amino acid sequence homologies with the copper-binding regions of azurins and plastocyanins (see Chapter 4, Fig. 4-6).

The structure and composition of laccase is compared with that of ceruloplasmin or ascorbate oxidase in Table 10-5. Laccase has been isolated not just from the latex of the lacquer tree, but also from fungi. The two forms differ a great deal in size, but almost all of this is probably due to the large amount of carbohydrate in the tree latex enzyme. Like ascorbate oxidase (which is probably a homodimer and has a similar-sized polypeptide unit), and like ceruloplasmin, there is a reactive center of three or four copper atoms responsible for the reduction of dioxygen in the different reactions catalyzed. [The most recent work on ascorbate oxidase (see below), amine oxidase (see Chapter 4), and ceruloplasmin (see Chapter 4) suggests that it is a trinuclear copper center, composed of one type 2 and two type 3 Cu atoms, with at least one blue copper for delivery of electrons.] There is considerable amino acid sequence homology between the C-terminal third of ceruloplasmin and laccase from the fungus

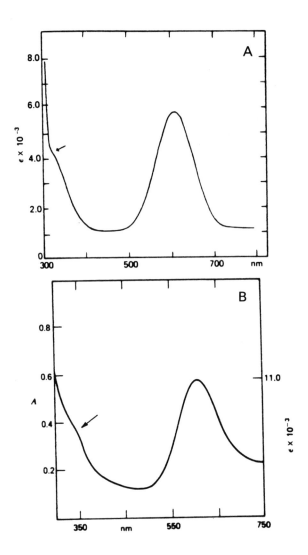
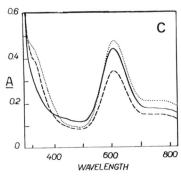

Figure 10-9. Comparative UV–visible absorption spectra for laccase, ceruloplasmin, and ascorbate oxidase. (A) Laccase and (B) ceruloplasmin; arrows indicate the 330-nm absorption peak. Reprinted, with permission, from Solomon (1981). (C) Spectra for ascorbate oxidase, showing the native enzyme at pH 5.2 (·····), after treatment with EDTA dimethylglyoxime (———), and reconstitution (----), Reprinted, with permission, from Mondovi and Avigliano (1984).

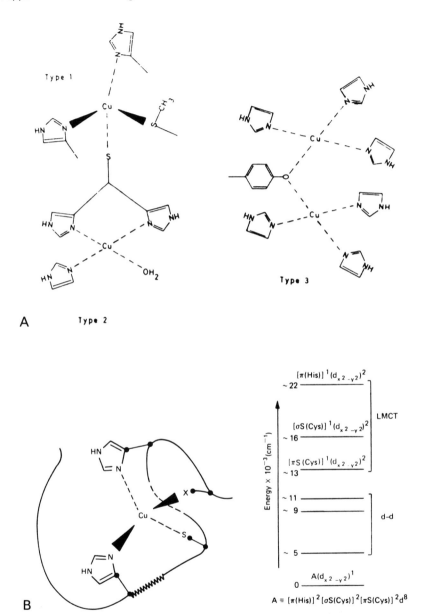

Figure 10-10. Model for the copper sites in laccase and for blue copper centers in proteins. (A) Model for laccase. Reprinted, with permission, from Reinhammar and Malmstrom (1981). (B) Generalized model for blue copper centers in proteins. Reprinted, with permission, from Gray and Solomon (1981). Both models are taken from Spiro (1981).

Table 10-5. Structure and Composition of Laccase, Ascorbate Oxidase, and Ceruloplasmin[a]

	Molecular weight	No. of subunits	No. of amino acids	Copper (g-atom/mol)[b]				Carbohydrate (%)	Ratio of absorbance, 280:610 nm
				Type 1	Type 2	Type 3	Total		
Laccase									
Rhus	120,000 (about 62,000 polypeptide)	1	540	1 (394)	1 (365)	2 (434)	4	49	
Fungal	70,000? (66,000 peptide)	1	570[c]	1 (785)	1	2 (782)	4	12	
Ascorbate oxidase	140,000[d] (pH ⩾ 7) (63,000 each polypeptide)[d]	2[d]	550	3	2[e]	4	8–10	10 (or 2.3)	0.040
Ceruloplasmin (human)	132,000 (121,000 peptide)	1	1046	2 (490)	1	4[f]	6–7	7	0.045

[a] References are mainly Reinhammar and Malmström (1981) and Ryden (1988).
[b] Values in parentheses are Redox potentials (mV).
[c] Lerch and Germann (1988).
[d] Mondovi and Avigliano (1984).
[e] Messerschmidt et al. (1988).
[f] Tentative data.

Neurospora crassa (Lerch and Germann, 1988). A 114-amino acid unit can be aligned that shows an exact correpondence between 27 amino acids and also 12 isofunctional substitutions. The sequence of ascorbate oxidase is also homologous.

The functions of laccase are only partly understood and would seem to be different in the lacquer tree as compared with the fungus. Being part of the latex, laccase is released when the tree is damaged, exposing the latex to air. Specific phenols in the lacquer (urushiol and laccol; Fig. 10-11) are then oxidized by laccase, forming free-radical compounds that polymerize. This solidifies or clots the latex, sealing the "wound". In the case of the fungus, the enzyme is expressed in certain basidiomycetes that can decompose lignin, which is a polyphenol (Fig. 10-11), and absent in others that cannot do this (Reinhammar and Malmström, 1981).

As regards the similarities with tyrosinase, they are not structural, as tyrosinase (also known as catechol oxidase and cresolase) is descended from the same ancestor as the hemocyanins (see later) and contains only type 3 copper. However, both plant enzymes have a very similar substrate specificity and are particularly active against diphenols (Mayer and Harel,

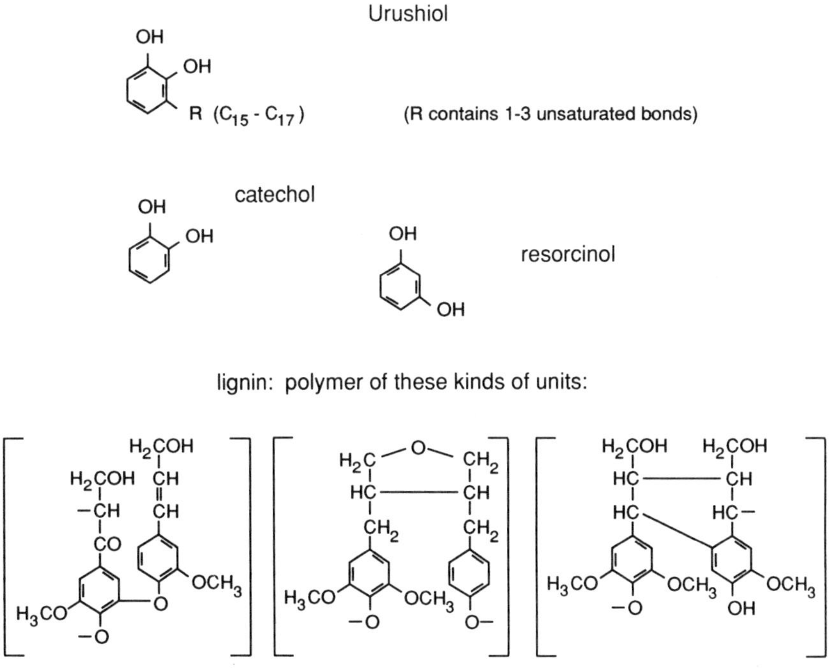

Figure 10-11. Structures of plant substances that may be substrates for blue copper enzymes (urushiol, catechol, resorcinol, and lignin).

1979). Laccase will not, however, oxidize tyrosine (or *p*-cresol). Laccase may also be involved in the polymerization of phenols that form the humic acids of soil humus, being prevalent especially in soil fungi, like *Asco-* and *Basidio-mycetes*, as well as *Fungi imperfecti* (Mayer and Harel, 1979). Apart from the genus *Rhus*, there are only limited reports of laccase in higher plants, although it has been detected in the secretory duct of many Anacardiaceae, including the mango (Joel *et al.*, 1978), in peaches (Lehmann *et al.*, 1974), in *Aesculus* (Wosilait *et al.*, 1954; Mayer and Harel, 1979), and, more recently, in sycamore trees (Bligny *et al.*, 1986). In most cases, laccase seems to be extracellular, although there is some soluble intracellular enzyme in certain fungi (especially *Podospora*). The extracellular nature of most the laccases fits with the fact that they have attached carbohydrate. This is also consistent with the nature of the other blue copper oxidases, ascorbate oxidase and ceruloplasmin. In this connection, it is tempting to think that the laccases might serve another function that has been identified with ceruloplasmin, namely, that of scavenging free radicals. This might serve to protect the exterior of plant (or fungal cells) from oxidative damage. However, that might also interfere with some of its catalytic functions.

In the fungus *Neurospora crassa*, laccase is produced in conjunction with tyrosinase (catechol oxidase) during sexual differentiation of the organism (Huber and Lerch, 1987; Hirsch, 1954), where tyrosinase is thought to be responsible for forming the melanin pigment of the fruiting bodies. Both of these enzymes can be induced together in cultured *N. crassa* cells by addition of cycloheximide (Huber and Lerch, 1987), and the increase in activity (but not protein) is much greater when copper is added as well. The mechanism of cycloheximide induction is not understood, but it has been speculated that it involves inhibition of the synthesis of a fast-turnover protein that represses expression of both enzymes. The reason for laccase induction during fruiting is not clear, although, from its substrate specificity, it seems possible that it could assist in the formation of a distinctive melanin. Melanin formation involves oxidation of tyrosine and subsequent phenols and quinones (see Fig. 6-20 and Chapter 6). This occurs intracellularly, where some laccase may also be present, although much of it is secreted. Again, an extracellular protective function is possible, as may be others involving substrates still to be discovered.

10.5.4 Ascorbate Oxidase

Ascorbate oxidase has most recently been reviewed by Mondovi and Avigliano (1984). This enzyme is widely distributed in plants and also found in some fungi (White and Krupka, 1965) and bacteria (Dawson, 1966). While capable of acting on many of the same substrates as laccase (and ceruloplasmin), it is notable for its ascorbate oxidase activity (EC 1.10.3.3). The most common forms of the enzyme that have been isolated are from the crookneck and zucchini squashes and from the cucumber. As already indicated, there appears to be considerable sequence homology with laccase (Messerschmidt et al., 1988; Lerch and Germann, 1988), as well as with the C-terminal third of ceruloplasmin. The X-ray structure of the zucchini protein is being resolved (Messerschmidt et al., 1988), and the initial findings indicate that the protein is a dimer, each subunit being of about the same size as the laccase polypeptide chain and half the size of ceruloplasmin (Table 10-5). It also has some carbohydrate and more copper atoms than even ceruloplasmin. The ratio of absorbance at 610 to that at 280 nm is 0.040 (Mondovi and Avigliano, 1984), or very close to that of human ceruloplasmin (0.045) and laccase. At pHs below 7, the purified protein aggregates to form larger structures.

According to the 2.5 Å X-ray map of the crystallized zucchini protein, the individual subunits appear to be identical (Messerschmidt et al., 1988). Each subunit seems to have two plastocyanin-type folding domains, each with a single copper, bound in a fashion analogous to that in plastocyanin and azurin. Halfway between these domains is a trinuclear copper cluster, probably made up of types 2 and 3. In one of the subunits of the dimeric molecule, only one of the type 1 copper sites seems to be occupied, giving a total of nine Cu atoms per total molecule. It is thought that the type 1 coppers at each end of the subunit accept electrons from ascorbate and that the type 3 (coupled) coppers (in the trinuclear site) are responsible for the oxygen binding. [Earlier work by Dawson et al. (1980) with fluoride suggested that the type 2 copper (in the trinuclear site) might also be involved in the ascorbate binding.] The two type 1 sites in each subunit are about 13 Å distant from the trinuclear copper cluster.

The function of ascorbate oxidase is not really known. Studies with plants indicate that it is mostly attached to the cell walls (Halloway et al., 1970), but it is also in the cytoplasm (Bidwell, 1979), and it is most prevalent in regions of tissue that are growing rapidly, as, for example, in pea seedling roots (Suzuki and Ogiso, 1973). It is also found in fruits, such as squashes and cucumbers, that also grow quite rapidly.

Ascorbate oxidase is capable of oxidizing a number of compounds that may be found in plant tissues, including catechols, flavonoids, and hydroxycinnamic acids (Fig. 10-11). How this might be useful is not clear. The fact that synthesis of the enzyme is positively affected by exposure of sprouts (cotyledons) to infrared light (in the case of mustard plants; Attridge, 1974) should be some kind of a clue, but one we have yet to understand. When vegetables or fuits are sliced and darken, the latter process may be mediated by ascorbate oxidase catalyzing the reaction of oxygen with o-diphenolic compounds that are present (Mondovi and Avigliano, 1984). Ascorbate, of course, inhibits this process by competing for the enzyme. Indeed, one could rationalize the ascorbate specificity of the enzyme on the basis that, when plants are injured, certain compounds are better protected from oxidative action of the inherent ascorbate oxidase, which normally has another function but can inadvertently cause inappropriate oxidation when circumstances like cell disruptions occur, unless "controlled" by the presence of ascorbate. Alternatively, the darkening reaction itself may be protective and form a polymeric protective layer.

10.5.5 Tyrosinase (Phenolase, Catechol Oxidase)

The structure of mammalian tyrosinase (involved in melanin formation) has already been reviewed in Chapter 6. This enzyme has as its primary substrates tyrosine and DOPA, and its function in these reactions is relatively well understood. The evolutionarily related enzyme in plants, which is widely distributed in different tissues and species of this phylum (and is also found in other nonvertebrate organisms), has a broader substrate specificity and probably catalyzes other kinds of reactions. Consequently, it has variously been called phenolase, catechol oxidase, cresolase, and, more officially, o-diphenol oxidoreductase (EC 1.10.3.1). (It also has been reported to have ferroxidase activity.) It is distinguished from ascorbate oxidase primarily in being unable to oxidize p-diphenols and physiologically in terms of its intracellular location, ascorbate oxidase being mainly in the cell wall and cytoplasm, while tyrosinase/phenolase is in the thylakoid membrane of chloroplasts and perhaps in the mitochondria (Walker and Webb, 1981). Mayer and Harel (1979) did an extensive review of both tyrosinase (catechol oxidase) and ascorbate oxidase of plants under the heading of "polyphenol oxidases." [A general review of plant copper proteins also appeared in 1981 (Walker and Webb, 1981).] No further comprehensive reviews of the plant forms hae appeared since, although new data on the amino acid sequence of the fungal (and bacterial) polyphenol oxidases have become available that are important for placing the enzyme in evolutionary connection with mammalian (mouse) tyrosinase and with the hemocyanins (Lerch and Germann, 1988). Information on the structure of the plant enzymes is much more scant, and no X-ray data on any form of tyrosinase are available. All of the plant, fungal, and bacterial tyrosinases/phenolases (as well as the hemocyanins) have a pair of coupled type 3 coppers as their reactive center (or a binuclear copper center). At least in the case of the fungal and bacterial tyrosinase (but probably in all), each copper (termed Cu_A and Cu_B) is associated with three histidine imidazoles (Lerch and Germann, 1988) that come from a pair of α-helices in the protein structure.

As regards the overall structure of plant catechol oxidases, there have been numerous reports of various molecular weights ranging from 25–30 kDa up to 120–150 kDa, with "subunits" of various sizes. Overall, there is the impression that the basic functional unit might be on the order of 60 kDa perhaps comprising "heavy" and "light" chains of about 42 and 13 kDa, respectively (Meyer and Harel, 1979; Strothkamp *et al.*, 1976; Lerch, 1976), with a pair of copper atoms. Dimers of such basic units probably also occur, and it is still not entirely clear whether both coppers may be on the same (heavy or light) polypeptide chain. [The gene structure suggests that they would both be on the heavier unit (Lerch and Germann, 1988).] In general, the amino acid composition appears to be high in basic amino acids and very low in sulfur.

As already mentioned, the enzyme is very widely distributed and catalyzes the oxidation of monophenols and *o*-diphenols to quinones. Like ascorbate oxidase, the enzyme is induced in some plants by cycloheximide (see earlier) and appears in rapidly growing tissues. As in the case of germinating seeds (Taneja and Sachar, 1977), it also appears to be induced (or its activity increased) by wounding (Bastin, 1968) or infection (Balasubramani *et al.*, 1971). Its distribution among different plant parts is not, however, restricted, and it is found in all sorts of places, including pollen grains and guard cells. In cells, it is found largely in the cell fluid, although it also appears on the inside of the chloroplast lamellae, as determined histologically (Parish, 1971, 1972; Mayer and Harel, 1979), and in the mitochondria.

The function(s) of this enzyme is still remarkably unclear. In the fungi (where it is almost always detected), its appearance is associated with development of the fruiting bodies, and added substrates like tyrosine may even help induce fruiting. Numerous suggestions have been made about possible roles in plants; these are reviewed in detail, and generally found wanting, by Mayer and Harel (1979). Two kinds of possibilities that may be viable are a role in making seed coats less permeable to water and in tissue darkening upon wounding and oxygen exposure, which may also have a protective function. It seems unlikely that the enzyme is involved in providing quinones for electron transport in chloroplasts or mitochondria. Another role may be in fighting infective organisms via production of potentially toxic quinones. This fits with numerous observations that resistance to infection often (but not always) correlates with catechol oxidase presence and activity. All kinds of infections and assaults on a plant can increase its catechol oxidase activity, yet the exact reasons for this response (and its consequences) are still not clear.

10.5.6 Low-Molecular-Weight Copper Ligands

Reports of metallothionein-like proteins in plants have already been mentioned in connection with root function and copper tolerance (see Section 10.4.2). The "Cu-thionein" first isolated, however, was found to have a molecular weight of only 1700 (Rauser, 1984). Later work has shown that the protein is not actually a metallothionein, but rather a peptide composed of two to ten linear repeats of γ-glutamylcysteine attached to a C-terminal glycine (Grill et al., 1985a,b) (see Fig. 10-4). This peptide is widely distributed in higher plants, binds a variety of metal ions, including copper, and accumulates in plant cells in culture in response to heavy-metal exposure. It is therefore likely that this "phytochelatin" is responsible for the limited resistance plants may have (particularly in their roots) to copper toxicity. The same kind of peptide probably also protects plants against other toxic heavy metals, such as Cd(II) (Grill and Zenk, 1989; Loeffler et al., 1990). The facts that it is made up principally of γ-glutamylcysteine, that there is a γ-peptidyl link, and that it is very close in structure to glutathione indicate that it cannot be a gene-blueprinted polypeptide product. Indeed, it appears to be formed from glutathione by a specific γ-glutamyl cysteine dipeptidyl transpeptidase (Loeffler et al., 1990). Induction of the molecule by heavy metals is dependent on glutathione (Grill et al., 1986a,b), and its synthesis is self-limiting, as the inducing metals are bound by the phytochelatin product. The molecular weights of the resulting polymers range from 559 to 1931, and they can be separated in reversed-phase (C_{18}) high-pressure liquid chromatography (HPLC). Results for extracts from a plant (Fig. 10-12A) show that the $n=4$ form predominates. [In others, it can be the $n=2$ or $n=3$ forms (Grill et al., 1987).] In the fission yeast (Fig. 10-12B) (Grill et al., 1986a), the $n=2$ and $n=3$ forms (of 1250 Da) were most abundant and varied with the growth phase, the $n=2$ form predominating in early logarithmic growth. In both the yeasts and plants, the phytochelatins can be rapidly induced by heavy metals (Fig. 10-13), and, as already indicated, this induction is partly dependent upon glutathione. The smaller forms seem to appear more rapidly, and, with time, larger polymers also form. The roster of metal ions that are able to induce appearance of the phytochelatins is larger than that for genuine metallothioneins (Grill et al., 1986b) and includes Bi(II), AsO_4(III), Sb(III), W(IV), Sn(II), and Ag(I) [as well as

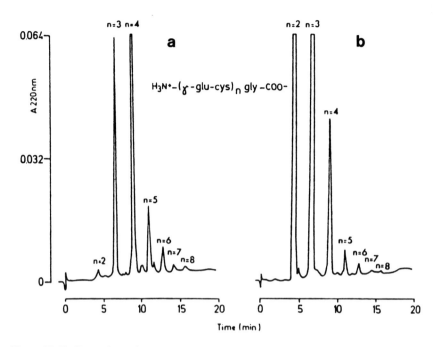

Figure 10-12. Separation of various sizes of phytochelatins by HPLC: (a) from plants; (b) from fission yeast. Reprinted, with permission, from Grill et al. (1986).

Cd(II), Cu(II) and Zn(II)]. Induction has been shown in roots (as well as in suspension cultures). The roots are where accumulation of excess copper (and probably other metals) is usually found, because metal ions usually enter through the roots.

Still another form of multi-metal-binding peptide has been partially purified from lettuce leaves (Walker and Welch, 1987) by DEAE-cellulose and size exclusion chromatography. It has an apparent molecular weight of 1250 (or close to that of phytochelatins), also contains sulfur, but (on first approximation at least) also contains a variety of other amino acids. There have also been reports of "chelatins" in the alga *Euglena gracilis*, chelatins in this case being thought of as cysteine-rich proteins about the size of metallothioneins, but with a lower sulfur amino acid content and some aromatic amino acids. [Metallothioneins generally lack aromatic amino acids (see Chapter 6).] The big question is whether they might be variants of the "phytochelatins" of Grill et al., already discussed, or whether they are actual metallothioneins that have been altered during purification (see Chapter 6). Two high-cysteine proteins of about 8000 Da, binding Cd(II), Zu(II), and some Cu(II), have been partially purified from *Euglena* by

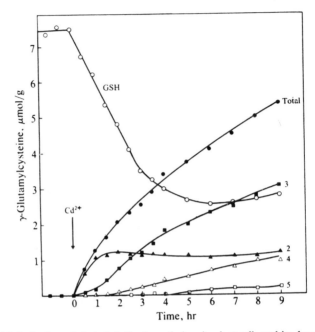

Figure 10-13. Induction of phytochelatins by cadmium in plant cells and its dependence upon glutathione (GSH). Numbers to the right of curves indicate the number of repeating γ-glutamylcysteine residues per molecule. Reprinted, with permission, from Grill et al. (1987).

Petering and colleagues. They do not react with antibody against rat liver MT and have quite different cadmium binding/dissociation characteristics than MT (Gingrich et al., 1986). The apparent size and the detection of sulfide in the protein (Weber et al., 1987) suggest that it is not the polymeric γ-glutamylcysteine-glycine peptide of Grill et al.. Two copper-containing polypeptides have also been isolated by Piccini et al. (1985). Both are reported to have a rather peculiar amino acid composition, with 28 and 54% aspartate, respectively, and 13–20% cysteine. Apparent molecular weights are 7400 and 3700. If glutamate rather than aspartate were so high, these would fit with the γ-glutamylcysteinyl glycine "phytochelatins," although the molecular weights would appear to be higher. Although plants (and schizosaccharomyces) were the sources of the phytochelatins of Grill et al., and algae might be expected to be like plants (and indeed also appear to have no metallothionein), algae may have evolved still another form of metal detoxification. This might explain the differences reported. Apart from aiding in detoxification, these compounds may play a role in storage or transport, a matter that needs to be further examined.

10.6 Copper in Mollusks and Arthropods

Mollusks and arthropods probably contain many or even most of the common copper-dependent proteins that we recognize in animal species (and to some degree in other creatures, as well), but they also use copper in a unique way, by producing hemocyanins to carry oxygen to their tissues. Hemocyanin, like catechol oxidase (tyrosinase), is a protein that has a binuclear copper center made up of a pair of type 3 copper atoms. Evolutionarily, evidence of Lerch and Germann (1988) suggests that the molluscan hemocyanin may be more closely related to the tyrosinases than to arthropodan hemocyanin (Fig. 10-14), the Cu_A site being added to the tyrosinase mollusk branch, while the Cu_B site is doubled in the case of the arthropods (Lerch and Huber, 1986). Certainly, except for a stretch of 42 amino acids (Drexel et al., 1987), the amino acid sequence of the less

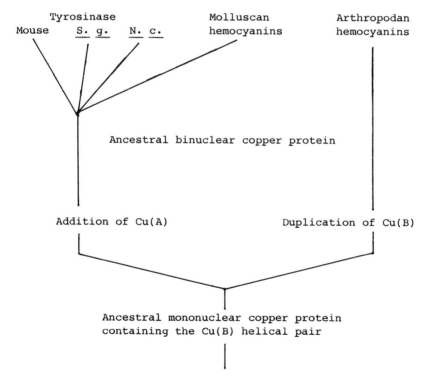

Figure 10-14. Evolutionary relationships between tyrosinases and hemocyanins (S.g. = Saccharomyces glaucescens; N.c. = neurospora crassa). Redrawn from Lerch and Germann (1988).

studied molluscan protein is completely different from that of the arthropodan protein (Volbeda and Hol, 1988), and so is the quaternary structure (as determined largely by electron microscopy; Bruggen et al., 1981; Ellerton et al., 1983). Molluscan hemocyanin has 10 or 20 subunits made up of repeated polypeptide units, to form a cylinder with a total molecular weight of about 9×10^6. [There are about 640 amino acid residues per binuclear copper cluster (or O_2 binding site).] The arthropod proteins are probably built of hexameric units linked together to form dimers, tetramers, and higher aggregates. The total molecular weights are not as high as in the mollusk. The largest form (from the horseshoe crab) is about 3.5×10^6 Da, the smallest about 460,000 (Van Holde and Brenowitz, 1981). Based on the 3.2-Å X-ray data for the deoxy form of hemocyanin from the spiny lobster (*Panularis*) (Volbeda and Hol, 1988), the basic 460,000-Da hexamer has 3:2 symmetry, or is a trimer of very tighty coupled subunit dimers. Each ca. 75,000-Da subunit in the structure (subunits show microheterogeneity) has three domains. The N- and

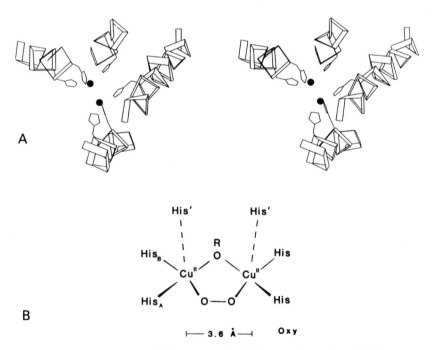

Figure 10-15. X-ray structure of the copper site of hemocyanins and model for oxygen binding. (A) Stereo diagram of binuclear copper site. Reprinted, with permission, from Volbeda and Hol (1988). (B) Model for oxygen binding to binuclear site, showing the histidine configuration. Reprinted, with permission, from Reed (1985).

C-terminal domains have carbohydrate and disulfide bonds, respectively. The functional, central domain has seven α-helices (Fig. 10-15A). Two of the histidines come out of one helix and are separated by three other amino acids. The proposed configuration of the binuclear copper center in these proteins, when bound to O_2, is shown in Fig. 10-15B. The reaction would be

$$Cu(I)Cu(I) + O_2 \rightarrow Cu(II)O_2^- Cu(II)$$

In the X-ray structure, no oxygen ligand between the coppers has been seen as yet, but, as suggested in the drawing, some of the histidines are much closer to the coppers than others. [The coppers are about 3.5 Å apart, with two of the histidines only about 2.0 Å away.]

The function of hemocyanins is exactly the same as that of hemoglobin, namely, oxygen transport. Each O_2 unit is bound as indicated, and, as in hemoglobin, there is positive binding cooperativity among the subunits. However, in contrast to hemoglobin, this cooperativity does not extend to carbon monoxide (Reed, 1985). Also, hemocyanins are not sequestered in cells, but are directly dissolved in what is called the hemolymph of these creatures, which makes up their circulatory systems.

In mollusks, the kidney is one of the major sites where metals accumulate. In the case of the quahog [*Mercenaria mercenara* (*L.*)], the kidneys contain both intra- and extracellular granules, where most of the Fe, Mg, and Zn (as well as Ca and P) are found (Sullivan *et al.*, 1988). In contrast, most of the copper is in the cytoplasm (at least in the case of creatures taken from unpolluted waters), where it may be bound to metallothionein and SOD (although this last has not been examined).

Appendixes

Appendix A. Copper Contents of Foods

Food	Copper concentration μg/g or μg/ml[a]	Mean ± SD (N)[b]
Meats, fish, dairy products		
Beef, raw or corned	0.5, 0.8, 0.9	0.8 ± 0.5 (22)
Cheddar cheese	1.1	1.1 ± 0.7 (12)
Chicken	0.4–0.5, 1.8	
Cod (fish)	0.4, 2.2, 4.7	1.8 ± 2.2 (12)
Cottage cheese	0.13, 0.2	
Duck	4.1, 4.1, 5.0	
Egg		
Chicken	0.5, 1.0, 4.0	1.0 ± 0.7 (42)
Yolk	2.4, 3.1, 3.5	3.1 ± 2.6 (75)
White	0.5, 1.3, 1.7	0.5 ± 0.5 (69)
Haddock (fish)	1.9, 2.3	1.9 ± 1.3 (27)
Halibut (fish)	1.6–2.3	
Herring (fish)	1.9, 2.5, 4.4	1.9 ± 1.5 (5)
Milk		
Cow	0.1, 0.2, 0.35	0.35 ± 0.26 (426)
Human	0.4, 0.5, 0.7	0.52 ± 0.15 (14)
Oysters	12, 17, 37	17 ± 42 (430)
Pork (muscle meat)	3.1, 3.9, 4.1	
Ham	0.3–0.9	
Salmon	2.0	2.0 ± 1.6 (5)
Shrimp	0.8, 4.3, 6.0	6.0 ± 4.0 (8)
Trout (fish)	1.7–3.3	1.7 ± 1.6 (5)
Turkey	1.0–1.1	
Light meat	0.8, 1.0	
Dark meat	1.5, 1.2	
Veal	1.2, 2.5	1.2 ± 0.9 (5)

(continued)

Appendix A. (Continued)

Food	Copper concentration μg/g or μg/ml[a]	Mean ± SD (N)[b]
Fruits		
Apple	0.4, 0.9 (1.0)*	0.9 ± 0.7 (69)
Apricot	0.9, 1.1, (1.5)*	
Avocado	2.1–2.6	
Banana	1.0, 1.3, 1.7	
Blackberry	1.4	1.6 ± 0.5 (8)
Blueberry	0.6, 1.5 (1.1)*	
Cherry	0.9, 1.0, 1.2	1.2 ± 0.5 (17)
Cranberry	0.6, 1.1 (1.4–2.6)*	1.1 ± 0.5 (11)
Fig	0.7, 0.7, 0.7	
Grapefruit	0.4, 0.4, 0.5	0.4 ± 0.1 (15)
Orange	0.5, 0.6 (0.7)*	0.6 ± 0.4 (31)
Peach	0.5, 0.7, 0.9	
Pear	0.9, 1.1, 1.5	1.5 ± 0.7 (41)
Pineapple	0.6, 0.8, 1.1	
Plum	0.4, 1.0 (0.9)*	1.0 ± 0.4 (21)
Raspberry	0.7, 1.8 (1.4)*	
Strawberry	0.5, 0.7 (1.2)*	0.7 ± 0.5 (23)
Nuts		
Almond	8, 9, 14	
Brazil nut	13, 15	
Cashew	37	
Coconut	4.6, 33?	
Chestnut	2.3, 4.2	
Hazelnut (Filbert)	13, 13, 22	
Peanut	6–11	
Walnut	9, 14, 19	13.9 ± 10.8 (5)
Grains and pasta		
Barley	3.0, 8.2	8.2 ± 2.4 (17)
Milled (pearled)	2.0	
Maize (all kinds)	0.7–2.5, 3.3	
Yellow	1.4	1.4 ± 0.8 (51)
Grits	0.8	
Germ	10.1	
Millet	8.5	
Oats	3.4–5.3	
Pasta (wheat flour based)	0.7–2.8	
Rice		
Brown	2.4, 2.8, 4.1	2.8 ± 1.4 (10)
Polished	1.3, 1.7, 2.0	2.0 ± 0.9 (12)
Rye		
Whole	4.0, 6.0, 7.2	7.2 ± 2.2 (9)
Germ	7.5	

Appendix A. (Continued)

Food	Copper concentration μg/g or μg/ml[a]	Mean ± SD (N)[b]
Wheat		
Whole	4.1–4.6, 9.3	9.3 ± 7.3 (122)
Germ	6.2, 9.5, 23.9	23.9 ± 20.8 (5)
Bran	15–16	14.5 ± 4.0 (8)
Patent flour (white)	1.5, 2.0	2.0 ± 0.7 (19)
Vegetables		
Artichoke	0.7 (3.2)*	
Asparagus	1.1–1.6	1.1 ± 0.4 (30)
Beans		
Green	0.7 (1.4)*	1.4 ± 1.2 (35)
Kidney (dry)	9.6	8.4 ± 1.7 (3)
Beets	0.8, 1.5 (1.9)*	1.5 ± 0.7 (34)
Broccoli	0.5, 0.3	
Brussels sprouts	0.5, 0.7 (0.9)*	
Cabbage	0.2, 0.8 (0.6)*	0.8 ± 0.5 (51)
Carrots	0.5, 1.0 (0.8)*	1.0 ± 0.5 (72)
Cauliflower	0.3, 0.7	0.7 ± 0.6 (5)
Cucumber	0.4, 0.6 (0.9)*	0.6 ± 0.3 (6)
Kale	0.9–2.4	
Lentils	3.5, 6.4 (6.6)*	
Lettuce	0.2, 0.9 (0.5)*	0.9 ± 0.7 (308)
Mushrooms	4.0, 4.9, 10	10 ± 5.5 (4)
Onions	0.4 (0.8)*	1.0 ± 0.5 (37)
Peas		
Green	1.8 (3.8)*	2.2 ± 0.2 (5)
Chick	3.4	
Soy	4.3, 11.7	11.7 ± 3.4 (7)
Potato		
White	2.1, 2.6 (1.5)*	2.1 ± 1.0 (121)
Sweet (yam)	1.7, 1.8, 2.2	
Radish	0.4 (1.4)*	0.9 ± 0.7 (5)
Spinach	1.1–1.3	1.1 ± 0.5 (272)
Squash	0.6–1.0	
Tomato	0.8–0.9	1.1 ± 0.5 (469)
Miscellaneous		
Beer	0.09, 0.7 (0.4)*	
Bread		
White	2.2, 2.3	
Whole wheat	4.2	
Chocolate		
Bitter	27	
Milk	4.9	

(continued)

Appendix A. *(Continued)*

Food	Copper concentration	
	μg/g or μg/ml[a]	Mean ± SD (N)[b]
Cocoa		
Dry	39, 36	36 ± 11 (10)
Mix	2.9	
Coffee, ground	1.3	
Honey	0.4, 0.9?	
Sugar		
Cane	1.0	
Raw beet	3.4, 4.2	
White	0.2, 0.6	0.2 ± 0.1 (13)
Brown	3.5	3.5 ± 2.5 (7)
Molasses	6.8, 14.2	14.2 ± 9.6 (64)
Wine		
White	0.2 (0.7)*	
Red	0.2 (0.8–1.0)*	
Port	0.5 (1.0)*	
All varieties	1.1	1.1 ± 1.3 (412)
Yeast, Brewer's (dry)	33, 50	50 ± 33 (9)

[a] Values from USDA Handbooks, as well as Schroeder (1973) and Souci *et al.* (1981); starred values in parentheses are for Europe.
[b] From Pennington and Calloway (1973).

Appendix B. Copper Content of Human and Animal Tissues

Tissue	Copper concentration (μg/g wet wt.)[a]			
	Human	Rat	Pig	Other
Adipose	0.2, 0.3 (2)	0.35 (1)	0.7, 0.8 (2)	2.4 (1) (mouse)
Adrenal	1.6 ± 0.4 (11)	0, 2.1 (2)		1.2 (1) (whale)
Aorta	1.6 ± 0.4 (9)	2.7 (1)		
Aqueous humor	0.14 (1)			
Bile	4.0 ± 1.9 (5)[b]	1.8 ± 0.8 (3)	0.2–3.4 (1)	8.1 (1) (chick)
Bladder	0.8 ± 0.3 (3)			0.1, 2 (2) (sheep)
Blood	1.1 ± 0.1 (5)		1.0, 0.9 (2)[b]	1.23 (1)[b] (chicken)
Bone	3.9 ± 1.3 (8)	2.5 ± 0.6 (3)	1.4, 2.4 (2)	2.3–6.5 (1) (sheep)
Brain	5.2 ± 1.1 (10)	3.1 ± 1.2 (10)	3.9 ± 1.5 (4)	4.0 ± 2.1 (4) mouse)
Breast	0.7 ± 0.3 (5)			
Cartilage	0.4–0.7 (1)			
Cerebrospinal fluid	5 ± 2 (4)			
Cervix	0.6 (1)			
Chorioamniotic membrane	83? (1)			
Cornea	1.2, 2.1 (2)	0.3, 0.8 (2)		
Epididymus		1.0 (1)		
Esophagus	0.9 ± 0.3 (3)			
Eye	0.6–1.6 (1)			
Fallopian tube	0.9–1.2 (1)			
Secretions	1.1 (1)			
Hair	20 ± 6 (21)	11.3 (1)[c]		
Heart	4.8 ± 1.9 (14)	4.8 (2)[c]	4.5 (1)	4.6 (1) (cow)
Hypothalamus	4.4 ± 0.7 (3)	4.9 (1)		
Intestine	1.0, 3.0 (2)	1.7, 2.1 (2)		1.7 (1) (mouse)
Kidney	12 ± 7 (19)	7.9 ± 5.5 (14)	7.3 ± 4.5 (4)	4.4 ± 1.1 (3) (mouse)
Cortex	3.6 (1)[d]			
Larynx	1.0, 1.8 (2)			
Liver	6.2 ± 0.8 (9)	4.6 ± 1.1 (23)	5.2 ± 0.7 (5)	
Lung	1.3 ± 0.4 (11)	1.8 ± 0.6 (5)	1.2, 1.4 (2)	3.9 (2) (mouse)
				2.6 (1) (dog)

(continued)

Appendix B. *(Continued)*

Tissue	Copper concentration (μg/g wet wt.)[a]			
	Human	Rat	Pig	Other
Lymph	1.2 (1)			
Lymph nodes	0.7 (1)	1.4 (1)		
Marrow (red)	0.3–1.1 (1)	5.0 (1)	1.2–2.6 (1)	
Muscle	0.9 ± 0.3 (7)	1.0 ± 0.4 (5)	1.0 ± 0.6 (3)	3.7 (1) (cow)
				0.5–0.9 (3) (beef)
Nails				
Finger	20 ± 17 (10)			
Toe	8, 8 (2)			
Ovary	1.3 ± 0.4 (5)			
Pancreas	1.9 ± 0.6 (7)	2.7–4.3 (1)	1.2–1.9 (1)	
Parotid	1.3 (1)			
Pituitary				
(see Hypothalamus)				
Placenta	2.2 ± 0.8 (3)	0.2–21 (1)[e]		0.5–1.5[e] (1) (sheep)
Prostate	1.3 ± 0.2 (4)			
Saliva	0.22 ± 0.08 (4)	0.2–3.0 (1)		
Semen	0.5, 1.5 (2)			
Skin	0.8 ± 0.4 (9)	1.7 ± 0.8 (4)	1.0, 1.5 (2)	0.4 (1) (cow)
				0.7 (1) (calf)
Stomach	2.2 ± 0.7 (7)		1.6 (1)	
Submaxillary gland	3.4 (1)			
Sweat				
Men	0.6 (1)			
Women	1.5 (1)			
Synovial fluid	0.2, 0.5 (2)			
Teeth	9.2 ± 1.4 (3)	0.8 1.7 (1)		
Children	2.5–2.9 (1)[f]			
Testis	1.1 ± 0.7 (4)	1.9 ± 0.7 (6)	1.7 (1)	5 (1) (calf)
Thymus	0.8–2.7 (1)	1.5 (1)		
Thyroid	2.3 ± 0.8 (5)			
Tongue		1.3 (1)	2.1 (1)	3.4–4.8 (1) (cow)
Trachea	0.9, 1.0 (2)			
Trophoblast	1.9, 3.2 (2)			
Uterus (total content)	1.3 ± 0.6 (5)	2.9–3.7 (1)		
Vagina	0.7 (1)			
Vas deferens		1.8–2.2 (1)		
Vena cava	1.0–2.5 (1)			

[a] Means, or mean of mean literature values (±SD; N), for number of reports indicated in parentheses [from data gathered by Owen (1982a)]. Dry weight values corrected to wet weight values by factors given by Owen (1982a–e), where necessary.
[b] Summarized by Underwood (1971).
[c] Linder and Munro (1973).
[d] Piscator and Lind (1972).
[e] Changes during gestation.
[f] Pinchin et al. (1978).

References

Abraham, P. A., 1971. Hematologic, enzymic, and connective tissue changes associated with copper depletion and repletion in the growing rat, Ph.D. thesis, Rutgers University, *Diss. Abstr. Int. B* 32:5005–5006.

Abrahams, I. L., Bremner, I., Diakun, G. P., Garner, C. D., Hasnain, S. S., Ross. I., and Vasak, M., 1986. Structural study of the copper and zinc sites in metallothionein by using EXAFS, *Biochem. J.* 236:585–589.

Abu-Farsakh, F. A., Thajeel, A. H., Itani, S. M., Al-Khalily, A. S., and Abu-Farsakh, N. A., 1989. Interactions of copper with some analytes in human serum, *Clin. Chem.* 35:335.

Adham, N. D., 1980. Effect of calcium and copper on zinc absorption in the rat, *Nutr. Metab.* 24:281.

Adham, N. F., Song, M. K., and Rinderknecht, H., 1977. Binding of zinc to alpha-2-macroglobulin and its role in enzyme binding activity, *Biochim. Biophys. Acta* 495:212–219.

Aiginger, P., Kolarz, G., and Willvonseder, R., 1978. Copper in ankylosing spondylitis and rheumatoid arthritis, *Scand. J. Rheumatol.* 7:75–78.

Aisen, P., and Morell, A. G., 1965. Physical and chemical studies on ceruloplasmin. III. A stabilizing copper–copper interaction in ceruloplasmin, *J. Biol. Chem.* 240:1974–1978.

Aisen, P., Morell, A. G., Alpert, S., and Sternlieb, I., 1964. Biliary excretion of caeruloplasmin copper, *Nature* 203:873–874.

Albergoni, V., and Cassini, A., 1974. A cupro-zinc protein with superoxide dismutase activity from horse liver. Isolation and properties, *Comp. Biochem. Physiol.* 47B:767–777.

Albergoni, V., Cassini, A., Favero, N., and Rocco, G. P., 1975. Effect of penicillamine on some metals and metalloproteins in the rat, *Biochem. Pharmacol.* 24:1131–1132.

Aldred, A. R., Grimes, A., Schreiber, G., and Mercer, J. F. B., 1987. Rat ceruloplasmin: Molecular cloning and gene expression in liver, choroid plexus, yolk sac, placenta and testis, *J. Biol. Chem.* 2875–2878.

Aleinikova, T. D., Vasil'ev, V. B., Monakhov, N. K., and Shavlovskii, M. M., 1987. Synthesis and secretion of ceruloplasmin by isolated rat hepatocytes, *Biochem. S.S.R.* 52:1382–1389.

Alessandri, G., Raju, K., and Gullino, P. M., 1983. Mobilization of capillary endothelium *in vitro* induced by effectors of angiogenesis *in vivo*, *Cancer Res.* 43:1790–1797.

Alexander, J., and Aaseth, J., 1980. Biliary excretion of copper and zinc in the rat as influenced by diethylmaleate, selenite and diethyldithiocarbamate, *Biochem. Pharmacol.* 29:2129–2133.

Alfaro, B., and Heaton, F. W., 1973. Relationships between copper, zinc and iron in the plasma soft tissues and skeleton of the rat during Cu deficiency, *Br. J. Nutr.* 29:73–85.

Alfaro, B., and Heaton, F. W., 1974. The subcellular distribution of copper, zinc and iron in liver and kidney. Changes during copper deficiency in the rat, *Br. J. Nutr.* 32:435–445.

Alias, A. G., 1971. The effects of ACTH and of cortisol on serum ceruloplasmin in rabbits, *FEBS Lett.* 18:308–310.

Allen, K. G. D., and Klevay, L. M., 1978. Cholesterolemia and cardiovascular abnormalities in rats caused by copper deficiency, *Atherosclerosis* 29:81–93.

Allen, K. G. D., and Klevay, L. M., 1980. Hyperlipoproteinemia in rats due to copper deficiency, *Nutr. Rep. Intl.* 22:295–299.

Allerton, S. E., and Linder, M. C., 1985. Copper absorption: Effects of vitamin C, *Fed. Proc.* 44: Abstract No. 3415.

Allison, R. V., Bryan, D. C., and Hunter, J. H., 1927. The stimulation of plant response on the raw peat soils of the Florida everglades through the use of copper sulphate and other chemicals, *Fla. Agric. Exp. Stn. Bull.* 190:35–80.

Alpern-Elran, H., and Brem, S., 1985. Angiogenesis in human brain tumors: Inhibition by copper depletion, *Surg. Forum* 36:498–500.

Al-Timini, D. J., and Dormandy, T. L., 1977. The inhibition of lipid autoxidation by human caeruloplasmin, *Biochem. J.* 168:283–288.

Ambler, R. P., 1971. Cytochrome c, in: *Recent Developments in the Chemical Study of Protein Structures* (A. Previero, J. F. Pechere, and M. A. Coletti-Preiero, eds.), INSERM, Paris, p. 289.

Ambler, R. P., and Tobari, J., 1985. The primary structure of *Pseudomonas* AM1 amicyanin and pseudoazurin, *Biochem. J.* 232:451–457.

Ameyama, M., Hayashi, M., Matsushita, K., Shinagawa, E., and Adachi, O., 1984. Microbial production of pyrroloquinoline, *Agric. Biol. Chem.* 48:561.

Andreasson, L. E., and Vanngard, T., 1970. Evidence of a specific copper (II) in human ceruloplasmin as a binding site for inhibitory anions, *Biochem. Biophys. Acta* 200:247–257.

Andrews, G. K., Adamson, E. D., and Gedamu, L., 1984. The ontogeny of expression of murine metallothionein: Comparison with the α-fetoprotein gene, *Dev. Biol.* 103: 294–303.

Andrews, G. K., Gallant, K. R., and Cherian, M. G., 1987. Regulation of the ontogeny of rat liver metallothionein mRNA by zinc, *Eur. J. Biochem.* 166:527–531.

Andrews, G. S., 1979. Studies of plasma zinc, copper, caeruloplasmin, and growth hormone, *J. Clin. Pathol.* 32:325–333.

Antholine, W. E., Petering, D. H., and Pickart, L., 1989. ESR studies of the interaction of copper(II)GHL, histidine, and Ehrlich cells, *J. Inorg. Biochem.* 35:215–224.

Antonini, E., Brunori, M., Colosimo, A., Greenwood, C., and Wilson, M. T., 1977. Oxygen "pulsed" cytochrome c oxidase: Functional properties and catalytic relevance, *Proc. Natl. Acad. Sci. USA* 74:3128–2132.

Aoyagi, Y., Ikenaka, T., and Ichida, F., 1978. Copper(II)-binding ability of human alpha-fetoprotein, *Cancer Res.* 38:3483–3486.

Apelgot, S., Coppey, J., Grisvard, J., Guille, E., and Sissoeff, I., 1981. Distribution of copper-64 in control mice and in mice bearing ascitic Krebs tumor cells, *Cancer Res.* 41:1502–1507.

Appleton, D. W., and Sarkar, B., 1971. The absence of specific copper(II)-binding in dog albumin: A comparative study of the human and dog albumin, *J. Biol. Chem.* 246:5040–5046.

Argentu-Ceru, M. P., and Autuori, F., 1985. Localization of diamine oxidase in animal tissues, in *Structure and Functions of Amine Oxidases* (B. Mondovi, ed.), CRC Press, Boca Raton, Florida, pp. 90–104.

Arnon, D. I., and Stout, P. R., 1939. The essentiality of certain elements in minute quantity for plants with special reference to copper, *Plant Physiol.* 14:371–374.

Asada, K., Urano, M., and Takahashi, M., 1973. Subcellular location of superoxide dismutase in spinach leaves and preparation and properties of crystalline spinach superoxide dismutase, *Eur. J. Biochem.* 36:257–266.

Ash, D. E., Papadopoulos, N. J., Colombo, G., and Villafranca, J. J., 1984. Kinetic and spectroscopic studies of the interaction of copper with dopamine β-hydroxylase, *J. Biol. Chem.* 259:3395–3398.

Ashida, J., and Nakamura, H., 1959. Role of sulfar metabolism in copper-resistance of yeast, *Plant Cell Physiol.* 1:71–79.

Ashwell, G., and Morell, A. G., 1974. The role of surface carbohydrates in hepatic recognition and transport of circulating glycoproteins, *Adv. Enzymol.* 41:99–128.

Attridge, T. H., 1974. Phytochrome-mediated synthesis of ascorbic acid oxidase in mustard cotyledons, *Biochim. Biophys. Acta* 372:258.

Audran, R., Blatrix, C., Amouch, P., Drouet, J., and Steinbuch, M., 1974. Etude de l'élimination plasmatique chez l'homme normal de la céruloplasmine et d'apocéruloplasmines, *C. R. Acad. Sci.* 278:D2227–2230.

Auer, D. E., Ng, J. C., Steele, D. P., and Seawright, A. A., 1988. Monthly variation in the plasma copper and zinc concentration of pregnant and non-pregnant mares, *Aust. Vet. J.* 65:61–62.

August, D., Janghorbani, M., and Young, V. R., 1989. Determination of zinc and copper absorption at three dietary zinc–copper ratios by using stable isotope methods in young adult and elderly subjects, *Am. J. Clin. Nutr.* 50:1457–1463.

Azzi, A., 1980. Cytochrome *c* oxidase. Towards clarification of its structure, interactions and mechanism, *Biochim. Biophys. Acta* 594:231–252.

Azzi, A., Casey, R. P., and Nalecz, M. J., 1984. The effect of *N. N'*-dicyclohexylcarbodiimide on enzymes of bioenergetic relevance, *Biochim. Biophys. Acta* 768:209–226.

Babior, B. M., 1978a. Oxygen-dependent microbial killing by phagocytes (part one), *N. Engl. J. Med.* 298:659–668.

Babior, B. M., 1978b. Oxygen-dependent microbial killing by phagocytes (part two), *N. Engl. J. Med.* 298:721–725.

Bajpayee, D. P., 1975. Significance of plasma copper and ceruloplasmin concentrations in rheumatoid arthritis, *Ann. Rheum. Dis.* 34:162–165.

Balasubramani, K. A., Deverall, B. J., and Murphy, J. V., 1971. Changes in respiratory rate, polyphenoloxidase, and polygalacturonase activity in and around lesions caused by Botrytis in leaves of Vitia faba, *Physiol. Plant Pathol.* 1:105.

Banaszkiewicz, W., and Chodera, A., 1962. The effect of some corticoids on the level of copper in the blood serum of rabbits, *Acta Physiol. Pol.* 13:415–419.

Banerjee, D., Onosaka, S., and Cherian, M. G., 1982. Immunohistochemical localization of metallothionein in cell nucleus and cytoplasm of rat liver and kidney, *Toxicology* 24:95–105.

Banford, J. C., Brown, D. H., Hazelton, R. A., McNeil, C. J., Sturrock, R. D., and Smith, W. E., 1982. Serum copper and erythrocyte superoxide dismutase in rheumatoid arthritis, *Ann. Rheum. Dis.* 41:458–462.

Bannister, J. V., and Bannister, W. H., 1985. The production of oxygen-centered radicals by neutrophils and macrophages as studied by electron spin resonance (esr), *Environ. Health Perspect.* 104:37.

Bannister, J. V., Bannister, W. H., Hill, H. A. O., Mahood, J. F., Wilson, R. L., and Wolfendon, B. S., 1980. Does ceruloplasmin dismutate superoxide? No. *FEBS Lett.* 118:127–129.

Bannister, J. V., Bannister, W. H., and Rotilio, G., 1987. Aspects of the structure, function, and applications of superoxide dismutase, *CRC Crit. Rev. Biochem.* 22:111–180.

Barber, E. F., and Cousins, R. J., 1986. Induction of ceruloplasmin by interleukin-1 and retinoic acid., *Fed. Proc.* 45:994 (Abstract 3410).

Barber, R. S., Braude, R., Mitchell, K. G., and Cassidy, J., 1955. High copper mineral mixture for fattening pigs, *Chem. Ind. (London)* 74:601.

Barbosa, J., Seal, U. S., and Doe, R. P., 1971. Effects of anabolic steroids on haptoglobin, orosomucoid, plasminogen, fibrinogen, transferrin, ceruloplasmin, alpha.1-antitrypsin, beta.-glucuronidase, and total serum proteins, *J. Clin. Endocrinol. Metab.* 33:388–398.

Barnea, A., and Cho, G., 1987. Copper amplification of prostaglandin E_2 stimulation of the release of luteinizing hormone-releasing hormone is a post receptor event, *Proc. Natl. Acad. Sci. USA* 84:580–584.

Barnea, A., Cho, G., and Hartter, D. E., 1988a. A correlation between the ligand specificity for ^{67}copper uptake and for copper-prostaglandin E_2 stimulation of the release of gonadotropin-releasing hormone from median eminence explants, *Endocrinology* 122:1505–1510.

Barnea, A., Colombani-Vidal, M., Cho, G., and Hartter, D. E., 1988b. Evidence for synergism between copper and prostaglandin E_2 in stimulating the release of gonadotropin-releasing hormone from median eminence explants: Na^+/Cl^- requirements, *Mol. Cell. Endocrin.* 56:11–19.

Barnes, G., and Frieden, E., 1984. Ceruloplasmin receptors of erythrocytes, *Biochem. Biophys. Res. Commun.* 125:157–162.

Barnes, R., Connelly, J. L., and Jones, O. T. G., 1972. The utilization of iron and its complexes by mammalian mitochondria, *Biochem. J.* 128:1043–1055.

Barra, D., Martini, F., Bannister, J. V., Schinina, M. E., Rotilio, G., Bannister, W. H., and Bossa, F., 1980. The complete amino acid sequence of human Cu/Zn superoxide dismutase, *FEBS Lett.* 120:53–57.

Barra, R., 1987. Effects of glycyl-histidyl-lysine on Morris hepatoma 7777 cells, *Cytobios* 52:99–107.

Barrow, L., and Tanner, M. S., 1988. Copper distribution among serum proteins in paediatric liver disorders and malignancies, *Eur. J. Clin. Invest.* 18:555–560.

Bastin, M., 1968. Effect of wounding on the synthesis of phenols, phenoloxidase, and peroxidase in the tuber tissue of Jerusalem artichoke, *Can. J. Biochem.* 46:1339–1343.

Batey, R. G., Lai, C. F. P., Shamir, S., and Sherlock, S., 1980. A non-transferrin-bound serum iron in idiopathic hemocromatosis, *Dig. Dis. Sci.* 25:340–346.

Battersby, S., and Chandler, J. A., 1977. Correlation between elemental composition and motility of human spermatozoa, *Fert. Steril.* 28:557–561.

Bauer, H. C., Parent, J. B., and Olden, K., 1985. Role of carbohydrate in glycoprotein secretion by human hepatoma cells, *Biochem. Biophys. Res. Commun.* 128:368–374.

Baumann, H., Richards, C., and Gauldie, J., 1987. Interaction among hepatocyte-stimulating factors, interleukin 1, and glucocorticoids for regulation of acute phase plasma proteins in human hepatoma (HepG2) cells, *J. Immunol.* 139:4122–4128.

Beaman, B. L., Scates, S. M., Moring, S. E., Deem, R., and Misra, H. P., 1983. Purification and properties of a unique superoxide dismutase from *Nocardia asteroides*, *J. Biol. Chem.* 258:91–96.

Bearn, A. G., 1960. A genetic analysis of thirty families with Wilson's disease, *Ann. Hum. Genet.* 24:33.

Bearn, A. G., and Kunkel, H. G., 1954. Localization of ^{64}Cu in serum fractions following oral administration: An alteration in Wilson's disease. *Proc. Soc. Exp. Biol. Med.* 85:44–48.

Bearn, A. G., and Kunkel, H. G., 1955. Metabolic studies in Wilson's disease using Cu^{64}, *J. Lab. Clin. Med.* 45:623–631.

Beaumier, D. L., Caldwell, M. A., and Holbein, B. E., 1984. Inflammation triggers

References

hypoferremia and de novo synthesis of serum transferrin and ceruloplasmin in mice, *Infect. Immun.* 46:489–494.

Behari, J. R., and Tandon, S. K., 1980. Effect of zinc on cadmium, copper and zinc contents in cadmium exposed rats, *Toxicol. Lett.* 5:151–154.

Beisel, W. R., 1991. Nutrition and infection, in *Nutritional Biochemistry and Metabolism*, 2nd ed. (M. C. Linder, ed.), Elsevier, New York (Chapter 17).

Bender, C. L., and Cooksey, D. A., 1986. Indigenous plasmids in *Pseudomonas syringae* pv. *tomato*: Conjugative transfer and role in copper resistance, *J. Bacteriol.* 165:534–541.

Benne, R., Van Den Burg, J., Brakenhoff, J. P. J., Sloof, P., VanBoom, J. H., and Tromp, M. C., 1986. Major transcript of the frameshifted cox II gene from trypanosome mitochondria contains four nucleotides that are not encoded in the DNA, *Cell*, 46:819–826.

Bennetts, H. W., Beck, A. B., and Harley, R., 1948. *Aust. Vet. J.* 24:237.

Bennetts, H. W., and Chapman, F. E., 1937. Copper deficiency in sheep in Western Australia: A preliminary account of the etiology of enzooic ataxia of lambs and an anemia of ewes, *Aust. Vet. J.* 13:138–149.

Beoku-Betts, D., Chapman, S. K., Knox, C. V., and Sykes, A. G., 1984. Kinetic studies on electron-transfer reactions involving blue copper proteins. II. Effects of pH, competitive inhibition, and chromium (II) modification on the reaction of plastocyanin with cytochrome *f*, *Inorg. Chem.* 24:1677–1681.

Berkenbosch, F., Van Oers, J., del Rey, A., Tilders, F., and Besedovsky, H., 1987. Corticotropin-releasing factor-producing neurons in the rat activated by interleukin-1, *Science* 238:524–526.

Bermingham-McDonogh, O., Gralla, E. B., and Valentine, J. S., 1988. The copper, zinc-superoxide dismutase gene of *Saccharomyces cerevisiae*: Cloning, sequencing, and biological activity, *Proc. Natl. Acad. Sci. USA* 85:4789–4793.

Bernton, E. W., Beach J. E., Hoaday, J. W., Smallridge, R. C., and Fein, H. G., 1987. Release of multiple hormones by a direct action of interleukin-1 on pituitary cells, *Science* 238:519–521.

Berthon, G., and Kayali, A., 1982. Histamine as a ligand in blood plasma. Part 5. Computer simulated distribution of metal histamine complexes in normal blood plasma and discussion of the implications of a possible role of zinc and copper in histamine catabolism, *Agents Actions* 12:398–407.

Bettger, W. J., Fish, T. J., and O'Dell, B. L., 1978. Effects of copper and zinc status of rats on erythrocyte stability and superoxide dismutase activity, *Proc. Soc. Exp. Biol. Med.* 158:279–282.

Beutler, B., and Cerami, A., 1987. Cachectin: More than a tumor necrosis factor, *N. Engl. J. Med.* 316:379–385.

Beutler, B. D., Greenwals, D., Holmes, J. D., Chang, M., Pan, Y.-C. E., Mathison, J., Ulevitch, R., and Cerami, A., 1985. Identity of tumor necrosis factor and the macrophage-secreted factor cachetin, *Nature (London)* 316:552–554.

Beveridge, T. J., and Fyfe, W. S., 1985. Metal fixation by bacterial cell walls, *Can. J. Earth Sci.* 22:1893–1898.

Bhasker, K. R., and Barnea, A., 1988. Progesterone augments copper-prostaglandin E2 stimulation of the release of gonadotropin-releasing hormone from explants of the median eminence of immature female rats: an estrogen-dependent process, *Endocrinology* 122:2143–2149.

Bhathena, S. J., Recant, L., Voyles, N. R., Timmers, K. I., Reiser, S., Smith, J. C., Jr., and Powell, A. S., 1986. Decreased plasma enkephalins in copper deficiency in man, *Am. J. Clin. Nutr.* 43:42–46.

Bhathena, S. J., Reiser, S., Smith, Jr., J. C., Revett, K., Kennedy, B. W., Powell, A. S., Voyles, N. R., and Recant, L., 1988. Increased insulin receptors in carbohydrate-sensitive subjects: A mechanism for hyperlipaemia in these subjects, *Eur. J. Clin. Nutr.* 42:465–472.

Bicknell, R., and Vallee, B. L., 1989. Angiogenin stimulates endothelial cell prostacyclin secretion by activation of phospholipase A_2, *Proc. Natl. Acad. Sci. USA* 86:1573–1577.

Bidwell, R. G. S., 1979. *Plant Physiology*, 2nd ed., Macmillan, New York.

Biesold, D., and Gunther, J., 1972. Improved methods for investigation of copper metabolism in patients with Wilson's disease using ^{64}Cu, *Clin. Chim. Acta* 42:353–359.

Bingle, C. D., Srai, S. K. S., Whiteley, G. S. W., and Epstein, O., 1988. Neonatal and adult copper-64 metabolism and Wilson's disease, *Biol. Neonate* 54:294–300.

Bingle, C. D., Epstein, O., Srai, S. K. S., 1990c. A developmental shift from low- to high-Mr copper binders in guinea pig serum, *Biochem. Soc. Trans.* 18:645.

Bingle, C. D., Srai, S. K. S., and Epstein, O., 1990a. Developmental changes in hepatic copper proteins in guinea pig, *J. Hepatol.* 10:138–143.

Bingle, C. D., Gitlin, J. D., Epstein, O., and Srai, S. K. S., 1990b. Transcriptional regulation of caeruloplasmin gene expression in the developing guinea pig liver, in *Trace Elements in Man and Animals* (TEMA-7), (B. Momcilovic, ed.), in press.

Birbeck, M. S. C., Cartwright, P., and Hall, J. G., 1970. The transport of immunoglobulin A from blood to bile visualized by autoradiography and electron microscopy, *Immunology* 37:477–484.

Bird, T. A., and Levene, C. I., 1982. Lysyl oxidase: Evidence that pyridoxal phosphate is a cofactor, *Biochem. Biophys. Res. Commun.* 108:1172–1180.

Bisson, R., Schiavo, G., and Papini, E. 1984. Cytochrome c oxidase, *Eur. Bioenerg. Conf. Rep.* 3A:61–62.

Blackett, P. R., Lee, D. M., Donaldson, D. L., Fesmire, J. D., Chan, W. Y., Holcombe, J. H., and Rennert, O. M., 1984. Studies of lipids, lipoproteins, and apolipoproteins in Menkes' disease, *Pediatr. Res.* 18:864–870.

Blake, D. R., Hall, N. D., Treby, D. A., Halliwell, B., and Gutteridge, J. M. C., 1981. Protection against superoxide and hydrogen peroxide in synovial fluid from rheumatoid patients, *Clin. Sci.* 61:483–486.

Blakely, B. R., and Hamilton, D. L., 1987. The effect of copper deficiency on the immune response in mice, *Drug Nutr. Inter.* 5:103–111.

Blalock, T. L., Dunn, M. A., and Cousins, R. J., 1988. Metallothionein gene expression in rats: Tissue-specific regulation by dietary copper and zinc, *J. Nutr.* 118:222–228.

Blenk, H., Hofstetter, A., Bowering, R., Buttler, R., Hartmann, M., and Marx, F. J., 1974. Immunoelektrophore des Ejakulats, *Munch Med. Wochenschr.* 116:35–38.

Bigny, R., Gaillard, J., and Douce, R., 1986. Excretion of laccase by sycamore (*Acerpseudoplatanus L.*) cells. Effects of copper deficiency, *Biochem. J.* 237:583–588.

Bloomer, L. C., and Sourkes, T. L., 1974. Hepatic cuproproteins in copper-loaded rats, in *Trace Element Metabolism in Animals-2* (W. G. Hoekstra, J. W. Suttie, H. E. Ganther, and W. Mertz, eds.), University Park Press, Baltimore, pp. 479–481.

Boer, P. H., and Gray, M. W., 1988. Transfer RNA genes and the genetic code in *Chlamydomonas rienhardtii* mitochondria, *Curr. Genet.* 14:583–590.

Bogden, J. D., Lintz, D. I., Joselow, M. M., Charles, J., and Salaki, J. S., 1977. Effect of pulmonary tuberculosis on blood concentrations of copper and zinc, *Am. J. Clin. Pathol.* 67:251–256.

Bogden, J. D., Thind, I. S., Louria, D. B., and Caterini, H., 1978. Maternal and cord blood metal concentrations and low birth weight—a case-control study, *Am. J. Clin. Nutr.* 31:1181–1187.

References

Bohner, H., and Boger, P., 1978. Reciprocal formation of cytochrome c-553 and plastocyanin in *Scenedesmus*, *FEBS Lett.* 85:337–339.

Bombardieri, G., Milani, A., and Rossi, L., 1985. Copper-dependent amine oxidases; clinical aspects, in *Structure and Functions of Amine Oxidases* (B. Mondovi, ed.), CRC Press, Boca Raton, Florida, pp. 195–204.

Bombelka, E., Richter, F.-W., Stroh, A., and Kadenbach, B., 1986. Analysis of the Cu, Fe, and Zn contents in cytochrome c oxidases from different species and tissues by proton-induced X-ray emission (PIXE), *Biochem. Biophys. Res. Comm.* 140:1007–1013.

Bonaccorsi di Patti, M. C., Musci, G., Giartosio, A., D'Allesio, S., and Calabrese, L., 1990. The multidomain structure of ceruloplasmin from calorimetric and limited proteolysis studies, *J. Biol. Chem.* 265:21016–21022.

Bonewitz, R. F., and Howell, R. R., 1981. Synthesis of a metallothionein-like protein in cultured human skin fibroblasts: Relation to abnormal copper distribution in Menkes' disease, *J. Cell Physiol.* 106:339–348.

Bonta, I. L., 1977. Endogenous substances as modulators of inflammation, in *inflammation: Mechanisms and Their Impact on Therapy* (I. L. Bonta, I. J. Thompson, and K. Brune, eds.). Birkhauser, Basel, pp. 121–131.

Boulanger, Y., Goodman, C. M. Forte, C. P., Fesik, S. W., and Armitage, I. M., 1983. Model for mammalian metallothionein structure, *Proc. Natl. Acad. Sci. USA* 80:1501.

Bowman, B. H., and Yang, F., 1987. DNA sequencing and chromosomal locations of human plasma protein genes, in *The Plasma Proteins*, Vol. V, 2nd ed. (F. W. Putnam, ed.), Academic Press, New York, pp. 1–48.

Bowman, B. H., Yang, F., Brune, J. L., Naylor, S. L., Barnett, D. R., McGill, J. R., Moore, C. M., Lum, J. B., and McCombs, J., 1985. Organization and chromosomal locations of genes encoding human plasma proteins, *Protides Biol. Fluids* 33:15–20.

Box, B. M., and Mogenson, G. J., 1982. Blood pressure differences in spontaneously hypertensive rats, *Nutr. Res.* 2:619.

Boyer, R. F., and Schori, B. E., 1983. The incorporation of iron into apoferritin as mediated by ceruloplasmin, *Biochem. Biophys. Res. Commun.* 116:244–250.

Boyle, E., Freeman, P. C., Goudie, A. C., Magan, F. R., and Thomson, M., 1976. Role of copper in preventing gastrointestinal damage by acidic anti-inflammatory drugs, *J. Pharm. Pharmacol.* 28:865.

Boyne, R., and Arthur, J. R., 1981. Effects of selenium and copper deficiency on neutrophil function in cattle, *J. Comp. Pathol.* 91:271–276.

Bradshaw, R. A., and Peters, T., 1969. The amino acid sequence of peptide (1-24) of rat and human serum albumin, *J. Biol. Chem.* 244:5582–5589.

Brady, F. O., 1975. Tryptophan-2-3-dioxygenase: a review of the roles of the heme and copper cofactors in catalysis, *Bioinorg. Chem.* 5:167–182.

Brady, F. O., Monaco, M. E., Forman, H. J., Schutz, G., and Feigelson, P., 1972. On the role of copper in activation of and catalysis by tryptophan-2,3-dioxygenase, *J. Biol. Chem.* 247:7915–7922.

Braganza, J. M., and Sturniolo, G., 1980. Evidence that pancreatic secretions influence copper metabolism in man, *J. Physiol.* 305:51P.

Bray, R. C., Cockle, S. A., Fielden, E. M., Roberts, P. B., Rotilio, G., and Calabrese, L., 1974. Reduction and inactivation of superoxide dismutase by hydrogen peroxide, *Biochem. J.* 139:43–48.

Brem, S., 1976. The role of vascular proliferation in the growth of brain tumors, *Clin. Neurosurg.* 23:440–453.

Brem, S., Alpern-Elran, H., Tanaka, Y., and Zagzag, D., 1986. Control of neoplastic development in the brain by angiosuppression with copper depletion, *Canadian J. Neurolog. Sci.* 13:162–163.

Bremner, I., 1974. Copper and zinc proteins in ruminant liver, in *Trace Element Metabolism in Animals—2* (W. G. Hoekstra, J. W. Suttie, H. E. Ganther, and W. Mertz, eds), University Park Press, Baltimore, pp. 489–492.

Bremner, I., 1976. The relationship between the zinc status of pigs and the occurrence of copper- and Zn-binding proteins in liver, *Br. J. Nutr.* 35:245–252.

Bremner, I., 1979. Factors influencing the occurrence of copper thioneins in tissues, in *Metallothionein* (J. H. R. Kägi and M. Nordberg, eds.), Birkhauser, Basel, pp. 273–280.

Bremner, I., 1980. Absorption, transport and distribution of copper, *Ciba Found. Symp.* (1979) *Exerpta. Med.* 79:23–48.

Bremner, I., 1987. Nutritional and physiological significance of metallothionein, *Experientia* (Suppl.) 52:81–107.

Bremner, I., and Beattie, J. H., 1990. Metallothionein and the trace minerals, *Annu. Rev. Nutr.* 10:63–84.

Bremner, I., and Davies, N. T., 1974a. Studies on the appearance of a copper-binding protein in rat liver, *Biochem. Soc. Trans.* 2:245–427.

Bremner, I., and Davies, N. T., 1974b. Studies on the appearance of a hepatic copper binding protein in normal and zinc-deficient rats, *Br. J. Nutr.* 36:101–112.

Bremner, I., and Price, J., 1985. Effects of dietary iron supplements on copper metabolism in rats, in *Trace Elements in Man and Animals* (TEMA-5) (C. F. Mills, I. Bremner, and J. K. Chesters, eds.), Commonwealth Agricultural Bureaux, Slough, UK, pp. 374–376.

Bremner, I., Hoekstra, W. G., Davies, N. T., and Williams, R. B., 1978. Renal accumulation of (copper, zinc)-thioneins in physiological and pathological states, in *Trace Element Metabolism in Man and Animals—3* (M. Kirchgessner, ed.), Arbeitskreis für Tierernährungsforschung Weihenstephan, Freising, Germany, pp. 44–51.

Bremner, I., Young, B. W., and Mills, C. F., 1979. The effect of ammonium tetrathiotungstate on the absorption and distribution of copper in the rat, *Biochem. Soc. Trans.* 7:677–678.

Bremner, I., Mehra, R. K., Morrison, J. N., and Wood, A. M., 1986. Effects of dietary copper supplementation of rats on the occurrence of metallothionein-I in liver and its secretion into blood, bile and urine, *Biochem. J.* 235:735–759.

Bremner, I., Humphries, W. R., Phillippo, M., Walker, M. J., and Morrice, P. C., 1987a. Iron-induced copper deficiency in calves: Dose–response relationships and interactions with molybdenum and sulphur, *Anim. Prod.* 45:403–414.

Bremner, I., Morrison, J. N., Wood, A. M., and Arthur, J. R., 1987b. Effects of changes in dietary zinc, copper and selenium supply and of endotoxin administration on metallothionein-I concentrations in blood cells and urine in the rat, *J. Nutr.* 117:1595–1602.

Breslow, E., 1964. Comparison of cupric ion-binding sites in myoglobin derivatives and serum albumin, *J. Biol. Chem.* 239:3252–3259.

Brewer, G., 1986. Presentation at International Symposium of the Society for Trace Element Research in Humans, Palm Springs, California (December).

Brewer, G. J., Yuzbasiyan, V. A., Iyengar, V., Hill, G. M., Dick, R. D., and Prasad, A. S., 1988. Regulation of copper balance and its failure in humans, *Curr. Top. Nutr. Dis. (Essent. Toxic Trace Elem. Hum. Health Dis.)* 18:95–103.

Bridges, K. R., and Hoffman, K. E., 1986. The effects of ascorbic acid on the intracellular metabolism of iron and ferritin, *J. Biol. Chem.* 261:14273–14277.

Briggs, J., Chambers, I. R., Finch, P., Slaiding, I. R., and Weigel, H., 1981. Thin layer chromatography on cellulose impregnated with tungstate: A rapid method resolving mixtures of some commonly occurring carbohydrates, *Carbohydr. Res.* 78:365–367.

Briggs, J., Finch, P., Matulewicz, M. C., and Weigel, H., 1981. Complexes of copper(II), calcium, and other metal ions with carbohydrates: Thin-layer ligand-exchange

chromatography and determination for relative stabilities of complexes, *Carbohyd. Res.* 97:181–188.
Briggs, M. Austin, J., and Staniford, M., 1970. Oral contraceptives and copper metabolism, *Nature* 225:81.
Broadley, C., and Hoover, R. L., 1989. Ceruloplasmin reduces the adhesion and scavenges superoxide during the interaction of activated polymorphonuclear leukocytes with endothelial cells, *Am. J. Pathol.* 135:647–655.
Bronner, F., and Yost, J. H., 1985. Saturable and nonsaturable copper and calcium transport in mouse duodenum, *Am. J. Physiol.* 249:G108–112.
Bronson, R. E., Calaman, S. D., Traish, A. M., and Kagan H. M., 1987. Stimulation of lysyl oxidase (EC 1.4.3.13) activity by testosterone and characterization of androgen receptors in cultured calf aorta smooth-muscle cells, *Biochem. J.* 244:317–323.
Brouwer, M., Winge, D. R., and Gray, W. R., 1989. Structural and functional diversity of copper-metallothioneins from the American lobster *Homarus americanus*, *J. Inorg. Biochem.* 35:289–303.
Brown, D. A., Chatel, K. W., Chan, A. Y., and Knight, B., 1980. Cytosolic levels and distribution of cadmium, copper and zinc in pretumorous livers from diethylnitrosamine-exposed mice and in non-cancerous kidneys from cancer patients, *Chem.-Biol. Interact.* 32:13–27.
Brown, D. H., Dunlop, J., Smith, W. E., Teape, J., and Lewis, A. J., 1980. Total serum copper and ceruloplasmin levels following administration of copper aspirinate to rats and guinea pigs, *Agents Actions* 10:465–470.
Brown, J. C. W., and Strain, J. D., 1990. Effect of dietary homocysteine on copper status in rats, *J. Nutr.* 120:1068–1074.
Brown, J. R., Low, T., Beherns, P., Sepulveda, P., Parker, K., and Blakeney, E., 1971. Amino acid sequence of bovine and porcine serum albumin, *Fed. Proc.* 30:Abstract 1104.
van Bruggen, E. F. J., Schutter, W. G., van Breemen, J. F. L., Bijlholt M. C. C., and Wichertjes, T., 1981. Subunits of hemocyanins, *Biol. Chem. Hoppe-Seyler* 362:617–635.
van Bruggen, E. F. J., Schutter, W. G., van Breemen, J. F. L., Bijlholt, M. M. C., and Wichertjes, T., 1981. Arthropodan and molluscan haemocyanins, in *Electron Microscopy of Proteins* (M. Harris, ed.), Academic Press, New York, pp. 1–38.
Brunori, M., and Chance B. (eds.), 1989. *Cytochrome Oxidase, Structure, Function, and Physiopathology*, Vol. 1. CRC Press, Boca Raton, FL.
Brunori, M., Silvestrini, M. C., Wilson, M. T., and Weiss, H., 1983. Transient kinetic studies of *Neurospora crassa* cytochrome *c* oxidase, *Biochem. J.* 215:425–427.
Bryan, S. E., Vizard, D. L., Beary, D. A., LaBiche, R. A., and Hardy, K. J., 1981. Partitioning of zinc and copper within subnuclear nucleoprotein particles, *Nucleic. Acid Res.* 9:5811–5823.
Bryan, S. E., LeGros, L., Brown, J., Byrne, C., and Re, R. N., 1985. Copper-rich nucleoprotein generated by micrococcal nuclease, *Biol. Trace Elem. Res.* 8:219–229.
Brzozowaska, A., Sicinska, A., Witkowska, J., and Roszkowski, W., 1989. Effect of protein quality and dietary levels of iron, zinc and copper on the apparent absorption and tissue concentration of trace elements in rats, *Spec. Publ.-R. Soc. Chem. (Nutr. Availability: Chem. Biol. Aspects)* 72:206–208.
Buchman, C., Skroch, P., Dixon, W., Tullius, T. D., and Karin, M., 1990. A single amino acid change in CuP2 alters its mode of DNA binding, *Mol. Cell. Biol.* 10:4778–4887.
Budesinsky, M., Budzikiewicz, H., Prochazka, Z., Ripperger, H., Romer, A., Scholz, G., and Schreiber, K., 1980. Nicotianamine, a possible phytosiderophore of general occurence, *Phytochemistry* 19:2295–2297.
Burbelo, P. D., Kagan, H. M., and Chichester, C. O., 1985. Immunological characterization of bovine lysyl oxidase, *Comp. Biochem. Physiol.* 81B:845–849.

Burch, R. E., Williams, R. V., Hahn, H. K. J., Jetton, M. M., and Sullivan, J. F., 1985. Serum and tissue enzyme activity and trace-element content in response to zinc deficiency in the pig, *Clin. Chem.* 21:568–577.

Burrows, A., and Pekala, B., 1971. Serum copper and ceruloplasmin in pregnancy, *Am. J. Obstet. Gynecol.* 109:907–909.

Burrows, G. H., and Barnea, A., 1982. Copper stimulates the release of luteinizing hormone releasing hormone from isolated hypothalamic granules, *Endocrinology* 110:1456–1458.

Buse, C., Steffens, G. M. C., Steffens, C. J., Meinecke, L., Biewald, R., and Erdweg, M., 1982. Subunits of cytochrome *c* oxidase, *Eur. Bioenerg. Conf. Rep.* 2:163–164.

Bush, J. A., Mahoney, J. P., Markowitz, H., Gubler, C. J., Cartwright, G. E., and Wintrobe, M. M., 1955. Studies on copper metabolism. XVI. Radioactive copper studies in normal subjects and in patients with hepatolenticular degeneration, *J. Clin. Invest.* 34:1766–1778.

Bush, J. A., Jensen, W. N., Athens, J. W., Ashenbrucker, H., Cartwright, G. E., and Wintrobe, M. M., 1956a. Studies on copper metabolism. XIX. The kinetics of iron metabolism and erythrocyte life-span in copper-deficient swine, *J. Exp. Med.*, 103:701–712.

Bush, J. A., Mahoney, J. P., Gubler, C. J., Cartwright, G. E., and Wintrobe, M. M., 1956b. Studies on copper metabolism. XXI. The transfer of radiocopper between erythrocytes and plasma, *J. Lab. Clin. Med.* 47:898–906.

Buskirk, H. H., Crim, J. A., Van Giessen, G. J., and Petering, H. G., 1973. Rapid *in vitro* method for determining cytotoxicity of antitumor agents, *J. Natl. Cancer Inst.* 51:135–138.

Butkus, D. E., and Jones, F. T., 1982. Reversal of copper-induced inhibition of vasopressin responsiveness by reducing agents, *Biochim. Biophys. Acta* 685:203–206.

Butler, E. J., 1963. The influence of pregnancy on the blood, plasma, and ceruloplasmin copper levels of sheep, *Comp. Biochem. Physiol.* 9:1–12.

Butler, E. J., and Curtis, M. J., 1977. The effect of *Escherichia coli* endotoxin on plasma histaminase activity in the domestic fowl and the involvement of ceruloplasmin, *Res. Vet. Sci.* 22:267–270.

Butler, E. J., and Newman, G. E., 1956. The urinary excretion of copper and its concentration in the blood of normal human adults, *J. Clin. Pathol.* 9:157–161.

Butt, T. R., Stenberg, E. J., Gorman, J. A., Clark, P., Hamer, D., Rosenberg, M., and Crooke, S. T., 1984. Copper metallothionein of yeast, structure of the gene, and regulation of expression, *Proc. Natl. Acad. Sci. USA* 81:3332–3336.

Caffrey, Jr., J. M., Smith, H. A., Schmitz, J. C., Merchant, A., and Frieden, E., 1990. Hemolysis of rabbit erythrocytes in the presence of copper ions: Inhibition by albumin and ceruloplasmin, *Biol. Trace Elem. Res.* 25:11–19.

Cain, K., and Holt, D. E., 1979. Metallothionein degradation: Metal composition as a controlling factor, *Chem.-Biol. Interact.* 28:91–106.

Calabrese, L. C., and Carbonaro, M., 1986. An e.p.r. study of the non-equivalence of the copper sites of ceruloplasmin, *Biochem. J.* 238:291–295.

Calabrese, L. C., Carbonaro, M., and Musci, G., 1980. Chicken ceruloplasmin. Evidence in support of a trinuclear cluster involving type 2 and 3 copper centers, *J. Biol. Chem.* 263:6480–6483.

Calabrese, L. C., Malatesta, F., and Barra, D., 1981. Purification and properties of bovine caeruloplasmin, *Biochem. J.* 199:667–673.

Calabrese, L. C., Carbonaro, M., and Musci, G., 1987. Chicken ceruloplasmin, *J. Biol. Chem.* 263:6480–6483.

Calabrese, L. C., Carbonaro, M., and Musci, G., 1988. Presence of coupled trinuclear copper cluster in mammalian ceruloplasmin is essential for efficient electron transfer to oxygen, *J. Biol. Chem.* 264:6183–6187.

Calviello, G., Palozza, P., Bossi, D., Piccioni, E., Cittadini, A., and Bartoli, G. M., 1988. Impairment of microsomal calcium sequestration activity upon superoxide dismutase depletion in rat liver, *Biochem. Biophys. Acta* 964:289–292.

Camakaris, J., 1987. Copper transport, absorption and storage, in *Copper in Animals and Man*, Vol. 1 (J. M. Howell and J. M. Gawthorne, eds.), CRC Press, Boca Raton, Florida, pp. 63–77.

Camakaris, J., Mann, J. R., and Danks, D. M., 1979. Copper metabolism in mottled mouse mutants, copper concentrations in tissues during development, *Biochem. J.* 180:597–604.

Camakaris, J., Danks, D. M., Ackland, L., Cartwright, E., Borger, P., and Cotton, R. G. H., 1980. Altered copper metabolism in cultured cells from human Menkes' syndrome and mottled mutants, *Biochem. Genet.* 18:117–127.

Camerman, N., Camerman, A., and Sarkar, B., 1976. Molecular design to mimic the copper(II) transport site of human albumin: The crystal and molecular structure of copper(II)-glycylglycyl-L-histidine-N-methyl amide monoaquo complex, *Can. J. Chem.* 54:1309–1316.

Campbell, C. H., Brown, R., and Linder, M. C., 1981. Circulating ceruloplasmin is an important source of copper for normal and malignant cells, *Biochim. Biophys. Acta* 678:27–38.

Campbell, C. H., Solgonick, R., and Linder, M. C., 1989. Translational regulation of ferritin synthesis in rat spleen: Effects of iron and inflammation, *Biochem. Biophys. Res. Commun.* 160:453–459.

Cann R. L., Brown, W. M., and Wilson, A. C., 1984. Polymorphic sites and the mechanism of evolution in human mitochondrial DNA, *Genetics* 106:479–499.

Capaldi, R. A., 1990. Structure and assembly of cytochrome *c* oxidase, *Arch. Biochem. Biophys.* 280:252–262.

Capaldi, R. A., Malatesta, F., and Darley-Usmar, V. M., 1983. Structure of cytochrome *c* oxidase, *Biochim. Biophys. Acta* 726:135–148.

Caple, I. W., and Heath, T. J., 1978. Regulation of excretion of copper in bile of sheep: Effect of anaesthesia and surgery, *Comp. Biochem. Physiol.* 61:503–507.

Carlton, W. W., and Henderson, W., 1963. Cardiovascular lesions in experimental copper deficiency in chickens, *J. Nutr.* 81:200.

Carlton, W. W., and Henderson, W., 1965. Studies in chickens fed a copper-deficient diet supplemented with ascorbic acid, *J. Nutr.* 85:67–72.

Carrico, R. J., and Deutsch, H. F., 1969. Isolation of human hepatocuprein and cerebrocuprein, *J. Biol. Chem.* 244:6087–6093.

Carrico, R. J., and Deutsch, H. F., 1970. The presence of zinc in human cytocuprein and some properties of the apoprotein, *J. Biol. Chem.* 245:723–727.

Carrico, R. J., Deutsch, H. F., Beinert, H., and Orme-Johnson, W. H., 1969. The properties of an apoceruloplasmin-like protein in human serum, *J. Biol. Chem.* 244:4141–4146.

Cartei, G., Meani, A., and Causarano, D., 1970. Rise of serum transferrin, serum iron and ceruloplasmin in men due to "Esoestrolo" and "Clorotrianisene," *Nutr. Rep. Int.* 2:343–350.

Cartwright, G. E., and Wintrobe, M. M., 1964. Copper metabolism in normal subjects, *Amer. J. Clin. Nutr.* 14:224–232.

Cartwright, G. E., Gubler, C. J., Bush, J. A., and Wintrobe, M. M., 1956. Studies on copper metabolism XVII. Further observations on the anemia of copper deficiency in swine, *Blood*, 11:143.

Cartwright, G. E., Markowitz, H., Shields, G. S., and Wintrobe, M. M., 1960. Studies on copper metabolism. XIX. A critical analysis of serum copper and ceruloplasmin concentration in normal subjects, patients with Wilson's disease and relatives of patients with Wilson's disease, *Am. J. Med.* 28:555–563.

Carville, D. G. M., and Strain, J. J., 1989. The effect of copper deficiency on blood antioxidant enzymes in rats fed sucrose or sucrose and lactose diets, *Nutr. Rep. Intl.* 39:25–33.
Caster, W. O., and Doster, J. M., 1979. Effect of the dietary zinc/copper ratio on plasma cholesterol., *Nutr. Rep. Int.* 19:773–775.
Cattell, F. C. R., and Scott, W. D., 1978. Copper in aerosol particles produced by the ocean, *Science* 202:429–430.
Cederbaum, A. I., and Wainio, W. W., 1972. Binding of iron and copper to bovine heart mitochondria, *J. Biol. Chem.* 247:4593–4603.
Cerutti, P. A., 1985. Prooxidant states and tumor promotion, *Science* 227:375–381.
Cerveny, J., Fass, D. N., and Mann, K. G., 1984. Synthesis of coagulation factor V by cultured aortic endothelium, *Blood* 63:1467–1474.
Chai, M., and Liu, H., 1989. Effects of zinc supplement to weaning feed on absorption of copper, *Shipin Kexue* 119:20–23 (CAS Abstr.).
Chakravarty, P. K., Ghosh, A., and Chowdhury, J. R., 1984a. Sub-cellular distribution of copper and caeruloplasmin in chemically-induced tumour tissue, *J. Comp. Pathol.* 94:607–609.
Chakravarty, P. K., Ghosh, A., and Chowdhury, R. R., 1984b. Serum copper and caeruloplasmin activity in experimental malignancies, *J. Comp. Pathol.* 94:603–606.
Chan, S. I., and Li, P. M., 1990. Cytochrome c oxidase: Understanding nature's design of a proton pump, *Biochemistry* 29:1–12.
Chan, W.-Y., Cushing, W., Coffman, M. A., and Rennert, O. M., 1978. Genetic expression of Wilson's disease in cell culture: A diagnostic marker, *Science* 208:299–300.
Chan, W.-Y., Garnica, A. D., and Rennert, O. M., 1978. Cell culture studies of Menkes' kinky hair disease, *Clin. Chim. Acta* 88:495–507.
Chan, W.-Y., Richichi, J., Griesmann, G. F., Cushing, W., Kling, O. R., and Rennert, O. M., 1980. Copper and ceruloplasmin activity in human amniotic fluid, *Am. J. Obstet. Gynecol.* 138:257–259.
Chang, E. C., and Kosman, D. J., 1990. Oxygen-dependent methionine auxotrophy in copper–zinc superoxide dismutase-deficient mutants of *Saccharomyces cerevisiae*, *J. Bacteriol.* 172:1840–1845.
Chang, I. C., Miholland, D. C., and Matrone, G., 1976. Controlling factors in the development of ceruloplasmin in pigs during the neonatal growth period. *J. Nutr.* 106:1343–1350.
Chang, J.-Y., Owman, C., and Steinbusch, H. W. M., 1988. Evidence for coexistence of serotonin and noradrenaline in sympathetic nerves supplying brain vessels of guinea pig, *Brain Res.* 438:237–246.
Chapman, D. B., and Way, E. L., 1980. Metal ion interaction with opiates, *Annu. Rev. Pharmacol. Toxicol.* 20:553–579.
Chapman, H. L. J., and Bell, M. C., 1963. Relative absorption and excretion by beef cattle of copper from various sources, *J. Anim. Sci.* 22:82–85.
Chaudhry, F. M., and Loneragan, J. F., 1970. Effects of nitrogen, copper and zinc fertilizers on the copper and zinc nutrition of wheat plants, *Austr. J. Agric. Res.* 21:865.
Chayen, J., Bitensky, L., Butcher, R. G., and Poulter, L. W., 1969. Redox control of lysosomes in human synovia, *Nature* 222:281–282.
Cheek, D. B., Powell, G. K., Reba, R., and Feldman, M., 1966. Manganese, copper and zinc in rat muscle and liver cells and in thyroid and pituitary insufficiency, *Bull. Jons Hopkins Hosp.* 118:338–348.
Cheh, A., and Neilands, J. B., 1973. Zinc, as essential metal ion for beef liver δ-levulinate dehydratase, *Biochem. Biophys. Res. Commun* 55:1060–1063.
Chen, M. L., and Failla, M. L., 1988. Metallothionein metabolism in the liver and kidney of the streptozotocin-diabetic rat, *Comp. Biochem. Physiol.* 90B:439–445.
Cherian, M. G., Templeton, D. M., Gallant, K. R., and Banarjee, D., 1987. Biosynthesis and

metabolism of metallothionein in rat during perinatal development, in *Experientia Supplementum*, Vol. 52 (J. H. R. Kagito, and Y. Kojiuma, eds.), Birkhauser Verlag, Basel, pp. 499–505.

Chevreul, --, 1868. Sur la présence du cuivre dans les êtres organisés, *C. R. Acad. Sci.* 66:567–568. (Obtained from Owen, 1982c.)

Chez, R. A., Henkin, R. I., and Fox, R., 1978. Amniotic fluid copper and zinc concentrations in human pregnancy. *Obstet. Gynecol.* 52:125–127.

Chikvaidze, E. N., 1990. Complexes of copper(II) ions with albumins, *Soobschch. Akad. Nauk. Gruz. SSR* 137:153–156 (CAS Abstr.).

Chothia, C., and Lesk, A., 1982. Evolution of proteins formed by β-sheets. I. Plastocyanin and azurin, *J. Mol. Biol.* 160:309–323.

Chu, F.-F., and Olden, J., 1985. The expression of ceruloplasmin, an angiogenic glycoprotein, by mouse embryonic fibroblasts, *Biochem. Biophys. Res. Commun.* 126:15–24.

Church, W. R., Jernigan, R. L., Toole, J., Hewick, R. M., Knopf, J., Knutson, G. J., Nesheim, M. E., Mann, K. G., and Fass, D. N., 1984. Coagulation factors V and VIII and ceruloplasmin constitute a family of structurally related proteins, *Proc. Natl. Acad. Sci. USA* 81:6934–6937.

Cikrt, M., 1972. Biliary excretion of ^{203}Hg, ^{64}Cu, ^{52}Mn, and ^{210}Pb in the rat, *Br. J. Ind. Med.* 29:74–80.

Cikrt, M., 1973. Enterohepatic circulation of ^{64}Cu, ^{52}Mn and ^{203}Hg in rats, *Arch. Toxicol.* 31:51–59.

Cikrt, M., and Tichy, M., 1972. Polyacrylamide gel disc electrophoresis of rat bile after intravenous administration of ^{52}MnCl$_2$, ^{64}CuCl$_2$, ^{203}HgCl$_2$, ^{210}Pb(NO$_3$)$_2$, *Experientia* 28:383–384.

Cikrt, M., Havrdova, J., and Tichy, M., 1974. Changes in the binding of copper and zinc in the rat bile during 24 hours after application, *Arch. Toxicol.* 32:321–329.

Cikrt, M., Tichy, M., and Holusa, R., 1975. Biliary excretion of ^{64}Cu, ^{65}Zn and ^{203}Hg in the rat with liver injury induced by CCl$_4$, *Arch. Toxicol.* 34:227–236.

Clark, I. A., Thumwood, C. M., Chaudhri, G., Cowden, W. B., and Hunt, N. H., 1988. Tumor necrosis factor and reactive oxygen species: Implications for free radical-induced tissue injury, in *Proceedings of Oxygen Radicals and Tissue Injury Symposium* (B. Halliwell, ed.), Fed. Amer. Soc. Exp. Biol., Bethesda, Maryland.

Cloez, I., and Bourre, J.-M., 1987. Copper, manganese and zinc in the developing brain of control and quaking mice, *Neurosc. Lett.* 83:118–122.

Cobley, J. G., and Haddock, B. A., 1975. The respiratory chain of *thiobacillus ferroxidans*: the reduction of cytochromes by Fe^{2+} and the preliminary characterization of rusticyanin, a novel 'blue' copper protein, *FEBS Lett.* 60:29.

Cohen, A. M., Teitelbaum, A., Miller, E., Ben-Tor, V., Hirt, R., and Fields, M., 1982. Effect of copper on carbohydrate metabolism in rats, *Isr. J. Med. Sci.* 18:840.

Cohen, D. I., Illowsky, B., and Linder, M. C., 1979. Altered copper absorption in tumor bearing and estrogen treated rats, *Am. J. Physiol.* 236:E309–E315.

Cohen, N. L., Keen, C. L., Lonnerdal, B., and Hurley, L. S., 1985. Effects of varying iron on the expression of copper deficiency in the growing rat: Anemia, ferroxidase I and II, tissue trace elements, ascorbic acid, and xanthine dehydrogenase, *J. Nutr.* 115:633–649.

Cohn, J. R., and Emmett, E. A., 1978. The excretion of trace metals in human sweat, *Ann. Clin. Lab. Sci.* 8:270–275.

Collins, T., Lapierre, L. A., Fiers, W., Stominger, J. L., and Pober, J. S., 1986. Recombinant human tumor necrosis factor increases mRNA levels and surface expression of HLA-A,B antigens in vascular endothelial cells and dermal fibroblasts *in vitro*, *Proc. Natl. Acad. Sci. USA* 83:446–450.

Colombo, G., Rajashekhar, B., Giedroc, D. P., and Villafranca, J. J., 1984. Mechanism-based inhibitors of dopamine β-hydoxylase: Inhibition by 2-bromo-3-(p-hydroxylphenyl)-1-propene, *Biochemistry* 23:3590–3598.

Colombo, G., Papadopoulos, N. J., Ash, D. E., and Villafranca, J. J., 1987. Characterization of highly purified dopamine beta-hydroxylase, *Arch. Biochem. Biophys.* 252:71–80.

Conforti, A., Franco, L., Milanino, R., and Velo, G. P., 1982. Copper and ceruloplasmin (Cp) concentrations during the acute inflammatory process in the rat, *Agents Actions* 12:303–307.

Conforti, A., Franco, L., Milanino, R., Totorizzo, A., and Velo, G. P., 1983a. Copper metabolism during acute inflammation: Studies on liver and serum copper concentrations in normal and inflamed rats, *Br. J. Pharmacol.* 79:45–52.

Conforti, A., Franco, L., Menegale, G., Milanino, R., Piemonte, G., and Velo, G. P., 1983b. Serum copper and ceruloplasmin levels in rheumatoid arthritis and degenerative joint disease and their pharmacolgical implications, *Pharmacol. Res. Commun.* 15:859–867.

Cordano, A., Baertl, J. M., and Graham, G. G., 1964. Copper deficiency in infancy, *Pediatrics* 34:324–336.

Cordano, A., Placko, R. P., and Graham, G. G., 1966. Hypocupremia and neutropenia in copper deficiency, *Blood* 28:280–283.

Cornatzer, W. E., and Klevay, L. M., 1986. Fructose lowers cardiac copper in rats, *Fed. Proc.* 45:357 (Abstract).

Corrigan, Jr., J. J., Jeter, M. A., Burck, D., and Feinberg, W. M., 1990. Histidine-rich glycoprotein levels in children: The effect of age, *Thromb. Res.* 59:681–686.

Coulson, W. F., and Carnes, W. H., 1967. Cardiovascular studies on copper-deficient swine. IX. Repair of vascular defects in deficient swine treated with copper, *Am. J. Pathol.* 50:861–868.

Cousins, R. J., 1985. Transport and hepatic metabolism of copper and zinc: Special reference to metallothionein and ceruloplasmin, *Physiol. Rev.* 65:238–309.

Cousins, R. J., and Barber, E. F., 1987. Regulation of ceruloplasmin synthesis by retinoic acid and interleukin-1, in *Biology of Copper Complexes* (J. R. J. Sorenson, ed.), Humana Press, Clifton, New-Jersey, pp. 17–38.

Cousins, R. J., and Swerdel, M. R., 1985. Ceruloplasmin and metallothionein induction by zinc and 13-*cis*-retinoic acid in rats with adjuvant inflammation, *Proc. Soc. Exp. Biol. Med.* 179:168–172.

Cox, D. R., and Palmiter, R. D., 1982. Assignment of the mouse metallothionein-I (MTI) gene to chromosome 8: Implications for human Menkes' disease, *Pediatr. Res.* 16:190A.

Cox, J. C., and Boxer, D. H., 1978. The purification and some properties of rusticyanin, a blue copper protein involved in iron(II) oxidation from *Thiobacillus ferrooxidans*, *Biochem. J.* 174:497–502.

Crampton, R. F., Matthews, D. M., and Poisner, R., 1965. Observations on the mechanism of absorption of copper by the small intestine, *J. Physiol.* 178:111–126.

Crane, F. L., Glenn, J. L., and Green, D. E., 1956. Studies on the electron transfer system IV. The electron transfer particle, *Biochim. Biophys. Acta* 22:475–487.

Crane, I. J., and Hunt, D. M., 1983. A study of intestinal copper-binding proteins in mottled mice, *Chem.-Biol. Interact.* 45:113–124.

Cronlund, A. L., Smith, B. D., and Kagan, H. M., 1985. Binding of lysyl oxidase to fibrils of type I collagen, *Connect. Tissue Res.* 14:109–119.

Culotta, V. C., Hsu, T., Hu, S., Fuerst, P., and Hamer, D., 1989. Copper and the ACE1 regulatory protein reversibly induce yeast metallothionein gene transcription in a mouse extract, *Proc. Natl. Acad. Sci. USA* 86:8377–8381.

Cummings, J. N., 1951. The effects of B.A.L. in hepatolenticular degeneration, *Brain* 74:10–22.

Cunnane, S. C., 1988. The profile of long-chain fatty acids in serum phospholipids: A possible indicator of copper status in humans, *Am. J. Clin. Nutr.* 48:1475–1478.
Cunningham, I. J., 1931. Some biochemical and physiological aspects of copper in animal nutrition, *Biochem. J.* 25:1267–1294.
Curzon, G., 1961. Some properties of coupled iron–caeruloplasmin oxidation systems, *Biochem. J.* 79:656–663.
Curzon, G., and O'Reilly, S., 1960. A coupled iron–caeruloplasmin oxidation system, *Biochem. Biophys. Res. Commun.* 2:284–286.
Curzon, G., and Young, S. N., 1972. The ascorbate oxidase activity of caeruloplasmin, *Biochem. Biophys. Acta* 268:41–48.
Cymbaluk, N. F., Schryver, H. F., Hintz, H. F., and Lowe, J. E., 1981. Influence of dietary molybdenum in copper metabolism in ponies, *J. Nutr.* 111:96–106.
Czaja, M. J., Weiner, F. R., Schwarzenberg, S. J., Sternlieb, I., Scheinberg, I. H., Van Thiel, D. H., LaRusso, N. F., Glambrone, M.-A., Kirschner, R., Koschinsky, M. L., MacGillivray, R. T. A., and Zern, M. A., 1987. Molecular studies of ceruloplasmin deficiency in Wilson's desease, *J. Clin. Invest.* 80:1200–1204.
Dallman, P. R., 1967. Cytochrome oxidase repair during treatment of copper deficiency: Relation to mitochondrial turnover, *J. Clin. Invest.* 46:1819–1827.
Dallman, P. R., and Goodman, J. R., 1970. Enlargement of mitochondrial compartment in iron and copper deficiency, *Blood*, 35:496–505.
D'Amato, R. J., Benham, D. F., and Snyder, S. H., 1987. Characterization of the binding of N-methyl-4-phenylpyridine, the toxic metabolite of the Parkinsonian neurotoxin N-methyl-4-phenyl-1,2,3,6-tetrahydropyridine, to neuromelanin, *J. Neurochem.* 48:653–658.
Dameron, C. T., and Harris, E. D., 1987a. Regulation of aortic CuZn-superoxide dismutase with copper, *Biochem. J.* 248:663–668.
Dameron, C. T., and Harris, E. D., 1987b. Regulation of aortic CuZn-superoxide dismutase with copper. Ceruloplasmin and albumin re-activate and transfer copper to the enzyme in culture. *Biochem. J.* 248:669–675.
D'Amore, P. A., and Thompson, R. W., 1987. Mechanisms of angiogenesis, *Annu. Rev. Physiol.* 49:453–464.
Danks, D. M., 1983. Hereditary disorders of copper metabolism in Wilson's disease and Menkes' disease, in *The Metabolic Basis of Inherited Disease* (Stanbury, J. B., Wyngaarden, J. B., Fredrickson, D. S., Goldstein, J., and Brown, M., eds.) McGraw-Hill, New York.
Danks, D. M., 1987. Copper deficiency in infants with particular reference to Menkes' disease, in *Copper in Animals and Man*, Vol. II (J. M. Howell and J. M. Gawthorne, eds.), CRC Press, Boca Raton, Florida, pp. 29–51.
Danks, D. M., 1988. Copper deficiency in humans, *Ann. Rev. Nutr.* 8:235–257.
Danks, D. M., and Camakaris, J., 1983. Mutations affecting trace elements in humans and animals. A genetic approach to an understanding of trace elements, in *Advances in Human Genetics*, Vol. 13 (H. Harris and K. Hirschhorn, eds.), Plenum Press, New York, p. 149.
Danks, D. M., Campbell, P. E., Walker-Smith, J., Stevens, B. J., Gillespie, J. M., Blomfield, J., and Turner, B., 1972a. Menkes' kinky-hair syndrome, *The Lancet* i:1100–1102.
Danks, D. M., Campbell, P. E., Stevens, B. J., Maynes, V., and Cartwright, E., 1972b. Menkes' kinky hair syndrome. An inherited defect in copper absorption with widespread effects, *Pediatrics* 50:188–201.
Danks, D. M., Cartwright, E., Stevens, B. J., and Townley, R. R. W., 1973. Menkes' kinky hair disease: Further definition of the defect in copper transport, *Science* 179:1140–1142.
Darwish, H. M., Hoke, J. E., and Ettinger, M. J., 1983. Kinetics of Cu(II) transport and

accumulation by hepatocytes from copper-deficient mice and the brindled mouse model of Menkes' disease, *J. Biol. Chem.* 258:13621–13626.

Darwish, H. M., Schmitt, R. C., Cheney, J. C., and Ettinger, M. J., 1984a. Copper efflux kinetics from rat hepatocytes, *Am. J. Physiol.* 246:G48–G55.

Darwish, H. M., Cheney, J. C., Schmitt, R. C., and Ettinger, M. J., 1984b. Mobilization of copper(II) from plasma components and mechanism of hepatic copper transport, *Am. J. Physiol.* 246:G72–G79.

Davidson, I. W. F., and Burt, R. L., 1973. Physiologic changes in plasma chromium of normal and pregnant women: Effect of a glucose load, *Am. J. Abst. Gynecol.* 116:601–608.

Davies, M. T., and Williams, R. N., 1976. Effects of pregnancy on uptake and distribution of copper in the rat, *Proc. Nutr. Soc.* (Abstr.) 34:4A–5A.

Davies, N. T., and Campbell, J. K., 1977. The effect of cadmium on intestinal copper absorption and binding in the rat, *Life Sci.* 20:955–960.

Davis, G. K., and Mertz, W., 1987. Copper, in *Trace Elements in Human and Animal Nutrition*. Vol. 1, Academic Press, New York, pp. 301–364.

Davis, N. R., and Anwar, R. A., 1970. On mechanism of formation of desmosine and isodesmosine cross-links of elastin, *J. Am. Chem. Soc.* 92:3778–3782.

Davis, P. N., Norris, L. C., and Kratzer, F. H., 1962. Interference of soybean proteins with utilization of trace minerals, *J. Nutr.* 77:217–223.

Dawson, C. R., 1966. In *The Biochemistry of Copper* (J. Peisach, P. Aisen, and W. W. Blumberg, eds.), Academic Press, New York, p. 305.

Dawson, C. R., Strothkamp, K. G., and Krul, K. G., 1975. Ascorbate oxidase and related copper proteins, *Ann. N.Y. Acad. Sci.* 258:209–220.

Dawson, J. H., Dooley, D. M., Clark, R., Stephens, P. J., and Gray, H. B., 1979. Spectroscopic studies of ceruloplasmin. Electronic structures of the copper sites, *J. Am. Chem. Soc.* 101:5046–5051.

Dawson, J. H., Dooley, D. M., and Gray, H. B., 1980. Coordination environment and fluoride binding of type 2 copper in the blue copper protein ascorbate oxidase, *Proc. Natl. Acad. Sci. USA* 77:5028–5031.

Dawson, J. H., Dooley, D. M., and Gray, H. B., 1983. Coordination environment and fluoride binding of type 2 copper in the blue copper oxidase ceruloplasmin, *Proc. Natl. Acad. Sci. USA* 75:4078–4081.

Day, F. A., Panemangalore, M., and Brady, F. O., 1981. *In vivo* and *ex vivo* effects of copper on rat liver metallothionein, *Proc. Soc. Exp. Biol. Med.* 168:306–310.

Dayer, J. M., Beutler, B., and Cerami, A., 1985. Cachectin/tumor necrosis factor stimulates collagenase and prostaglandin E_2 production by human synovial cells and dermal fibroblasts, *J. Exp. Med.* 162:2163–2168.

Deatherage, J. F., Henderson, R., and Capaldi, R. A., 1982. Relationship between membrane and cytoplasmic domains in cytochrome *c* oxidase by electron microscopy in media of different density, *J. Mol. Biol.* 158:501–514.

Deeming, S. B., and Weber, C. W., 1978. Hair analysis of trace minerals in human subjects as influenced by age, sex, and contraceptive drugs, *Am. J. Clin. Nutr.* 31:1175–1180.

Deinum, J., van Vanngard, T., 1973. The stoichiometry of the paramagnetic copper and the oxidation–reduction potentials of type I copper in human ceruloplasmin, *Biochim. Biophys. Acta* 310:321–330.

Delhaize, E., Loneragan, J. F., and Webb, J., 1987. Copper in plants, its relation to soils and availability to animals, in *Copper in Animals and Man*, Vol. 1 (J. McC. Howell and J. M. Gawthorne, eds). CRC Press, Boca Raton, Florida, pp. 1–20.

Del Principe, D., Menichelli, A., and Colistra, C., 1989. The ceruloplasmin and transferrin system in cerebrospinal fluid of acute leukemia patients, *Acta Paediatr. Scand.* 78:327–328.

Denis, M., 1986. Structure and function of cytochrome-*c*-oxidase, *Biochemistry* 68:459–470.
Denko, C. W., 1979. Protective role of ceruloplasmin in inflammation, *Agents Actions* 9/(4):333–336.
Denny-Brown, D., and Porter, H., 1951. The effect of BAL (2,3-dimer captopropanol) on hepatolenticular degeneration (Wilson's disease), *N. Engl. J. Med.* 245:917–925.
Desideri, A., Hartmann, H. J., Morante, S., and Weser, U., 1990. An EXAFS study of the copper accumulated by yeast cells, *Biol. Met.* 3:45–47.
Desiderio, M. A., Sessa, A., and Perin, A., 1982. Induction of diamine oxidase activity in rat kidney during compensatory hypertrophy, *Biochim. Biophys. Acta*, 714:243.
Desiderio, A., Falconi, M., Parisi, V., Morante, S., and Rotilio, G., 1988. Is the activity-linked electrostatic gradient of bovine Cu, Zn superoxide dismutases conserved in homologous enzymes irrespective of the number and distribution of charges?, *Free Radical Biol. Med.* 5:313–317.
Dick, A. T., Dewey, D. W., and Gawthorne, J. M., 1975. Thiomolybdates and the copper–molybdenum–sulphur interaction in ruminant nutrition, *J. Agric. Sci.* 85:567.
Dickson, E. R., McCall, J. T., and Baggenstoss, A. H., 1975. Determination of copper in needle biopsies of the liver, *Gastroenterology* 69:279–280.
Diez, M., Cerdan, F. J., Arroyo, M., and Balibrea, J. L., 1989. Use of the copper/zinc ratio in the diagnosis of lung cancer, *Cancer* 63:726–730.
Dinarello, C. A., 1984. Interleukin-1 and the pathogenesis of the acute-phase response, *N. Engl. J. Med.* 311:1413–1418.
Dinarello, C. A., Cannon, J. G., Wolff, S. M., Bernheim, H. A., Beutler, B., Cerami, A., Figari, I. S., Palladino, M. A., Jr., and O'Connor, J. V., 1986. Tumor necrosis factor (cachectin) is an endogenous pyrogen and induced production of interleukin 1, *J. Exp. Med.* 163:1433–1450.
Dipaolo, R. V., Kanfer, J. N., and Newberne, P. M., 1974. Copper deficiency and the central nervous system, *J. Neuropathol. Exp. Neurol.* 33:226–236.
DiSilvestro, R. A., 1986. Plasma levels of immunoreactive ceruloplasmin and other acute phase proteins during lactation (42415), *Proc. Soc. Exp. Biol. Med.* 183:257–261.
DiSilvestro, R. A., 1988. Influence of copper intake and inflammation on rat serum superoxide dismutase activity levels, *J. Nutr.* 118:474–479.
DiSilvestro, R. A., and David, E. A., 1986. Enzyme immunoessay for ceruloplasmin: application to cancer patients serum, *Clin. Chim. Acta* 158:287–292.
DiSilvestro, R. A., and Cousins, R. J., 1984. Translational regulation of rat liver metallothionein levels by glucagon, *Fed. Proc.* 43:3310 (Abstract).
DiSilvestro, R. A., and Harris, E. D., 1981. A postabsorption effect of L-ascorbic acid on copper metabolism in chicks, *J. Nutr.* 111:1964–1968.
DiSilvestro, R. A., and Harris, E. D., 1985. Purification and partial characterization of ceruloplasmin from chicken serum, *Arch. Biochem. Biophys.* 241:438–446.
Dive, C., 1971. Les protéines de la bile, Maloine Librarie, Paris.
Dixon, J. S., Greenwood, M., and Lowe, J. R., 1988. Caeruloplasmin concentration and oxidase activity in polyarthritis, *Rheum. Int.* 8:11–14.
Dixon, J. W., and Sarkar, B., 1974. Isolation, amino acid sequence and copper(II)-binding properties of peptide (1-24) of dog serum albumin, *J. Biol. Chem.* 249:5872–5877.
Dooley, D. M., Cote, C. E., Coolbaugh, T. S., and Jenkins, P. L., 1981. Characterization of bovine ceruloplasmin, *FEBS Lett.* 131:363–365.
Dreosti. I. E., and Quicke, G. V., 1966. *S. Afr. J. Agric. Sci.* 9:365.
Drews, L. M., Keis, M. C., and Fox, H. M., 1979. Effect of dietary fiber on copper, zinc and magnesium utilization by adolescent boys, *Am. J. Clin. Nutr.* 32:1893–1897.
Drexel, R., Seigmund, S., Schneider, H. J., Kinzen, B., Gielens, C., Preaux, G., Lontie, R., Kellerman, J., and Lottspeich, F., 1987. Complete amino-acid sequence of a functional

unit from a molluscan hemocyanin (*Helix pomatia*), *Biol. Chem. Hoppe-Seyler* 368:617–635.

Drouliscos, N. J., Bowland, J. P., and Elliott, J. I., 1970. Influence of supplemental dietary copper on copper concentration of pig blood, selected tissues and digestive tract contents, *Can. J. Anim. Sci.* 50:113–120.

Dunlap, W. M., James, G. W., III, and Hume, D. M., 1974. Anemia and neutropenia caused by copper deficiency, *Ann. Int. Med.* 80:470–476.

Durnam, D. M., and Palmiter, R. D., 1981. Transcriptional regulation of the mouse metallothionein-I gene by heavy metals, *J. Biol. Chem.* 256:5712–5716.

Dyke, B., Hegenauer, J., and Saltman, P., 1987. Isolation and characterization of a new zinc-binding protein from albacore tuna plasma, *Biochemistry* 26:3228–3234.

Earl, C. J., Moulton, M. J., and Selverstone, B., 1954. Metabolism of copper in Wilson's disease and in normal subjects: Studies with Cu-64, *Am. J. Med.* 17:205–213.

Eaton, D. L., and Toal, B. F., 1982. Evaluation of the Cd/hemoglobin affinity assay for the rapid determination of metallothionein in biological tissues, *Toxicol. Appl. Pharmacol.* 66:134–142.

Egan, D. A., and Murrin, M. P., 1973. Copper concentration and distribution in the livers of equine fetuses, neonates and foals, *Res. Vet. Sci.* 15:147–148.

Ehlers, N., and Bulow, N., 1977. Clinical copper metabolism parameters in patients with retinitis pigmentosa and other tapeto-retinal degenerations, *Br. J. Opthalmol.* 61:595–596.

Eipper, B. A., and Mains, R. E., 1988. Peptide α-amidation, *Annu. Rev. Physiol.* 50:333–344.

Eipper, B. A., Mains, R. E., and Glembotski, C. C., 1983. Identification in pituitary tissue of a peptide alpha-amidation activity that acts on glycine-extended peptides and requires molecular oxygen, copper, and ascorbic acid, *Neurobiology* 80:5144–5148.

Elek, G., Lapis, K., and Rockenbauer, A., 1979. Electron spin resonance (ESR) spectrum of paraffin embedded human liver tissue. A possible method for the estimation of copper content, *Histochemistry* 61:233–237.

El-Hag, A., Lipsky, P. E., Bennett, M., and Clark R. A., 1986. Immunomodulation by neutrophil myeloperoxidase and peroxide: Differential susceptibility of human lymphocyte functions, *J. Immunol.* 136:3420.

Ellerson, D. L., and Hilker, D. M., 1985. The effect of high dietary thiamine on copper metabolism in rats, *Nutr. Rep. Int.* 32:419–424.

Ellerton, H. D., Ellerton, N. F., and Robinson, H. A., 1983. Hemocyanin—a current perspective, *Prog. Biophys. Mol. Biol.* 41:143–248.

El-Shobaki, F. A., and Rummel, W., 1979. Binding of copper to mucosal transferrin and inhibition of intestinal iron absorption in rats, *Res. Exp. Med. (Berlin)* 174:187–195.

Elvehjem, C. A., Steenbok, H., and Hart, E. B., 1929. The effect of diet on the copper content of milk, *J. Biol. Chem.* 83:27–34.

Endo, M., Suzuki, K., Schmidt, K., Fournet, B., Karamanos, Y., Montrueil, J., Dorland, L., Van Halbeek, H., and Vliegenthart, J. F. G., 1982. *J. Biol. Chem.* 257:8755–8760.

Engel, R. W., Price, N. O., and Miller, F., 1967. Copper, manganese, cobalt, molybdenum balance in preadolescent girls, *J. Nutr.* 92:197–204.

Epstein, C. J., Avraham, K. B., Lovett, M., Smith, S., Elroy-Stein, O., Rotman, G., Bry, C., and Groner, Y., 1987. Transgenic mice with increased Cu/Zn-superoxide dismutase activity: Animal model of dosage effects in Down's syndrome, *Proc. Natl. Acad. Sci. USA* 84:8044–8048.

Epstein, O., and Sherlock, S., 1981. Is Wilson's disease caused by a controller gene mutation resulting in perpetuation of the fetal mode of copper metabolism into childhood? *Lancet* 7:303–305.

Erardi, F. X., Failla, M. L., and Falkinham, J. O., III, 1987. Plasmid-encoded copper

resistance and precipitation by *Mycobacterium scrofulaceum*, *Appl. Environ. Microbiol.* 53:1951–1954.
Ettinger, M. J., 1987. Cellular biochemistry of copper and inherited defects of copper metabolism, *Life Chem. Rep.* 5:169–186.
Ettinger, M. J., and Kosman, D. J., 1981. Chemical and catalytic properties of galactose oxidase, in *Copper Proteins* (T. G. Spiro, ed.), John Wiley and Sons, New York, pp. 220–261.
Ettinger, M. J., Darwish, H. M., and Schmitt, R. C., 1986. Mechanism of copper transport from plasma to hepatocytes, *Fed. Proc.* 45:2800–2804.
Etzel, K. R., and Cousins, R. J., 1981. Hormonal regulation of liver metallothionein zinc: Independent and synergistic action of glucagon and glucocorticoids, *Proc. Soc. Exp. Biol. Med.* 167:233–236.
Etzel, K. R., Shapiro, S. G., and Cousins, R. K., 1979. Regulation of liver metallothionein and plasma zinc by the glucocorticoid dexamethasone, *Biochem. Biophys. Res. Commun.* 89:1120–1126.
Evans, D. J., and Fritze, K., 1969. The identification of metal–protein complexes by gel chromatography and neutron activation analysis. *Anal. Chim. Acta* 44:117.
Evans, G. W., 1973. Copper homeostasis in the mammalian system, *Physiol. Rev.* 53:535–570.
Evans, G. W., and Cornatzer, W. E., 1971. Biliary copper excretion in the rat (35349), *Proc. Soc. Exp. Biol. Med.* 136:719–721.
Evans, G. W., and Wiederanders, R. E., 1967a. Pituitary-adrenal regulation of ceruloplasmin, *Nature* 215:766–768.
Evans, G. W., and Wiederanders, R. E., 1967b. Copper distribution in the neonatal rat, *Am. J. Physiol.* 213:1177–1182.
Evans, G. W., and Wiederanders, R. E., 1967c. Blood copper variation among species, *Amer. J. Physiol.* 213:1183–1185.
Evans, G. W., and Wiederanders, R. E., 1968. Effect of hormones on ceruloplasmin and copper concentrations in the plasma of the rat, *Am. J. Physiol.* 214:1152–1154.
Evans, G. W., and Reis, B. O., 1978. Impaired copper homeostasis in neonatal male and adult female brindled mice, *J. Nutr.* 108:554–560.
Evans, G. W., Cornatzer, N. F., and Cornatzer, W. E., 1970a. Mechanism for hormone-induced alterations in serum ceruloplasmin, *Am. J. Physiol.* 218:613–615.
Evans, G. W., Majors, P. F., and Cornatzer, W. E., 1970b. Ascorbic acid interaction with metallothionein, *Biochem. Biophys. Res. Comm.* 41:1244–1247.
Evans, G. W., Myron, D. R., Cornatzer, N. F., and Cornatzer, W. E., 1970c. Age-dependent alterations in hepatic subcellular copper distribution and plasma ceruloplasmin, *Am. J. Physiol.* 218:298–302.
Evans, G. W., Myron, D. R., and Wiederanders, R. E., 1969. Effect of protein synthesis inhibitors on plasma ceruloplasmin in the rat, *Am. J. Physiol.* 216:340–342.
Evans, J. L., and Abraham, P. A., 1973. Anemia, iron storage and ceruloplasmin in copper nutrition in the growing rat, *J. Nutr.* 103:196–201.
Evans, R. W., Madden, A. D., Patel, K. J., Gibson, J. F., and Wrigley, S. K., 1985. Type-2 copper in ceruloplasmin, *Biochem. Soc. Trans.* 13:627–629.
Everett, J. L., Day, C. L., and Bergel, F., 1964. Analysis of August rat liver for calcium, copper, iron, magnesium, manganese, molybdenum, potassium, sodium and zinc, *J. Pharm. Pharmacol.* 16:85–90.
Failla, M. L., and Cousins, R. J., 1978. Zinc accumulation and metabolism in primary cultures of rat liver cells: Regulation by glucocorticoids, *Biochim. Biophys. Acta* 543:293–304.
Failla, M. L., and Kiser, R. A., 1981. Altered tissue content and cytosol distribution of trace elements in experimental diabetes. *J. Nutr.* 111:1900–1909.

Failla, M. L., and Kiser, R. A., 1983. Hepatic and renal metabolism of copper and zinc in the diabetic rat, *Am. J. Physiol.* 244:E115–E121.

Failla, M. L., Benedict, C. D., and Weinberg, E. D., 1976. Accumulation and storage of Zn^{2+} by *Candida utilis*, *J. Gen. Microbiol.* 94:23–36.

Failla, M. L., Van De Veerdonk, M., Morgan, W. T., and Smith, Jr., J. C., 1982. Chæacterization of zinc-binding proteins of plasma in familial hyperzincemia, *J. Lab. Clin. Med.* 100:943–952.

Failla, M. L., Babu, U., and Seidel, K. E., 1988. Use of immunoresponsiveness to demonstrate that the dietary requirement for copper in young rats is greater with dietary fructose than dietary starch, *J. Nutr.* 118:487–496.

Farrer, P. A., and Mistilis, S. P., 1968. Copper metabolism in the rat. Studies of the biliary excretion and intestinal absorption of ^{64}Cu-labeled copper, *Birth Defects* 4:14–22.

Farver, O., and Pecht, I., 1981. Electron transfer processes of blue copper proteins, in *Copper Proteins* (T. G. Spiro, ed.), John Wiley and Sons, New York, pp. 153–192.

Farver, O., and Pecht, I., 1988. Preferred sites and pathways for electron transfer in blue copper proteins, in *Oxidases and Related Redox Systems* (T. E. King, H. S. Mason, and M. Morrison, eds.), Alan R. Liss, New York, pp. 269–283.

Fay, J., Cartwright, G. E., and Wintrobe, M. M., 1949. Studies on free erythrocyte protoporphyrin, serum iron, serum iron-binding capacity, and plasma copper during normal pregnancy, *J. Clin. Invest.* 28:487.

Fee, J. A., 1975. Copper proteins. Systems containing the blue copper center, *Struct. Bonding* 23:1–60.

Fee, J. A., 1981. In *Oxygen and Oxy-Radicals in Chemistry and Biology* (M. A. J. Rodgers and E. L. Powers, eds). Academic Press, New York, pp. 205–239.

Fee, J. A., 1982. Is superoxide important in oxygen poisoning? *Trends Biochem. Sci.* 7:84–86.

Fee, J. A., Choc, M. G., Findling, K. L., Lorence, R., and Yoshida, T., 1980. Properties of a copper-containing cytochrome c1aa3 complex: A terminal oxidase of the extreme thermophil *Thermus thermophilus HB8*, *Proc. Natl. Acad. Sci. USA* 77:147–151.

Feldman, B. F., Keen, C. L., Kaneko, J. J., and Farver, T. B., 1981. Anemia of inflammatory disease in the dog: Measurement of hepatic superoxide dismutase, hepatic nonheme iron, copper, zinc, and ceruloplasmin and serum iron, copper and zinc, *Am. J. Vet. Res.* 42:1114–1117.

Feldman, S. L., Failla, M. L., and Cousins, R. J., 1978. Degradation of liver metallothioneins *in vitro*, *Biochim. Biophys. Acta* 544:638–646.

Feldman, G., Abramowitz, C., Sarmini, H., and Rousselet, F., 1972. Répartition du cuivre dans les fractions subcellulaires hépatiques après intoxication subaiguë par le cuivre chez le rat, *Biologie Gastro-Entérologie* 5:37–46.

Felix, E. L., 1927. Correction of unproductive muck by the addition of copper, *Phytopathol.* (Abstr.) 17:49–50.

Fell, B. F., 1987. The pathology of copper deficiency in animals, in *Copper in Animals and Man*, Vol. II (J. M. Howell and J. M. Gawthorne, eds.), CRC Press, Boca Raton, Florida, pp. 1–28.

Fell, B. F., King, T. P., and Davies, N. T., 1982. Pancreatic atrophy in copper-deficient rats: Histochemical and ultrastructural evidence of a selective effect on acinar cells. *Histochem. J.* 14:665–680.

Fell, F., Farmer, L. J., Farquharson, C., Bremner, I., and Graca, D. S., 1985. Observations on the pancreas of cattle deficient in copper, *J. Comp. Path.* 95:573–590.

Feller, D. J., and O'Dell, B. L., 1981. Dopamine and norepinephrine in discrete areas of the copper-deficient rat brain. *J. Neurochem.* 34:1259–1263.

Ferrera, R., Faris, B., Mogayzel, P. J., Jr., Gonnerman, W. A., and Franzblau, C., 1982. A micromethod for the purification of lysyl oxidase, *Anal. Biochem.* 126:312–317.

Fett, J. W., Strydom, D. J., Lobb, R. R., Alderman, E. M., Bethune, J. L., Riordan, J. F., and Vallee, B. L., 1985. Isolation and characterization angiogenin, angiogenic protein from human carcinoma cells, *Biochemistry* 24:5480–5486.
Fett, J. W., Bethune, J. L., and Vallee, B. L., 1987. Induction of angiogenesis by mixtures of two angiogenic proteins, angiogenin and acidic fibroblast growth factor, in chick chorioallantoic membrane, *Biochem. Biophys. Res. Commun.* 146:1122–1131.
Fielden, E. M., Roberts, P. B., Bray, R. C., Lowe, D. J., Mautner, G. N., Rotilio, G., and Calabrese, L., 1974. The mechanism of action of superoxide dismutase from pulse radiolysis and electron paramagnetic resonance, *Biochem. J.* 139:49–60.
Fields, M., Ferretti, R. J., Smith, J. C., and Reiser, S., 1983a. Effects of copper deficiency on metabolism and mortality of rats fed sucrose or starch diets, *J. Nutr.* 113:1335–1345.
Fields, M., Fields, M., Reiser, S., and Smith, J. C., Jr., 1983b. Effect of copper or insulin in diabetic copper-deficient rats, *Proc. Soc. Exp. Biol. Med.* 173:137–139.
Fields, M., Ferretti, R. J., Smith, J. C., and Reiser, S., 1984a. Impairment of glucose tolerance in copper-deficient rats: Dependency on the type of carbohydrate, *J. Nutr.* 114:393.
Fields, M., Ferretti, R. J., Smith, Jr., J. C., and Reiser, S., 1984b. The severity of copper deficiency is determined by the type of dietary carbohydrate, *Proc. Soc. Exp. Biol. Med.* 175:530–537.
Fields, M., Craft, N., Lewis, C. G., Holbrook, J., Rose, A., Reiser, S., and Smith, J. C., 1986a. Contrasting effects of the stomach and small intestine of rats on copper absorption, *J. Nutr.* 116:2219–2228.
Fields, M., Holbrook, J., Scholfield, D., Smith, J. C., Jr., and Reiser, S., and the Los Alamos Medical Research Group, 1986b. Effect of fructose or starch on copper-67 absorption and excretion by the rat, *J. Nutr.* 116:625–632.
Fields, M., Holbrook, J., Scholfield, D., Powell, A. S., Rose, A. J., Reiser, S., and Smith, J. C., 1986c. Development of copper deficiency in rats fed fructose or starch: Weekly measurements of copper indices in blood, *Proc. Soc. Exp. Biol. Med.* 181:120–124.
Fields, M., Lewis, C. G., Scholfield, D. J., Powell, A. S., Rose, A. J., Reiser, S., and Smith, J. C., 1986d. Female rats are protected against the fructose-induced mortality of copper deficiency, *Proc. Soc. Exp. Biol. Med.* 183:145–149.
Fields, M., Lewis, C. G., Beal, T., Scholfield, D., Patterson, K., Smith, J. C., and Reiser, S., 1987. Sexual differences in the expression of copper deficiency in rats, *Proc. Soc. Exp. Biol. Med.* 186:183–187.
Fields, M., Lewis, C. G., and Beal, T., 1989. Accumulation of sorbitol in copper deficiency: Dependency on gender and type of dietary carbohydrate, *Metab. Clin. Exp.* 38:371–375.
Fields, M., Lewis, C. G., and Lure, M. D., 1991. Anemia aggravates the severity of copper deficiency in experimental animals, in *Trace Elements in Man and Animals* (TEMA-7). (B. Momcilovic, ed.), in press.
Fields, M., Lewis, C. G., and Lure, M. D., 1991a. Anemia plays a major role in myocardial hypertrophy of copper deficiency, *Metab. Clin. Exp.* 40:1–3.
Fields, M. Lewis, C. G., Lure, M. D., Burns, W. A., and Antholine, W. E., 1991b. The severity of copper deficiency can be ameliorated by deferoxamine, *Metab. Clin. Exp.* 40:105–109.
Fink, P. S., Whitford, T., Leffak, M., and Prochaska, L. J., 1987. Homology between bacterial DNA and bovine mitochondrial DNA encoding cytochrome *c* oxidase subunit III, *FEBS Lett.* 214:75–80.
Fischer, P. W. F., and L'Abbé, M. R., 1985. Copper transport by intestinal brush border membrane vesicles from rats fed high zinc or copper deficient diets, *Nutr. Res.* 5:759–767.
Fischer, P. W. F., Giroux, A., Belonje, B., and Shah, B. G., 1980. The effect of dietary copper and zinc on cholesterol metabolism, *Am. J. Clin. Nutr.* 33:1019–1025.

Fischer, P. W. F., Giroux, A., and L'Abbé, M. R., 1981. The effect of dietary zinc on intestinal copper absorption, *Am. J. Clin. Nutr.* 34:1670–1675.

Fischer, P. W. F., Giroux, A., and L'Abbé, M. R., 1983. Effects of zinc on mucosal copper binding and on the kinetics of copper absorption, *J. Nutr.* 113:462–469.

Fisher, D. B., Kirkwood, R., and Kaufman, S., 1972. Rat liver phenylalanine hydroxylase, an iron enzyme, *J. Biol. Chem.* 247:5161–5167.

Fisher, G. L., and Shifrine, M., 1978. Hypothesis for the mechanism of elevated serum copper in cancer patients, *Oncology* 35:22–25.

Fishman, J. H., and Fishman, J., 1988. Copper and endogenous mediators of estradiol action, *Biochem. Biophys. Res. Commun.* 152:783–788.

Fishman, J. B., Rubin, J. B., Handrahan, J. V., Connor, J. R., and Fine, R. E., 1987. Receptor-mediated transcytosis of transferrin across the blood–brain barrier, *J. Neurosci. Res.* 18:299–304.

Flanagan, P. R., Haist, J., MacKenzie, I., and Valberg, L. S., 1984. Intestinal absorption of zinc: Competitive interactions with iron, cobalt, and copper in mice with sex-linked anemia (sla), *Can. J. Physiol. Pharmacol.* 62:1124–1128.

Flatmark, T., and Romslo, I., 1975. Energy-dependent accumulation of iron by isolated rat liver mitochondria. Requirement of reducing equivalents and evidence for a unidirectional flux of Fe(II) across the inner membrane, *J. Biol. Chem.* 250:6433–6438.

Fleming, C. R., Dickson, E. R., Baggenstoss, A. H., and McCall, J. T., 1974. Copper and primary biliary cirrhosis, *Gastroenterology* 67:1182–1187.

Fleming, R. E., and Gitlin, J. D., 1990. Primary structure of rat ceruloplasmin and analysis of tissue-specific gene expression during development, *J. Biol. Chem.* 265:7701–7707.

Flynn, A., Loftus, M. A., and Finke, J. H., 1984. Production of interleukin-1 and interleukin-2 in allogenic mixed lymphocyte cultures under copper, magnesium and zinc deficient conditions, *Nutr. Res.* 4:673–679.

Fogel, S., and Welch, J. S., 1982. Tandem gene amplification mediates copper resistance in yeast, *Proc. Natl. Acad. Sci. USA* 79:5342–5346.

Folkman, J., 1987. Angiogenic factors, *Science* 235:442.

Forker, E. L., 1969. The effect of estrogen on bile formation in the rat, *J. Clin. Invest.* 48:654–663.

Forth, W., and Rummel, W., 1971. Absorption of iron and chemically related metals *in vitro* and *in vivo*: Specificity of the iron binding system in the mucosa of the jejunum, in *Intestinal Absorption of Metal Ions, Trace Elements and Radionuclides* (S. C. Skoyyna and D. Waldron-Edward, eds.), Pergamon Press, Oxford, pp. 173–191.

Foulkes, E. C., (ed.), 1982 *Biological Roles of Metallothionein*, Elsevier/North-Holland, Amsterdam.

Foulkes, E. C. (ed.), 1985. *Biological Roles of Metallothionein*. Elsevier/North-Holland, Amsterdam.

Frank, L., Wood, D. L., and Roberts, R. J., 1978. Effect of diethyldithiocarbamate on oxygen toxicity and lung enzyme activity in immature and adult rats, *Biochem. Pharmacol.* 27:251–254.

Frater-Schroder, M., Risau, W., Hallmann, R., Gautschi, P., and Bohlen, P., 1987. Tumor necrosis factor type α, a potent inhibitor of endothelial cell growth, *in vitro*, is angiogenic, *in vivo*, *Proc. Natl. Acad. Sci. USA* 84:5277–5281.

Freedman, J. A., and Chan, S. H. P., 1984. Interactions in cytochrome oxidase: functions and structure, *J. Bioenerg. Biomembr.* 16:75–100.

Freedman, J. H., Weiner, R. J., and Peisach, J., 1986. Resistance to copper toxicity of cultured hepatoma cells. Characterization of resistant cell lines, *J. Biol. Chem.* 261:11840–11848.

Freedman, J. H., Ciriolo, M R., and Peisach, J., 1989. The role of glutathione in copper metabolism and toxicity, *J. Biol. Chem.* 264:5598–5606.

Freeman, B. M., Manning, A. C. C., and Pole, D. S., 1973. Factors affecting ceruloplasmin activity in *Gallus domesticus*, *Comp. Biochem. Physiol.* 45A:689–698.
Freeman, H. C., and Martin, R.-P., 1969. Potentiometric study of equilibria in aqueous solution between copper(II) ions, L (or D)-histidine and L-threonine and their mixtures, *J. Biol. Chem.* 244:4823–4830.
Freeman, S., and Daniel, E., 1973. Dissociation and reconstitution of human ceruloplasmin, *Biochemistry* 12:4806–4810.
Frey, T. G., Costello, M. J., Karlsson, B., Haselgrove, J. C., and Leigh, J. S., 1982. Structure of the cytochrome *c* oxidase dimer. Electron microscopy of two-dimensional crystals, *J. Mol. Biol.* 162:113–130.
Fridovich, I., 1975. Superoxide dismutases, *Annu. Rev. Biochem.* 44:147–159.
Frieden, E., 1970. Ceruloplasmin, a link between copper and iron metabolism, *Nutr. Rev.* 28:87–91.
Frieden, E., 1971. Ceruloplasmin, a link between copper and iron metabolism, in *Bioinorganic Chemistry* (R. F. Gould, ed.), American Chemical Society, Washington, D. C., pp. 292–231.
Frieden, E., 1976. Copper and iron metalloproteins, *Trends Biochem. Sci.* 1976 (Dec.): 273–274.
Frieden, E., 1979. Ceruloplasmin: the serum copper transport protein with oxidase activity, in *Copper in the Environment*. Part II (J. O. Nriagu, ed.), John Wiley and Sons, New York, pp. 241–284.
Frieden, E., 1980. Ceruloplasmin: a multifunctional metalloprotein of vertebrate plasma, in *Biological Roles of Copper, Ciba Foundation Symposium 79 (New Series)*, Excerpta Medica, Amsterdam, p. 93.
Frieden, E., 1981a. The evolution of copper proteins, in *Metal Ions in Biological Systems*, Vol. 13 (H. Sigel, ed.), Marcel Dekker, New York, pp. 1–14.
Frieden, E., 1981b. Ceruloplasmin: A multifunctional metalloprotein of vertebrate plasma, in *Metal Ions in Biological Systems*, Vol. 13 (H. Sigel, ed.), Marcel Dekker, New York, pp. 117–142.
Frieden, E., 1983. The copper connection, *Sem. Hematol.* 20:114–117.
Frieden, E., 1986. Perspectives on copper biochemistry, *Clin. Physiol. Biochem.* 4:11–19.
Frieden, E., and Hsieh, H. S., 1976. Ceruloplasmin: The copper transport protein with essential oxidase activity, in *Advances in Enzymology and Related Areas of Molecular Biology*, Vol. 44 (A. Meister, ed.), John Wiley and Sons, New York, pp. 187–236.
Friedman, R. L., Manly, S. P., McMahon, M., Kerr, I. M., and Stark, G. R., 1984. Transcriptional and posttranscriptional regulation of interferon-induced gene expression in human cells, *Cell* 38:745–755.
Friere, E., 1990. Thermodynamic studies of the insertion and folding of membrane proteins: Yeast cytochrome *c* oxidase, *Thermochim. Acta* 163:47–56.
Frommer, D. J., 1971. The binding of copper by bile and serum, *Clin. Sci.* 41:485–493.
Frommer, D. J., 1972a. Studies on copper metabolism, Thesis, University of London.
Frommer, D. J., 1972b. The measurement of biliary copper secretion in humans, *Clin. Sci.* 42:26P.
Frommer, D. J., 1974. Defective biliary secretion of copper in Wilson's disease, *Gut* 15:125–129.
Frommer, D. J., 1977. Biliary copper excretion in man and the rat, *Digestion* 15:390–396.
Frydman, M., Bonne-Tamir, B., Farrer, L. A., Conneally, P. M., Magazanik, A., Askhel, S., and Goldwitch, Z., 1985. Assignment of the gene for Wilson's disease to chromosome 13: Linkage to the esterase D locus, *Proc. Natl. Acad. Sci. USA* 82:1819–1821.
Fujita, T., Ohtani, N., Aihara, M., and Nishioka, K., 1987. Comparison of the effects of iron

and copper on prostaglandin synthesis in rabbit kidney medulla slices, *J. Pharm. Pharmacol.* 39:230–233.

Fuller, S. D., 1981. The three-dimensional structure of cytochrome *c* oxidase, Doctoral dissertation, University of Oregon.

Fuller, S. D., Capaldi, R. A., and Henderson, R., 1979. Structure of cytochrome *c* oxidase in deoxycholate-derived two-dimensional crystal, *J. Mol. Biol.* 134:305–327.

Funckes, C. G., Chvapil, M., Carroll, R. W., and Bressler, R., 1976. Effect of a long-acting contraceptive drug, norethindrone enanthate, on serum zinc and copper in human volunteers, *Contraception* 14:291–295.

Funk, F., and Schneider, W., 1989. The dichotomy of mechanisms of copper accumulation in yeast induced by enhanced levels of copper as well as menadione in growth media, *Chem. Speciation Bioavailability* 1:113–138.

Furey, W. F., Robbins, A. H., Clancy, L. L., Winge, D. R., Wang, B. C., and Stout, C. D., 1986. Crystal structure of Cd, Zn metallothionein, *Science* 231:704–710.

Furey, W. F., Robbins, A. H., Clancy, L. L., Winge, D. R., Wang, B. C., and Stout, C. D., 1987. Crystal structure of cadmium–zinc metallothionein, *Experientia* (Suppl.) 52:139–148.

Furst, P., Hu, S., Hackett, R., and Hamer, D., 1988. Copper activates metallothionein gene transcription by altering the conformation of a specific DNA binding protein, *Cell* 55:705–717.

Fushimi, H., Hamison, C. R., and Ravin, H. A., 1971. Two new copper proteins from human brain; isolation and properties, *J. Biochem.* 69:1041–1054.

Gacheru, S. N., Trackman, P. C., and Kagan, H. M., 1988. Evidence for a functional role for histidine in lysyl oxidase catalysis, *J. Biol. Chem.* 263:16704–16708.

Gagnon, J. Palmiter, R. D., and Walsh, K. A., 1978. Comparison of the NH_2-terminal sequence of ovalbumin as synthesized *in vitro* and *in vivo*, *J. Biol. Chem.* 253:7464–7468.

Gaitskhoki, V. S., L'vov, V. M., Puchkova, L. V., Scwartzman, A. L., and Neifakh, S. A., 1981. Highly purified ceruloplasmin messenger RNA from rat liver, physico-chemical and functional characteristics, *Mol. Cell. Biochem.* 35:171–182.

Gaitskhoki, V. S., Voronina, O. V., Denezhkina, V. V., Pliss, M. G., Puchova, L. V., Shvartsman, A. L., and Neifakh, S. A., 1990. Ceruloplasmin gene expression in different organs of rats, *Biokhimiya* 55:927–937.

Gallagher, C. H., and Reeve, V. E., 1971. Copper deficiency in the rat. Effect on adenine nucleotide metabolism, *Aust. J. Exp. Biol. Med. Sci.* 49:445–451.

Gallagher, C. H., Judah, J. D., and Rees, K. R., 1956. The biochemistry of copper deficiency. I. Enzymological disturbances, blood chemistry and excretion of amino-acids, *Proc. R. Soc. London, Ser. B* 145:134–149.

Gallant, K. R., and Cherian, M. G., 1987. Changes in dietary zinc result in specific alterations of metallothionein concentrations in newborn rat liver, *J. Nutr.* 117:709–716.

Gallop, P. M., Paz, M. A., Fluckinger, R., and Kagan, H. M., 1989. PQQ, the elusive coenzyme, *Trends Biochem. Soc.* 14:343–346.

Galston, A. W., 1983. Polyamines as modulators of plant development, *BioScience* 33:382.

Gao, L., Li, R., and Wang, K., 1989. Kinetic studies of mobilization of copper(II) from human serum albumin with chelating agents, *J. Inorg. Biochem.* 36:83–92.

Gardiner, P. E., Rosick, E., Rosick, U., Bratter, P., and Kynast, G., 1982. The application of gel filtration immunonephelometry and electrothermal atomic absorption spectrometry to the study of the distribution of copper-, iron- and zinc-bound constituents in human amniotic fluid, *Clin. Chim. Acta* 120:103–117.

Gardiner, P. E., Ottaway, J. M., and Fell, G. S., 1981. The application of gel filtration and electrothermal atomic absorption spectroscopy to the speciation of protein bound zinc and copper in human blood serum. *Anal. Chim. Acta* 124:281–294.

Gardiner, P. E., Roesick, E., Roesick, U., Braettner, P., and Kynast, G., 1982. The application of gel filtration, immunonephelometry and electrothermal atomic absorption spectrometry to the study of the distribution of copper-, iron-, and zinc-bound constituents in human amniotic fluid, *Clin. Chim. Acta* 120:103–117.

Garnier, A., Tosi, L., and Steinbuch, M., 1981. Ferroxidase II. The essential role of copper in enzymatic activity, *Biochem. Biophys. Res. Commun.* 98:66–71.

Garrett, I. R., and Whitehouse, M. W., 1987. Copper and inflammation, in *Copper in Animals and Man*, Vol. 2, (J. McC. Howell and J. M. Gawthorne, eds), CRC Press, Boca Raton, Florida, pp. 107–122.

Gartrell, J. W., 1981. Distribution and correction of copper deficiency in crops and pastures, in *Copper in Soils and Plants* (J. F. Loneragan, A. D., Robson, and R. D., Graham, eds.), Academic Press, New York, pp. 313–349.

Garvey, J. S., and Chang, C. C., 1981. Detection of circulating metallothionein in rats injected with zinc or cadmium, *Science* 214:805–807.

Gavriel, P., and Kagan, H. M., 1988. Inhibition by heparin of the oxidation of lysine in collagen by lysyl oxidase, *Biochemistry* 27:2811–2815.

Gavrilla, I., Feticu, M.., and Ardeleanu, R., 1968. Valoarea sideremeie si cupremiei, precum si a transaminazelor in diagnosticul icterelor. (Determination of blood iron and copper levels and serum transaminase in the diagnosis of jaundices), *Med. Intern. Bucur.* 20:943–954.

Geiger, J. D., Seth, P. K., Klevay, L. M., and Parmar, S. S., 1984. Receptor-binding changes in copper deficient rats, *Pharmacology* 28:196–202.

Geller, B. L., and Winge, D. R., 1982a. Metal binding sites of rat liver Cu-thionein, *Arch. Biochem. Biophys.* 213:109–117.

Geller, B. L., and Winge, D. R., 1982b. Rat liver CuZn superoxide dismutase. Subcellular location in lysosomes, *J. Biol. Chem.* 257:8945–8952.

George, G. N., Winge, D. R., Sout, C. T., and Cramer, S. P., 1986. X-ray absorption studies of the copper-beta domain of rat liver metallothionein, *J. Inorg. Biochem.* 27:213–220.

George, G. N., Byrd, J., and Winge, D. R., 1988. X-ray absorption studies of yeast copper metallothionein, *J. Biol. Chem.* 263:8199–8203.

George, G. N., Prince, R. C., Frey, T. G., and Cramer, S. P., 1989. Oriented x-ray absorption spectroscopy of membrane bound metalloproteins, *Physica B* 158:81–83.

Germann, U. A., and Lerch, K., 1986. Isolation and partial nucleotide sequence of the laccase from *neurospora crassa*: Amino acid sequence homology of the protein to human ceruloplasmin, *Proc. Natl. Acid. Sci. USA* 83:8854–8858.

Getzoff, E. D., Tainer, J. A., Weiner, P. K., Kollman, P. A., Richardson, J. S., and Richardson, D. C., 1983. Electrostatic recognition between superoxide and copper, zinc superoxide dismutase, *Nature (London)* 306:287–290.

Gibbs, K., and Walshe, J. M., 1977. The effect of certain chelating compounds on the urinary excretion of copper by the rat: Observations on their clinical significance, *Clin. Sci. Mol. Med.* 53:317–320.

Gibbs, K., and Walshe, J. M., 1979. A study of the ceruloplasmin concentrations found in 75 patients with Wilson's disease, their kinships and various control groups, *Q. J. Med. New Series XLVIII* 191:447–463.

Gibbs, K., and Walshe, J. M., 1980. Biliary excretion of copper in Wilson's disease, *Lancet* 6:538–539.

Giclas, P. C., Manthei, U., and Strunk, R. C., 1985. The acute phase response of C3, C5, ceruloplasmin, and C-reactive protein induced by turpentine pleurisy in the rabbit, *Am. J. Pathol.* 120:146–156.

Gillespie, J. M., 1973. Keratin structure and changes in copper deficiency, *Aust. J. Dermatol.* 14:127.

Gingrich, D. J., Weber, D. N., Shaw, C. F., Garvey, J. S., and Petering, D. H., 1986. Characterization of a highly negative and labile binding protein induced in *Euglena gracilis* by cadmium, *Eniron. Health Perspect.* 65:77–85.

Gipp, W. F., Pond, W. G., Kallfelz, F. A., Tasker, J. B., Van Campen, D. R., Krook, L., van Visek, W. J., 1974. Effect of dietary copper, iron and ascorbic acid levels on hematology, blood and tissue copper, iron and zinc concentrations and ^{64}Cu and ^{59}Fe metabolism in young pigs, *J. Nutr.* 504:532–541.

Giroux, E. L., 1975. Determination of zinc distribution between albumin and alpha 2-macroglobulin in human serum, *Biochem. Med.* 12:258–266.

Gitlin, D., and Biasucci, A., 1969. Development of γG, γA, γM, $\beta IC/\beta IA$, C'1 esterase inhibitor, ceruloplasmin, transferrin, hemopexin, haptoglobin, fibrinogen, plasminogen, α_1-antitrypsin, orosomucoid, β-lipoprotein, α_2-macroglobulin, and prealbumin in the human conceptus, *J. Clin. Invest.* 48:1433–1446.

Gitlin, J. D., 1988a. Transcriptional regulation of ceruloplasmin gene expression during inflammation, *J. Biol. Chem.* 263:6281–6287.

Gitlin, J. D., 1988b. Ceruloplasmin gene expression in human peripheral blood monocytes and macrophages, Pediatr. Res. 23: Abstract no. 826.

Gjedde, A., 1986. The selective barrier between blood and brain, *Trends Biochem. Sci.* 11:525–527.

Glembotski, C. C., Eipper, B. A., and Mains, R. E., 1984. Characterization of a peptide alpha-amidation activity from anterior pituitary, *J. Biol. Chem.* 259:6385–6392.

Glennon, J. D., and Sarkar, B., 1982. Nickel(II) transport in human blood serum, *Biochem. J.* 203:15–23.

Goebel, K.-M., Storck, U., and Neurath, F., 1981. Intrasynovial fluid therapy in rheumatoid arthritis, *Lancet* 1:1015.

Goka, T. J., Stevenson, R. E., Hefferan, P. M., and Howell, R. R., 1976. Menkes' disease: A biochemical abnormality in cultured human fibroblasts, *Proc. Natl. Acad. Sci. USA* 73:604–606.

Goldfischer, S., and Bernstein, J., 1969. Lipofuscin (aging) pigment granules of the newborn human liver, *J. Cell. Biol.* 42:253–261.

Goldfischer, S., and Sternlieb, I., 1969. Cytochemical study of copper in Wilson's disease and newborn liver, *Gastroenterology* 56:403.

Goldfischer, S., Popper, H., and Sternlieb, I., 1980. The significance of variations in the distribution of copper in liver disease, *Am. J. Pathol.* 99:715–730.

Goldstein, I. M., and Charo, I. F., 1982. Ceruloplasmin: An acute phase reactant and antioxidant, *Lymphokines* 8:373–411.

Goldstein, I. M., Kaplan, H. B., Edelson, H. S., and Weissmann, G., 1979. Ceruloplasmin: A scavenger of superoxide anion radicals, *J. Biol. Chem.* 254:4040–4045.

Gollan, J. L., 1975. Studies on the nature of complexes formed by copper with human alimentary secretions and their influence on copper absorption in the rat, *Clin. Sci. Mol. Med.* 49:237–245.

Gollan, J. L., and Deller, D. J., 1973. Studies on the nature and excretion of biliary copper in man, *Clin. Sci.* 44:9–15.

Gollan, J. L., Davis, P. S., and Deller, D. J., 1971a. A radiometric assay of copper binding in biological fluids and its application to alimentary secretions in normal subjects and Wilson's disease, *Clin. Chim. Acta* 31:197–204.

Gollan, J. L., Davis, P. S., and Deller, D. J., 1971b. Copper content of alimentary secretions, *Clin. Biochem.* 4:42–44.

Gollan, J. L., Davis, P. S., and Deller, D. J., 1971c. Binding of copper by human alimentary secretions, *Am. J. Clin. Nutr.* 24:1025–1027.

Gollub, E. G., and Dayan, J., 1985. Regulation by heme of the synthesis of cytochrome c oxidase subunits V and VII in yeast, *Biochem. Biophys. Res. Comm.* 128:1447–1454.

Goode, C. A., Dinh, C. T., and Linder, M. C., 1989. Mechanism of copper transport and delivery in mammals: Review and recent findings, *Adv. Exp. Med. Biol.* 258:131–144.

Goodman, J. R., and Dallman, P. R., 1969. Role of copper in iron localization in developing erythrocytes, *Blood* 34:747–753.

Gooneratne, R., and Christensen, D., 1985. Gestation age and maternal-fetal liver copper levels in the bovine, in *Trace Elements in Man and Animals* (C. F. Mills, I. Bremner, and J. K., Chesters, eds.), Commonwealth Agric. Bureaux, Farnham Royal, UK, p. 334–336.

Goormaghtigh, E., Brasseur, R., and Ruysschaert, J. M., 1982. Adriamycin inactivates cytochrome c oxidase by exclusion of the enzyme from its cardiolipin essential environment, *Biochem. Biophys. Res. Commun.* 104:314–320.

Gordon, D. T., Leinart, A. S., and Cousins, R. J., 1987. Portal copper transport in rats by albumin, *Am. J. Physiol.* 252:E327–333.

Goresky, C. A., Holmes, T. H., and Sass-Kortsak, A., 1968. The initial uptake of copper by the liver in the dog, *Can. J. Physiol. Pharmacol.* 46:771–784.

Gorin, M. B., Cooper, D. L., Eiferman, F., van de Rijn, P., and Tilghman, S. M., 1981. The evolution of alpha-fetoprotein and albumin. I. A comparison of the primary amino acid sequences of mammalian alpha-fetoprotein and albumin, *J. Biol. Chem.* 256:1954–1959.

Gorman, J. A., Clark, P. E., Lee, M. C., Debouck, C., and Rosenberg, M., 1986. Regulation of the yeast metallothionein gene, *Gene* 48:13–22.

Gottschall, D. W., Dietrich, R. F., Benkovic, S. J., and Shiman, R., 1982. Phenylalanine hydroxylase. Correlation of the iron content with activity and the preparation and reconstitution of the apoenzyme, *J. Biol. Chem.* 257:845.

Graham, R. D., 1981. Absorption of copper by plant roots, in *Copper in Soils and Plants* (J. F. Loneragan, A. D. Robson, and R. D. Graham, eds.), Academic Press, Sydney, Chapter 7.

Grassman, E., and Kirchgessner, M., 1974. On the metabolic availability of absorbed copper and iron, in *Trace Element Metabolism in Animals-2* (W. G. Hoekstra, J. W. Suttie, H. E. Ganther, and W. Mertz, eds.), University Park Press, Baltimore, pp. 523–526.

Graul, R. S., Epstein, O., Sherlock, S., and Scheurer, P. J., 1982. Immunocytochemical identification of caeruloplasmin in hepatocytes of patients with Wilson's disease, *Liver* 2:207–211.

Gray, H. B., and Solomon, E. I., 1981. Electronic structures of blue copper centers in proteins, in *Copper Proteins* (T. G. Spiro, ed.), John Wiley and Sons, New York, pp. 3–39.

Greco, M. A., Hrab, D. I., Magner, W., and Kosman, D. J., 1990. Copper, zinc superoxide dismutase and copper deprivation and toxicity in Saccharomyces cerevisiae, *J. Bacteriol.* 172:317–325.

Greenberger, N. J., Balcerzak, S. P., and Ackerman, G. A., 1969. Iron uptake by isolated intestinal brush borders: Changes induced by alterations in iron stores, *J. Lab. Clin. Med.* 73:711–721.

Greengard, O., 1970. Developmental formation of enzymes in rat liver, in *Biochemical Actions of Hormones*, Vol. 1, Academic Press, New York, pp. 53–85.

Greger, J. L., and Buckley, S., 1977. Menstrual loss of zinc, copper, magnesium and iron by adolescent girls, *Nutr. Rep. Int.* 16:639–647.

Greger, J. L., Bennett, O. A., Buckley, S., and Baligar, P., 1978. Zinc, copper and manganese balance in adolescent girls, in *Trace Element Metabolism in Man and Animals-3* (M. Kirchgessner, ed.), Arbeitskreis für Tierernährungsforschung Weihenstephan, Feising, Germany, pp. 300–303.

Gregor, J. L., and Snedeker, S. M., 1980. Effect of dietary protein and phosphorus levels on the utilization of zinc, copper and manganese, *J. Nutr.* 110:2243–2253.
Gregoriadis, G., and Sourkes, T. L., 1968. Role of protein in removal of copper from the liver, *Nature* 218:290–291.
Gregoriadis, G., and Sourkes, T. L., 1970. Regulation of hepatic copper in the rat by the adrenal gland, *Can. J. Biochem.* 48:160–163.
Gregoriadis, G., Morell, A. G., Sternlieb, I., and Scheinberg, H., 1970. Catabolism of desialylated ceruloplasmin in the liver, *J. Biol. Chem.* 245:5833–5837.
Gregory, E. M., Goscin, S. A., and Fridovich, I., 1974. Superoxide-dismutase and oxygen toxicity in a eukaryote, *J. Bacteriol.* 117:456–460.
Grenouillet, P., Martin, R.-P., Rossie, A., and Ptak, M., 1973. Interactions between copper(II) ions and L-threonine, L-*allo*-threonine and L-serine in aqueous solution, *Biochim. Biophys. Acta* 322:185–194.
Griffith, G. C., Butt, E. M., and Walker, J., 1954. The inorganic element content of certain human tissues, *Ann. Intern. Med.* 41:501–509.
Grill, E., Winnacker, E.-L., and Zenk, M. H., 1985a. Phytochelatins: The principle heavy-metal complexing peptides of higher plants, *Science* 230:674–676.
Grill, E., Zenk, M. H., and Winnacker, E.-L., 1985b. Induction of heavy-metal sequestering phytochelatin by cadmium in cell cultures of *Rauvolfia serpentina*, *Naturwissenschaften* 72:432–434.
Grill, E., Winnacker, E.-L., and Zenk, M. H., 1986a. Synthesis of seven different homologous phytochelatins in metal-exposed *Schizosaccharomyces pombe* cells, *FEBS Lett.* 197:115–120.
Grill, E., Winnacker, E.-L., and Zenk, M. H., 1987. Phytochelatins, a class of heavy-metal-binding peptides from plants, are functionally analogous to metallothioneins, *Proc. Natl. Acad. Sci* 84:439–443.
Gross, A. M., and Prohaska, J. R., 1990. Copper-deficient mice have higher cardiac norepinephrine turnover, *J. Nutr.* 120:88–96.
Grover, W. D., and Henkin, R. I., 1976. Trichopoliodystrophy (TPD): A fetal disorder of copper metabolism, *Pediatr. Res.* 10:448.
Grover, W. D., Henkin, R. I., Schwartz, M., Brodsky, N., Hobdell, E., and Stolk, J. M., 1982. A defect in catecholamine metabolism in kinky-hair disease, *Ann. Neurol.* 12:263–270.
Gubler, C. J., Lahey, M. E., Ashenbrucker, H., Cartwright, G. E., and Wintrobe, M. M., 1952a. Studies on copper metabolism. I. A method for the determination of copper in whole blood, red blood cells, and plasma, *J. Biol. Chem.* 196:209–220.
Gubler, C. J., Lahey, M. E., Chase, M. S., Cartwright, G. E., and Wintrobe, M. M., 1952b. Studies on copper metabolism III. The metabolism of iron in copper deficient swine, *Blood*, 7:1075.
Gubler, C. J., Lahey, M. E., Cartwright, C. E., and Wintrobe, M. M., 1953. Studies on copper metabolism. IX. The transportation of copper in blood, *J. Clin. Invest.* 14:224.
Gubler, C. J., Cartwright, G. E., and Wintrobe, M. M., 1957. Studies on copper metabolism. XX. Enzyme activities and iron metabolism in copper and iron deficiencies, *J. Biol. Chem.* 224:533–546.
Gullino, P. M., 1978. Angiogenesis and oncogenesis, *J. Natl. Cancer Inst.* 61:639–643.
Gupte, S. S., and Hackenbrock, C. R., 1988. The role of cytochrome c diffusion in mitochondrial electron transport, *J. Biol. Chem.* 263:5248–5253.
Gurtler, H., Wolf, H., and Heilmann, P., 1973. Konzentration and subzellulare Verteilung von Eisen and Kupfer in der Leber von Schweinen. I. Konzentration und subzellulare Verteilung in den ersten 10 Lebenstagen. *Acta Biol. Med. Ger.* (No. 4) 30:447–456.
Guss, J. M., Merritt, E. A., Phizackerley, R. P., Hedman, B., Murata, M., Hodgson, K. O.,

and Freeman, H. C., 1988. Phase determination by multiple-wavelength x-ray diffraction: crystal structure of a basic "blue" copper protein from cucumbers, *Science* 241: 806–816.
Guthans, S. L., and Morgan, W. T., 1982. The interaction of zinc, nickel and cadmium with serum albumin and histidine-rich glycoprotein assessed by equilibrium dialysis and immunoadsorbent chromatography, *Arch. Biochem. Biophys.* 218:320–328.
Gutteridge, J. M. C., 1984. Copper-phenanthroline-induced site-specific oxygen-radical damage to DNA. *Biochem. J.* 218:983–985.
Gutteridge, J. M. C., 1985. Inhibition of the Fenton reaction by the protein caeruloplasmin and other copper complexes. Assessment of ferroxidase and radical scavenging activities, *Chem.-Biol. Interact.* 56:113–120.
Gutteridge, J. M. C., 1986. Antioxidant properties of the proteins caeruloplasmin, albumin and transferrin. A study of their activity in serum and synovial fluid from patients with rheumatoid arthritis, *Biochim. Biophys. Acta* 869:119–127.
Gutteridge, S., Winter, D. B., Bruyninckx, W. J., and Mason, H. S., 1977. Localization of Cu and heme *a* on cytochrome *c* oxidase polypeptides, *Biochem. Biophys. Res. Commun.* 78:945–951.
Gutteridge, J. M. C., Richmond, R., and Halliwell, B., 1980. Oxygen free-radicals and lipid peroxidation: Inhibition by the protein caeruloplasmin, *FEBS Lett.* 112:269–272.
Gutteridge, J. M. C., Hill, C., and Blake, D. R., 1984. Copper stimulated phospholipid membrane peroxidation: Antioxidant activity of serum and synovial fluid from patients with rheumatoid arthritis, *Clin. Chim. Acta* 139:85–90.
Gutteridge, J. M. C., Winyard, P. G., Blake, D. R., Lunec, J., Brailsford, S., and Halliwell, B., 1985. The behaviour of caeruloplasmin in stored human extracellular fluids in relation to ferroxidase II activity, lipid peroxidation and phenanthroline-detectable copper, *Biochem. J.* 230:517–523.
Hagenfeldt, K., 1972a. Intrauterine contraception with the copper-T device. 1. Effect on trace elements in the endometrium, cervical mucus and plasma, *Contraception* 6:37–54.
Hagenfeldt, K., 1972b. Studies on the mode of action of copper T device, *Acta Endocrinol.* 71 (Suppl 169):1–37.
Hagenfeldt, K., Plantin, L.-O., and Diczfalusy, E., 1970. Trace elements in the human endometrium. 1. Zinc, copper, manganese, sodium and potassium concentrations at various phases of the normal menstrual cycle, *Acta Endocrinol.* 65:541–551.
Hager, L. G., and Palmiter, R. D., 1981. Transcriptional regulation of mouse liver metallothionein-1 gene by glucocorticoids, *Nature* (London) 291:340–342.
Hagen, C. J., and Evans, G. W., 1975. Absorption of trace metals in the zinc-deficient rat, *Am. J. Physiol.* 228:1020–1023.
Hahn, N., Paschen, K., and Holler, J., 1972. Das Verhalten von Kupfer, Eisen, Magnesium, Calcium und Zink bei Frauen mit normalen Menstruationscyclus, unter Einnahme von Ovulationshemmern und in der Graviditat, *Arch. Gynakol.* 213:176–186.
Hall, G. A., and Howell, J. Mc., 1969. The effect of copper deficiency on rat reproduction, *Br. J. Nutr.* 23:41–45.
Hall, A. C., Young, B. W., and Bremner, I., 1979. Intestinal metallothionein and the mutual antagonism between copper and zinc in the rat, *J. Inorg. Biochem.* 11:57–66.
Hallaway, M., Phethean, P. D., and Laggart, J., 1970. A critical study of the intracellular distribution of ascorbate oxidase and a comparison of the kinetics of the soluble and cell-wall systems, *Phytochemistry* 9:935.
Halliwell, B., and Gutteridge, J. M. C., 1985. The importance of free radicals and catalytic metal ions in human diseases, *Mol. Aspects Med.* 8:89–193.
Halliwell, B., and Gutteridge, J. M. C., 1986. Oxygen free radicals and iron in relation to

biology and medicine: Some problems and concepts, *Arch. Biochem. Biophys.* 246:501–514.
Hallman, P. S., Perrin, D. D., and Watt, A. E., 1971. The computed distribution of copper(II) and zinc(II) ions among seventeen amino acids present in human blood plasma, *Biochem. J.* 121:549–555.
Hambidge, K. M., and Droegemüeller, W., 1974. Changes in plasma and hair concentrations of zinc, copper, chromium, and manganese during pregnancy, *Obstet. Gynecol* 44:666–672.
Hamer, D. H., 1986. Metallothionein, *Annu. Rev. Biochem.* 55:913–951.
Hamer, D. H., Thiele, D. J., and Lemontt, J. E., 1985. Function and autoregulation of yeast copperthionein, *Science* 228:685–690.
Hamilton, G. A., 1981. Oxidases with monocopper reactive sites, in *Copper Proteins* (T. G. Spiro, ed.), John Wiley and Sons, New York, pp. 193–218.
Hamilton, G. A., Adolf, P. K., de Jersey, J. DuBois, G. C., Dyrkacz, G. R., and Libby, R. D., 1978. Trivalent copper, superoxide, and galactose oxidase, *J. Am. Chem Soc.* 100:1899–1912.
Hampton, J. K., Rider, L. J., Goka, T. J., and Preslock, J. P., 1972. The histaminase activity of ceruloplasmin, *Proc. Soc. Exp. Biol. Med.* 141:974–977.
Hanaichi, T., Kidokoro, R., Hayashi, H., and Sakamoto, N., 1984. Electron probe X-ray analysis on human hepatocellular lysosomes with copper deposits: Copper binding to a thiol-protein in lysosomes, *Lab. Invest.* 51:592–597.
Hanlon, D. P., and Shuman, S., 1975. Copper ion binding and enzyme inhibitory properties of the antithyroid drug methimazole, *Experientia* 31:1005–1006.
Hansen, J. S., Heegaard, P. M. H., Jensen, S. P., Norgaard-Pederson, B., and Bog-Hansen, T. C., 1988. Heterogeneity in copper and glycan content of ceruloplasmin in human serum differs in health and disease, *Electrophoresis* 9:273–278.
Hardy, K. J., and Bryan, S. E., 1975. Localization and uptake of copper into chromatin, *Toxicol. Appl. Pharmacol.* 33:62–69.
Harless, E., 1847. *Arch. Anat. Physiol.*, p. 148.
Harris, E. D., 1976. Copper-induced activation of aortic lysyl oxidase *in vivo, Proc. Natl. Acad. Sci. USA* 73:371–374.
Harris, E. D., 1979. Localization of lysyl oxidase in hen oviduct: Implications in egg shell membrane formation and composition, *Science* 208:55–56.
Harris, E. D., 1986. Biochemical defect in chick lung resulting from copper deficiency, *J. Nutr.* 116:252.
Harris, E. D., and Dameron, C. T., 1987. Incorporation of copper into CuZn SOD in culture, in *Trace Elements in Man and Animals-6* (L. Hurley, C. L. Keen, B. Lönnerdal, and R. B. Rucker, eds.), Plenum Press, London, pp. 135–138.
Harris, E. D., and DiSilvestro, R. A., 1981. Correlation of lysyl oxidation activation with the *p*-phenylenediamine oxidase activity (ceruloplasmin) in serum, *Proc. Soc. Exp. Med.* 166:528–531.
Harris, E. D., and Percival, S. S., 1988. Insights into ceruloplasmin mediated transport of copper into cells, *FASEB J.* 2:Abstract 6537.
Harris, D. I. M., and Sass-Kortsak, A., 1967. The influence of amino acids on copper uptake by rat liver slices, *J. Clin. Invest.* 46:659–667.
Harris, E. D., Gonnerman, W. A., Savage, J. E., and O'Dell, B. L., 1974. Connective tissue amine oxidase. II. Purification and partial characterization of lysyl oxidase from chick aorta, *Biochim. Biophys. Acta* 341:332–344.
Harris, E. D., Rayton, J. K., and DeGroot, J. E., 1977. A critical role for copper in aortic elastin structure and synthesis, in *Proceedings of an International Conference on Elastin and Elastic Tissue*, Plenum Press, New York, pp. 543–559.

Harris, E. D., Blount, J. E., and Leach, Jr., R. M., 1980. Localization or lysyl oxidase in hen oviduct: Implications in egg shell membrane formation and composition, *Science* 208:55–56.

Harrison, F. L., 1987. The impact of increased copper concentrations on fresh water ecosystems, personal communication.

Hart, B. A., Voss, G. W., and Garvey, J. S., 1989. Induction of pulmonary metallothionein following oxygen exposure, *Environ. Res.* 50:269–278.

Hart, E. B., Steenbock, H., Waddell, J., and Elvehjem, C. A., 1928. Iron in nutrition. VII. Copper as a supplement to iron for hemoglobin building in the rat, *J. Biol. Chem.* 77:797–812.

Hartmann, C., and Klinman, J. P., 1988. Pyrroloquinoline quinone: A new redox cofactor in eukaryotic enzymes, *Biofactors* 1:41–49.

Hartter, D. E., and Barnea, A., 1988. Brain tissue accumulates ^{67}copper by two ligand-dependent saturable processes. A high affinity, low capacity and a low affinity, high capacity process, *J. Biol. Chem.* 263:799–805.

Harvey, P. W., and Allen, K. G. D., 1981. Decreased plasma lecithin: cholesterol acyltransferase activity in copper-deficient rats, *J. Nutr.* 111:1855–1858.

Hashimoto, Y., Iijima, H., Nozaki, Y., and Shudo, K., 1986. Functional analogues of bleomycin: DNA clevage by bleomycin and hemin-intercalators, *Biochemistry* 25:5103–5110.

Hassel, C. A., Marchello, J. A., and Lei, K. Y., 1983. Impaired glucose tolerance in copper-deficient rats, *J. Nutr.* 113:1081–1083.

Hassel, C. A., Carr, T. P., Marchello, J. A., and Lei, K. Y., 1988. Apolipoprotein E-rich HDL binding to liver plasma membranes in copper-deficient rats, *Proc. Soc. Exp. Biol. Med.* 187:296–308.

Havivi, E., and Guggenheim, K., 1966. The effect of copper and fluoride on the bone of mice treated with parathyroid hormone, *J. Endocrinol.* 36:357–361.

Hayaishi, O., 1974. *Molecular Mechanisms of Oxygen Activation*, Academic Press: New York.

Hayaishi, O., Ishimura, Y., Fujisawa, H., and Nozaki, M., 1973. Prostaglandin F synthetase, a dual function enzyme, in *Oxidases and Related Redox Systems* (T. E. King, H. S. Mason and M. Morrison, eds.), University Park Press, Baltimore, pp. 577–590.

Haywood, S., 1985. Renal copper accumulation and excretion in copper overloaded rats, in *Trace Elements in Man and Animal* (TEMA-5) (C. F. Mills, I. Bremner, and J. K. Chesters, eds.) Farnham Royal, Slough, United Kingdom, pp. 190–192.

Hazelrig, J. B., Owen, C. A., Jr., and Ackerman, E., 1966. A mathematical model for copper metabolism and its relation to Wilson's disease, *Am. J. Physiol.* 211:1075–1081.

Hearing, V. J., Nicholson, J. M., Montague, P. M., Ekel, T. M., and Tomecki, K. J., 1978. Mammalian tyrosinase, structural and functional interrelationship of isozymes, *Biochim. Biophys. Acta* 522:327–339.

Hedges, J. D., and Kornegay, E. T., 1973. Interrelationship of dietary copper and iron as measured by blood parameters, tissue stores and feedlot performance of swine, *J. Anim. Sci.* 37:5:1147–1154.

Hegenauer, J., and Saltman, P., 1976. Bioavailable copper and iron in rat diets, *Am. J. Clin. Nutr.* 29:936–938.

Heikila, R. E., Cabbat, F. S., and Cohen, G., 1976. *In vivo* inhibition of superoxide dismutase in mice by diethyldithiocarbamate, *J. Biol. Chem.* 251:2182–2185.

Heilmann, P., Gurtler, H., and Wolf, H., 1975. Konzentration und subzellulare Verteilung von Eisen und Kupfer in der Leber von Schweinen. II. Untersuchungen an der Leber von Laufer und Schlachtschweinen unter besonderer Berücksichtigung eines Zusatzes von Kupfersulfat zur Futterration, *Acta Biol. Med. Ger.* 34:1589–1601.

Heimburger, N., Haupt, H., Kranz, T., and Baudner, S., 1972. Humanserumproteine mit hoher Affinität zu Carboxymethylcellulose, II, *Hoppe-Seyler's Z. Physiol. Chem.* 353:1133–1140.

Henkin, R. I., 1971. Newer aspects of copper and zinc metabolism, in *Newer Trace Elements in Nutrition* (W. Mertz and W. E. Cornatzer, eds.), Marcel Dekker, New York, p. 256.

Henkin, R. I., 1974a. Metal–albumin–amino acid interactions: Chemical and physiological interrelationships, in *Protein–Metal Interactions* (M. Friedman, ed.), Plenum Press, New York, pp. 299–328.

Henkin, R. I., 1974b. On the role of adrenocorticosteroids in the control of zinc and copper metabolism, in *Trace Element Metabolism in Animals–2* (W. G. Hoekstra, J. W. Suttie, H. E. Ganther, and W. Mertz, eds.), University Park Press, Baltimore, pp. 647–651.

Henkin, R. I., 1974c. In *Trace Element Metabolism in Animals–2* (W. G. Hoekstra, J. W. Suttie, H. E. Ganther, and W. Mertz, eds.), University Park Press, Baltimore, pp. 652–655.

Henkin, R. I., Meret, S., and Jacobs, J. B., 1969. Steroid-dependent changes in copper and zinc metabolism, *J. Clin. Invest.* 48:38a.

Henkin, R. I., Schulman, J. D., Schulman, C. B., and Bronzart, D. A., 1973. Changes in total, nondiffusible, and diffusible plasma zinc and copper during infancy, *J. Pediatr.* 82:831–837.

Herd, S. M., Camakaris, J., Christofferson, R., Wookey, P., and Danks, D. M., 1987. Uptake and efflux of copper-64 in Menkes' disease and normal continuous lymphoid cell lines, *Biochem. J.* 247:341–347.

Herlyn, M., Mancianti, M. L., Jambrosic, J., Bolen, J. B., and Koprowski, H., 1988. Regulatory factors that determine growth and phenotype of normal human melanocytes, *Exp. Cell Res.* 179:322–331.

Hernandez, O., Aznar, R., Hicks, J. J., Ballesteros, L. M., and Rosado, A., 1975. Subcellular distribution of trace metals in the normal and in the copper treated human secretory endometrium, *Contraception* 11:451–464.

Herzyk, E., Lee, D. C., Dunn, R. C., Bruckdorfer, K. R., and Chapman, D., 1987. Changes in the secondary structure of apolipoprotein B-100 after copper (2+)-catalysed oxidation of human low-density lipoproteins monitored by Fourier transform infrared spectroscopy, *Biochim. Biophys. Acta* 922:145–154.

Heydorn, K., Damsgaard, E., Horn, N., Mikkelsen, M., Tygstrup, I., Vestermark, S., and Weber, J., 1975. Extra-hepatic storage of copper. A male foetus suspected of Menkes' disease, *Humangenetik* 29:171–175.

Hien, P. V., Kovacs, K., and Matkovics, B., 1974. Properties of enzymes, I. Study of superoxide dismutase activity change in human placenta of different ages, *Enzyme* 18:341–347.

Hilewicz-Grabska, M., Zgirski, A., Krajewski, T., and Plonka, A., 1988. Purification and partial characterization of goose ceruloplasmin, *Arch. Biochem. Biophys.* 260:18–27.

Hill, C. H., 1969. The role of copper in elastin formation, *Nutr. Rev.* 27:99–100.

Hill, C. H., and Starcher, B., 1965. Effect of reducing agents on copper deficiency in the chick, *J. Nutr.* 85:271–274.

Hill, G. M., Brewer, G. J., Hogikyan, N. D., and Stellini, M. A., 1984. The effect of depot parenteral zinc on copper metabolism in the rat, *J. Nutr.* 114:2283–2291.

Hillman, L. S., 1981. Serial serum copper concentrations in premature and SGA infants during the first 3 months of life, *J. Pediatr.* 98:305–308.

Hirsch, E., Graybiel, A. M., and Agid, Y. A., 1988. Melanized dopaminergic neurons are differentially susceptible to degeneration in Parkinson's disease, *Nature* 334:345p-348p.

Hirsch, H. M., 1954. *Physiol. Plant.* 7:72–97.

Hoagland, D. R., 1932. Mineral nutrition in plants, *Annu. Rev. Biochem.* 1:618–636.

Hohnadel, D. C., Sunderman, Jr., F. W., Nechay, M. W., and McNeely, M. D., 1973. Atomic absorption spectrometry of nickel, copper, zinc and lead in sweat collected from healthy subjects during sauna bathing, *Clin. Chem.* 19:1288–1292.

Holbein, B. E., 1980. Iron-controlled infection with *Neisseria meningitis* in mice. *Infect. Immun.* 29:886–891.

Holbrook, J., Fields, M., Smith, J. C., Jr., Reiser, S., and the Los Alamos Medical Research Group, 1986. Tissue distribution and excretion of copper-67 intraperitoneally administered to rats fed fructose or starch, *J. Nutr.* 116:831–838.

Hollister, L. E., Cull, V. L., Gonda, V. A., and Kolb, F. O., 1960. Hepatolenticular degeneration. Clinical, biochemical and pathologic study of a patient with fulminant course aggravated by treatment with BAL and vesenate, *Am. J. Med.* 28:623–630.

Holmberg, C. G., and Laurell, C.-B., 1947. Investigations in serum copper. II. Isolation of the copper containing protein and a description of some of its properties, *Acta Chem. Scand.* 2:550–556.

Holmberg, C. G., and Laurell, C.-B., 1948. Investigations in serum copper. I. Nature of serum copper and its relation to the iron-binding protein in human serum, *Acta Chem. Scand.* 1:944–950.

Holme, E., Bankel, L., Lundberg, P. A., and Waldeenstroem, J., 1980. Determination of human copper-containing superoxide dismutase in biological fluids with a radioimmunoassay, *Dev. Biochem.* 11B:262–270.

Holt, D., Magos, L., and Webb, M., 1980. The interaction of cadmium induced rat renal metallothionein with bivalent mercury *in vitro*, *Chem-Biol. Interact.* 32:125–135.

Holtzman, N. A., and Gaumnitz, B. M., 1970a. Studies on the rate of release and turnover of ceruloplasmin and apoceruloplasmin in rat plasma, *J. Biol. Chem.* 245:2354–2358.

Holtzman, N. A., and Gaumnitz, B. M., 1970b. Identification of an apoceruloplasmin-like substance in the plasma of copper-deficient rats, *J. Biol. Chem.* 245:2350–2353.

Holtzman, N. A., Naughton, M. A., Iber, F. L., and Gaumnitz, B. M., 1967. Ceruloplasmin in Wilson's disease, *J. Clin. Invest.* 46:993–1002.

Hon-nami, K., and Oshima, T., 1980. Cytochrome oxidase from an extreme thermophile, *Thermus thermophilus HB8*, *Biochem. Biophys. Res. Commun.* 92:1023–1029.

Hon-nami, K., and Oshima, T., 1984. Purification and characterization of cytochrome *c* oxidase from *Thermus thermophilus HB8*, *Biochemistry* 23:454–460.

Hoogenraad, T. U., Dekker, A. W., and van den Hamer, C. J. A., 1985. Copper responsive anemia, induced by oral zinc therapy in a patient with acrodermatitis enteropathica, *Sci. Total Environ.* 42:37–43.

Horio, T., 1958. Terminal oxidation system inbacteria. I. Purification of cytochromes from *pseudomonas aeruginosa*, *J. Biochem.* (Tokyo) 45:195.

Hormel, S., Adam, E., Walsch, K. A., Beppu T., and Titani, K., 1986. The amino acid sequence of the blue copper protein of *Alcaligenes faecalis*, *FEBS Lett.* 197:301–304.

Horn, N., 1976. Copper incorporation studies on cultured cells for prenatal diagnosis of Menkes' disease, *Lancet* 1:1156–1158.

Horn, N., 1981. Menkes' X-linked disease; Prenatal diagnosis of hemizygous males and heterozygous females, *Prenat. Diag.* 1:107–121.

Horn, N., 1984. Copper metabolism in Menkes' disease, in *Metabolism of Trace Metals in Man*, Vol. II, *Genetic Implications* (O. M. Rennert and W. Y. Chan, eds.), CRC Press, Boca Raton, Florida, p. 25.

Horn, N., Heydorn, K., Damsgaard, E., Tygstrup, I., and Vestermark, S., 1978. Is Menkes syndrome a copper storage disorder? *Clin. Genet.* 14:186–187.

Howell, J. Mc., and Davison, A. N., 1959. The copper content and cytochrome oxidase activity of tissues from normal and swayback lambs, *Biochem. J.* 72:365–368.

Howell, J. Mc., and Gawthorne, J. M., (eds.), 1987. *Copper in Animals and Man*, Vol. II. CRC Press, Boca Raton, Florida.

Howell, J. Mc., and Hall, G. A., 1969. Histological observations on foetal resorption in copper-deficient rats, *Br. J. Nutr.* 23:47–52.

Howell, J. S., 1958. The effect of copper acetate on *p*-dimethylaminoazobenzene carcinogenesis in the rat, *Br. J. Cancer* 12:594–608.

Hsieh, S., and Frieden, E., 1975. Evidence for ceruloplasmin as a copper transport protein, *Biochem. Biophys. Res. Commun.* 67:1326–1331.

Huang, S. S., and Lee, D. M., 1979. A novel method for converting apolipoprotein B, the major protein moiety of human plasma low-density lipoprotein, into a water-soluble protein, *Biochim. Biophys. Acta* 577:424–441.

Huber, C. T., and Frieden, E., 1970. Copper biosystems. 34. Substrate activation and the kinetics of ferroxidase, *J. Biol. Chem.* 245: 3973–3978.

Huber, M., and Lerch, K., 1987. The influence of copper on the induction of tyrosinase and laccase in *Neurospora crassa*, *FEBS Lett.* 219:335–338.

Huber, M., and Lerch, K., 1988. Identification of two histidines as copper ligands in *Streptomyces glaucescens* tyrosinase, *Biochemistry* 27:5610–5615.

Huber, W., and Saifer, M. G. P., 1977. Orgotein, the drug version of bovine Cu-Zn superoxide dismutase. 1. A summary account of safety and pharmacology in laboratory animals, in *Superoxide and Superoxide Dismutases* (A. M. Michelson, J. M. McCord, and I. Fridovich, eds.), Academic Press, London, p. 517.

Hunsaker, H. A., Morita, M., and Allen, K. G. D., 1984. Marginal copper deficiency in rats, aortal morphology of elastin and cholesterol values in first-generation adult males, *Athersclerosis* 51:1–19.

Hunt, C. E., and Carlton, W. W., 1965. Cardiovascular lesions associated with experimental copper deficiency in the rabbit, *J. Nutr.* 87:385–393.

Hunt, C. E., Carlton, W. W., and Newberne, P. M., 1970. Interrelationships between copper deficiency and dietary ascorbic acid in the rabbit, *Br. J. Nutr.* 24:61–69.

Hunt, C. E., Landesman, J., and Newberne, P. M., 1970. Copper deficiency in chicks: Effects of ascorbic acid on iron, copper, cytochrome oxidase activity, and aortic mucopolysaccharides, *Br. J. Nutr.* 24:607.

Hunt, D. M., 1974. Primary defect in copper transport underlies mottled mutants in the mouse, *Nature* 249:852–854.

Hunt, D. M., 1980. Copper and neurological function, in *Biological Roles of Copper* (D. Evered and G. Lawrenson, eds.), Excerpta Medica, Amsterdam, p. 247–266.

Hunt, D. M., and Johnson, D. R., 1972. An inherited deficiency in noradrenaline biosynthesis in the brindled mouse, *J. Neurochem.* 19:2811–2819.

Hunt, D. M., Wake, S. A., Mercer, J. F. B., and Danks, D. M., 1986. A study of the role of metallothionein in the inherited copper toxicosis of dogs, *Biochem. J.* 236:409–415.

Husain, M., and Davidson, L., 1986. Properties of *paracoccus dinitrificans* amicyanin, *Biochemistry* 25:2431–2436.

Ichikawa, Y., and Yamano, T., 1970. Cytochrome b_5 and co-binding cytochromes in the Golgi membranes of mammalian livers, *Biochem. Biophys. Res. Commun.* 40:297–305.

Iguchi, H., and Sano, S., 1985. Cadmium- or zinc-binding to bone lysyl oxidase and copper replacement. *Connect. Tissue Res.* 14:129–139.

Iijima, Y., Fukushima, T., Bhuiyan, L. A., Yamada, T., Kosaka, F., and Sato, J. D., 1990. Synergistic and additive induction of metallothionein in Chang liver cells. A possible mechanism of marked induction of metallothionein by stress, *FEBS Lett.* 269: 218–220.

Ikemi, Y., and Ohmori, K., 1990. Change of serum ceruloplasmin level by acute ozone exposure, *Igaku to Seibutsugaku* 121:37–39 (CAS Abstr.).
Imahara, H., Wakatsuki, T., Kitamura, T., and Tanaka, H., 1978. Effect of copper on growth of yeast, *Agric. Biol. Chem.* 42:1173–1179.
Imbert, J., Zafarullah, M., Culotta, V. C., Gedamu, L., and Hamer, D., 1989. Transcription factor MBF-1 interacts with metal regulatory elements of higher eukaryotic metallothionein genes, *Mol. Cell. Biol.* 9:5315–5323.
Iodice, A. A., Richert, D. A., and Schulman, M. P., 1958. Copper content of purified δ-aminolevulinic acid dehydrase, *Fed. Proc.* 17:248 (Abstract).
Irie, S., and Tavassoli, M., 1986. Liver endothelium desialates ceruloplasmin, *Biophys. Res. Commun.* 15:94–100.
Irie, S., and Tavassoli, M., 1988. Analysis of microheterogeneities of the glycan chains of rat and human transferrin and human ceruloplasmin by lectin affinity chromatographies, *Biochem. Int.* 17:1079–1085.
Irie, S., Minguell, J. J., and Tavassoli, M., 1988. Analysis of the microheterogeneity of the glycan chain of rat transferrin, *Biochem. Int.* 17:1079–1085.
Irie, S., Minguell, J. J., and Tavassoli, M., 1989. Relationship of the glycan structure of glycoproteins to the desialylation process by rat liver endothelium, *Biochem. Int.* 19:345–352.
Ishihara, N., and Matsushiro, T., 1986. Biliary and urinary excretion of metals in humans, *Arch. Environ. Health* 41:324–330.
Ishihara, N., Yoshida, A., and Koizumi, M, 1987. Metal concentrations in human pancreatic juice, *Arch. Environ. Health* 42:356–360.
Ishimura, Y., and Hayaishi, O., 1973. Non-involvement of copper in the L-tryptophan-2,3-dioxygenase reaction, *J. Biol. Chem.* 248:8610–8612.
Ishimura, Y., and Hayaishi, O., 1976. On the prosthetic groups of L-tryptophan-2,3-dioxygenase from pseudomonas: Evidence for non-involvement of copper in the reaction, In *Advances in Experimental Medicine and Biology*, Vol. 74, (K. T. Yasunobu, H. F. Mower, and O. Hayaishi, eds.), Plenum Press, New York, pp. 363–373.
Ishimura, Y., Nozaki, M., Hayaishi, O., Nakamura, T., Tamura, M., and Yamazaki, I., 1970. The oxygenated form of L-tryptophan-2,3-dioxygenase as reaction intermediate, *J. Biol. Chem.* 245:3593–3602.
Itoh, O., Torikai, T., Satoh, M., Okumura, O., and Osawa, T., 1981. Antitumor and toxohormone-neutralizing activities of human ceruloplasmin, *Gann* 72:370–376.
Iyengar, B., and Misra, R. S., 1987. Reaction of dendritic melanocytes in vitiligo to the substrates of tyrosine metabolism, *Acta Anat.* 129:203–205.
Iyengar, V., Brewer, G. J., Dick, R. D., and Owyang, C., 1988. Studies of cholecystokinin-stimulated biliary secretions reveal a high molecular weight copper-binding substance in normal subjects that is absent in patients with Wilson's disease, *J. Lab. Clin. Med.* 111:267–274.
Jabusch, J. R., Farb, D. L., Kerschensteiner, D. A., and Deutsch, H. F., 1980. Some sulfydryl properties and primary structure of human erythrocyte superoxide dismutase, *Biochem.* 19:2310–2316.
Jackman, M. P., Sinclair-Day, J. O., Sisley, M. J., Sykes, A. G., Denys, L. A., and Wright, P. E., 1987. Kinetic studies on 1:1 electron-transfer reactions involving blue copper proteins, 15. The reactivity of *Anabaena variabilis* plastocyanin with inorganic complexes and related NMR studies, *J. Am. Chem. Soc.* 109:6443–6449.
Jackson, G. E., 1978. The action of chelating agents in the removal of copper from ceruloplasmin, *FEBS Lett.* 90:173–177.
Jacob, R. A., Skala, J. H., Omaye, S. T., and Turnland J. R., 1987. Effect of varying ascorbic-

acid intakes on copper absorption and ceruloplasmin levels in young men. *J. Nutr.* 117:2109–2115.

Jacobs, A., 1977. Low molecular weight intracellular iron transport compounds, *Blood* 50:433–439.

Jahroudi, N., Foster, R., Price-Haughey, J., Beitel, G., and Gedamu, L., 1990. Cell-type specific and differential regulation of the human metallothionein genes. Correlation with DNA methylation and chromatin structure, *J. Biol. Chem.* 265–6506–6511.

Jain, S. K., and Williams, D. M., 1988. Copper deficiency anemia: Altered red blood cell lipids and viscosity in rats, *Am. J. Clin. Nutr.* 48:637–640.

Jamison, M. H., Sharma, H., Case, R. M., and Braganza, J. M., 1981. Pancreatic secretions assist bile in limiting copper absorption in the rat, *Gut* 22:A866–A867.

Jamison, M. H., Sharma, H., Braganza, J. M., and Case, R. M., 1983. The influence of pancreatic juice on ^{64}Cu absorption in the rat, *Br. J. Nutr.* 50:113–119.

Janes, S. M., Mu, D., Wemmer, D., Smith, A. J., Kaur, S. Maltby, D., Burlingame, A. L., and Klinman, J. P., 1990. A new redox cofactor in eukaryotic enzymes: 6-hydroxy dopa at the active site of bovine serum amine oxidase, *Science* 248:981–987.

Janssens, A. R., and van den Hamer, C. J. A., 1982. Kinetics of ^{64}copper in primary biliary cirrhosis, *Hepatology* 2:822–827.

Jarausch, J., and Kadenbach, B., 1985. Structure of the cytochrome c oxidase complex of rat liver. 2. Topological orientation of polypeptides in the membrane as studied by proteolytic digestion and immunoblotting, *Eur. J. Biochem.* 146:219–225.

Jarvis, S. C., 1984. The effects of nitrogen supply on the absorption and distribution of copper in red clover (*Trifolium ptratense L.*) grown in flowing solution culture with a low, maintained concentration of copper, *Ann. Bot.* 53:153.

Jecht, E. W., and Bernstein, G. S., 1973. Influence of copper on the motility of human spermatozoa, *Contraception* 7:381–340.

Jenkinson, S. G., Lawrence, R. A., Burk, R. F., and Williams, D. M., 1982. Effects of copper deficiency on the activity of the selenoenzyme glutathione peroxidase and on excretion and tissue retention of 75SeO2-, *J. Nutr.* 112:197–204.

Jenkinson, S. G., Lawrence, R. A., Grafton, W. D., Gregory, P. E., and McKinney, M. A., 1984. Enhanced pulmonary toxicity in copper-deficient rats exposed to hyperoxia, *Fund. Appl. Toxicol.* 4:170–177.

Jeunet, F., Richterich, R., and Aebi, H., 1962. Bile et céruloplasmine; étude *in vitro* à l'aide de la perfusion du foie de rat isolé, *J. Physiol. Paris* 54:729–737.

Joel, D., Marbach, I., and Mayer, A. M., 1978. Laccase in anacardiaceae, *Phytochemistry* 17:796.

Johansen, J. T., Overballe-Petersen, C., Martin, B., Hasemann, V., and Svendsen, I., 1979. The complete amino acid sequence of copper, zinc superoxide dismutase from *Saccharomyces cerevisiae*, *Carlsberg Res. Commun.* 44:201–217.

Johnson, C. A., 1976. The determination of some toxis metals in human liver as a guide to normal levels in New Zealand. Part 1. Determination of Bi, Cd, Cr, Co, Cu, Pb, Mn, Ni, Ag, Ti, and Zn, *Anal. Chim. Acta* 81:69–74.

Johnson, D. A., Osaki, S., and Frieden, E., 1967. A micromethod for the determination of ferroxidase (ceruloplasmin) in human serum, *Clin. Chem.* 13:142–150.

Johnson, G. F., Morell, A. G., Stockert, R. J., and Sternlieb, I., 1981. Hepatic lysosomal copper protein in dogs with an inherited copper toxicosis, *Hepatology* 2:243–248.

Johnson, N. C., Kheim, T., and Kountz, W. B., 1959. Influence of sex hormone on total serum copper, *Proc. Soc. Exp. Biol. Med.* 102:98–99.

Johnson, P. E., 1982. A mass spectrometric method for use of stable isotopes as tracers in studies of iron, zinc, and copper absorption in human subjects, *J. Nutr.* 112:1414–1424.

Johnson, P. E., 1989. Factors affecting copper absorption in humans and animals, *Adv. Exp. Med. Biol.* 258:71–79.

Johnson, P. E., and Canfield, W. K., 1989. Stable zinc and copper absorption in free-living infants fed breast milk or formula, *J. Trace Elem. Exp. Med.* 2:285–295.

Johnson, W. T., and Evans, G. W., 1980. Age-dependent variation of copper in tissue and proteins of neonatal rat small intestine, *Proc. Soc. Exp. Biol. Med.* 165:496–501.

Johnson, W. T., and Nordlie, R. C., 1977. Differential effects of Cu^{2+} on carbomoyl phosphate: glucose phosphotransferase and glucose-6-phosphate phosphohydrolase activities of multifunctional glucose-6-phosphatase, *Biochemistry* 16:2458–2466.

Johnson, W. T., and Saari, J. T., 1990. Dietary supplementation with t-butylhydroquinone reduces cardiac hypertrophy and anemia associated with copper deficiency in rats, *Nutr. Res.* 9:1355–1662.

Johnson, W. T., Nordlie, R. C., and Klevay, L. M., 1984. Glucose-6-phosphatase activity in copper-deficient rats, *Biol. Trace Elem. Res.* 6:369–378.

Jones, D. G., 1984. Effects of dietary copper depletion on acute and delayed inflammatory responses in mice, *Res. Vet. Sci.* 37:205–210.

Jones, D. G., and Suttle, N. R., 1981. Some effects of copper deficiency on leucocyte function in sheep and cattle, *Res. Vet. Sci.* 31:151–156.

Jones, D. G., and Suttle, N. R., 1983. The effect of copper deficiency on the resistance of mice to infection with *Pasteurella haemolytica*, *J. Comp. Pathol.* 93:143–149.

Judd, E. S., Dry, T. J., and Bledsoe, M. S., 1935. The significance of iron and copper in the bile of man, *J. Lab. Clin. Med.* 20:609–615.

Julshamn, K., Utne, F., and Braekkan, O. R., 1977. Interactions of cadmium with copper, zinc and iron on different organs and tissues of the rat, *Acta Pharmacol. Toxicol.* 41:515–524.

Julshamn, K., Andersen, K., Ringdal, O., and Brenna, J., 1988. Effect of dietary copper on the hepatic concentration and subcellular distribution of copper and zinc in the rainbow trout, *Aquaculture* 73:143–155.

Juurlink, B. H. J., 1988. Lysosomal accumulation of neuromelanin-like material in astrocytes after exposure to norepinephrine, *J. Neuropathol. Exp. Neurol.* 47:48–53.

Kaar, K., Rhen, K., and Tarkkila, T., 1984. Short-term effects of desogestrel and ethinyloestradiol on serum proteins in women, *Scand. J. Clin. Lab. Invest.* 44:623–627.

Kadenbach, B., Ungibauer, M., Jarausch, J., Buge, U., and Kuhn-Nentwig, L., 1983a. The complexity of respiratory complexes, *Trends Biochem. Sci.* 8:398–400.

Kadenbach, B., Jarausch, J., Hartmann, R., and Merle, P., 1983b. Separation of mammalian cytochrome c oxidase into 13 polypeptides by a sodium dodecyl sulfate-gel electrophoretic procedure, *Anal. Biochem.* 129:517–521.

Kadenbach, B., Kuhn-Nentwig, L., and Buge, U., 1987. Evolution of a regulatory enzyme: Cytochrome-c-oxidase (complex IV), *Curr. Top. Bioenerg.* 15:113–161.

Kadenbach, B., Reimann, A., Stroh, A., and Huther, F. J., 1988. Evolution of cytochrome c oxidase, in *Oxidases and Related Redox Systems* (T. E. King, H. S. Mason, and M. Morrison, eds.), Alan R. Liss, New York, pp. 653–668.

Kagan, H. M., 1986. Lysyl oxidase, in *Biology of Extracellular Matrix* (R. P. Mecham, ed.), Academic Press, New York, pp. 322–398.

Kagan, H. M., Sullivan, K. A., Olsson, T. A., III, and Cronlund, A. L., 1979. Purification and properties of four species of lysyl oxidase from bovine aorta, *Biochem. J.* 177:203–214.

Kagan, H. M., Tseng, L., Trackman, P. C., Okamoto, K., Rapaka, S., and Urry, D. W., 1980. Repeat polypeptide models of elastin as substrates for lysyl oxidase, *J. Biol. Chem.* 255:3656–3659.

Kagan, H. M., Williams, M. A., Williamson, P. R., and Anderson, J. M., 1984. Influence of sequence and charge on the specificity of lysyl oxidase toward protein and synthetic peptide substrates, *J. Biol. Chem.* 259:11203–11207.

Kagan, H. M., Vaccaro, C. A., Bronson, R. E., Tang, S. S., and Brody, J. S., 1986. Ultrastructural immunolocalization of lysyl oxidase in vascular connective tissue, *J. Cell. Biol.* 103:1121–1128.

Kageyama, G. H., and Wong-Riley, M. T. T., 1982. Histochemical localization of cytochrome oxidase in the hippocampus: Correlation with specific neuronal types and afferent pathways, *Neuroscience* 7:2337–2361.

Kagi, J. H. R., and Kojima, Y. (eds.), 1987. *Metallothionein II*, Birkhauser Verlag, Basel.

Kagi, J. H. R., and Nordberg, M. (eds.), 1979. *Metallothionein*, Birkhauser Verlag, Basel.

Kagi, J. H. R., Vasak, M., Lerch, K., Gilg, D. E. O., Hunziker, P., Bernard, W. R., and Good, M., 1984. Structure of mammalian metallothionein, *Environ. Health Perspect.* 54:93–103.

Kahn, M. I., Yasmin, R., Fahmy, M. A., Riaz-Ul-Haq, Zia-Ur-Rehman, and Shakoori, A. R., 1988. Studies on cell survival and induction of metallothionein in yeast, *Pak. J. Zool.* 20:275–281.

Kanagy, C., Vanderkooi, J. M., and Bonner, W. D. Jr., 1988. Luminescence from the carbon monoxide derivative of *Agaricus bispora* tyrosinase, *Arch. Biochem. Biophys.* 267:668–675.

Kang, H. K., Harvey, P. W., Valentine, J. L., and Swendseid, M. E., 1977. Zinc, iron, copper, and magnesium concentrations in tissues of rats fed various amounts of zinc, *Clin. Chem.* 23:1834–1837.

Karabelas, D. S., 1972. Copper metabolism in the adjuvant-induced arthritic rat, Ph.D. thesis, New York University, *Diss. Abstr. Int.* B 33:2776.

Karcioglu, Z. A., Sarper, R. M., Van Rinsvelt, H. A., Guffey, J. A., and Fink, R. W., 1978. Trace element concentrations in renal cell carcinoma, *Cancer* 42:1330–1340.

Karin, M., 1985. Metallothioneins: Proteins in search of functions, *Cell* 41:9–10.

Karin, M., and Richards, R. I., 1982. Human metallothionein genes—primary structure of the metallothionein II gene and related processed gene, *Nature* 299:797–802.

Karin, M., Najarian, R., Haslinger, A., Valenzuela, P., Welch, J., and Fogel, S., 1984. Primary structure and transcription of an amplified genetic locus: the *CUP1* locus of yeast, *Proc. Natl. Acad. Sci. USA* 81:337–341.

Karlsson, K., and Marklund, S. L., 1987. Heparin-induced release of extracellular superoxide dismutase to human blood plasma, *Biochem. J.* 242:55–59.

Karp, B. I., Roboz, M., and Linder, M. C., 1986. Regulation of ceruloplasmin and copper turnover, by estrogens and tumors, in the rat, *J. Nutr. Growth. Cancer* 3:47–55.

Karpel, J. T., and Peden, V. H., 1972. Copper deficiency in long term parenteral nutrition, *J. Pediatr.* 80:32–36.

Kasper, C. B., and Deutsch, H. F., 1963a. Physicochemical studies of human ceruloplasmin, *J. Biol. Chem.* 238:2325–2337.

Kasper, C. B., and Deutsch, H. F., 1963b. Immunochemical studies of crystalline human ceruloplasmin and derivatives, *J. Biol. Chem.* 238:2343–2350.

Kataoka, M., and Tavassoli, M., 1985. Identification of ceruloplasmin receptors on the surface of human blood monocytes, granulocytes, and lymphocytes, *Exp. Hematol.* 13:806–810.

Katoh, S., 1960. A new copper protein from *Chlorella ellipsoidea*, *Nature* (London) 186:533.

Katoh, S., and Takamiya, A., 1961. A new leaf copper protein "plastocyanin", a natural Hill oxidant, *Nature* 189:665.

Katz, B. M., and Barnea, A., 1990. The ligand specificity for uptake of complexed copper-67 by brain hypothalamic tissue is a function of copper concentration and copper: ligand molar ratio, *J. Biol. Chem.* 265:2017–2021.

Kayali, A., and Berthon, G., 1980. Le histamine comme ligand dans le plasma sanguin. Détermination des constantes d'équilibre de ses complexes avec Co(II), Ni(II), Cu(II), Ag(II), Cd(II) et simulation quantitative de leur distribution à divers taux plasmatiques significatifs de troubles cliniques, *J. Chim. Phys.* 77:333–341.

Keen, C. L., and Hurley, L. S., 1979. Developmental patterns of copper and zinc concentrations in mouse liver and brain: Evidence that the gene crinkled (cr) is associated with an abnormality in copper metabolism, *J. Inorg. Biochem.* 11:269–277.

Keen, C. L., and Hurley, L. S., 1980. Developmental changes in concentrations of iron, copper, and zinc in mouse tissues, *Mech. Ageing Dev.* 13:161–176.

Keen, C. L., Lonnerdal, B., Sloan, M. V., and Hurley, L. S., 1980. Effect of dietary iron, copper and zinc chelates of nitrilotriacetic acid (NTA) on trace metal concentrations in rat milk and maternal and pup tissues, *J. Nutr.* 110:897–906.

Keen, C. L., Lonnerdal, B., and Fisher, G. L., 1981. Age-related variations in hepatic iron, copper, zinc and selenium concentrations in beagles, *Am. J. Vet. Res.* 42:1884–1887.

Kelly, W. A., Kesterson, J. W., and Carlton, W. W., 1974. Myocardial lesions in the offspring of female rats fed a copper deficient diet, *Exp. Med. Pathol.* 20:40–56.

Kelsay, J. L., Jacob, R. A., and Prathers, E. S., 1979. Effect of fiber from fruits and vegetables on metabolic responses of human subject. III. Zinc, copper and phosphorus balances, *Am. J. Clin. Nutr.* 32:2307–2311.

Kensler, T. W., Bush, D. M., and Kozumbo, W. J., 1983. Inhibition of tumor promotion by a biomimetic superoxide dismutase, *Science* 221:77–77.

Kern, S. R., Smith, H. A., Fontaine, D., and Bryan, S. E., 1981. Partitioning of zinc and copper in fetal liver subfractions: Appearance of metallothionein-like proteins during development, *Toxicol. Appl. Pharmacol.* 59:346–354.

Kessel, D., and Allen, J., 1975. Elevated plasma sialyltransferase in the cancer patient, *Cancer Res.* 35:670–672.

Kesseru, E., and Leon, F., 1974. Effect of different solid metals and metallic pairs on human sperm motility, *Int. J. Fert.* 19:81–84.

Ketcheson, M. R., Barron, G. P., and Cox, D. H., 1969. Relationship of maternal dietary zinc during gestation and lactation to development and zinc, iron and copper content of the postnatal rat, *J. Nutr.* 98:303–311.

Keyhani, E., and Keyhani, J., 1980. Identification of porphyrin present in apo-cytochrome c oxidase of copper-deficient yeast cells, *Biochim. Biophys. Acta,* 633:221–227.

Killgore, J., Schmidt, C., Duich, L., Romero-Chapman, N., Tinker, D., Reiser, K., Melko, M., Hyde, D., and Rucker, R. B., 1989. Nutritional importance of pyrroloquinoline quinone, *Science* 245:850–852.

Kiiholma, P., Gronoroos, M., Erkkola, R., Pakarinen, P., and Nanto, V., 1984a. The role of calcium, copper, iron and zinc in preterm delivery and premature rupture of fetal membranes, *Gynecol. Obstet. Invest.* 17:194–201.

Kiiholma, P., Gronoroos, M., Liukko, P., Pakarinen, P., Hyora, H., and Erkkola, R., 1984b. Maternal serum copper and zinc concentrations in normal and small-for-date pregnancies, *Gynecol. Obstet. Invest.* 18:212–216.

Kimber, C. L., Mukherjee, T., and Deller, D. J., 1973. In vitro iron attachment to the intestinal brush border. Effect of iron stores and other environmental factors, *Am. J. Dig. Dis.* 18:781–791.

Kimoto, E., Tanaka, H., Gyotoku, J., Morishage, F., and Pauling, L., 1982. Enhancement of antitumor activity of ascorbate against Ehrlich ascites tumor cells by the copper: glycylglycylhistidine complex, *Cancer Res.* 43:824.

King, J. C., and Wright, A. L., 1985. In *Trace Elements in Man and Animals* (C. F. Mills, I. Bremner, and J. K. Chesters, eds.), Commonwealth Agric. Bureaux, Farnham Royal, U.K., p. 318.

King, J. C., Reynolds, W. L., and Margen, S., 1978. Absorption of stable isotopes of iron, copper, and zinc during oral contraceptive use, *Am. J. Clin. Nutr.* 31:897–906.

King, T. E., Mason, H. S., and Morrison, M. (eds.), 1988. *Oxidases and Related Redox Systems*, Alan R. Liss, New York.

Kingston, I. B., Kingston, B. L., and Putnam, F. W., 1977. Chemical evidence that proteolytic cleavage causes the heterogeneity present in human ceruloplasmin preparations, *Proc. Natl. Acad. Sci. USA* 74:5377–5381.

Kirchgessner, M., and Grassman, E., 1970. The dynamics of copper absorption, in *Trace Element Metabolism in Animals* (C. F. Mills, ed.), Livingstone, Edinburgh, pp. 277–286.

Kirchgessner, M., Spoerl, R., and Schneider, U. A., 1978. Studies on the super-retention of trace elements (Cu, Zn, Mn, Ni, Fe) during gravidity, in *Trace Element Metabolism in Man and Animals—3* (M. Kirchgessner, ed.), Arbeitskreis für Tierernahrungsforschung Weihenstephan, Freising, Germany, pp. 440–443.

Kishore, V., Latman, N., Roberts, D. W., Barnett, J. B., and Sorenson, J. R. J., 1984. Effect of nutritional copper deficiency on adjuvant arthritis and immunocompetence in the rat, *Agents Actions* 14:274–282.

Kishore, V., Wokocha, B., and Fourcade, L., 1990. Effect of nutritional copper deficiency on carrageenin edema in the rat: A quantitative study, *Biol. Trace Elem. Res.* 23:97–107.

Kitano, S., 1980. Membrane and contractile properties of rat vascular tissue in copper-deficient conditions, *Circ. Res.* 46:681.

Kizer, J. S., Bateman, R. C., Jr., Miller, C. R., Humm, J., Busby, W. H., Jr., and Youngblood, W. W., 1986. Purification and characterization of a peptidyl glycine monooxygenase from porcine pituitary, *Endocrinology* 118:2262–2267.

Klaassen, C. D., 1973. Effect of alteration in body temperature on the biliary excretion of copper, *Proc. Soc. Exp. Biol. Med.* 144:8–12.

Kleyay, L. M., 1973. Hypercholesterolemia in rats by an increase in the ratio of zinc to copper ingested, *Am. J. Clin. Nutr.* 26:1060–1068.

Klevay, L. M., 1977. The role of copper and zinc in cholesterol metabolism, in *Advances in Nutritional Research*, Vol. 1 (H. H. Draper, ed.), Plenum Press, New York, pp. 227–252.

Klevay, L. M., 1981. Interactions of copper and zinc in cardiovascular disease, *Ann. N.Y. Acad. Sci.* 355:140.

Klevay, L. M., 1982. Metabolic interactions among cholesterol, cholic acid and copper, *Nutr. Rept. Intl.* 26:405–414.

Klevay, L. M., 1983. Copper and ischemic heart disease, *Biol. Trace Elem. Res.* 5:245.

Klevay, L. M., 1985. Atrial thrombosis, abnormal electrocardiograms and sudden death in mice due to copper deficiency, *Atherosclerosis*, 54:213–224.

Klevay, L. M., 1986. Hypertension in rats deficient in copper, *Fed. Proc.* (Abstr.) 45:235.

Klevay, L. M., Inman, L., Johnson, L. K., Lawler, M., Mahalko, J. R., Milne, D. B., Lukaski, H. C., Bolonchuk, W., and Sandstead, H. H., 1984b. Increased cholesterol in plasma in a young man during experimental copper depletion, *Metabolism* 33:1112–1118.

Klevay, L. M., Canfield, W. K., Gallagher, S. K., Henriksen, L. K., Lukaski, H. C., Bolonchuk, W., Johnson, L. K., Milne, D. B., and Sanstead, H. H., 1986. Decreased glucose tolerance in two men during experimental copper depletion, *Nutr. Rep. Intl.* 33:371–382.

Klinman, J. P., and Brenner, M., 1988. Role of copper and catalytic mechanism in the copper monooxygenase, dopamine β-hydroxylase (DBH), in *Oxidases and Related Redox Systems* (T. E. King, H. S. Mason, and M. Morrison, eds.), Alan R. Liss, New York, pp. 227–248.

Klinman, J. P., Krueger, M., and Edmondson, D. E., 1984. Evidence for two copper atoms/subunit in dopamine β-monooxygenase catalysis, *J. Biol. Chem.* 259:3399–3402.

Kloeckener-Gruissem, B., McEwen, J. E., and Poyton, R. O., 1988. Identification of a third nuclear protein-coding gene required specifically for post-transcriptional expression of the mitochondrial COX3 gene in *Saccharomyces cerevisiae*, *J. Bacteriol.* 170:1399–1402.

Klug, D., Rabani, J., and Fridovich, I., 1972. A direct demonstration of the catalytic action of superoxide dismutase through the use of pulse radiolysis, *J. Biol. Chem.* 247:4839–4842.

Klug-Roth, D., Fridovich, I., and Rabani, J., 1973. Pulse radiolytic investigations of superoxide catalyzed disproportionation mechanism for bovine superoxide dismutase, *J. Am. Chem. Soc.* 95:2786–2790.

Klyavinya, D. R., Ozolinya, G. R., and Tauchius, B. A., 1982. Accumulation of copper-containing protein in leaves with increase in the concentration of copper in them, *Sov. Plant Physiol.* 30:731.

Kobayashi, T., and Kobayashi, T., 1965. A study of mechanisms by which copper salts induced ovulation in the rabbit, *Endocrinol. Jap.* 12:277–288.

Kobayashi, S., Imano, M., and Kimura, M., 1982. Turnover of metallothionein in mammalian cells, in *Biological Roles of Metallothionein* (E. C. Foulkes, ed.), Elsevier North Holland, Amsterdam, pp. 305–322.

Kobayashi, Y., Asada, M., and Osawa, T., 1984. Mechanism of phorbol myristate acetate-induced lymphotoxin production by a human T cell hybridoma, *J. Biochem.* 95:1775–1782.

Koenig, W., Richter, F. W., Meinel, B., and Bode, J. C., 1979. Variation of trace element contents in a single human liver, *J. Clin. Chem. Clin. Biochem.* 17:23–27.

Koide, T., Odani, S., and Ono, T., 1982. The N-terminal sequence of human plasma histidine-rich glycoprotein homologous to antithrombin with high affinity for heparin, *FEBS Lett.* 141:222–224.

Koide, T., Foster, D., Yoshitake, S., and Davie, E. W., 1986. Amino acid sequence of human histidine-rich glycoprotein derived from the nucleotide sequence of its cDNA, *Biochemistry* 25:2220–2225.

Kolberg, J., Michaelsen, T. E., and Jantzen, E., 1983. Interactions of purified human ceruloplasmin with *Lathyrus odoratus, Lens culinaris*, and *Canavalia ensiformis* lectins, *Hoppe-Seyler's Z. Physiol. Chem.* 364:111–117.

Koller, L. D., Mulhern, S. A., Frankel, N. C., Steven, M. G., and Williams, J. R., 1987. Immune dysfunction in rats fed a diet deficient in copper, *Am. J. Clin. Nutr.* 45:997–1006.

Koo, S. I., and Lee, C. C., 1989. Copper deficiency and altered cholesterol metabolism, presentation at ACS Symposium on copper, *Bioavailability and metabolism*, April 9–14, Dallas, Texas.

Koo, S. I., Lee, C. C., and Norvell, J. E., 1988. Effect of copper deficiency on the lymphatic absorption of cholesterol, plasma chylomicron clearance, and postheparin lipase activities, *Proc. Soc. Ecp. Biol. Med.* 188:410–419.

Kopp, S. J., Klevay, L. M., and Feliksik, J. M., 1983. Physiological and metabolic characterization of a cardiomyopathy induced by chronic copper deficiency, *Am. J. Physiol.* 245H:855.

Korner, A. M., and Pawelek, J., 1980. Dopachrome conversion: A possible control point in melanin biosynthesis, *J. Invest. Dermatol* 75:192–195.

Korte, J. J., and Prohaska, J. R., 1987. Dietary copper deficiency alters protein and lipid composition of murine lymphocyte plasma membranes, *J. Nutr.* 117:1076–1084.

Koschinsky, M. L., 1988. Characterization of the human ceruloplasmin cDNA and gene, *Diss. Abstr. Int. B* 50:65.

Koschinsky, M. L., Funk, W. D., van Oost, B. A., and MacGillivray, R. T. A. 1986. Complete cDNA sequence of human preceruloplasmin, *Proc. Natl. Acad. Sci. USA* 83:5086–5090.

Koschinsky, M. L., Chow, B. K. C., Schwartz, J., Hamerton, J. L., and MacGillivray, R. T. A., 1987. Isolation and characterization of a processed gene for human ceruloplasmin. *Biochemistry* 26:7760–7767.

Kosman, D. J., and Driscoll, J. J., 1988. Solvent exchangeable protons and the activation of molecular oxygen: The galactose oxidase reaction, in *Oxidases and Related Redox Systems* (T. E. King, H. S. Mason and M. Morrison, eds.), Alan R. Liss, New York, pp. 251–267.

Kosman, D. J., Ettinger, M. J., Weiner, R., and Massaro, E. J., 1974. The molecular properties of the copper enzyme galactose oxidase, *Arch. Biochem. Biophys.* 165:456.

Kramer, T. R., Johnson, W. T., and Briske-Anderson, M., 1988. Influence of iron and the sex of rats on hematological, biochemical and immunological changes during copper deficiency, *J. Nutr.* 118:214–221.

Kressner, M. S., Stockert, R. J., Morell, A. G., and Sternlieb, I., 1984. Origins of biliary copper, *Hepatology* 4:867–870.

Kuivaniemi, H., 1985. Partial characterization of lysyl oxidase from several human tissues, *Biochem. J.* 230:639–643.

Kuivaniemi, H., Savolainen, E.-R., and Kivirikko, K. I., 1984. Human placental lysyl oxidase: Purification, partial characterization, and preparation of two specific antisera to the enzyme, *J. Biol. Chem.* 259:6996–7002.

Kuivaniemi, H., Korhonen, R.-M., Vaheri, A., and Kivirikko, K. I., 1986. Deficient production of lysyl oxidase in cultures of malignantly transformed human cells, *FEBS Lett.* 261–264.

Kumagai, H., and Yamada, H., 1985. Bacterial and fungal amine oxidases, in *Structure and Functions of Amine Oxidases* (B. Mondovi, ed). Boca Raton, Florida, CRC Press, pp. 37–49.

Kumar, C., Naqui, A., and Chance, B., 1984a. The identity of pulsed cytochrome oxidase, *J. Biol. Chem.* 259:2073–2076.

Kumar, C., Naqui, A., and Chance, B., 1984b. Peroxide interaction with pulsed cytochrome oxidase. Optical and EPR studies, *J. Biol. Chem.* 259:11668–11671.

Kunapuli, S. P., Verde, A., Riley, L. A., and Kumar, A., 1986. Isolation of cDNA clone for human ceruloplasmin, *Fed. Proc.* 45:944 (Abstract No. 3411).

Kwon, B. S., Haq, A. K., Pomerantz, S. H., and Halaban, R., 1987. Isolation and sequence of a cDNA clone for human tyrosinase that maps at the mouse c albina locus, *Proc. Natl. Acad. Sci. USA*, 84:7473–7477.

Kwon, B. S., Haq, A. K., Pomerantz, S. H., and Halaban, R., 1988. Cloning and characterization of a human tyrosinase cDNA, *Progr. Clin. Biol. Res.* 256:273–282.

Laemmli, U. K., Lebkowski, J. S., and Lewis, C. D., 1981. Evidence for a structural role of copper in histone-depleted chromosomes and nuclei, *ICN-UCLA Symp. Mol. Cell. Biol.* 23:275–292.

Lahey, M. E., Gubler, C. J., Chase, M. S., Cartwright, G. E., and Wintrobe, M. M., 1952. Studies on copper metabolism II. Hematologic manifestations of copper deficiency in swine, *Blood* 7:1053–1074.

Lahey, M. E., Gubler, C. J., Cartwright, G. E., and Wintrobe, M. M., 1953. Studies on copper metabolism. VII. Blood copper in pregnancy and various pathological states, *J. Clin. Invest.* 32:329–339.

Lai, J. C. K., Chan, A. W. K., Minski, M. J., and Lim, L., 1985. Roles of metal ions in brain development and aging, in *Metal Ions in Neurology and Psychiatry* (S. Gabay, J. Harris, and B. T. Ho, eds.), Alan R. Liss, New York, pp. 49–67.

Lal, S., and Sourkes, T. L., 1971a. Intracellular distribution of copper in the liver during chronic administration of copper sulfate to the rat, *Toxicol. Appl. Pharmacol.* 18:562–572.

Lal, S., and Sourkes, T. L., 1971b. Deposition of copper in rat tissues. The effect of dose and duration of administration of copper sulfate, *Toxicol. Appl. Pharm.* 20:269–283.

Lampi, K. J., Mathias, M. M., Rengers, B. D., and Allen, K. G. D., 1988. Dietary copper and copper dependent superoxide dismutase in hepatic prostaglandin synthesis by rat liver homogenates, *Nutr. Res.* 8:1191–1202.

Langmyhr, F. J., Eyde, B., and Jonsen, J., 1979. Determination of the total content and distribution of cadmium, copper and zinc in human parotid salvia, *Anal. Chim. Acta* 107:211–218.

Lapin, C. A., Morrow, G., III, Chvapil, M., Belke, D. P., and Fisher, R. S., 1976. Hepatic trace elements in sudden death syndrome, *J. Pediatr.* 89:607–608.

LaRusso, N. F., Dickson, E. R., and Pineda, A. A., 1980. Plasmaperfusion and studies of copper metabolism in benign recurrent cholestasis; novel procedures in a rare syndrome, *Mayo Clin. Proc.* 55:450–454.

Lastowski-Perry, D., Otto, E., and Maroni, G., 1985. Nucleotide sequence and expression of a *Drosophila* metallothionein, *J. Biol. Chem.* 260:1527–1530.

Lau, B. W. C., and Klevay, L. M., 1981. Plasma lecithin:cholesterol actyltransferase in copper-deficient rats, *J. Nutr.* 111:1698–1703.

Lau, B. W. C., and Klevay, L. M., 1982. Postheparin plasma lipoprotein lipase in copper-deficient rats, *J. Nutr.* 112:928–933.

Lau, S., and Sarkar, B., 1971. Ternary coordination complex between human serum albumin, copper(II) and L-histidine, *J. Biol. Chem.* 246:5938–5943.

Lau, S., and Sarkar, B., 1981a. The interaction of copper(II) and glycyl-L-histidyl-lysine, a growth modulating tripeptide from plasma, *Biochem. J.* 199:649–656.

Lau, S., and Sarkar, B., 1981b. A critical examination of the interaction between copper(II) and glycylglycyl-L-histidine, *J. Chem. Soc., Dalton Trans.* 1981:491–494.

Lau, S., and Sarkar, B., 1984. Comparative studies of manganese(II)-, nickel(II)-, zinc(II)-, copper(II)-, cadmium(II)-, and iron(III)-binding components in human cord and adult sera, *Can. J. Biochem. Cell. Biol.* 62:449–455.

Lau, M. L., Strickland, G. T., and Yeh, S. J., 1971. Tissue copper, zinc, and manganese levels in Wilson's disease: Studies with the use of neutron activation analysis, *J. Lab. Clin. Med.* 77:438–444.

Lau, S. J., Laussac, J. P., and Sarkar, B., 1989. Synthesis and copper(II)-binding properties of the N-terminal peptide of human alpha-fetoprotein, *Biochem. J.* 257:745–750.

Laurie, S. H., and Pratt, D. E., 1986. Copper-albumin: What is its functional role?, *Biochem. Biophys. Res. Commun*, 135:1064.

Laurin, D. E., and Klasing, K. C., 1990. Roles of synthesis and degradation in the regulation of metallothionein accretion in a chicken macrophage cell line, *Biochem. J.* 268:459–463.

Laussac, J.-P., and Sarkar, B., 1984. Characterization of the copper(II)- and nickel(II)-transport site of human serum albumin. Studies of copper(II) and nickel(II) binding to peptide 1-24 of human serum albumin by ^{13}C and ^{1}H NMR spectroscopy, *Biochemistry* 23:2832–2838.

Lawrence, G. D., and Sawyer, D. T., 1979. Potentiometric titrations and oxidation–reduction potentials of manganese and copper zinc superoxide dismutases, *Biochemistry* 18:3045–3050.

Lee, G. R., Cartwright, G. E., and Wintrobe, M. M., 1968a. Heme biosynthesis in copper deficient swine, *Proc. Soc. Exp. Biol. Med.* 127:977.

Lee, G. R., Nacht, S., Lukens, J. N., and Cartwright, G. E., 1968b. Iron metabolism in copper deficient swine, *J. Clin. Invest.* 47:2058.

Lee, G. R., Williams, D. M., and Cartwright, G. E., 1976. Role of copper in iron metabolism and heme biosynthesis, In *Trace Elements in Human Health and Disease*, Vol. 1, *Zinc and Copper* (A. S. Prasad and D. Oberleas, eds.), Academic Press, New York, pp. 373–393.

Lee, R. E., and Lands, W. E. M., 1972. Cofactors in biosynthesis of prostaglandins F1-alpha and F2-alpha, *Biochim. Biophys. Acta* 260:203–211.

Lee, R. E., and Lands, W. E. M., 1973. Cofactors in the biosynthesis of prostaglandin $F_{1\alpha}$ and $F_{2\alpha}$, *Biochim. Biophys. Acta* 260:203–211.

Lee, S., Lancey, R., Montaser, A., Madani, N., and Linder, M. C., 1991. Transfer of copper from mother to fetus during the latter part of gestation in the rat. Submitted.

Lefevre, M., Keen, C. L., Lonnerdal, B., Hurley, L. S., and Schneeman, B. O., 1985. Different effects of zinc and copper deficiency on composition of plasma high density lipoproteins in rats, *J. Nutr.* 115:359–368.

Le Guilly, Y., Launois, B., Lenoir, P., and Bourel, M., 1973. Production of serum proteins by subcultures of adult human liver, *Biomed. Express* 19:361–364.

Lehmann, H. P., Schosinsky, K. H., and Beeler, M. F., 1974. Standardization of serum ceruloplasmin concentrations in international enzyme units with *o*-dianisidine dihydrochloride as substrate, *Clin. Chem.* 20:1564–1567.

Lei, K. Y., 1977. Cholesterol metabolism in copper-deficient rats, *Nutr. Rep. Int.* 15:597–605.

Lei, K. Y., Prasad, A. S., Bowersox, E., and Oberleas, D., 1976. Oral contraceptives, norethindrone and mestranol: Effects on tissue levels of minerals, *Oral Contracept. Min. Metab.* 231:98–103.

Leibetseder, J., Kment, A., Skalicky, M., Haider, I., and Helali, I. A., 1977. Copper metabolism in pigs, *Nutr. Metab.* 21 (Suppl. 1):211–214.

Leone, A., Pavlakis, G. N., and Hamer, D. H., 1985. Menkes' disease: Abnormal metallothionein gene regulation in response to copper, *Cell* 40:301–309.

Lerch, K., 1976. Neurospora tyrosinase: Molecular weight, Cu content and spectral properties, *FEBS Lett.* 69:157–160.

Lerch, K., 1980. Copper metallothionein, a copper binding protein from *Neurospora crassa*, *Nature* 284:368–370.

Lerch, K., 1981. Copper monooxygenases: Tyrosinase and dopamine β-monooxygenase, in *Metal Ions in Biological Systems*. Vol. 13, *Copper Proteins* (H. Sigel, ed.), Marcel Dekker, New York, pp. 144–180.

Lerch, K., and Germann, U. A., 1988. Evolutionary relationships among copper proteins containing coupled binuclear copper sites, in *Oxidases and Related Redox Systems* (T. E. King, H. S. Mason, and M. Morrison, eds.), Alan R. Liss, New York, pp. 331–348.

Lerch, K., and Huber, M., 1986. Different origins of metal binding sites in binuclear copper proteins, tyrosinase and hemocyanin, *J. Inorg. Biochem.* 26:213–217.

Lerch, K., Deinum, J., and Reinhammar, B., 1978. The state of copper in neurospora laccase, *Biochim. Biophys. Acta* 534:7–14.

Lerch, K., Ammer, D., and Olafson, R. W., 1982. Crab metallothionein, primary structures of metallothioneins 1 and 2, *J. Biol. Chem.* 257:2420–2426.

Lesna, M., 1987. Presentation at *Symposium on Metabolism of Minerals and Trace Elements in Human Diseases*, New Delhi, Aligarh, and Srinagar, India, September, 1987.

Letendre, E. D., and Holbein, B. E., 1984. Ceruloplasmin and regulation of transferrin iron during *Neisseria meningitidis* infection in mice, *Infect. Immunol* 45:133–138.

Leu, M. L., Strickland, G. T., and Yeh, S. J., 1971. Tissue copper, zinc, and manganese levels in Wilson's disease: Studies with the use of neutron activation analysis, *J. Lab. Clin. Med.* 77:438–444.

Leung, L. L. K., Harpel, P. C., Nachman, R. L., and Rabellino, E. M., 1983. Histidine-rich glycoprotein is present in human platelets and is released following thrombin stimulation, *Blood* 62:1016–1021.

Leung, M. K. M., Fessler, L. I., Greenberg, D. B., and Fessler, J. H., 1979. Separate amino and carboxyl procollagen peptidases in thick embryo tendon, *J. Biol. Chem.* 254:224–232.

Leuthauser, S. W. C., Oberley, L. W., Oberley, T. D., Sorenson, J., and Ramakrishna, K., 1981. Antitumor effect of a copper-coordination compound with superoxide dismutase-like activity, *J. Natl. Cancer Inst.* 66:1077–1081.

Levanon, D., Lieman-Hurwitz, J., Dafni, N., Wigderson, M., Sherman, L., Banstein, Y., Laver-Rudich, Z., Daniger, E., Stein, O., and Groner, Y., 1985. Architecture and anatomy of the chromosomal locus in human chromosome 21 encoding the Cu/Zn superoxide dismutase, *EMBO J.* 4:77–84.

Leverton, R. M., and Binkley, E. S., 1944. The copper metabolism and requirement of young women, *J. Nutr.* 27:43–53.

Levinson, W., Oppermann, H., and Jackson, J., 1980. Transition series metals and sulfhydryl reagents induce the synthesis of four proteins in eukaryotic cells, *Biochim. Biophys. Acta* 606:170–180.

Lewis, C. D., and Laemmli, U. K., 1982. Higher order metaphase chromosome structure: Evidence for metalloprotein interactions, *Cell* 29:171–181.

Lewis, C. G., Fields, M., Craft, N., Yang, C.-Y., and Reiser, S., 1987. Alteration of pancreatic enzyme activities in small intestine of rats fed a high fructose, low-copper diet, *J. Nutr.* 117:1447–1452.

Lewis, C. G., Fields, M., Craft, N., Yang, C.-Y., and Reiser, S., 1988. Changes in pancreatic enzyme specific activities of rats fed a high-fructose, low-copper diet, *J. Am. Coll. Nutr.* 7:27–34.

Lewis, T. L., Hunt, L. T., and Barker, W. C., 1989. Striking sequence similarity among sialic acid-binding lectin, pancreatic ribonucleases, and angiogenin: Possible structural and functional relationships, *Protein Seq. Data Anal.* 2:101–105.

Lewis, C. G., Fields, M., and Beal, T., 1990. Effect of changing the type of dietary carbohydrate or copper level of copper-deficient, fructose-fed rats on tissue sorbitol concentrations, *J. Nutr. Biochem.* 1:160–166.

Lewis, J. A., and Tata, J. R., 1973. A rapidly sedimenting fraction of rat liver endoplasmic reticulum, *J. Cell. Sci.* 13:447–459.

Lewis, K. O., 1973. The nature of the copper complexes in bile and their relationship to the absorption and excretion of copper in normal subjects and in Wilson's disease, *Gut* 14:221–232.

Li, P. M., Gelles, J., Chan, S. I., Sullivan, R. J., and Scott, R. A., 1987. Extended X-ray absorption fine structure of copper in Cu_A-depleted, p-(hydroxymercuri)benzoate-modified, and native cytochrome c oxidase, *Biochemistry* 26:2091–2095.

Li, T.-Y., Kraker, A. J., Shaw, C. F., III, and Petering, D. H., 1980. Ligand substitution reactions of metallothioneins with EDTA and apo-carbonic anhydrase, *Proc. Natl. Acad. Sci. USA* 77:6334–6338.

Lijima, Y., Fukushima, T., Bhulyan, L. A., Yamada, T., Kosaka, F., and Sato, J. D., 1990. Synergistic and additive induction of metallothionein in Chang liver cells. A possible mechanism of marked induction of metallothionein by stress, *FEBS Lett.* 269:218–220.

Lijnen, H. R., Hoylaerts, M., and Collen, D., 1980. Isolation and characterization of a human plasma protein with affinity for the lysine binding sites in plasminogen, *J. Biol. Chem.* 255:10214–10222.

Lin, C. M., and Kosman, D. J., 1990. Copper uptake in wild type and copper metallothionein-deficient *Saccharomyces cerevisiae*. Kinetics and mechanism, *J. Biol. Chem.* 265:9194–2000.

Lin, I. M., and Lei, K. Y., 1981. Cholesterol kinetic analyses in copper-deficient rats, *J. Nutr.* 111:450–457.

Lin, P. S., Kwock, L., Ciborowski, L., and Butterfield, C., 1978. Sensitization effects of diethyldithiocarbamate, *Radiat. Res.* 74:515.

Lin, P. S., Kwock, L., and Goodchild, N. T., 1980. Copper chelator enhancement of bleomycin cytotoxicity, *Cancer* 46:2360–2364.
Linder, M. C., 1978a. Functions and metabolism of trace elements, in *Perinatal Physiology* (U. Stave, ed.), Plenum Medical Book Co., New York, pp. 425–454.
Linder, M. C., 1978b. Iron and copper metabolism in cancer, as exemplified by changes in ferritin and ceruloplasmin in rats with transplantable tumors, *Adv. Exp. Med. Biol.* 92:643–664.
Linder, M. C., 1983. Changes in the distribution and metabolism of copper in cancer, *J. Nutr. Growth Cancer* 1:27–38.
Linder, M. C., 1985. *Nutritional Biochemistry and Metabolism*, Elsevier, New York.
Linder, M. C., 1985a. Human nutrition in context, in *Nutritional Biochemistry and Metabolism* (M. C. Linder, ed), Elsevier, New York, Chap. 1.
Linder, M. C., 1985b. Nutrition and metabolism of trace elements, in *Nutritional Biochemistry and Metabolism* (M. C. Linder, ed.), Elsevier, New York, Chap. 7.
Linder, M. C., 1985c. Nutrition and metabolism of vitamins, in *Nutritional Biochemistry and Metabolism*, (M. C. Linder, ed.), Elsevier, New York, Chap. 5.
Linder, M. C., 1991. *Nutritional Biochemistry and Metabolism*, 2nd ed., Elsevier, New York.
Linder, M. C., 1991a. Nutrition and metabolism of trace elements, in *Nutritional Biochemistry and Metabolism*, 2nd ed. (M. C. Linder, ed.), Elsevier, New York, Chap. 7.
Linder, M. C., 1991b. Nutrition and metabolism of fats, in *Nutritional Biochemistry and Metabolism*, 2nd ed. (M. C. Linder, ed.), Elsevier, New York, Chap. 3.
Linder, M. C., and Moor, J. R., 1977. Plasma ceruloplasmin: Evidence for its presence in and uptake by the heart and other organs of the rat, *Biochim. Biophys. Acta* 499:329–336.
Linder, M. C., and Moor, J. R., 1978. In *Proceedings of the Third International Symposium on Detection and Prevention of Cancer* (H. E. Nieburgs, ed.), Marcel Dekker, New York, pp. 191–207.
Linder, M. C., and Munro, H. N., 1973. Iron and copper metabolism during development, *Enzyme* 15:111–138.
Linder, M. C., and Munro, H. N., 1974. Iron and copper metabolism in development, in *Biochemical Bases of Development of Physiological Function* (O. Greengard, ed.), *Enzyme* 15:111–138.
Linder, M. C., and Munro, H. N., 1977. The mechanism of iron absorption and its regulation, *Fed. Proc.* 36:2017–2023.
Linder, M. C., and Munro, H. N., 1978. Ferritin: Structure, biosynthesis, and role in iron metabolism, *Physiol. Rev.* 58:317–396.
Linder, M. C., Bryant, R. B., Lim, S., Scott, L. E., and Moor, J. R., 1979a. Ceruloplasmin elevation and synthesis in rats with transplantable tumors, *Enzyme* 24:85–95.
Linder, M. C., Houle, P. A., Isaacs, E., Moor, J. R., and Scott, L. E., 1979b. Copper regulation of ceruloplasmin in copper deficient rats, *Enzyme* 24:23–35.
Linder, M. C., Moor, J. R., and Wright, K., 1981. Ceruloplasmin assays in diagnosis and treatment of human lung, breast and gastrointestinal cancer, *J. Natl. Cancer Inst.* 67:263–275.
Linder, M. C., Weiss, K. C., and Wirth, P. L., 1985. Copper transport within the mammalian organism, in *Trace Elements in Man and Animals* (TEMA-5) (C. F. Mills, I. Bremner, and J. C. Chesters, eds.) Commonwealth Agricultural Bureau, Farnham Royal, Slough, UK, pp. 323–328.
Linder, M. C., Roboz, M., and the Los Alamos Medical Radioisotope Research Group, 1986. Turnover and excretion of copper in rats as measured with ^{67}Cu, *Am. J. Physiol.* 251:E551–E555.
Linder, M. C., Weiss, K. C., and Vu, H. M., 1987a. Structure and function of transcuprein in

transport of copper by mammalian blood plasma, Proceedings of the 6th International Conference on Trace Elements in Man and Animals (TEMA-6), Asilomar, California.
Linder, M. C., and Goode, C. A., 1988. Evidence for transfer of copper from ceruloplasmin to the plasma membrane of rat brain cells during copper uptake. *FASEB J.* 2:Abstract 6538.
Linder, M. C., Goode, C. A., Weiss, K. C., Wirth, P.-L., and Vu, M. H., 1989. Mammalian copper transport: Review and recent findings, in *Metabolism of Minerals and Trace Elements in Human Disease* (M. Abdulla, H. Dashti, B. Sarkar, H. Al-Sayer, and N. Al-Naqeeb, eds.), Smith-Gordon, Nishimura, London, pp. 219–229.
Linder, M. C., Lee, S. H., Lancey, R. W., and Madani, N., 1990. Transport of copper to the rat fetus on ceruloplasmin, albumin, and transcuprein, in *Trace Elements in Man and Animals* (TEMA-7) (B. Momcilovic, ed.), in press.
Lipman, C. B., and Mackinney, J., 1931. *Plant Physiol.* 6:593.
Lipsky, P. E., 1981. Modulation of human antibody production *in vitro* by penicillamine and $CuSO_4$, *J. Rheumatol.* 8 (Suppl. 7):69–73.
Lipsky, P. E., 1982. Modulation of lymphocyte functions by copper and thiols, in *Inflammatory Disease and Copper*, J. R. J. Sorenson (ed.), Human Press, Clifton, New Jersey, p. 581–592.
Lipsky, P. E., 1984. Immunosuppression by D-penicillamine *in vitro*. *J. Clin. Invest.* 73:56.
Lipsky, P. E., and Ziff, M., 1980. Inhibition of human helper T cell function *in vitro* by D-penicillamine and $CuSO_4$, *J. Clin. Invest.* 65:1069–1076.
Liu, C. C. F., and Medeiros, D. M., 1986. Excess diet copper increases systolic blood pressure in rats, *Biol. Trace Element Res.* 9:15–24.
Liukko, P., Erkkola, R., and Bergink, E. W., 1988. Progestagen-dependent effect on some plasma proteins during oral contraception, *Gynecol. Obstet. Invest.* 25:118–122.
Locke, J., Boase, D. R., and Smalldon, K. W., 1979. The quantitative multi-element analysis of human liver tissue by spark-source mass spectrometry, *Anal. Chim. Acta* 104:233–244.
Loeffler, S., Hochberger, A., Grill, E., Winnacker, E. L., and Zenk, M. H., 1989. Termination of the phytochelatin synthase reaction through sequestration of heavy metals by the reaction product, *FEBS Lett.* 258:42–46.
Lolkema, P. C., Donker, M. H., Schouten, A. J., and Ernst, W. H. O., 1984. The possible role of metallothioneins in copper tolerance of *Silence cucubalus*, *Planta* 162:174.
Loneragan, J. F., 1981. Distribution and movement of copper in plants, in *Copper in Soils and Plants* (J. F. Loneragan, A. D. Robson, and R. D. Graham, eds), Academic Press, Sydney, Chapter 8.
Long, G., Zhou, J., and Zhang, W., 1989. Serum transferrin and ceruloplasmin determination during normal pregnancy, *Hunan Yixue* 6:323–324 (CAS Abstr.).
Lonnerdal, B., Keen, C. L., and Hurley, L. S., 1981. Iron, copper, zinc, and manganese in milk, *Annu. Rev. Nutr.* 1:149–174.
Lorentz, K., and Jaspers, G., 1970. Untersuchungen zur bilaren Elimination von Enzymen. I. Verhalten von Coeruloplasmin, Acylcholinhydrolase, Benzoylcholinhydrolase, Alkalischer Phosphatase, Ornithin-carbamyl-transferase, Glutamathydrogenase and Glucose-6-phosphat-dehydrogenase, *Klin. Wochenschr.* 48:215–218.
Loschen, G., Azzi, A., Richter, C., and Flobe. L., 1974. Superoxide radicals as precursors of mitochondrial hydrogen peroxide, *FEBS Lett.* 42:68–72.
Lovstad, R. A., 1978. Thesis, University of Oslo, pp. 5–56.
Lovstad, R. A., 1981. The protective action of ceruloplasmin on Fe^{2+}-stimulated lysis of rat erythrocytes, *Int. J. Biochem.* 13:221–224.
Lovstad, R. A., 1982. The protective action of ceruloplasmin on copper ion stimulated lysis of rat erythrocytes, *Int. J. Biochem.* 14:585–589.

Lovstad, R. A., 1983. Iron ion induced haemolysis: Effect of caeruloplasmin, albumin and ascorbate (vitamin C), *Int. J. Biochem.* 15:1067–1071.

Lovstad, F. A., 1986. Hemin-induced lysis of rat erythrocytes. Protective action of ceruloplasmin and different serum albumins, *Int. J. Biochem.* 18:171–173.

Lown, J. W., and Sim, S., 1977. The mechanism of the bleomycin-induced cleavage of DNA, *Biochem. Biophys. Res. Commun.* 77:1150–1157.

Lu, S., Wa, W., Lin, Y., and Pu, C., 1987. Measurement of zinc and copper concentration in the amniotic fluid, *Shengzhi Yu Biyun* 7:72–73 (CAS Abstr.).

Lucky, A. W., and Hsia, Y. E., 1979. Distribution of ingested and injected radiocopper in two patients with Menkes' kinky hair disease, *Pediatr. Res.* 13:1280.

Lukasewycz, O. A., and Prohaska, J. R., 1982. Immunization against transplantable leukemia impaired in copper-deficient mice, *J. Natl. Cancer. Inst.* 69:489–493.

Lukasewycz, O. A., and Prohaska, J. R., 1983. Lymphocytes from copper-deficient mice exhibit decreased mitogen reactivity, *Nutr. Res.* 3:335–341.

Lukasewycz, O. A., Prohaska, J. R., Meyer, S. G., Schmidtke, J. R., Hatfield, S. M., and Marder, P., 1985. Alterations in lymphocyte subpopulations in copper-deficient mice, *Infect. Immunol.* 48:644–647.

Lukasewycz, O. A., Kolquist, K. L., and Prohaska, J. R., 1987. Splenocytes from copper deficient mice are low responders and weak stimulators in mixed lymphocyte reactions, *Nutr. Res.* 7:43–52.

Lukasewycz, O. A., and Prohaska, J. R., 1990. The immune response in copper deficiency, *Ann. N.Y. Acad. Sci.* (*Micronutr. Immune Funct./Cytokines Metab.*) 587:147–159.

Lunec, J., Wickens, D. G., Graff, T. L., and Dormandy, T. L., 1982. Copper, free radicals and rheumatoid arthritis, in *Inflammatory Diseases and Copper* (J. R. J. Sorenson, ed.), Humana Press, Clifton, New Jersey, pp. 231–240.

Lykins, L. F., Akey, C. W., Christian, E. G., Duval, G. E., and Topham, R. W., 1977. Dissociation and reconstitution of human ferroxidase II, *Biochemistry* 16:693–698.

Mack, U., Cooksley, W. G. C., Ferris, R. A., Halliday, J. W., and Powell, L. W., 1981. Regulation of plasma ferritin by the isolated perfused rat liver, *Br. J. Haematol.* 47:403–412.

Mackellar, W. C., and Crane, F. L., 1982. Iron and copper in plasma membranes, *J. Bioenerg. Biomembr.* 14:241–247.

Mackiewicz, A., 1989. Studies of mechanisms regulating glycosylation of acute phase proteins. Studies in vitro, *Immunol. Pol.* 14:103–127 (CAS Abstr.).

Mackiewicz, A., Ganapathi, M. K., Schultz, D., and Kushner, I., 1987. Monokines regulate glycosylation of acute-phase proteins, *J. Exp. Med.* 166:253–258.

Mackiewicz, A., Ganapathi, M. K., Schultz, D., Samols, D., Reese, J., and Kushner, I., 1988. Regulation of rabbit acute phase protein biosynthesis by monokines, *Biochem. J.* 253:851–857.

Madapallimatam, G., and Riordan, J. R., 1977. Antibodies to the low molecular weight copper binding protein from liver, *Biochem. Biophys. Res. Commun.* 77:1286–1293.

Maddox, I. S., 1973. The role of copper in prostaglandin synthesis, *Biochim. Biophys. Acta* 306:74–81.

Maeshima, M., and Asahi, T., 1978. Purification and characterization of sweet potato cytochrome *c* oxidase, *Arch. Biochem. Biophys.* 187:423.

Magdoff-Fairchild, B. S., Lovell, F. M., and Low, B. W., 1969. X-ray crystallographic study of ceruloplasmin Determination of molecular weight, *J. Biol. Chem.* 244:3497–3499.

Magnus-Levy, A., 1910. *Biochem. Z.* 24:363.

Mahoney, J. P., Bush, J. A., Gubler, C. J., Moretz, W. H., Cartwright, G. E., and Wintrobe, M. M., 1955. Studies on copper metabolism. XV. The excretion of copper by animals, *J. Clin. Invest.* 46:702–708.

Makino, R., and Ishimura, Y., 1976. Negligible amount of copper in hepatic L-tryptophan 2,3-dioxygenase, *J. Biol. Chem.* 251:7722–7725.

Malatesta, F., Antonini, G., Sarti, P., Vallone, B., and Brunori, M., 1990. Structure and function of cytochrome *c* oxidase, *Gazz. Chim. Ital.* 120:475–484.

Malinowska, A., 1986. Changes in the content of zinc, copper, and ceruloplasmin in biological fluids and tissues of sows and their fetuses during pregnancy, *Med. Weter.* 42:368–372.

Mallet, B., and Aquaron, R., 1983. Isolation and purification of ceruloplasmin in oculocutaneous albinism, Menkes' disease, Wilson's disease and pregnant women, *Clin. Chim. Acta* 132:245–256.

Mallette, L. E., and Henkin, R. I., 1976. Altered copper and zinc metabolism in primary hyperparathyroidism, *Am. J. Med. Sci.* 272:167–174.

Malmstroem, B. G., 1990. Cytochrome oxidase: Some unresolved problems and controversial issues, *Arch. Biochem. Biophys.* 280:233–241.

Mann, J. R., Camakaris, J., and Danks, D. M., 1979. Copper metabolism in mottled mouse mutants, distribution of ^{64}Cu in brindled (Mo^{br}) mice, *Biochem. J.* 180:613–619.

Mann, J. R., Camakaris, J., and Danks, D. M., 1980. Copper metabolism in mottled mutants. Defective placental transfer of ^{64}Cu to foetal brindled (Mo^{br}) mice, *Biochem. J.* 186:629–631.

Mann, J. R., Camakaris, J., Francis, N., and Danks, D. M., 1981. Copper metabolism in mottled mouse mutants: Studies of blotchy (Mo^{blo}) mice and comparison with brindled (Mo^{br}) mice, *Biochem. J.* 196:81–89.

Mann, K. G., Lawler, C. M., Vehar, G. A., and Church, W. R., 1984. Coagulation factor V contains copper ions, *J. Biol. Chem.* 259:12949–12951.

Mann, K. G., Nesheim, M. E., and Tracy, P. B., 1986. Nonenzymic cofactors: factor V, *New Comp. Biochem.* 13:15–34.

Mann, T., and Keilin, D., 1938. Haemocuprein and hepatocuprein, copper-protein compounds of blood and liver in mammals, *Proc. R. Soc. London, Ser. B.* 126:303–315.

Maquart, F. X., Pickart, L., Laurent, M., Gillery, P., Monboisse, J. C., and Borel, J. P., 1988. Stimulation of collagen synthesis in fibroblast cultures by the tripeptide-copper complex glycyl-L-histidyl-L-lysine-copper(2+), *FEBS Lett.* 238:343–346.

Marceau, N., Aspin, N., 1972. Distribution of ceruloplasmin-bound ^{67}Cu in the rat, *Am. J. Physiol.* 222:106–110.

Marceau, N., and Aspin, N., 1973. The intracellular distribution of the radiocopper derived from ceruloplasmin and from albumin, *Biochim. Biophys. Acta* 328:338–350.

Marceau, N., Aspin, N., and Sass-Kortsak, A., 1970. Absorption of copper 64 from gastro-intestinal tract of the rat, *Am. J. Physiol.* 218:377–383.

Margoshes, M., and Vallee, B. L., 1957. A cadmium protein from equine kidney cortex, *J. Am. Chem. Soc.* 79:4813–4814.

Markkanen, T., and Aho, A. J., 1972. Metabolic aspects of trace metal content of gallstones and gallbladders, *Acta Chim. Scand.* 138:301–305.

Marklund, S. L., 1982. Human copper-containing superoxide dismutase of high molecular weight, *Proc. Natl. Acad. Sci. USA* 79:7634–7638.

Marklund, S. L., 1984a. Extracellular superoxide dismutase and other superoxide dismutase isoenzymes in tissues from nine mammalian species, *Biochem. J.* 222:649–655.

Marklund, S. L., 1984b. Properties of extracellular superoxide dismutase from human lung, *Biochem. J.* 220:269–272.

Marklund, S. L., 1984c. Extracellular superoxide dismutase in human tissues and human cell lines, *J. Clin. Invest.* 74:1398–1403.

Marklund, S. L., 1984d. Extracellular superoxide dismutase and other superoxide dismutase isozymes in tissues from nine mammalian species, *Biochem. J.* 222:649–655.

Marklund, S. L., 1985. Product of extracellular-superoxide dismutase catalysis, *FEBS Lett.* 184:237–239.

Marklund, S. L., 1986. Ceruloplasmin, extracellular-superoxide dismutase, and scavenging of superoxide anion radicals, *J. Free Radicals Biol. Med.* 2:255–260.

Marklund, S. L., Holme, E., and Hellner, L., 1982. Superoxide dismutase in extracellular fluids, *Clin. Chim. Acta* 126:41–51.

Marmor, M. F., Nelson, J. W., and Levin, A. S., 1978. Copper metabolism in American retinitis pigmentosa patients, *Br. J. Ophthalmol.* 62:168–171.

Marshall, L. E., Graham, D. R., Reich, K. A., and Sigman, D. S., 1981. Cleavage of DNA by 1,10-phenanthroline-cuprous complex. Hydrogen peroxide requirement and primary and secondary structure specificity, *Biochemistry* 20:244.

Martin, M. T., Licklider, K. F., Jacobs, F. A., and Brushmiller, J. G., 1981. Detection of low molecular weight copper(II) and zinc(II) binding ligands in ultrafiltered milks—the citrate connection, *J. Inorg. Biochem.* 15:55–65.

Martin, M. T., Jacobs, F. A., and Brushmiller, J. G., 1984. Identification of copper-and zinc-binding ligands in human and bovine milk, *J. Nutr.* 114:869–879.

Martin, M. T., Jacobs, F. A., and Brushmiller, J. G., 1986. Low molecular weight copper-binding ligands in human bile, *Proc. Soc. Exp. Biol. Med.* 181:249–255.

Martinez, J. H., Solano, F., Arocas, A., Garcia-Borron, J. C., Iborra, J. L., and Lozano, J. A., 1987. The existence of apotyrosinase in the cytosol of Harding–Passey mouse melanoma melanocytes and characteristics of enzyme reconstitution by Cu(II), *Biochim. Biophys. Acta* 923:413–420.

Marzullo, G., and Hine, B., 1980. Opiate receptor function may be modulated through an oxidation–reduction mechanism, *Science* 208:1171–1173.

Mas, A., and Sarkar, B., 1988. The metabolism of metals in rat placenta, *Biol. Trace Elem. Res.* 18:191–199.

Maslinski, C., Bieganski, T., Fogel, W. A., and Kitler, M. E., 1985. Diamine oxidase in developing tissues, in *Structure and Functions of Amine Oxidases* (B. Mondovi, ed.), CRC Press, Boca Raton, Florida, pp. 153–166.

Mason, H. S., North, J. C., and Vanneste, M., 1965. Microsomal mixed-function oxidations: The metabolism of xenobiotics, *Fed. Proc.* 24:1172–1180.

Mason, J., 1982. The putative role of thiomolybdates in the pathogenesis of Mo-induced hypocupraemia and molybdenosis: Some recent developments, *Ir. Vet. J.* 36:164.

Mason, K. E., 1979. A conspectus of research on copper metabolism and requirements of man, *J. Nutr.* 109:1979–2066.

Mason, R., Brady, F. O., and Webb, M., 1981. Metabolism of zinc and copper in the neonate: Accumulation of Cu in the gastrointestinal tract of the newborn rat, *Br. J. Nutr.* 45:391–399.

Masters, D. G., Keen, C. L., Lonnerdal, B., and Hurley, L. S., 1983. Comparative aspects of dietary copper and zinc deficiencies in pregnant rats, *J. Nutr.* 113:1448–1451.

Mathew, B. M., Salahuddin, M., Ahmad, M., Kumar, S., Seth, T. D., Mahdi, S. Q., and Jamil, S. A., 1978. Temporal profile of changes in myocardial copper after isoproterenol induced cardiac necrosis, *Jpn. Circ. J.* 42:695–699.

Matsuba, Y., and Takahashi, Y., 1970. Spectrophotometric determination of copper with N, N, N', N'-tetraethylthiuram disulfide and an application of this method for studies of subcellular distribution of copper in rat brain, *Anal. Biochem.* 36:182–191.

Matsuda, I., Pearson, T., and Holtzman, N. A., 1974. Determination of apoceruloplasmin by radioimmunoassay in nutritional copper deficiency, Menkes' kinky hair syndrome, Wilson's disease and umbilical cord blood, *Pediat. Res.* 8:821–824.

May, P. M., Linder, P. W., and Williams, D. R., 1976. Ambivalent effect of protein binding

on computed distributions of metal ions complexed by ligands in blood plasma, *Experientia* 32:1492–1494.
May, P. M., Linder, P. W., and Williams, D. R., 1977. Computer simulation of metal-ion equilibria in biofluids: Models for the low-molecular-weight complex distribution of calcium(II), magnesium(II), manganese(II), iron(III), copper(II), zinc(II), and lead(II) ions in human blood plasma, *J. Chem. Soc.* 1977:588–594.
Mayer, A. M., and Harel, E., 1968. Interconversion of subunits of catechol oxidase from apple chloroplasts *Phytochemistry* 7:199–204.
Mayer, A. M., and Harel, E., 1979. Polyphenol oxidases in plants, *Phytochemistry* 18:193–215.
Mayo, K. E., and Palmiter, R. D., 1982. Glucocorticoid regulation of the mouse metallothionein 1 gene is selectively lost following amplification of the gene, *J. Biol. Chem.* 257:3061–3067.
McArdle, H. J., and Erlich, R., 1990. Copper uptake and transfer to the mouse during pregnancy, *J. Nutr.* (in press).
McArdle, H. J., Guthrie, J. R., Ackland, M. L., and Danks, D. M., 1987a. Albumin has no role in the uptake of copper by human fibroblasts, *J. Inorg. Biochem.* 31:123–131.
McArdle, H. J., Gross, S. M., and Danks, D. M., 1988a. The uptake of copper by mouse hepatocytes, *J. Cell. Physiol.* 136:373–378.
McArdle, H. J., Gross, S. M., and Danks, D. M., 1988b. The role of albumin in copper uptake by hepatocytes and fibroblasts, in *Trace Elements in Man and Animals* (TEMA-6) (L. S. Hurley, C. L. Keen, B. Lonnerdal, and R. B. Rucker, eds.), Plenum Press, New York, pp. 139–162.
McArdle, H. S., Gross, S. M., Creaser, I., Sargeson, A. M., and Danks, D. M., 1989a. Effect of chelators on copper metabolism and copper pools in mouse hepatocytes, *Am. J. Physiol.* 256:G667–672.
McArdle, H. J., Gross, S. M., Vogel, H. M., Ackland, M. L., and Danks, D. M., 1989b. The effect of tetrathiomolybdate on the metabolism of copper by hepatocytes and fibroblasts, *Biol. Trace Elem. Res.* 22:179–188.
McArdle, H. J., Gross, S. M., Danks, D. M., and Wedd, A. G., 1990a. Role of albumin's copper binding site in copper uptake by mouse hepatocytes, *Am. J. Physiol.* 258:G988–G991.
McArdle, H. J., Kyriakou, P., Grimes, A., Mercer, J. F. B., and Danks, D. M., 1990b. The effect of D-penicillamine on metallothionein mRNA levels and copper distribution in mouse hepatocytes, *Chem.-Biol. Interact.* 75:315–324.
McAuslan, B. R., Hannan, G. N., Reilly, W., Whittaker, R. G., and Florence, M., 1980. Reappraisal of evidence for the role of copper in angiogenesis. The nature of an active copper complex, in *CSIRO Symposium on the Importance of Copper in Biology and Medicine* (B. R. McAuslan, ed.), Commonwealth Scientific and Industrial Research Organisation, Canberra, pp. 42–49.
McCall, J. T., and Davis, G. K., 1961. Effect of dietary protein and zinc on the absorption and liver deposition of radioactive and total copper, *J. Nutr.* 74:45–50.
McCall, J. T., Goldstein, N. P., and Smith, L. H., 1971. Implications of trace metals in human diseases, *Fed. Proc.* 30:1011–1015.
McCord, J. M., 1974. Free radicals and inflammation: Protection of synovial fluid by superoxide dismutase, *Science* 185:529–531.
McCord, J. M., and Fridovich, I., 1969. Superoxide dismutase—an enzymic function for erythrocuprein (hemocuprein), *J. Biol. Chem.* 244:6049–6055.
McCracken, J., Desai, P. R., Papadopoulos, N. J., Villafranca, J. J., and Peisach, J., 1988a. Electron spin-echo studies of the copper(II) binding sites in dopamine β-hydroxylase, *Biochemistry* 27:4133–4137.

McCracken, J., Pember, S., Benkovic, S. J., Villafranca, J. J., Miller, R. J., and Peisach, J., 1988b. Electron spin-echo studies of the copper binding site in phenylalanine hydroxylase from *Chromobacterium violaceum*, *J. Am. Chem. Soc.* 110:1069–1074.

McGinnis, J., Sinclair-Day, J. D., and Sykes, A. G., 1985. Active site protonations of blue copper proteins, in *Biological and Inorganic Copper Chemistry* (K. D. Karlin and J. Zubieta, eds.), Adenine Press, Guilderland, New York, pp. 11–22.

McHargue, J. S., 1925. The occurrence of copper, manganese, zinc, nickel, and cobalt in soils, plants, and animals, and their possible function as vital factors, *J. Agric. Res.* 30:193–196.

McHargue, J. S., 1926. Mineral constituents of the cotton plant, *J. Am. Soc. Agronomy.* 18:1076–1083.

McHargue, J. S., 1927a. The proportion and significance of copper, iron, manganese and zinc in some mollusks and crustaceans, *Trans. Ky. Acad. Sci.* 2:46–52.

McHargue, J. S., 1927b. Significance of the occurrence of manganese, copper, zinc, nickel, and cobalt in Kentucky bue grass, *Indust. Eng. Chem.* 19:274–276.

McKee, D. J., and Frieden, E., 1971. Binding of transition metal ions by ceruloplasmin (ferroxidase), *Biochemistry* 10:3880–3883.

McKeel, D. W., and Jarett, L., 1970. Preparation and characterization of a plasma membrane fraction from isolated fat cells, *J. Cell. Biol.* 44:417–425.

McNatt, E. N., Campbell, W. G., Jr., and Callahan, B. C., 1971. Effects of dietary copper loading on livers of rats. I. Changes in subcellular acid phosphatases and detection of an additional acid *p*-nitrophenylphosphatase in the cellular supernatant during copper loading, *Am. J. Pathol.* 64:123–144.

McNaughton, J. L., Day, E. J., Dilworth, B. C., and Lott, B. D., 1974. Iron and copper availability from various sources, *Poultry Sci.* 53:1325–1330.

Mearrick, P. T., and Mistilis, S. P., 1969. Excretion of radiocopper by the neonatal rat, *J. Lab. Clin. Med.* 74:421–426.

Medeiros, D. M., Lin, K., Liu, C., and Thorne, B. M., 1984. Pre-gestation dietary copper restriction and blood pressure in the Longs-Evans rat, *Nutr. Rep. Intl.* 30:559.

Mehra, R. K., and Bremner, I., 1983. Development of a radioimmunoassay for rat liver metallothionein-1 and its application to the analysis of rat plasma and kidneys, *Biochem. J.* 213:459.

Mehra, R. K., and Bremner, I., 1984. Metallothionein-I in the plasma and liver of neonatal rats, *Biochem. J.* 217:859–862.

Mehra, R. K., and Bremner, I., 1985. Studies on the metabolism of rat liver copper-metallothionein, *Biochem. J.* 227:903–938.

Mehta, S. W., and Eikum, R., 1989. Effect of estrogen on serum and tissue copper and zinc levels in the female rat, *Nutr. Rep. Int.* 40:1101–1106.

Meinel, B., Bode, J. C., Koenig, W., and Richter, F.-W., 1979. Contents of rare elements in the human liver before birth, *Biol. Neonate* 36:225–232.

Meissner, W., 1916. Versuche über den Kupfer-Gehalt einiger Pflanzeaschen, *Schweigger's Jahrb. Chemie* 17:340–354, 436–448.

Menard, P., McCormick, C. C., and Cousins, R. J., 1981. Regulation of intestinal metallothionein biosynthesis in rats by dietary zinc, *J. Nutr.* 111:1353–1361.

Menke, K. H., Lantzsch, H.-J., and Schenkel, H., 1978. A new method for estimation of Zn and Cu status by chelating agents, in *Trace Element Metabolism in Man and Animals—3* (M. Kirchgessner, ed.), Arbeitskreis für Tierernährungsforschung Weihenstephan, Freising, Germany, pp. 456–459.

Menkes, J. H., Alter, M., Steigleder, G., Weakley, D. R., and Sung, J. H., 1962. A sex-linked recessive disorder with retardation of growth and peculiar hair, and focal cerebral and cerebellar degeneration, *Pediatrics* 29:764–779.

Mercer, J. F. B., and Grimes, A., 1986. Isolation of a human ceruloplasmin cDNA clone that includes the N-terminal leader sequence, *FEBS Lett.* 203:185–190.

Merle, P., and Kadenbach, B., 1982. Kinetic and structural differences between cyhtochrome *c* oxidases from beef liver and heart, *Eur. J. Biochem.* 125:239–244.

Messerschmidt, A., Rossi, A., Ladenstein, R., Huber, R., Bolognesi, M., Marchesini, A., Petruzelli, R., and Finazzi-Agro, A., 1988. Preliminary x-ray crystal structure and partial cDNA-sequence of ascorbate oxidase from *Zucchini*, in *Oxidases and Related Redox Systems* (T. E. King, H. S. Mason, and M. Morrison, eds.), Alan R. Liss, New York, pp. 285–288.

Messerschmidt, A., and Huber, R., 1990. The blue oxidases, ascorbate oxidase, laccase and ceruloplasmin. Modeling and structural relationships, *Eur. J. Biochem.* 187:341–352.

Metzler, D. E., 1977. *Biochemistry. The Chemical Reactions of Living Cells.* Academic Press, New York.

Micera, G., Decock, P., Kozlowski, H., Pettit, L. D., and Pusino. A., 1989. Different copper(II)-binding ability of amino sugars, *Chitin Chitosan: Sources. Chemical, Biochemical, Physical Properties Applications* (G. Skaak-Break, T. Anthonsen, and P. A. Sandford, eds.) (Proc. Intl. Conf.), 4th. Elsevier, London, pp. 487–490.

Mignatti, P., Tsuboi, R., Robbins, E., and Rifkin, D. B., 1989. In vitro angiogenesis on the human amniotic membrane: Requirement for basic fibroblast growth factor-induced proteinases, *J. Cell Biol.* 108:671–682.

Milanino, R., and Velo, G. P., 1981. Multiple actions of copper in control of inflammation: Studies in copper-deficient rats, in *Trace Elements in the Pathogenesis and Treatment of Inflammation* (K. D. Rainsford, K. Brune, and M. W. Whitehouse, eds.), Birkhauser, Basel, pp. 209–230.

Milanino, R., Mazzli, S., Passarella, E., Tarter, G., and Velo, G. P., 1978a. Carrageenan oedema in copper-deficient rats, *Agents Actions* 8:618–622.

Milanino, R., Passarella, E., and Velo, G. P., 1978b. Adjuvant arthritis in young copper-deficient rats, *Agents Actions* 8:623–628.

Milanino, R., Passarella, E., and Velo, G. P., 1979. Copper and the inflammatory process, in *Advances in Inflammation Research* (G. Weissman, B., Samuelsson, and R. Paoletti, eds.), Raven Press, New York, pp. 281–291.

Milanino, R., Franco, A., Conforti, A., Marella, M., and Velo, G. P., 1984. Copper metabolism in acute inflammatory process and its possible significance for a novel approach to the therapy on inflammation, *Acts of 1st World Conference on Inflammation*, Venice, April 16–18.

Milanino, R., Conforti, A., Franco, L., Marrella, M., and Velo, G., 1985. Review: Copper and inflammation—a possible rationale for the pharmacological manipulation of inflammatory disorders, *Agents Actions* 16:504–513.

Milici, A. J., Watrous, N. E., Stukenbrok, H., and Palade, G. E., 1987. Transcytosis of albumin in capillary endothelium. *J. Cell Biol.* 105:2603–2612.

Miller, D. S., and O'Dell, B. L., 1987. Milk and casein-based diets for the study of brain catecholamines in copper-deficient rats, *J. Nutr.* 117:1890–1897.

Miller, E. B., Kanabrocki, E. L., Case, L. F., Graham, L. A., Fields, T., Oester, Y. T., and Kaplan, E., 1967. Nondialyzable manganese, copper and sodium in human bile, *J. Nucl. Med.* 8:891–895.

Miller, L. L., Hanavan, H. R., Titthasiri, N., and Chowdhury, A., 1964. Dominant role of the liver in the biosynthesis of the plasma proteins with special reference to the plasma mucoproteins seromucoid, ceruloplasmin, and fibrinogen, *Adv. Chem. Ser.* 44:17–40.

Millett, F., deJong, C., Paulson, L., and Capaldi, R. A., 1983. Identification of specific carboxylate groups on cytochrome *c* oxidase that are involved in binding cytochrome *c*, *Biochemistry* 22:546–552.

Mills, C. F., Dalgarno, A. C., and Wenham, G., 1976. Biochemical and pathological changes in tissues of freisian cattle during the experimental induction of copper deficiency, *Br. J. Nutr.* 35:309.

Mills, C. F., El-Gallad, T. T., Bremner, I., and Wenham, G., 1981. Copper and molybdenum absorption by rats given ammonium tetrathiomolybdate, *J. Inorg. Biochem.* 14:163–175.

Milne, D. B., and Matrone, G., 1970, Forms of ceruloplasmin in developing piglets, *Biochim. Biophys. Acta* 212:43–49.

Milne, D. B., and Weswig, P. H., 1968. Effect of supplementary copper on blood and liver copper-containing fractions in rats, *J. Nutr.* 95:429–433.

Minghetti, P. P., Ruffner, D. E., Kuang, W. J., Dennison, O. E., Hawkins, J. W., Beattie, W. G., and Dugaiczyk, A., 1986. Molecular structure of the human albumin gene is revealed by nucleotide sequence within q11–22 of chromosome 4, *J. Biol. Chem.* 261:6747–6757.

Mishima, Y., and Imokawa, G., 1983. Selective aberration and pigment loss in melanosomes of malignant melanoma cells *in vitro* by glycosylation inhibitors: Premelanosomas as glycoprotein, *J. Invest. Dermatol.* 81:106–114.

Mistilis, S. P., and Farrer, P. A., 1968. The absorption of biliary and non-biliary radiocopper in the rat, *Scand. J. Gastroenterol* 3:586–592.

Mistilis, S. P., and Mearrick, P. T., 1969. The absorption of ionic, biliary, and plasma radiocopper in neonatal rats, *Scand. J. Gastroenterol.* 4:691–696.

Mo, S. C., Choi, D. S., and Robinson, J. W., 1988. A study of the uptake by duckweed of aluminum, copper, and lead from aqueous solution, *J. Environ. Sci. Health* A23:139–156.

Moffitt, Jr., A. E., and Murphy, S. D., 1973. Effect of excess and deficient copper intake on rat liver microsomal enzyme activity, *Biochem. Pharmacol.* 22:1463–1476.

Mohanakrishnan, P., and Chignell, C. F., 1984. Copper and nickel binding to canine serum albumin. A circular dichroism study, *Comp. Biochem. Physiol.* 79:321–323.

Mokdad, R., Debec, A., and Wegnez, M., 1987. Metallothionein genes in *Drosophila melanogaster* constitute a dual system, *Proc. Natl. Acad. Sci. USA* 84:2658–2662.

Mondovi, B. (ed.), 1985. *Structure and Functions af Amine Oxidases*, CRC Press, Boca Raton, Florida.

Mondovi, B., and Avigliano, L., 1984. Ascorbate oxidase, in *Copper Proteins and Copper Enzymes*, Vol. III, (R. Lontie, ed.), CRC Press, Boca Raton, Florida, pp. 101–118.

Mondovi, B., and Riccio, P., 1985. Animal intracellular amine oxidases, in *Structure and Functions of Amine Oxidases* (B. Mondovi, ed.), CRC Press, Boca Raton, Florida, pp. 64–76.

Moog, R. S., McGuirl, M. A., Cote, C. E., and Dooley, D. M., 1986. Evidence for methoxatin (pyrroloquinolinequinone) as the cofactor in bovine plasma amine oxidase from resonance Raman spectroscopy, *Proc. Natl. Acad. Sci. USA* 83:8435–8439.

Moore, C. V., 1951. *Harvey Lect.* 55:67. Cited by E. J. Underwood, in *Trace Elements in Human and Animal Nutrition*, 3rd ed., Academic Press, New York, 1971, p. 34.

Moore, J. M., Case, D. A., Chazin, W. J., Gippert, G. P., Havel, T. F., Powls, R., and Wright, P., 1988. Three-dimensional solution structure of plastocyanin from the green alga *Scenedesmus obliquus*, *Science* 240:314–318.

Moore, R. J., and Klevay, L. M., 1988. Effect of copper deficiency on blood pressure and plasma and lung agiotensin-converting enzyme activity in rats, *Nutr. Res.* 8:489–497.

Morell, A. G., and Scheinberg, I. H., 1958. Preparation of an apoprotein from ceruloplasmin by reversible dissociation of copper, *Science* 127:588–590.

Morell, A. G., and Scheinberg, I. H., 1960. Heterogeneity of human ceruloplasmin, *Science* 131:930–932.

Morell, A. G., Aisen, P., Blumberg, W. E., and Scheinberg, I. H., 1964. Physical and chemical studies on ceruloplasmin II. Molecular oxygen and the blue color of ceruloplasmin, *J. Biol. Chem.* 239:1042–1047.

Morell, A. G., Irvine, R. A., Sternlieb, I., Scheinberg, I. H., and Ashwell, G., 1968. Physical and chemical studies on ceruloplasmin. V. Metabolic studies on sialic acid-free ceruloplasmin *in vivo*, *J. Biol. Chem.* 243:155–159.

Morgan, R. F., and O'Dell, B. L., 1977. Effect of copper deficiency on the concentrations of catecholamines and related enzyme activities in the rat brain, *J. Neurochem.* 28:207–213.

Morgan, W. T., 1978. Human serum histidine-rich glycoproteins. I. Interactions with heme, metal ions and organic ligands, *Biochim. Biophys. Acta* 533:319–333.

Morgan, W. T., 1981. Interactions of the histidine-rich glycoprotein of serum with metals, *Biochemistry* 20:1054–1061.

Morgan, W. T., 1985. The histidine-rich glycoprotein of serum has a domain rich in histidine, proline, and glycine that binds heme and metals, *Biochemistry* 24:1496–1501.

Morgan, W. T., Sutor, R. P., Muller-Eberhard, U., and Koskelo, P., 1975. Interactions of rabbit hemopexin with copro-and uroporphyrins, *Biochim. Biophys. Acta* 400:415–422.

Morgan, W. T., Koskelo, P., Koenig, H., and Coneay, E. P., 1978. Human histidine-rich glycoprotein. II. Serum levels in adults, pregnant women and neonates, *Proc. Soc. Exp. Biol. Med.* 158:647–651.

Morgan, W. T., Deaciuc, V., and Riehm, J. P., 1989. A heme- and metal-binding hexapeptide from the sequence of rabbit plasma histidine-rich glycoprotein, *J. Mol. Recognit.* 2:122–126.

Morimoto, A., Sakata, Y., Watanabe, T., and Murakami, N., 1989. Characteristics of fever and acute-phase response induced in rabbits by IL-1 and TNF, *Am. J. Physiol.* 256:R35–R41.

Morisawa, M., and Mohri, H., 1972. Heavy metals and spermatozoan motility, I., Distribution of iron, zinc, and copper in sea urchin spermatozoa, *Exp. Cell Res.* 70:311.

Morley, C. G. D., and Bezkorovainy, A., 1985. Cellular iron uptake from transferrin: Is endocytosis the only mechanism? *Int. J. Biochem.* 17:553–564.

Morrison, J. N., Wood, A. M., and Bremner, I., 1988. Effects of inflammatory stress on metallothionein-I concentrations in blood cells and plasma of rats, *Biochem. Soc. Trans.* 16:820–821.

Moshkov, K. A., Lakatos, S., Hajdu, J., Zavodszky, P., and Neifakh, S. A., 1979. Proteolysis of human ceruloplasmin. Some peptide bonds are particularly susceptible to proteolytic attack, *Eur. J. Biochem.* 94:127–134.

Moshkov, K. A., Vagin, A. A., and Zaitsev, V. N., 1988. Localization of active sites in human ceruloplasmin according to the data of intra- and intermolecular homology, *Mol. Biol.* 21:934–939.

Moss, B. R., Madsen, F., Hansard, S. L., and Gamble, C. T., 1974. Maternal-fetal utilization of copper by sheep, *J. Anim. Sci.* 38:475–479.

Mota de Freitas, D., Luchinat, C., Banci, L., Bertini, I., and Valentine, J. S., 1987. ^{31}P NMR study of the interaction of inorganic phosphate with bovine copper-zinc superoxide dismutase, *Inorg. Chem.* 26:2788–2791.

Movaghar, M., and Hunt, D. M., 1987. Tyrosinase activity and the expression of the agouti gene in the mouse, *J. Exp. Zool.* 243:473–480.

Mulhern, S. A., and Koller, L. D., 1988. Severe or marginal copper deficiency results in a graded reduction in immune system status in mice, *J. Nutr.* 118:1041–1047.

Mulhern, S. A., Raveche, E. S., Smith, H. R., and Lal, R. B., 1987. Dietary copper deficiency and autoimmunity in the NZB mouse, *Am. J. Clin. Nutr.* 46:1035–1039.

Muller, D., and Sarkar, B., 1981. Non-transferrin Fe(III)-binding fraction in human adult and cord blood serum, *Proc. Annu. Meet. Can. Fed. Biol. Soc.* 24:Abstract 470.

Muller, H. B., 1970. Der Einfluss kupferarmer Kost auf das Pankreas, *Virchows Archiv. Abt. A. Pathologische Anatomie* 350:353–367.

Muller, M., and Azzi, A., 1986. The role of cytochrome *a* in the proton pump of cytochrome-*c* oxidase, *Biochemistry* 68:401–406.

Muller, M., Schlapfer, B., and Azzi, A., 1988a. Cytochrome *c* oxidase from *Paracoccus denitrificans*: Both hemes are located in subunit I, *Proc. Natl. Acad. Sci. USA* 85:6647–6651.

Muller, M., Schlapfer, B., and Azzi, A., 1988b. Preparation of a one-subunit cytochrome oxidase from *Paracoccus denitrificans*: Spectral analysis and enzymatic activity, *Biochemistry* 27:7546–7551.

Munch-Peterson, S., 1950. On the copper content of mother's milk before and after intravenous copper administration, *Acta Paediatr.* 39:378.

Munro, H. N., and Linder, M. C., 1978. Ferritin: Structure, biosynthesis, and role in iron metabolism, *Physiol. Rev.* 58:317–396.

Murillo, C., 1985. Master's thesis, California State University, Fullerton, California.

Murphy, A. S. N., Mains, R. E., and Eipper, B. A., 1986. Purification and characterization of peptidylglycine α-amidating monooxygenase from bovine neuro-intermediate pituitary, *J. Biol. Chem.* 261:1815–1822.

Murthy, L., and Petering, H. G., 1976. Effect of dietary zinc and copper interrelationships on blood parameters of the rat, *J. Agric. Food Chem.* 24:808–811.

Musci, G., Carbonaro, M., Adriani, A., Lania, A., Galtieri, A., and Calabrese, L., 1990. Unusual stability properties of a reptilian ceruloplasmin, *Arch. Biochem. Biophys.* 279:8–13.

Myers, B. A., Dubick, M. A., Reynolds, R. D., and Rucker, R. B., 1985. Effect of vitamin B-6 (pyridoxine) deficiency on lung elastin cross-linking in perinatal and weanling rat pups, *Biochem. J.* 229:153–160.

Nakagawa, T., Maeshima, M., Muto, H., Kajiura, H., Hattori, H., and Asahi, T., 1987. Separation, amino-terminal sequence and cell-free synthesis of the smallest subunit of sweet potato cytochrome *c* oxidase, *Eur. J. Biochem.* 165:303–307.

Nalbandyan, R. M., 1983. Copper in brain, *Neurochem. Res.* 8:1211–1232.

Narasaka, S., 1938. Studies in the biochemistry of copper. XXXIII. Metabolism of copper and adrenaline in normal and thyroidless animals, *Jpn. J. Med. Sci. II. Biochem.* 4:93–95.

Narayanan, A. S., Siegel, R. C., and Martin, G. R., 1972. On the inhibition of lysyl oxidase by β-aminopropionitrile, *Biochem. Biophys. Res. Commun.* 46:745.

Narayanan, A. S., Page, R. C., Kuzan, F., and Cooper, C. G., 1978. Elastin cross-linking *in vitro*. Studies on factors influencing the formation of desmosines by lysyl oxidase action on tropoelastin, *Biochem. J.* 173:857–862.

Nartey, N. O., Cherian, M. G., and Banerjee, D., 1987a. Immunohistochemical localization of metallothionein in human thyroid tumors, *Am. J. Pathol.* 129:177–182.

Nartey, N. O., Banerjee, D., and Cherian, M. G., 1987b. Immunohistochemical localization of metallothionein in cell nucleus and cytoplasm of fetal human liver and kidney and its changes during development, *Pathology*, 19:233–238.

Naylor, S. L., Yang, F., Cutshaw, S., Barnett, D. R., and Bowman, B. H., 1985. Mapping ceruloplasmin cDNA to human chromosome 3, *Cytogenet. Cell Genet.* 40:711.

Neal, W. M., Becker, R. B., and Shealy, A. L., 1931. A natural copper deficiency in cattle rations, *Science* 74:418–419.

Nederbragt, H., and Lagerwerf, A. J., 1986. Strain-related patterns of biliary excretion and hepatic distribution of copper in the rat, *Hepatology* 6:601–607.

Neifakh, S. A., Monakhov, N. K., Shaposhnikov, A. M., and Zubzitski, Y. N., 1969. Localization of ceruloplasmin biosynthesis in human and monkey liver cells and its copper regulation, *Experientia* 25:337–344.
Nelson, L. S., Jr., Jacobs, F. A., and Brushmiller, J. G., 1985. Solubility of calcium and zinc in model solutions based on bovine and human milks, *J. Inorg. Biochem.* 254:255–265.
Nelson, L. S., Jr., Jacobs, F. A., Brushmiller, J. G., and Ames, R. W., 1986. Effect of pH on the speciation and solubility of divalent metals in human and bovine milks, *J. Inorg. Biochem.* 26:153–168.
Nemer, M., Wilkinson, D. G., Travaglini, E. C., Sternberg, E. J., and Butt, T. R., 1985. Sea urchin metallothionein sequence: Key to an evolutionary diversity, *Proc. Natl. Acad. Sci. USA* 82:4992–4994.
Nesheim, M. E., Karzmann, J. A., Tracy, P. B., and Mann, K. G., 1981. Factor V. *Methods Enzymol. (Proteolytic Enzymes, Pt. C)* 80:249–274.
Nestler, J. E., Clore, J. N., Failla, M. L., and Blackard, W. G., 1987. Effects of extreme hyperinsulinaemia on serum levels of trace metals, trace metal binding proteins, and electrolytes in normal females, *Acta Endocrinol.* 114:235–242.
Netsky, M. G., Harrison, W. W., Brown, M., and Benson, C., 1969. Tissue zinc and human disease. Relation of zinc content of kidney, liver and lung to atherosclerosis and hypertension, *Am. J. Clin. Pathol.* 51:358–365.
Neumann, P. Z., and Sass-Kortsak, A., 1967. The state of copper in human serum: Evidence for an amino acid-bound fraction, *J. Clin. Invest.* 46:646–658.
Neumann, P. Z., and Silverberg, M., 1966. Active copper transport in mammalian tissues—a possible role in Wilson's disease, *Nature* 210:414–416.
Newberne, P. M., Hunt, C. E., and Young, V. R., 1968. The role of diet and the reticuloendothelial system in the response of rats to *Salmonella typhimurium* infection, *Br. J. Exp. Pathol.* 49:448–457.
Newton, S. A., Yeo, K.-T., Yeo, T.-K., Parent, J. P., and Olden, K., 1986. Possible involvement of intracellular lectins in intracellular transport of glycoproteins, in *Vertebrate Lectins* (K. Olden, J. B. Parent, eds.), Van Nostrand Reinhold Co., New York, pp. 211–226.
Nielson, A. L., 1944b. Om serumkobber, *Nord. Med.* 23:1657–1659.
Nielson, K. B., and Winge, D. R., 1984. Preferential binding of copper to the beta domain of metallothionein, *J. Biol. Chem.* 259:4941–4946.
Nielson, K. B., and Winge, D. R., 1985. Independence of the domains of metallothionein metal binding, *J. Biol. Chem.* 260:8698–8701.
Nielson, K. B., Atkin, C. L., and Winge, D. R., 1985. Distinct metal-binding configurations in metallothionein, *J. Biol. Chem.* 260:5342–5350.
Nilsson, T., Copeland, R. A., Smith, P. A., and Chan, S. I., 1988. Conversion of Cu_A to a type II copper in cytochrome *c* oxidase, *Biochemistry* 27:8254–8260.
Nisbet, A. D., Saundry, R. H., Moir, A. J. G., Fothergill, L. A., and Fothergill, J. E., 1981. The complete amino-acid sequence of hen ovalbumin, *Eur. J. Biochem.* 115:335–345.
Nomoto, K., Mino, Y., Ishida, T., Yoshioka, H., Ota, N., Inoue, M., Takagi, S., and Takemoto, T., 1981. X-ray crystal structure of the copper(II) complex of mugineic acid, a naturally occurring metal chelator of graminaceous plants, *J. Chem. Soc. Chem. Commun.*, p. 338.
Norheim, G., and Soli, N. E., 1977. Chronic copper poisoning in sheep. II. The distribution of soluble copper-, molybdenum- and zinc-binding proteins from liver and kidney, *Acta Pharmacol. Toxicol.* 40:178–187.
Norton, D. S., and Heaton, F. W., 1980. Distribution of copper and zinc among protein fractions in the cytoplasm of rat tissues, *J. Inorg. Biochem.* 13:1–9.

Noyer, M., and Putnam, F. W., 1981. A circular dichroism study of undegraded human ceruloplasmin, *J. Am. Chem. Soc.* 20:3536–3542.

Nozdryukhina, L. R., 1978. The role of trace elements of the blood in distinguishing diagnostics of heart and liver diseases, in *Trace Element Metabolism in Man and Animals-3* (M. Kirchgessner, ed.), Arbeitskreis für Tierernahrungsforschung Weihenstephan, Freising, Germany, pp. 336–339.

Nozdryukhina, L. R., Petelin, S., Nadzharyan, T. L., Belsky, N. K., Kravchuk, I. V., and Povrovskaya, G. A., 1977. Blood trace elements during emergency conditions in clinical treatment of cardiovascular diseases, Abstracts of the Third International Symposium on Trace Element Metabolism in Man and Animals. Freising, Germany.

Nriagu, J. O., 1980. *Zinc in the Environment. Health Effects*, Wiley, New York, Pt. 2, p. 480.

Numata, Y., and Hayakawa, T., 1986. Lysyl-oxidase activity in the odontoblast-predentine layer isolated from bovine incisor teeth, *Arch. Oral Biol.* 31:67–68.

Nunoz, M. T., Cole, E. S., and Glass, J., 1983. The reticulocyte plasma membrane pathway of iron uptake as determined by the mechanism of α, α'-dipyridyl inhibition, *J. Biol. Chem.* 258:1146–1151.

Nusbaum, M. J., and Zettner, A., 1973. Content of calcium, magnesium, copper, iron, sodium and potassium in amniotic fluid from eleven to nineteen weeks' gestation, *Am. J. Obstet. Gynecol.* 115:219–226.

Oakes, B. W., Danks, D. M., and Campbell, P. E., 1976. Human copper deficiency: Ultrastructural studies of the aorta and skin in a child with Menkes' syndrome, *Exp. Mol. Pathol.* 25:82–98.

Oberley, L. W., and Buettner, G. R., 1973. The role of superoxide dismutase in cancer: A review. *Cancer Res.* 39:1141.

Oberley, L. W., and Buettner, G. R., 1979. The production of hydroxyl radical by bleomycin and iron(II), *FEBS Lett.* 97:47–49.

Oberley, L. W., Leuthauser, S. W. H. C., Buettner, G. R., Sorenson, J. R. J., Oberley, T. D., and Bize, I. B., 1982a. The use of superoxide dimutase in treatment of cancer, in *Pathology of Oxygen*, (A. P. Autor, ed.) Academic Press, New York, pp. 207–218.

Oberley, L. W., Leuthauser, S. W. C., Oberley, T. D., Sorenson, J. R. J., and Pasternack, R. F., 1982b. Antitumor activities of compounds with superoxide dismutase activity, in *Inflammatory Diseases and Copper* (J. R. J. Sorenson, ed.), Humana Press, Clifton, New Jersey, pp. 423–432.

Oberley, L. W., Rogers, K. L., Schutt, L., Oberley, T. D., Leuthauser, S. W. C., and Sorenson, J. R. K., 1983. Possible role of glutathione in the antitumor effect of a copper-containing synthetic superoxide dismutase in mice, *J. Natl. Cancer Inst.* 71:1089–1094.

O'Cuill, T., Hamilton, A. F., and Egan, D. A., 1970. *Ir. Vet. J.* 24:21.

O'Dell, B. L., and Prohaska, J. R., 1983. Biochemical aspects of copper deficiency in the nervous system, *Neurobiology of Trace Elements*, Vol. 1 (I. Dreoshi and R. M. Smith, eds.), Humana Press, Clifton, New Jersey 41–81.

O'Dell, B. L., Hardwick, B. C., and Reynolds, G., 1961a. Mineral deficiencies of milk and congenital malformations in the rat, *J. Nutr.* 73:151–156.

O'Dell, B. L., Hardwick, B. C., Reynolds, G., and Savage, J. E., 1961b. Connective tissue defect in the chick resulting from copper deficiency, *Proc. Soc. Exp. Biol. Med.* 108:402.

O'Dell, B. L., Kilburn, K. H., McKenzie, W. N., and Thurston, R. J., 1978. The lung of the copper-deficient rat, a model for developmental pulmonary emphysema, *Am. J. Pathol.* 91:413–423.

Oestreicher, P., and Cousins, R. J., 1985. Copper and zinc absorption in the rat: Mechanism of mutual antagonism, *J. Nutr.* 115:159–166.

Ogiso, T., Ogawa, N., and Miura, T., 1979. Inhibitory effect of high dietary zinc on copper absorption in rats. II. Binding of copper and zinc to cytosol proteins in the intestinal mucosa, *Chem. Pharm. Bull.* 27(2):515–521.

Oh, S. H., Deagan, J. T., Whanger, P. D., and Weswig, P. H., 1978. Biological function of metallothionein. IV. Biosynthesis and degradation of liver and kidney metallothionein in rats fed diets containing zinc or cadmium, *Bioinorg. Chem.* 8:245–254.

Ohchi, H., Jusuhara, T., Katayama, T., Ohara, K., and Kato, N., 1987. Effects of dietary xenobiotics on the metabolism of copper, α-tocopherol and cholesterol in rats, *J. Nutr. Sci. Vitaminol.* 33:281–288.

Ohguchi, S., Ichimiya, H., Yagi, A., Hayashi, H., and Sakamoto, N., 1988. Copper-induced hypercholesterolemia of golden hamsters: Enhanced synthesis of cholesterol in the liver, *Gastroentol. Jpn.* 23:629–632.

Ohlson, M. A., and Daum, K., 1935. A study of the iron metabolism of normal women, *J. Nutr.* 9:75–89.

Oimatov, M., Vagin, A. A., Nekrasov, Yu. V., and Moshkov, J. A., 1986. Rotational function for the trigonal human ceruloplasmin crystals at a resolution of 6.ANG, *Kristallografiya* 31:937–941.

Olatunbosun, D. A., Bolodeoku, J. O., Cole, T. O., and Adadevoh, B. K., 1976. Relationship of serum copper and zinc to human hypertension in Nigerians, *Bull. WHO* 53:134–135.

Oliver, J. W., 1975. Interrelationships between athyrotic and copper-deficient states in rats, *Am. J. Vet. Res.* 36:1649–1653.

Olson, K. B., Heggen, G. E., and Edwards, C. F., 1958. Analysis of 5 trace elements in the liver of patients dying of cancer and noncancerous disease, *Cancer* 11:554–561.

Omoto, E., and Tavassoli, M., 1989a. Purification and characterization of ceruloplasmin receptors from rat liver endothelium, *Clin. Res.* 37:

Omoto, E., and Tavassoli, M., 1989b. The role of endosomal traffic in the transendothelial transport of ceruloplasmin in the liver, *Biochem. Biophys. Res. Commun.* 162:1346–1350.

Omoto, E., and Tavassoli, M., 1990. Purification and partial characterization of ceruloplasmin receptors from rat liver endothelium, *Arch. Biochem. Biophys.* 282:34–38.

Onderka, H. K., and Kirksey, A., 1975. Influence of dietary lipids on iron and copper levels of rats administered oral contraceptives, *J. Nutr.* 105:1269–1277.

Onoda, M., and Djakie, D., 1990. Modulation of Sertoli cell secretory function by rat round spermatid protein(s), *Mol. Cell. Endocrinol.* 73:35–44.

O'Reilly, S., Weber, P. M., Oswald, M., and Shipley, L., 1971. Abnormalities of the physiology of copper in Wilson's disease, *Arch Neurol.* 25:28–32.

Orena, S. J., Goode, C. A., and Linder, M. C., 1986. Binding and uptake of copper from ceruloplasmin, *Biochem. Biophys. Res. Commun.* 139:822.

Orskov, I., and Orskov, F., 1973. Plasmid-determined hydrogen sulfide character in *Escherichia coli* and its relation to plasmid-carried raffinose fermentation and tetracycline resistance characters. Examination of 32 hydrogen sulfide-positive strains isolated during the years 1950 to 1971, *J. Gen. Microbiol.* 77:487–499.

Ortel, T. L., Takahashi, N., Bauman, R. A., and Putnam, F. W., 1983. The domain structure of the human ceruloplasmin molecule, *Protides Biol. Fluids* 31:243–248.

Orten, J. M., Orten, A. U., Sardesai, V. M., and Cheng, L. C., 1975. The role of copper in the biosynthesis of heme, in *Erythrocyte Structure and Function*, Alan R. Liss, New York, pp. 643–665.

Osaki, S., Johnson, D. A., and Frieden, E., 1966. The possible significance of the ferrous oxidase activity of ceruloplasmin in normal human serum, *J. Biol. Chem.* 241:2746–2751.

Osaki, S., and Johnson, D. A., 1969. Mobilization of liver iron by ferroxidase (ceruloplasmin), *J. Biol. Chem.* 244:5757–5761.

Osaki, S., Johnson, D. A., and Frieden, E., 1971. The mobilization of iron from the perfused mammalian liver by a serum copper enzyme, ferroxidase I, *J. Biol. Chem.* 9:3018–3023.

Osborne, S. B., Walshe, J. M., and Williams, R., 1972. Copper dynamics in Wilson's disease. *Ann. Biol. Chim.* 30:391–396.

Osterberg, R., and Malmensten, B., 1984. Methylamine-induced conformational change of α_2-macroglobulin and its zinc(II) binding capacity. An X-ray scattering study, *Eur. J. Biochem.* 143:541–544.

Owen, C. A., Jr., 1964. Absorption and excretion of Cu^{64}-labeled copper by the rat, *Am. J. Physiol.* 207:1203–1206.

Owen, C. A., Jr., 1965. Metabolism of radiocopper (Cu^{64}) in the rat, *Am. J. Physiol.* 209:900–904.

Owen, C. A., Jr., 1971. Metabolism of copper 67 by the copper-deficient rat, *Am. J. Physiol.* 221:1722–1727.

Owen, C. A., Jr., 1973. Effects of iron on copper metabolism and copper on iron metabolism in rats, *Am. J. Physiol.* 224:514–518.

Owen, C. A., Jr., 1975. Copper metabolism after biliary fistula, obstruction, or sham operation in rats, *Mayo Clin. Proc.* 50:412–415.

Owen, C. A., Jr., 1980. The effect of surgery and ether anesthesia on excretion of biliary copper by the rat, *Proc. Soc. Exp. Biol. Med.* 163:496–497.

Owen, Jr., C. A., 1982a. *Physiological Aspects of Copper. Copper in Organs and Systems*, Park Ridge, New Jersey, New Jersey, Noyes Publications.

Owen, Jr., C. A., 1982b. *Biochemical Aspects of Copper. Copper Proteins. Ceruloplasmin, and Copper Protein Binding*, Park Ridge, New Jersey, Noyes Publications.

Owen, Jr., C. A., 1982c. *Biochemical Aspects of Copper. Occurrence, Assay and Interrelationships*, Park Ridge, New Jersey, Noyes Publications.

Owen, Jr., C. A., 1982d. *Copper deficiency and Toxicity. Acquired and Inherited, in Plants, Animals, and Man*. Park Ridge, New Jersey, Noyes Publications.

Owen, Jr., C. A., 1982e. *Wilson's Disease. The Etiology, Clinical Aspects, and Treatment of Inherited Copper Toxicosis*. Park Ridge, New Jersey, Noyes Publications.

Owen, C. A., Jr., and Hazelrig, J. R., 1968. Copper deficiency and copper toxicity in the rat, *Am. J. Physiol.* 215:334–338.

Owen, C. A., Jr., Randall, R. V., and Goldstein, N. P., 1975. Effect of dietary D-penicillamine on metabolism of copper in rats, *Am. J. Physiol.* 228:88–91.

Owen, C. A., Jr., Dickson, E. R., Goldstein, N. P., Baggenstoss, A. H., and McCall, J. T., 1977. Hepatic subcellular distribution of copper in primary biliary cirrhosis. Comparison with other hyperhepatocupric states and review of the literature, *Mayo Clin. Proc.* 52:73–80.

Ozgunes, H., Beksac, M. S., Duru, S., and Kayakirilmaz, K., 1987. Instant effect of induced abortion on serum ceruloplasmin activity, copper and zinc levels, *Arch. Gynecol.* 240:21–25.

Pacht, E. R., and Davis, W. B., 1988. Decreased ceruloplasmin ferroxidase activity in cigarette smokers, *J. Lab. Clin. Med.* 111:661–668.

Page, E., Earley, J., McCallister, L. P., and Boyd, C., 1974. Copper content and exchange in mammalian hearts, *Circ. Res.* 35:67–76.

Palida, F., Mas, A., Lonergan, P., and Ettinger, M., 1987. Non-metallothionein, cytosolic Cu-binding components in normal, Cu-deficient and brindled-mouse cells, presentation at symposium on Trace Elements in Man and Animals, Asilomar, California, May 31–June 5, 1987.

Palida, F. A., Mas, A., Arola, L., Bethin, K., Lonergan, P. A., and Ettinger, M. J., 1990. Cytosolic copper-binding proteins in rat and mouse hepatocytes incubated continuously with copper(II), *Biochem. J.* 268:359–366.

References

Paliwal, V. K., Iversen, P. L., and Ebadi, M., 1990. Regulation of zinc metallothionein II mRNA level in rat brain, *Neurochem. Int.* 17:441–447.

Palmissano, D. J., 1974. Nutrient deficiencies after intensive parenteral alimentation, *N. Engl. J. Med.* 291:799.

Palumbo, A., d'Ischia, M., and Prota, G., 1987. Tyrosinase-promoted oxidation of 5,6-dihydroxyindole-2-carboxylic acid to melanin. Isolation and characterization of oligomer intermediates, *Tetrahedron* 43:4203–4206.

Palumbo, A., d'Ischia, M., Misuraca, G., Prota, G., and Schultz, T., 1988. Structural modifications in biosynthetic melanins induced by metal ions, *Biochim. Biophys. Acta* 964:193–199.

Pande, J., Vasak, M., and Kagi, J. H. R., 1985. Interaction of lysine residues with the metal thiolate clusters in metallothionein, *Biochemistry* 24:6717–6722.

Panemangalore, M., Banerjee, D., Onosaka, S., and Cherian, M. G., 1983. Changes in the intracellular accumulation and distribution of metallothionein in rat liver and kidney during postnatal development, *Dev. Biol.* 97:95–102.

Parent, J. B., Bauer, H. C., and Olden, K., 1985. Three secretory rates in human hepatoma cells, *Biochim. Biophys. Acta* 846:44–50.

Parent, J. B., Yeo, T. K., Yeo, K. T., and Olden, K., 1986. Differential effects of 1-deoxynojirimycin on the intracellular transport of secretory glycoproteins of human hepatoma cells in culture, *Mol. Cell. Biochem.* 72:21–23.

Parish, R. W., 1971. Intracellular location of phenoloxidases and peroxidase in stems of spinach beet (Beta Vulgaris), *Z. Pflanzenphysiol.* 66:176–188.

Parish, R. W., 1972. The intracellular location of phenoloxidases, peroxidase, and phosphatases in the leaves of spinach beet (*Beta Vulgaris* L. subspecies *vulgaris*), *Eur. J. Biochem.* 31:446–455.

Parisi, A. F., and Vallee, B. L., 1970. Isolation of a zinc α_2-macroglobulin from human serum, *Biochemistry* 9:2421–2426.

Patek, E., and Hagenfeldt, K., 1974. Trace elements in the human fallopian tube epithelium, *Int. J. Fertil.* 19:85–88.

Patterson, D. S. P., Foulkes, J. A., Sweasey, D., Glancy, E. M., and Terlecki, S., 1974. A neurochemical study of field cases of the delayed spinal form of swayback (enzootic ataxia) in lambs, *J. Neurochem.* 23:1245–1253.

Patterson, D. S. P., Brush, P. J., Foulkes, J. A., and Sweasey, D., 1974. Copper metabolism and the composition of wool in Border disease, *Vet. Rec.* 95:214–215.

Pawelek, J., Korner, A., Bergstrom, A., and Bologna, J., 1980. New regulators of melanin biosynthesis and the autodestruction of melanoma cells, *Nature* 286:617–619.

Paynter, D. I., 1984. Copper containing enzymes in the sheep: factors affecting their activities and requirements, in *Ruminant Physiology, Concepts and Consequences*, S. K. Baker, J. M. Gawthorne, J. B. MacIntosh, and D. B. Purser (eds.), University of Western Australia, Perth, p. 329.

Paynter, D. I., 1987. The diagnosis of copper insufficiency, in *Copper in Animals and Man*, Vol. 1 (J. M. Howell and J. M. Gawthorne, eds.), CRC Press, Boca Raton, Florida, pp. 101–120.

Paynter, D. I., Moir, R. J., and Underwood, E. J., 1979. Changes in activity of the Cu-Zn superoxide dismutase enzyme in tissues of the rat with changes in dietary copper, *J. Nutr.* 109:1570–1576.

Paynter, J. A., Camakaris, J., and Mercer, J. F. B., 1990. Analysis of hepatic copper, zinc, metallothionein, and metallothionein-1a mRNA in developing sheep, *Vet. Rec.* 95:214–215.

Pedroni, E., Bianchi, E., Ugazio, A. G., and Burgio, G. R., 1975. Immunodeficiency and steely hair, *Lancet* i:1303–1304.

Peeters-Joris, C., Vandevoorde, A.-M., and Baudhuin, P., 1975. Subcellular localization of superoxide dismutase in rat liver, *Biochem. J.* 150:31–39.

Pekarek, R. S., Powanda, M. C., and Wannemacher, R. W., Jr., 1972. Effect of leukocyte endogenous mediator (LEM) on serum copper and ceruloplasmin concentrations in the rat, *Proc. Soc. Exp. Biol. Med.* 141:1029–1031.

Pember, S. O., Villefranca, J. J., and Benkovic, S. J., 1986. Phenylalanine hydroxylase from *Chromobacterium violaceum* is a copper-containing monooxygenase. Kinetics of the reductive activation of the enzyme, *Biochemistry* 25:6611–6619.

Pember, S. O., Benkovic, S. J., Villafranca, J. J., Pasenkiewicz-Gierula, M., and Antholine, W. E., 1987. Adduct formation between the cupric site of phenylalanine hydroxylase from *Chromobacterium violaceum* and 6,7-dimethyltetrahydropterin, *Biochemistry* 26:4477–4483.

Percival, S. S., and Harris, E. D., 1989. Ascorbate enhances copper transport from ceruloplasmin into human K562 cells, *J. Nutr.* 119:779–784.

Pennington, J. T., and Calloway, D. H., 1973. Copper content of foods, *J. Am. Dietet. Assoc.* 63:143–153.

Perin, A., Sessa, A., and Desiderio, M. A., 1983. Polyamine levels and diamine oxidase activity in hypertrophic heart of spontaneously hypertensive rats and of rats treated with isoproterenol, *Biochim. Biophys. Acta* 755:344.

Perlmutter, D. H., Dinarello, C. A., Punsal, P. I., and Colten, H. R., 1986. Cachectin/tumor necrosis factor regulates hepatic acute-phase gene expression, *J. Clin. Invest.* 78:1349–1354.

Perrin, D. D., 1965. Multiple equilibria in assemblages of metal ions and complexing species: A model for biological systems, *Nature* 206:170–171.

Perrin, D. D., and Agarwal, R. P., 1973. Multimetal-multiligand equilibria: A model for biological systems, in *Metal Ions in Biological Systems* (H. Sigel, ed.), Marcel Dekker, New York, pp. 168–206.

Petering, D. H., 1977. Reaction of copper complexes with Ehrlich cells, *Adv. Exp. Med. Biol.* 91:179–197.

Petering, D. H., 1980. Carcinostatic copper complexes, in *Metal Ions in Biological Systems*, Vol. II (H. Sigel, ed.), Marcel Dekker, New York, pp. 198–229.

Petering, D. H., Krezoski, S., Lehn, D., Stone, D., and Loomans, H., 1980. Use of metal chelating agents to modulate Ehrlich cell growth- multiple metal-ligand interactions, *Trace Element Metabolism in Man and Animals*, Symposium presentation.

Petering, H. G., Murthy, L., Stemmer, K. L., Finelli, V. N., and Menden, E. E., 1986. Effects of copper deficiency on the cardiovascular system of the rat, *Biol. Trace Element Res.* 9:251–270.

Peters, R. A., Stocken, L. A., and Thompson, R. H. S., 1945. British antilewisite (BAL), *Nature* 156:616–619.

Peters, R., and Walshe, J. M., 1966. Studies on the toxicity of copper. I. The toxic action of copper *in vivo* and *in vitro*, *Proc. R. Soc. London, Ser. B* 166:273.

Peters, R., Shorthouse, M., and Walshe, J. M., 1966. Studies on the toxicity of copper. II. The behaviour of microsomal membrane ATPase of the pigeon's brain tissue to copper and some other metallic substances, *Proc. R. Soc. London, Ser. B* 166:285.

Peters, T., 1977. Serum albumin: Recent progress in the understanding of its structure and biosynthesis, *Clin. Chem.* 23:5–12.

Peters, T., and Blumenstock, F. A., 1967. Copper-binding properties of bovine serum albumin and its amino-terminal peptide fragment, *J. Biol. Chem.* 242:1574–1578.

Pettersson, G., 1985. Plasma amine oxidase, in *Structure and Functions of Amine Oxidases* (B. Mondovi, ed.), CRC Press, Boca Raton, Florida, pp. 105–125.

Pettit, L. D., and Formicka-Kozlowska, G., 1984. A suggested role for copper in the biological activity of neuropeptides, *Neurol. Sci. Lett.* 50:53–56.
Pfizenmaier, K., Scheurich, P., Schluter, C., and Kronke, M., 1987. Tumor necrosis factor enhances HLA-A, B, C and HLA-DR gene expression in human tumor cells, *J. Immunol.* 138:975–980.
Phillips, J. H., Allison, Y. P., and Morris, S. J., 1977. The distribution of calcium, magnesium, copper and iron in the bovine adrenal medulla, *Neuroscience* 2:147–152.
Piccinni, E., Coppellotti, O., and Guidolini, L., 1985. Chelatins in *Euglena gracilis* and *Ochromonas danica*, *Comp. Biochem. Physiol.* 82C:29–36.
Pickart, L., 1987. Iamin: a human growth factor with multiple wound-healing properties, in *Biology of Copper Complexes* (J. R. J. Sorenson, ed.), Humana Press, Clifton, New Jersey, pp. 273–285.
Pickart, L., and Lovejoy, S., 1987. Biological activity of human plasma copper-binding growth factor glycyl-L-histidyl-L-lysine, *Methods Enzymol. (Pept. Growth Factors. Pt. B)* 147:314–328.
Pickart, L., and Thaler, M. M., 1973. Tripeptide in human serum which prolongs survival of normal liver cells and stimulates growth in neoplastic liver, *Nature (London) New Biol.* 243:85–87.
Pickart, L., and Thaler, M. M., 1980. Growth-modulating tripeptide (glycylhistidyllysine): Association with copper and iron in plasma and stimulation of adhesiveness and growth of hepatoma cells in culture by tripeptide-metal ion complexes, *J. Cell. Physiol.* 102:129–139.
Pickart, L., Thaler, M. M., and Millard, M. M., 1979. Effect of transition metals on recovery of the growth-modulating tripeptide glycylhistidyl-lysine (GHL) from plasma, *J. Chromatogr.* 175:65–73.
Pickart, L., Freedman, J. H., Loker, W. J., Peisach, J., Perkins, C. M., Stenkamp, R. E., and Weinstein, B., 1980. Growth modulating plasma tripeptide may function by facilitating copper uptake into cells, *Nature (London)* 288:715–717.
Pierre, F., Baruthio, F., Diebold, F., Wild, P., and Goutet, M., 1988. Decreased serum ceruloplasmin concentration in aluminum welders exposed to ozone, *Int. Arch. Occup Environ. Health* 60:95–97.
Piez, K. A., 1968. Cross-linking of collagen and elastin, *Annu. Rev. Biochem.* 37:547–570.
Pinchin, M. J., Newham, J., and Thompson, R. P. J., 1978. Lead, copper and cadmium in teeth of normal and mentally retarded children, *Clin. Chim. Acta* 85:89–94.
Pinnell, S. R., and Martin, G. R., 1968. Cross-linking of collagen and elastin—enzymatic conversion of lysine in peptide linkage to α-aminoadipic-δ-semialdehyde (allysine) by an extract from bone, *Proc. Natl. Acad. Sci. USA* 61:708–716.
Piscator, M., 1964. Om kadmium i normala manniskonjurar samt redogorelse for isolering av metallothionein ur lever fran kadmiumexponerade kaniner, *Nord. Hyg. Tidskr.* 45:76–82.
Piscator, M., and Lind, B., 1972. Cadmium, zinc, copper, and lead in human renal cortex, *Arch. Environ. Health* 24:426–431.
Planas, J., and Frieden, E., 1973. Serum iron and ferroxidase activity in normal, copper-deficient, and estrogenized roosters, *Am. J. Physiol.* 225:423–428.
Plesnicar, M., and Bendall, D. S., 1970. The plastocyanin content of chloroplasts from some higher plants estimated by a sensitive enzymatic assay, *Biochim. Biophys. Acta* 216:192.
Poillon, W. N., and Bearn, A. G., 1957. The molecular structure of human ceruloplasmin: Evidence for subunits, *Biochim. Biophys. Acta* 127:407–427.
Poillon, W. N., Maeno, H., Koike, K., and Feigelson, P., 1969. Tryptophan oxygenase of *pseudomonas acidovorans*, *J. Biol. Chem.* 244:3447–3456.
Pond, W. G., Walker, E. F., Jr., Kirtland, D., and Rounsaville, T., 1978. Effect of dietary Ca,

Cu and Zn level on body weight gain and tissue mineral concentrations of growing pigs and rats, *J. Anim. Sci.* 47:1128–1134.

Porter, H., 1963. The intracellular distribution and chromatographic separation of copper proteins in Wilson's disease, *Trans. Am. Neurol. Assoc.* 88:159–164.

Porter, H., and Folch, J., 1957. Cerebrocuprein. 1. A copper-containing protein isolated from brain, *J. Neurochem.* 1:260–271.

Porter, H., Sweeney, M., and Porter, E. M., 1964a. Neonatal hepatic mitochondrocuprein. II. Isolation of the copper-containing subfraction from mitochondria of newborn human liver, *Arch. Biochem. Biophys.* 104:97–101.

Pos, O., van Dijk, W., Ladiges, N., Linthorst, C., Sala, M., Van Tiel, D., and Boers, W., 1988. Glycosylation of four acute-phase glycoproteins secreted by rat liver cells *in vivo* and *in vitro*. Effects of inflammation and dexamethasone, *Eur. J. Cell Biol.* 46:121–128.

Potts, A. M., and Au, P. C., 1976. The affinity of melanin for inorganic ions, *Exp. Eye Res.* 22:487–491.

Poulik, M. D., 1962. Electrophoretic and immunological studies on structural sub-units of human ceruloplasmin, *Nature* 194:842–844.

Poulik, M. D., and Weiss, M. L., 1975. Ceruloplasmin, in *The Plasma Proteins*, Vol. 2, 2nd Ed. (F. W. Putnam, ed.), Academic Press, New York, pp. 51–108.

Poyton, R. O., and Kavanagh, J., 1976. Regulation of mitochondrial protein synthesis by cytoplasmic proteins *Proc. Natl. Acad. Sci. USA* 73:3947–3951.

Pratt, C. W., and Pizzo, S. V., 1984. The effect of zinc and other divalent cations on the structure and function of human α_2-macroglobulin, *Biochim. Biophys. Acta* 791:123–130.

Prema, K., and Ramalakshmi, B. A., 1980. Serum copper and zinc in hormonal contraceptive users, *Fertil. Steril.* 33:267–271.

Premakumar, R., Winge, D. R., Wiley, R. D., and Rajagopalan, K. V., 1975a. Copper-induced synthesis of copper-chelatin in rat liver, *Arch. Biochem. Biophys.* 170:267–277.

Premakumar, R., Winge, D. R., Wiley, R. D., and Rajagopalan, K. V., 1975b. Copper-chelatin: Isolation from various eucaryotic sources, *Arch. Biochem. Biophys.* 170:278–288.

Pribyl, T., and Schreiber, V., 1977. Effect of oestrogen and ascorbic acid on the serum ceruloplasmin level in rats, *Physiol. Bohemoslov.* 26:325–330.

Prins, H. W., and Van den Hamer, C. J. A., 1978. Cu-content of the liver of young mice in relation to a defect in the Cu-metabolism, in *Trace Element Metabolism in Man and Animals*–3 (M. Kirchgessner, ed.), Arbeitskreis für Tierernährungsforschung Weihenstephan, Freising, Germany, pp. 397–400.

Prins, H. W., and Van den Hamer, C. J. A., 1980. Abnormal copper-thionein synthesis and impaired copper utilization in mutated brindled mice: Model for Menkes' disease, *J. Nutr.* 110:151–157.

Prochaska, L. J., and Fink, P. S., 1987. On the role of subunit III in proton translocation in cytochrome *c* oxidase, *J. Bioenerg. Biomembr.* 19:143–166.

Prochaska, L. J., and Reynolds, K. A., 1986. Characterization of electron-transfer and proton-translocation activities in bovine heart mitochondrial cytochrome *c* oxidase deficient in subunit III, *Biochemistry* 25:781–787.

Prohaska, J. R., 1980. Normal copper metabolism in quaking mice, *Life Sci.* 26:731–735.

Prohaska, J. R., 1981a. Comparison between dietary and genetic copper deficiency in mice: Copper-dependent anemia, *Nutr. Res.* 1:159–167.

Prohaska, J. R., 1981b. Changes in brain enzymes accompanying deficiencies of the trace elements, copper, selenium, or zinc, in *Trace Element Metabolism in Man and Animals (TEMA-4)* (M. J. Howell, J. M. Gawthorne, and C. L. White, eds.), Australian Academy of Science, Canberra, pp. 275–282.

Prohaska, J. R., 1983a. Comparison of copper metabolism between brindled mice and dietary copper-deficient mice using ^{67}Cu, *J. Nutr.* 113:1312–1220.

Prohaska, J. R., 1983b. Changes in tissue growth, concentrations of copper, iron, cytochrome oxidase and superoxide dismutase subsequent to dietary or genetic copper deficiency in mice, *J. Nutr.* 113:2148–2158.
Prohaska, J. R., 1984. Repletion of copper-deficient mice and brindled mice with copper or iron, *J. Nutr.* 114:422–430.
Prohaska, J. R., 1986. Genetic diseases of copper metabolism, *Clin. Physiol. Biochem.* 4:87–93.
Prohaska, J. R., 1987. Functions of trace elements in brain metabolism, *Physiol. Rev.* 67:858–901.
Prohaska, J. R., 1988. Biochemical functions of copper in animals, in *Essential and Toxic Trace Elements in Human Health and Disease* (A. S. Prasad, ed.), Alan R. Liss, New York, pp. 105–124.
Prohaska, J. R., 1989. Effect of diet on milk copper and iron content of normal and heterozygous brindled mice, *Nutr. Res.* 9:353–356.
Prohaska, J. R., 1990a. Development of copper deficiency in neonatal mice. *J. Nutr. Biochem.* 1:415–419.
Prohaka, J. R., 1990b. Biochemical changes in copper deficiency. *J. Nutr. Biochem.* 1:452–461.
Prohaska, J. R., and Cox, D. A., 1983. Decreased brain ascorbate levels in copper-deficient mice and in brindled mice, *J. Nutr.* 113:2623–2629.
Prohaska, J. R., and DeLuca, K. L., 1988. Norepinephrine and dopamine distribution in copper-deficient mice, in *Trace Elements in Man and Animals* (TEMA-6) (L. S. Hurley, C. L. Keen, B. Lonnerdal, and R. B. Rucker, eds.), Plenum, New York, pp. 109–111.
Prohaska, J. R., and Heller, L., 1982. Mechanical properties of the copper-deficient rat heart, *J. Nutr.* 112:2142.
Prohaska, J. R., and Lukasewycz, O. A., 1981. Copper deficiency suppresses the immune response of mice, *Science* 213:559–561.
Prohaska, J. R., and Lukasewycz, O. A., 1989a. Biochemical and immunological changes in mice following postweaning copper deficiency, *Biol. Trace Elem. Res.* 22:101–112.
Prohaska, J. R., and Lukasewycz, O. A., 1989b. Copper deficiency during perinatal development: Effects on the immune response of mice, *J. Nutr.* 119:922–931.
Prohaska, J. R., and Lukasewycz, O. A., 1990. Effects of copper deficiency on the immune system, *Adv. Exp. Med. Biol. (Antioxid. Nutr. Immune. Funct.)* 262:123–143.
Prohaska, J. R., and Smith, T. L., 1982. Effect of dietary or genetic copper deficiency on brain catecholamines, trace metals and enzymes in mice and rats, *J. Nutr.* 112:1706–1717.
Prohaska, J. R., and Wells, W. W., 1974. Copper deficiency in the developing rat brain: A possible model for Menkes' steely-hair disease, *J. Neurochem.* 23:91–98.
Prohaska, J. R., and Wells, W. W., 1975. Copper deficiency in the developing rat brain: Evidence for abnormal mitochondria, *J. Neurochem.* 25:221–228.
Prohaska, J. R., Downing, S. W., and Lukasewycz, O. A., 1983. Chronic dietary copper deficiency alters biochemical and morphological properties of mouse lymphoid tissues, *J. Nutr.* 113:1583–1590.
Prohaska, J. R., Cox, D. A., and Bailey, W. R., 1984. Ascorbic acid synthesis and concentrations in organs of copper-deficient and brindled mice, *Biol. Trace Elem. Res.* 6:441–453.
Prohaska, J. R., Bailey, W. R., and Cox, D. A., 1985a. Failure of iron injection to reverse copper-dependent anemia in mice, in *Trace Elements in Man and Animals* (TEMA-5) (C. F. Mills, I. Bremner, and J. K. Chesters, eds.), Commonwealth Agricultureal Bureau, Farnham Royal, Slough, UK, pp. 27–31.
Prohaska, J. R., Korte, J. J., and Bailey, W. R., 1985b. Serum cholesterol levels are not elevated in young copper-deficient rats, mice or brindled mice, *J. Nutr.* 115:1702–1707.
Prohaska, J. R., Wittmers, L. E., Jr., and Haller, E. W., 1988. Influence of genetic obesity, food intake and adrenalectomy in mice on selected trace element-dependent protective enzymes, *J. Nutr.* 118:739–746.

Prozorovski, V. N., Rashkovetsie, L. G., Shavolvski, M. M., Vasiliev, V. B., and Neifakh, S. A., 1982. Evidence that human ceruloplasmin molecule consists of homologous parts, *Int. J. Pept. Protein Res.* 19:40–53.

Puget, K., and Michelson, A. M., 1974. Isolation of a new copper-containing superoxide dismutase bacteriocuprein, *Biochem. Biophys. Res. Commun.* 58:830–838.

Puttner, I., Carafoli, E., and Malatesta, F., 1985. Spectroscopic and functional properties of subunit III-depleted cytochrome oxidase, *J. Biol. Chem.* 260:3719–3723.

Quinones, S. R., and Cousins, R. J., 1984. Augmentation of dexamethasone induction of rat liver metallothionein by zinc, *Biochem. J.* 219:959–963.

Ragan, H. A., Nacht, S., Lee, G. R., Bishop, C. R., and Cartwright, G. E., 1969. Effect of ceruloplasmin on plasma iron in copper deficient swine, *Am. J. Physiol.* 217:1320–1323.

Rainsford, J. D., 1982. Development and therapeutic actions of oral copper complexes of antiinflammatory drugs, in *Inflammatory Diseases and Copper* (J. R. J. Sorenson, ed.), Human Press, Clifton, New Jersey, pp. 375–384.

Rajan, K. S., Colburn, R. W., and Davis, J. M., 1976. Distribution of metal ions in the subcellular fractions of several rat brain areas, *Life Sci.* 18:423–432.

Raju, K. S., 1983. Isolation and characterization of copper-binding sites of human ceruloplasmin, *Mol. Cell. Biochem.* 56:81–88.

Raju, K. S., Alessandri, G., Ziche, M., and Gullino, P. N., 1982. Ceruloplasmin, copper ions and angiogenesis, *J. Natl. Cancer Inst.* 69:1183–1188.

Raju, K. S., Alessandri, G., and Gullino, P. M., 1984. Characterization of a chemoattractant for endothelium induced by angiogenesis effectors, *Cancer Res.* 44:1579–1584.

Ramadori, G., Rieder, H., Meuer, S., and Meyer zum Bueschenfelde, K. H., 1986. Interleukin-1: A new hormone, *Dtsch. Med. Wochenschr.* 111:1032–1038.

Ramharack, R., and Deeley, R. G., 1987. Structure and evolution of primate cytochrome *c* oxidase subunit II gene, *J. Biol. Chem.* 262:14014–14021.

Ran, Y., 1989. Forms and availability of copper in soil in loessial region, *Turang Tongbao* 20:232–234 (CAS Abstr.).

Rao, E. A., Saryan, L. A., Antholine, W. E., and Petering, D. H., 1980. Cytotoxic and antitumor properties of bleomycin and several of its metal complexes, *J. Med. Chem.* 23:1310–1318.

Rapoport, S. I., and Pettigrew, K. D., 1979. A heterogeneous pore-vesicle membrane model for protein transfer from blood to cerebrospinal fluid at the choroid plexus, *Microvasc. Res.* 18:105–119.

Rauser, W. E., 1984. Partial purification and characterization of copper-binding protein from roots of *Agrostis gigantea* Roth, *J. Plant Physiol.* 115:143.

Rauser, W. E., and Curvetto, N. R., 1980. Metallothionein occurs in roots of *Agrostis* tolerant to excess copper, *Nature (London)* 287:563.

Ravin, H. A., 1961. An improved colorimetric enzymatic assay of ceruloplasmin, *J. Lab. Clin. Med.* 58:161–168.

Rawlings, N. D., and Barrett, A. J., 1990. Evolution of proteins of the cystatin superfamily, *J. Mol. Evol.* 30:60–71.

Rayton, J. K., and Harris, E. D., 1979. Induction of lysyl oxidase with copper, *J. Biol. Chem.* 254:621–626.

Re, R., 1987. The myocardial intracellular renin-angiotensin system, *Am. J. Cardiol.* 59:56A–58A.

Recant, L., Voyles, N. R., Timmers, K. I., Zalenski, C., Fields, M., and Bhathena, S. J., 1986. Copper deficiency in rats increases pancreatic enkephalin-containing peptides and insulin, *Peptides* 7:1061–1069.

Rechenberger, J., 1957. Serumkupfer und Schilddrüsentätigkeit, *Dtsch. Z. Verdau. Stoffwechselkr.* 17:139–145.
Reed, C. A., 1985. Hemocyanin cooperativity: A copper coordination chemistry perspective, in *Biological and Inorganic Copper Chemistry* (K. D. Karlin and J. Zubieta, eds.), Adenine Press, Guilderland, New York, pp. 61–74.
Reed, D. W., Passon, P. G., and Hultquist, D. E., 1970. Purification and properties of a pink copper protein from human erythrocytes, *J. Biol. Chem.* 245:2954–2961.
Reed, G. A., and Madhu, C., 1987. Peroxide scavenging by Cu(II) sulfate and Cu(II) (3,5-diisopropylsalicylate)$_2$, in *Biology of Copper Complexes* (J. R. J. Sorenson, ed.), Humana: Clifton, New Jersey pp. 287–298.
Reich, K. A., Marshall, L. E., Graham, D. R., and Sigman, D. S., 1981. Cleavage of DNA by the 1,10-phenanthroline–copper ion complex. Superoxide mediates the reaction dependent on NADH and hydrogen peroxide, *J. Am. Chem. Soc.* 103:3582–3584.
Reid, L. S., Gray, H. B., Dalvit, C., Wright, P. E., and Saltman, P., 1987. Electron transfer from cytochrome b$_5$ to iron and copper complexes, *Biochemistry* 26:7102–7107.
Reid, T. C., McAllum, H. J. F., and Johnstone, O. D., 1980. Liver copper concentrations in red deer (*Cervus elaphus*) and wapiti (*C. canadensis*), *Res. Vet. Sci.* 28:261–262.
Reinhammar, B., 1984. Laccase, in *Copper Proteins and Copper Enzymes*, Vol. 3 (R. Lontie, ed.), CRC Press, Boca Raton, Florida, Chapter 1.
Reinhammar, B., and Malmström, B. G., 1981. "Blue" copper-containing oxidases, in *Copper Proteins* (T. G. Spiro, ed.), John Wiley and Sons, New York, pp. 111–149.
Reiser, S., Ferretti, R. J., Fields, M., and Smith, Jr., J. C., 1983. Role of dietary fructose in the enhancement of mortality and biochemical changes associated with copper deficiency in rats, *Am. J. Clin. Nutr.* 38:214–222.
Reiser, S., Smith, J. C., Mertz, W., Holbrook, J. T., Schonfield, D. J., Powell, A. S., Canfield, W. K., and Canary, J. J., 1985. Indices of copper status in humans consuming a typical American diet containing either fructose or starch, *Am. J. Clin. Nutr.* 42:242–251.
Reiser, S., Powell, A., Yang, C.-Y., and Canary, J. J., 1987. Effect of copper intake on blood cholesterol and its lipoprotein distribution in men, *Nutr. Rep. Intl.* 36:641–649.
Renston, R. H., Maloney, D. G., Jones, A. L., Hradek, G. T., Wong, K. Y., and Goldfine, I. D., 1980. Bile secretory apparatus: Evidence for a vesicular transport mechanism for proteins in the rat, using horseradish peroxidase and [^{125}I] insulin, *Gastroenterology* 78:1373–1388.
Rice, E. W., 1962. Standardization of ceruloplasmin activity in terms of international enzyme units. Oxidative formation of "Brandrowski's Base" from *p*-phenylenediamine by ceruloplasmin, *Anal. Biochem.* 3:452–456.
Rich, P. R., West, I. C., and Mitchell, P., 1988. The location of Cu$_A$ in mammalian cytochrome *c* oxidase, *FEBS Lett.* 233:25–30.
Richards, M. P., and Cousins, R. J., 1975a. Influence of parenteral zinc and actinomycin D on tissue zinc uptake and the synthesis of a zinc-binding protein, *Bioinorg. Chem.* 4:215–224.
Richards, M. P., and Cousins, R. J., 1975b. Mammalian zinc homeostasis: Requirements for RNA and metallothionein synthesis, *Biochem. Biophys. Res. Commun.* 64:1215–1223.
Richards, M. P., and Cousins, R. J., 1977. Isolation of an intestinal metallothionein induced by parenteral zinc, *Biochem. Biophys. Res. Commun.* 75:286–294.
Richards, R. I., Heguy, A., and Karin, M., 1984. Structural and functional analysis of the human metallothionein-IA gene: Differential induction by metal ions and glucocorticoids, *Cell* 37:263–272.
Ridlington, J. W., Chapman, D. C., Goeger, D. E., and Whanger, P. D., 1981. Metallothionein

and Cu-chelatin: Characterization of metal-binding proteins from tissues of four marine animals, *Comp. Biochem. Physiol.* 70B:93–104.

Rigo, A., Viglino, P., Calabrese, L., Cocco, D., and Rotilio, G., 1977a. The binding of copper ions to copper free bovine superoxide dismutase. Copper distribution in protein samples recombined with less than stoichiometric copper ion/protein ratios, *Biochem. J.* 161:27–30.

Rigo, A., Terenzi, M., Viglino, P., Calabrese, L., and Rotilio, G., 1977b. The binding of copper ions to copper free bovine superoxide dismutase. Properties of the protein recombined with increasing amounts of copper ions, *Biochem. J.* 161:31–35.

Rigo, A., Viglino, P., Bonori, M., Cocco, D., Calabrese, L., and Rotilio, G., 1978. The binding of copper ions to copper-free bovine superoxide dismutase, *Biochem. J.* 169:277–280.

Rinaldi, A., Floris, G., and Giartosio, A., 1985. Plant amine oxidases, in *Structure and Functions of Amine Oxidases* (B. Mondovi, ed.), CRC Press, Boca Raton, Florida, pp. 52–62.

Riordan, J. R., and Gower, I., 1975a. Small copper-binding proteins from normal and copper-loaded liver, *Biochim. Biophys. Acta* 411:393–398.

Riordan, J. R., and Gower, I., 1975b. Purification of low molecular weight copper proteins from copper loaded liver, *Biochem. Biophys. Res. Commun.* 66:678–686.

Riordan, J. R., and Jolicoeur-Pacquet, L., 1982. Metallothionein accumulation may account for intracellular copper retention in Menkes' disease, *J. Biol. Chem.* 257:4639–4645.

Riordan, J. R., and Richards, V., 1980. Human fetal liver contains zinc- and copper-rich forms of metallothionein, *J. Biol. Chem.* 255:5380–5383.

Riordan, J. R., and Zelinka, A., 1985. Copper entry into and exit from cultured human cells, *Fed. Proc.* 44:995.

Ritland, S., and Steinnes, E., 1975. Copper and liver disease, *Scand. J. Gastroenterol*, 10 (Suppl. 34):41–42.

Ritland, S., Eiliv, S., and Sverre, S., 1977. Hepatic copper content, urinary copper excretion, and serum ceruloplasmin in liver disease, *Scand. J. Gastroenterol.* 12:81–88.

Robb, D. A., 1984. Tyrosinase in *Copper Proteins*, Vol. 2 (R. Lontie, ed.), CRC Press, Boca Raton, Florida, pp. 207–241.

Roe, J. A., Butler, A., Scholler, D. M., Valentine, J. S., Marky, L., and Breslauer, K. J., 1988. Differential scanning calorimetry of Cu, Zn-superoxide dismutase, the apoprotein, and its zinc-substituted derivatives, *Biochemistry* 27:950–958.

Roeser, H. P., Lee, G. R., Nacht, S., and Cartwright, G. E., 1970. The role of ceruloplasmin in iron metabolism, *J. Clin. Invest.* 49:2408–2417.

Rogers, J. M., Keen, C. L., and Hurley, L. S., 1985. Zinc, copper and manganese deficiencies in prenatal and neonatal development, with special reference to the central nervous system, in *Metal Ions in Neurology and Psychiatry* (S. Gabay, J. Harris, and B. T. Ho, eds.), Alan R. Liss, New York, pp. 3–34.

Rosenthal, R. W., and Blackburn, A., 1974. Higher copper concentrations in serum than in plasma, *Clin. Chem.* 20:1233–1234.

Rouch, D., Lee, B. T. O., and Camakaris, J., 1989. Genetic and molecular basis of copper resistance in *Escherichia coli*, *UCLA Symp. Mol. Cell. Biol., New Ser. (Met. Ion Homeostasis)* 98:439–436.

Rouch, P. M., Camakaris, J., and Lee, B. T. O., 1989. Copper transport in *Escherichia coli*, *UCLA Symp. Mol. Cell. Biol., New Ser. (Met. Ion Homeostasis)* 98:469–477.

Roussanov, E., and Balevska, P., 1966. Copper content and distribution in subcellular fractions of liver from normal and copper sulphate treated rats, *Izv. Inst. Fiziol.* 10:163–169.

Rowan, L., and Meacham T. N., 1971. Effect of exogenous ACTH treatment on ceruloplasmin activity in rabbit serum, *Res. Div. Rep.* 135:64–66.

Rowe, D. W., McGoodwin, E. B., Martin, G. R., and Grahn, D., 1977. Decreased lysyl oxidase activity in the aneurysm-prone, mottled mouse, *J. Biol. Chem.* 252:939–942.

Royce, P. M., Camakaris, J., and Danks, D. M., 1980. Reduced lysyl oxidase activity in skin fibroblasts from patients with Menkes' syndrome, *Biochem. J.* 192:579–586.

Royle, N. J., Irwin, D. M., Koschinsky, M. L., MacGillivray, R. T. A., and Hamerton, J. L., 1987. Human genes encoding prothrombin and ceruloplasmin map to 11p11-q12 and 3q21-24, respectively, *Somatic Cell Molec. Genet.* 13:285–292.

Rubinfeld, Y., Maor, Y., Simon, D., and Modai, D., 1979. A progressive rise in serum copper levels in women taking oral contraceptives: A potential hazard?, *Fertil. Steril.* 32:599–601.

Rucker, R. B., Riggins, R. S., Laughlin, R., Chan, M. M., Chen, M., and Tom, K., 1975. Effects of nutritional copper deficiency on the biomechanical properties of bone and arterial elastin in the chick, *J. Nutr.* 105:1062.

Ruokonen, A., and Kaar, K., 1985. Effects of desogestrel, levonorgestrel and lynestrenol on serum sex hormone binding globulin, cortisol binding glodulin, ceruloplasmin and HDL-cholesterol, *Eur. J. Obstet. Gynecol. Reprod. Biol.* 20:13–18.

Ruppert, S., Mueller, G., Kwon, B., and Schutz, G., 1988. Multiple transcripts of the mouse tyrosinase gene are generated by alternative splicing, *Embo. J.* 7:2715–2722.

Rusinko, N., and Prohaska, J. R., 1985. Adenine nucleotide and lactate levels in organs from copper-deficient mice and brindled mice, *J. Nutr.* 115:936–943.

Russ, E. M., and Raymunt, J., 1956. Effects of estrogens on total serum copper and ceruloplasmin, *Proc. Soc. Exp. Biol. Med.* 92:465–466.

Rydén, L., 1971. Evidence for proteolytic fragments in commercial samples of human ceruloplasmin, *FEBS Lett.* 18:321–325.

Rydén, L., 1971a. Human ceruloplasmin as a polymorphic glycoprotein; analysis by chromatography on hydroxyapatite, *Int. J. Protein Res.* 3:131–138.

Rydén, L., 1971b. Human ceruloplasmin as a polymorphic glycoprotein. Tryptic glycopeptides from two forms of the protein, *Int. J. Protein Res.* 3:191–200.

Rydén, L., 1972. Single-chain structure of human ceruloplasmin, *Eur. J. Biochem.* 26:380–386.

Rydén, L., 1984. Structure and evolution of the small blue proteins, in *Copper Proteins and Copper Enzymes*, Vol. 1 (R. Lontie, ed.), CRC Press, Boca Raton, Florida, pp. 157–182.

Rydén, L., 1988. Evolution of blue copper proteins, in *Oxidases and Related Redox Systems* (T. E. King, H. S. Mason, and M. Morrison, eds.), Alan R. Liss, New York, pp. 349–366.

Rydén, L., and Deutsch, H. F., 1978. Preparations and properties of the major copper-binding component in fetal liver, *J. Biol. Chem.* 253:519–524.

Ryl'kov, V. V., Taras'ev, M. Y., and Moshkov, K. A., 1990. A new form of copper(2+) centers in human ceruloplasmin, *Biokhimiya* 55:1367–1374 (CAS Abstr.).

Sadee, W., Pfeiffer, A., and Herz, A., 1982. Opiate receptor: Multiple effects of metal ions, *J. Neurochem.* 39:659–667.

Sadhu, C., and Gedamu, L., 1989. Metal-specific posttranscriptional control of human metallothionein genes, *Mol. Cell. Biol.* 9:5738–5741.

Saggerson, E. D., Sooranna, S. R., and Evans, C. J., 1976. Insulin-like actions of nickel and other transition-metal ions in rat fat cells, *Biochem. J.* 154:349–357.

Saenko, E. L., and Yaropolov, A. I., 1990. Studies on receptor interaction of ceruloplasmin with human red blood cells, *Biochem. Int.* 20:215–225.

Saenko, E. L., Skorobogat'ko, O. V., and Yaropolov, A. I., 1990. The protective effect of normal and pathological (Wilson disease) ceruloplasmins on human erythrocytes, *Biochem. Int.* 20:463–469.

Sahu, S. K., Oberley, I. W., Stevens, R. H., and Riley, E. F., 1977. Superoxide dismutase activity of Ehrlich ascites tumor cells, *J. Natl. Cancer Inst.* 58:1125–1128.

Saigo, K., Doi, K., Adachi, M., Tatsumi, E., Ryo, R., and Yamaguchi, N., 1989. Inhibition of interleukin-2 receptor expression on T cells by histidine-rich glycoprotein, *Igaki no Ayumi* 151:627–628 (CAS Abstr.)

Saito, K., Saito, T., Draganac, P. S., Andrews, R. B., Lange, R. D., Etkin, L. D., and Farkas, W. R., 1985. Secretion of ceruloplasmin by a human clear cell carcinoma maintained in nude mice, *Biochem. Med.* 33:45–52.

Sakurai, T., and Nakahara, A., 1986. Distinction of two type I coppers in bovine ceruloplasmin, *Inorg. Chim. Acta* 123:217–220.

Salaspuro, M. P., Pikkarainen, P., Sipponen, P., Vuori, E., and Miettinen, T. A., 1981. Hepatic copper in primary biliary cirrhosis: Biliary excretion and response to penicillamine treatment, *Gut* 22:901–906.

Salaverry, G., Mendez, M., del C., Zipper, J., and Medel, M., 1973. Copper determination and localization in different morphologic components of human endometrium during the menstrual cycle in copper intrauterine contraceptive device wearers, *Am. J. Obstet. Gynocol.* 115:163–168.

Salgo, M. P., and Oster, G., 1974. Copper stimulation and inhibition of the rat uterus, *Fertil. Steril.* 25:113–120.

Salmenpera, L., Perheentupa, J., Pakarinem, P., and Siimes, M. A., 1986. Cu nutrition in infants during prolonged exclusive breast-feeding: Low intake but rising serum concentrations of Cu and ceruloplasmin, *Am J. Clin. Nutr.* 43:251–257.

Saltman, P., Alex, T., and McCormack, B., 1959. The accumulation of copper by rat liver slices, *Arch. Biochem. Biophys.* 83:538–547.

Saltzgaber-Muller, J., and Schatz, G., 1978. Heme is necessary for the accumulation and assembly of cytochrome *c* oxidase subunits in *Saccharomyces cerevisiae*, *J. Biol. Chem.* 253:305–310.

Samman, S., and Roberts, D. C. K., 1985. Dietary copper and cholesterol, *Nutr. Res.* 5:1021–1034.

Samokyszyn, V. M., Miller, D. M., Reif, D. W., and Aust, S. D., 1989. Inhibition of superoxide and ferritin-dependent lipid peroxidation by ceruloplasmin, *J. Biol. Chem.* 264:21–26.

Samsahl, K., Brune, D., and Wester, P. O., 1965. Simultaneous determination of 30 trace elements in cancerous and non-cancerous human tissue samples by neutron activation analysis, *Int. J. Appl. Radiat. Isotopes* 16:273–281.

Samuels, A. R., Freedman, J. H., and Bhargave, M. M., 1983. Purification and characterization of a novel abundant protein in rat bile that binds azo dye metabolites and copper, *Biochim. Biophys. Acta* 759:23–31.

Sanada, H., Shikata, J., Hamamoto, H., Ueba, Y., Yamamuro, T., and Takeda, T., 1978. Changes in collagen cross-linking and lysyl oxidase by estrogen, *Biochim. Biophys. Acta* 541:408–413.

Sandmann, G., 1986. Formation of plastocyanin and cytochrome *c*-553 in different species of blue-green algae, *Arch Microbiol.* 145:76–79.

Santamaria, P., Chordi, A., and Ortiz de Lanzaduri, E., 1967. Immunoelectrophoresis of human liver proteins, *Pathologia Europaea* 2:247–436.

Sarkar, B., 1970. State of iron(III) in normal human serum: Low molecular weight and protein ligands besides transferrin, *Can. J. Biochem.* 48:1339–1350.

Sarkar, B., and Kruck, T. P. A., 1966. Copper–amino acid complexes in human serum, in *Biochemistry of Copper* (J. Peisach, P. Aisen, and W. E. Blumberg, eds.), Academic Press, New York, pp. 183–196.

Sarkar, B., and Wigfield, Y., 1968. Evidence for albumin–Cu(II)–amino acid ternary complex, *Can. J. Biochem.* 46:601–607.

Sarkar, B., Laussac, J.-P., and Lau, S.-J. Y., 1983. Transport forms of copper in human serum, in *Biological Aspects of Metals and Metal-Related Diseases* (B. Sarkar, ed.), Raven Press, New York, pp. 23–40.

Sarzeau,--, 1830, Sur la présence du cuivre dans les végétaux et dans le sang. *J. Pharm. Sci. Accessoires* 16:505–518. (Obtained from Owen, 1982c)

Sass-Kortsak, A., 1975. Wilson's disease, a treatable disease in children, *Pediatr. Clin. N. Am.* 22:963–983.

Sass-Kortsak, A., Cherniak, M., Geiger, D. W., and Slater, R. J., 1959. Observations on ceruloplasmin in Wilson's disease, *J. Clin. Invest.* 38:1672–1682.

Sass-Kortsak, A., and Bearn, A. G., 1978. Hereditary diseases of copper metabolism: Wilson's disease (hepatolenticular degeneration) and Menkes' disease (kinky-hair or steely-hair syndrome), in *The Metabolic Basis of Inherited Diseases* (J. B. Steinberg, J. G. Wingaarden, and D. S. Frederickson, eds.), McGraw-Hill, New York, pp. 1098–1126.

Sato, M., and Bremner, I., 1984. Biliary excretion of metallothionein and a possible degradation product in rats injected with copper and zinc, *Biochem. J.* 223:475–479.

Sato, M., Schilsky, M. L., Stockert, R. J., Morell, A. G., and Sternlieb, I., 1990. Detection of multiple forms of human ceruloplasmin. A novel Mr 200,000 form, *J. Biol. Chem.* 265:2533–2537.

Sato, N., and Henkin, R. I., 1973. Pituitary-gonadal regulation of copper and zinc metabolism in the female rat, *Am. J. Physiol.* 225:508–512.

Sato, N., 1987. Recent progress in the research of angiogenic factors, *Jikken Igaku* 5:953–955 (CAS Abstr.).

Sauchelli, V., 1969. *Trace Elements in Agriculture*, Van Nostrand Reinhold, New York.

Sausville, E. A., Peisach, J., and Horwitz, S. B., 1978. Effect of chelating agents and metal ions on the degradation of DNA by bleomycin, *Biochemistry* 17:740–746.

Savin, M. A., and Cook, J. D., 1980. Mucosal iron transport by rat intestine, *Blood* 56:1029–1035.

Savina, P. N., 1971. Copper and ascorbic acid content in the blood of patients with diffuse toxic and different forms of endemic goiter, *Sov. Med.* 34:148–150.

Sawyer, D. T., and Valentine, J. S., 1981. How super is superoxide? *Acc. Chem. Res.* 14:393–400.

Saxena, A., and Fleming P. O. J., 1983. Isolation and reconstitution of the membrane bound form of dopamine β-hydroxylase, *J. Biol. Chem.* 258:4147–4152.

Scandhan, K. P., and Mazumdar, B. N., 1979. Semen copper in normal and infertile subjects, *Experientia* 15:877–878.

Schatz, G., and Mason, T. L., 1974. The biosynthesis of mitochondrial proteins, *Annu. Rev. Biochem.* 43:51–87.

Schechinger, T., Hartmann, H.-J., and Weser, U., 1986. Copper transport from Cu(I)-thionein into apocaeruloplasmin mediated by activated leucocytes, *Biochem. J.* 240:281–283.

Scheinberg, I. H., and Morell, A. G., 1973. Ceruloplasmin in *Organic Biochemistry* (G. I. Eichorn, ed.), Elsevier, New York, pp. 306–319.

Scheinberg, I. H., and Sternlieb, I., 1960. Copper metabolism. *Pharmacol. Rev.* 13:355–381.

Scheinberg, I. H., and Sternlieb, I., 1967. Copper metabolism and the central nervous system, in *Molecular Basis of Some Aspects of Mental Activity*, Vol. 2 (O. Walaas, ed.), Academic Press, New York, pp. 115–124.

Scheinberg, I. H., Cook, C. D., and Murphy, J. A., 1954. Concentration of copper and ceruloplasmin in maternal and infant plasma at delivery, *J. Clin. Invest.* 33:963 (Abstract).

Scheving, I. E., Pauly, J. E., Kanabrocki, E. L., and Kaplan, E., 1968. A 24-hour rhythm in serum copper and manganese levels of normal and adrenal medullectomized adult male rats, *Texas Rep. Biol. Med.* 26:341–347.

Schiff, J. M., Fisher, M. M., and Underdown, B. J., 1984. Receptor-mediated biliary transport of immunoglobulin A and asialoglycoprotein: Sorting and missorting of ligands revealed by two radiolabelling methods, *J. Cell Biol.* 98:79–89.

Schlesinger, D. H., Pickart, L., and Thaler, M. M., 1977. Growth-modulating serum tripeptide is glycyl-histidyl-lysine, *Experientia* 33:324–325.

Schmidt, C. J., and Hamer, D. H., 1983. Cloning and sequence analysis of two monkey metallothionein cDNAs, *Gene* 24:137–146.

Schmidt, C. J., Hamer, D. H., and McBride, O. W., 1984. Chromosomal location of human metallothionein genes: Implications for Menkes' disease, *Science* 224:1104–1106.

Schmitt, R. C., Darwish, H. M., Cheney, J. C., and Ettinger, M. J., 1983. Copper transport kinetics by isolated rat hepatocytes, *Am. Physiol. Soc.* 244:G183–G191.

Schoenemann, H. M., Failla, M. L., and Steele, N. C., 1990. Consequences of severe copper deficiency are independent of dietary carbohydrate in young pigs, *Am. J. Clin. Nutr.* 52:147–154.

Schraer, R., and Schraer, H., 1965. Changes in metal distribution of the avian oviduct during the ovulation cycles, *Proc. Soc. Exp. Biol. Med.* 119:937–942.

Schreiber, G., 1987. Synthesis, processing, and secretion of plasma proteins by the liver and other organs and their regulation, in *The Plasma Proteins*, Vol. V, 2nd ed. (F. W. Putnam, ed.), Academic Press, New York, pp. 293–361.

Schreiber, V., and Pribyl, T., 1977a. Effect of interaction of estrogen, testosterone and thyroid hormones on the serum ceruloplasmin level in rats, *Physiol. Bohemoslov.* 26:129–137.

Schreiber, V., and Pribyl, T., 1977b. Antioestrogenic action of the aldosterone antagonist canrenoate K in the rat (adenohypophysis, ceruloplasmin), *Physiol. Bohemoslov.* 26:385–395.

Schreiber, V., and Pribyl, T., 1978. The effect of hexose monophosphate shunt inhibitors on adenohypophyseal and ceruloplasmin reactions to oestrogen, *Physiol. Bohemoslov.* 27:301–307.

Schreiber, V., Pribyl, T., and Jahodova, J., 1978. Increase of ceruloplasmin level in blood of rats after ACTH, *Endocrinol. Exp.* 12:115–118.

Schroeder, H. A., 1973. *The Trace Elements and Man*, Devin-Adair, Old Greenwich, Connecticut.

Schroeder, H. A., 1974. The role of trace elements in cardiovascular diseases, *Med. Clin. N. Amer.* 58:381–396.

Schroeder, H. A. and Cousins, R. J., 1990. Interleukin 6 regulates metallothionein gene expression and zinc metabolism in hepatocyte monolayer cultures. *Proc. Nat. Acad. Sci. (USA)* 87:3137–3141.

Schultze, M. O., 1939. The effect of deficiencies in copper and iron on the cytochrome oxidase of rat tissues, *J. Biol. Chem.* 129:729.

Schultze, M. O., Elvehjem, C. A., and Hart, E. B., 1936. Further studies on the availability of copper from various sources as a supplement to iron in hemoglobin formation, *J. Biol. Chem.* 115:453–457.

Schwarz, F. J., and Kirchgessner, M., 1973. Wechselwirkungen bei der intestinalen Absorption von ^{64}Cu, ^{65}Zn, und ^{59}Fe nach Cu-, Zn- oder Fe-depletion, *Int. Z. Vit. Ern. Forsch.* 44:116–126.

Schwarz, F. J., and Kirchgessner, M., 1974. Intestinal absorption of copper, zinc, and iron after dietary depletion, in *Trace Element Metabolism in Animals—2* (W. G. Hoekstra, J. W. Suttie, H. E. Ganther, and W. Mertz, eds.), University Park Press, Baltimore, pp. 519–522.

Schwarz, F. J., Kirchgessner, M., and Spengler, M. A., 1981. Zur Absorption und endogenen Exkretion des Spurenelements Kupfer bei trachtigen und laktierenden Ratten, *Z. Tierphysiol, Tierernahrg. Futtermittelkde.* 46:207–213.

Schweigerer, L., 1988. Basic fibroblast growth factor and its relation to angiogenesis in normal and neoplastic tissue, *Klin. Wochenschr.* 66:340–345.

Schweigerer, L., Malerstein, B., and Gospodarowicz, D., 1987. Tumor necrosis factor inhibits the proliferation of cultured capillary endothelial cells, *Biochem. Biophys. Res. Commun.* 143:997–1004.

Scudder, P. R., Al-Timini, D., McMurray, W., White, A. G., Zoob, B. C., and Dormandy, T. L., 1978a. Serum copper and related variables in rheumatoid arthritis, *Ann. Rheum. Dis.* 37:67–70.

Scudder, P. R., McMuarray, W., White, A. G., and Dormandy, T. L., 1978b. Synovial fluid copper and related variables in rheumatoid and degenerative arthritis, *Ann. Rheum. Dis.* 37:71–72.

Seagrave, J., Hanners, J. L., Taylor, W., and O'Brien, H. A., 1986. Transfer of copper from metallothionein to nonmetallothionein proteins in cultured cells, *Biol. Trace Elem. Res.* 10:163–173.

Seaman, W. E., Gindhart, T. D., Blackman, M. A., Dalal, B., Talal, N., and Werb, Z., 1982. Suppression of natural killing *in vitro* by monocytes and polymorphonuclear leukocytes, *J. Clin. Invest.* 69:876–888.

Searle, P. F., Davison, B. L., Stuart, G. W., Wilkie, T. M., Norstedt, G., and Palmiter, R. D., 1984. Regulation, linkage and sequence of mouse metallothionein I and II genes, *Mol. Cell. Biol.* 4:1221–1230.

Searle, P. F., 1990. Zinc dependent binding of a liver nuclear factor to metal response element MRE-a of the mouse metallothionein-1 gene and variant sequences, *Nucleic Acids Res.* 18:4683–4690.

Seelig, A., and Seelig, J., 1985. Phospholipid composition and organization of cytochrome *c* oxidase preparations as determined by ^{31}P-nuclear magnetic resonance, *Biochim. Biophys. Acta* 815:153–158.

Segschneider, P., 1949. Das Verhalten des Serumkupfers bei der gesunden nichtschwangeren und schwangeren Frau. *Z. Geburtshilfe Gynakol.* 130:142–167.

Serfass, R. E., Park, K.-E., and Kaplan, M. L., 1988. Developmental changes of selected minerals in Zucker rats, *Proc. Soc. Exp. Biol. Med.* 189:229–239.

Sessa, A., Desiderio, M. A., and Perin, A., 1982. Diamine oxidase activity induction in regenerating rat liver, *Biochim. Biophys. Acta* 698:11.

Seymour, C. A., 1987. Copper toxicity in man, in *Copper in Animals and Man*, Vol. II (J. M. Howell and J. M. Gawthorne, eds.), CRC Press, Boca Raton, Florida, pp. 79–106.

Shaikh, Z. A., and Hirayama, K., 1979. Metallothionein in the extracellular fluids as an index of cadmium toxicity, *Environ. Health Perspect.* 28:267–271.

Shapiro, R., Strydom, D. J., Olson, K. A., and Vallee, B. L., 1987. Isolation of angiogenin from normal human plasma, *Biochemistry* 26:5141–5146.

Sharanov, B. P., Govorova, N. J., and Lyzlova, S. N., 1988. A comparative study of serum proteins ability to scavenge active oxygen species: O^- and OCl^-, *Biochem. Int.* 17:783–790.

Sharma, R. P., and McQueen, E. G., 1981. Effects of gold sodium thiomalate on cytosolic copper and zinc in the rat kidney and liver tissues, *Clin. Exp. Pharmacol. Physiol.* 8:591–599.

Sharma, S. C., Robinson, P., and Wilson, C. W. M., 1976. Effect of pharmacological doses of ascorbic acid on copper acetate-induced ovulation in the rabbit, *Proc. Nutr. Soc.* 35:10A–11A.

Sharoyan, S. G., Shaljian, A. A., Nalbandyan, R. M., and Buniatian, H. C., 1977. Two copper-containing proteins from white and gray matter of brain, *Biochim. Biophys. Acta* 493:478–487.

Shatzman, A. R., and Kosman, D. J., 1978. The utilization of copper and its role in the biosynthesis of copper-containing proteins in the fungus, *Dactylium dendroides*, *Biochim. Biophys. Acta* 544:163–179.

Shatzman, A. R., and Kosman, D. J., 1979a. Characterization of two copper-binding components of the fungus, *Dactylium dendroides*, *Arch. Biochem. Biophys.* 194:226–235.

Shatzman, A. R., and Kosman, D. J., 1979b. Biosynthesis and cellular distribution of the two superoxide dismutases of *Dactylium dendroides*, *J. Bacteriol.* 137:313–320.

Shearer, T. R., Lis, E. W., Johnson, K. S., Johnson, J. R., and Prescott, G. H., 1980. Copper and zinc in the amniotic fluid and serum from high-risk pregnant women, *Proc. Soc. Exp. Biol. Med.* 161:382–385.

Sherman, A. R., and Tschiember, N., 1979. Tissue copper in offspring of iron deficient rats, *Fed. Proc.* (Abstr.) 38:453.

Sherman, L., Levanon, D., Lieman-Hurwitz, J., Dafni, N., and Groner, Y., 1984. Human Cu/Zn superoxide dismutase gene: Molecular characterization of its two mRNA species, *Nucleic Acids Res.* 12:9349–9365.

Shibahara, S., Tomita, Y., Sakakura, T., Nager, C., Chaudhuri, B., and Muller, R., 1986. Cloning and expression of cDNA encoding mouse tyrosinase, *Nucleic Acids Res.* 14:2413–2427.

Shieh, J. J., and Yasunobu, K. T., 1976. Purification and properties of lung lysyl oxidase, a copper-enzyme, *Adv. Exp. Med. Biol.* 74:447–463.

Shields, G. S., Markowitz, H., Klassen, W. H., Cartwright, G. E., and Wintrobe, M. M., 1961. Studies on copper-metabolism. 31. Erythrocyte copper, *J. Clin. Invest.* 40:2007.

Shields, G. S., Coulson, W. F., Kimball, D. A., Carnes, W. H., Cartwright, G. E., and Wintrobe, M. M., 1962. Studies on copper metabolism. XXXII. Cardiovascular lesions in copper-deficient swine, *Am. J. Pathol.* 41:603.

Shils, M. E., and Randall, H. T., 1980. Diet and nutrition in the care of the surgical patient, in *Modern Nutrition in Health and Disease*. 6th ed. (R. S. Goodhart and M. E. Shils, eds.), Lea & Febiger, Philadelphia, pp. 1083–1124.

Shing, Y., 1988. Heparin–copper biaffinity chromatography of fibroblast growth factors, *J. Biol. Chem.* 263:9059–9062.

Shokeir, M. H. K., 1971. Investigations on the nature of caeruloplasmin deficiency in the neonate, *Clin. Genet.* 2:223–227.

Shokeir, M. H. K., and Shreffler, D. C., 1969. Cytochrome oxidase deficiency in Wilson's disease: A suggested ceruloplasmin function, *Proc. Natl. Acad. Sci. USA* 62:867–872.

Shreffler, D. C., Brewer, G. J., Gall, J. C., and Honeyman, M. S., 1967. Electrophoretic variation in human serum ceruloplasmin: A new genetic polymorphism, *Biochem. Genet.* 1:101–115.

Shvartsman, A. L., Voronina, O. V., Gaitskhoki, V. S., and Patkin, E. L., 1990. Expression of the ceruloplasmin gene in mammalian tissues studied by hybridization with complementary DNA probes, *Mol. Biol.* (*Moscow*) 24:657–662 (CAS Abstr.).

Siegel, R. C., 1979. Lysyl oxidase, in *International Review of Connective Tissue Research*, Vol. 8 (D. A. Hall and D. S. Jackson, eds.), Academic Press, New York, pp. 73–118.

Siegel, R. C., and Lian, J. B., 1975. Lysyl oxidase dependent synthesis of a collagen cross-link containing histidine, *Biochem. Biophys. Res. Commun.* 67:1353–1359.

Siegers, M. P., Kasperek, K., Heiniger, H. J., Iyengar, G. V., and Feinendegen, L. E., 1977. Trace element concentrations in eight organs of five inbred strains of mice, Abstracts, Third International Symposium on Trace Element Metabolism in Man and Animals, Freising, Germany.

Sigel, H., (ed.), 1981. *Metal Ions in Biological Systems*, Vol. 13, Marcel Dekker, New York.

Sigman, D. S., Graham, D. R., D'Aurora, V. D., and Stern, A. M., 1979. Oxygen-dependent cleavage of DNA by 1,10-phenanthroline-cuprous complex, *J. Biol. Chem.* 254:12269–12272.

References

Siimes, M. A., and Dallman, P. R., 1974. A new kinetic role for serum ferritin in iron metabolism. *Br. J. Haematol.* 28:7–18.

Silverberg, M., Neumann, P. Z., and Rotenberg, A. D., 1968. The role of amino acids in physiologic and pathologic copper transport: *in vitro* and *in vivo* studies, *Birth Defects, Orig. Article Ser.* 4:8–13.

Silverstone, B. Z., Nawratzki, E., Berson, D., and Yanko, L., 1986. Zinc and copper metabolism in oculocutaneous albinism in the Caucasian, *Metab. Pediatr. Syst. Ophthalmol.* 9:589–591.

Simpson, C. F., Jones, J. E., and Harms, R. H., 1967. Ultrastructure of aortic tissue in copper deficient and control chick embryos, *J. Nutr.* 91:283–291.

Simpson, C. F., Robbins, R. C., and Harms, R. H., 1971. Microscopic and biochemical observations of aortae of turkeys fed copper-deficient diets with and without ascorbic acid, *J. Nutr.* 101:1359–1366.

Singh, E. J., 1975. Effect of oral contraception and IUD's on the copper in human cervical mucus, *Obstet. Gynecol.* 45:328–330.

Singh, E. J., Baccarini, I. M., O'Neill, H. J., and Olwin, J. H., 1978. Effects of oral contraceptives on zinc and copper levels in human plasma and endometrium during the menstrual cycle, *Arch. Gynecol.* 226:303–306.

Singhal, A., Singh, M., Singh, G., and Sinha, S. N., 1982. Serum copper and ceruloplasmin as an index of foetal well being in abortions, *Indian. J. Path. Microbiol.* 25:242–244.

Sjollema, B., 1933. Kupfermangel als Ursache von Krankheiten bei Planzen und Tieren, *Biochem. Z.* 267:151–156.

Sjollema, B., 1938. Kupfermangel als Ursache von Tierkrankheiten, *Biochem. Z.* 295:372–376.

Skalicky, M., Kement, A., Haider, I., and Leibetseder, J., 1978. Effects of low and high copper intake on copper metabolism in pigs, in *Trace Element Metabolism in Man and Animals—3* (M. Kirchgessner, ed.), Arbeitskreis für Tierernährungsforschung Weihenstephan, Freising Germany, pp. 163–167.

Skandhan, K. P., and Mazumdar, B. N., 1979. Semen copper in normal and infertile subjects, *Experentia* 35:877–878.

Skinner, M. K., and Griswold, M. D., 1983. Sertoli cells synthesize and secrete a ceruloplasmin-like protein, *Biol. Reprod.* 28:1225–1229.

Slot, J. W., Geuze, H. J., Freeman, B. A., and Crapo, J. D., 1986. Intracellular localization of the copper–zinc and manganese superoxide dismutases in rat liver parenchymal cells, *Lab. Invest.* 55:363–371.

Smallwood, R. A., Williams, H. A., Rosenoer, V. M., and Sherlock, S., 1968. Liver copper levels in liver diseases: Studies using neutron activation analysis, *Lancet* ii:1310–1313.

Smith, A., and Morgan, W. T., 1985. Hemopexin-mediated heme transport to the liver. Evidence for a heme binding protein in liver plasma membranes, *J. Biol. Chem.* 260:8325–8329.

Smith, B. S. W., and Wright, H., 1973. Studies on the uptake of radioactive copper by sheep erythrocytes *in vitro*, *Biochim. Biophys. Acta* 307:590–598.

Smith, C. H., and Bidlack, W. R., 1980. Interrelationship of dietary ascorbic acid and iron on the tissue distribution of ascorbic acid, iron and copper in female guinea pigs, *J. Nutr.* 110:1398–1408.

Smith, P. A., Sunter, J. P., and Case, R. M., 1982. Progressive atrophy of pancreatic acinar tissue in rats fed a copper-deficient diet supplemented with D-penicillamine or triethylene tetramine: Morphological and physiological studies, *Digestion* 23:16–30.

Smith, R. M., Osborne-White, W. S., and O'Dell, B. L., 1976. Cytochromes in brain mitochondria from lambs with enzootic ataxia, *J. Neurochem.* 26:1145–1148.

Smith, W. E., Brown, D. H., Dunlop, J., Hazelton, R., Sturrock, R. D., and Lewis, A. J., 1980. The effect of therapeutic agents on serum copper levels and serum oxidase activities in the rat adjuvant model compared to analogous results from studies of rheumatoid arthritis in humans, *Inflammation: Mech. Treat.*, Proc. Int. Meet., 4th (D. A. Willoughby and J. P. Giroud, eds.), Univ. Park Press, Baltimore, Maryland, pp. 457–463.

Snedeker, S. M., Smith, S. A., and Greger, J. L., 1982. Effect of dietary calcium and phosphorus levels on the utilization of iron, copper and zinc by adult males, *J. Nutr.* 112:136–143.

Sobocinski, P. Z., Powanda, M. C., Canterbury, W. J., Matchotka, S. V., Walker, R. I., and Snyder, S. L., 1977. Role of zinc in the abatement of hepatocellular damage and mortality incidence in endotoxemic rats, *Infect. Immun.* 15:950–957.

Sobocinski, P. Z., Canterbury, Jr., W. J., Knutsen, G. L., and Hauer, E. C., 1981. Effect of adrenalectomy on cadmium- and turpentine-induced hepatic synthesis of metallothionein and α_2-macrofetoprotein in the rat, *Inflammation* 5:153–164.

Soli, N. E., and Rambaek, J. P., 1978. Excretion of intravenously injected copper-64 in sheep, *Acta Pharmacol. Toxicol.* 43:205–210.

Solomon, E. I., 1981. Binuclear copper active site. Hemocyanin, tyrosinase, and type 3 copper oxidases, in *Copper Proteins* (T. G. Spiro, ed.). Wiley: New York, pp. 41–108.

Solomon, E. I., 1988. Coupled binuclear copper proteins: Catalytic mechanisms and structure–reactivity correlations, in *Oxidases and Related Redox Systems* (T. E. King, H. S. Mason, and M. Morrison, eds.), Alan R. Liss, New York, pp. 309–329.

Solomon, E. I., Hare, J. W., Dooley, D. M., Dawson, J. H., Stephens, P. J., and Gray, H. B., 1980. Spectroscopic studies of stellacyanin, plastocyanin, and azurin. Electronic structure of the blue copper sites, *J. Am. Chem. Soc.* 102:168.

Solomon, E. I., Himmelwright, R. S., Eickman, N. C., LuBein, C. D., Schoeniger, L. O., and Lerch, K., 1981. Chemical and spectroscopic comparison of mollusc and arthropod hemocyanins with extensions to tyrosinase, in *Invertebrate Oxygen Binding Proteins* (J. Lamy and J. Lamy, eds.), Dekker, New York, pp. 553–569.

Sommer, A. L., 1930. Manganese, boron, zinc, and copper. Elements necessary in only small amounts for plant growth. Many have not heard of the experiments. Other materials may be essential, *Am. Fertilizer* 72:15–19.

Sonsma, T., Hixon, P., McWilliams, K., and Linder, M. C., 1981. Mechanism and regulation of intestinal copper absorption, in *Trace Element Metabolism in Man and Animals* (J. M. Howell. J. M. Grawthorne, and C. L. White, eds.), Australian Academy of Science, Canberra, pp. 145–147.

Sorenson, J. R. J., 1982. The anti-inflammatory activities of copper complexes, in *Metal Ions in Biological Systems* (H. Sigel, ed.), Marcel Dekker, New York, pp. 77–124.

Sorenson, J. R. J., 1977. Evaluation of copper complexes as potential anti-arthritic drugs, *J. Pharm. Pharmac.* 29:450–452.

Sorenson, J. R. J., 1978. Copper complexes—a unique class of antiarthritic drugs, *Progr. Medicinal Chem.* 15:211–260.

Sorenson, J. R. J., 1984. Copper complexes in biochemistry and pharmacology, *Chem. Britain*, 20:1110–1113.

Sorenson, J. R. J., 1985. Copper complexes: A physiologic approach to treatment of chronic diseases, *Comp. Ther.* 11:49–64.

Sorenson, J. R. J., 1987a. Antiinflammatory, antiulcer, and analgesic activities of copper complexes offer a physiological approach to treatment of rheumatoid arthritis, in *Proceedings* 3rd *International Conference* on *Bioinorganic Chemistry*, July 6–10, 1987, Noordwijkerhout, *Rec. Trav. Chim. Pays-Bas.* 106:378.

Sorenson, J. R. J., 1987b. Antiinflammatory and antiulcer activities of non-steroidal antiinflammatory agent copper complexes, in *Biology of Copper Complexes* (J. R. J. Sorenson, ed.), Humana Press, Clifton, New Jersey, pp. 243–252.

Sorenson, J. R. J., 1989. Copper complexes offer a physiological approach to treatment of chronic diseases, *Progr. Medicinal Chem.* 26:437–568.
Sorenson, J., Oberley, L. W., Oberley, T. D., Leuthauser, S. W. C., Ramakrishna, K., Vernino, L., and Kishore, V., 1982a. *Trace Substances in Environmental Health—XVI* (D. D. Hemphill, ed.), University of Missouri Press, Columbia, Missouri.
Sorenson, J. R. J., Ramakrishna, K., and Rolniak, T. M., 1982b. Antiulcer activities of D-penicillamine copper complexes, *Agents Actions*, 12:408–411.
Sorkina, D. A., Borodina, N. I., and Konoshenko, S. V., 1976. Study of the physiochemical properties and partial N-terminal amino acid sequence in the serum albumin of chickens, *Zh. Evol. Biokhim. Fiziol.* 12:422–427 (CAS Abstract).
Soroka, V. R., 1968. Trace element metabolism in the liver under the influence of adrenaline and insulin, *Mikroelem. Sel. Khoz. Med.* 4:200–206.
Souci, S. W., Fachmann, W., and Kraut, H., 1981. *Food Composition and Nutrition Tables 1981/1982*, Wissenschaftliche Verlagsgesellschaft mbH, Stuttgart.
Sourkes, T. C., 1970. Factors affecting the concentration of copper in the liver of the rat, in *Trace Element Metabolism in Animals* (C. F. Mills, ed.), E. and S. Livingstone, Edinburgh, pp. 247–256.
Sourkes, T. L., Lloyd, K., Birnbaum, H., 1968. Inverse relationship of hepatic copper and iron concentrations in rats fed deficient diets, *Can. J. Biochem.* 46:267–271.
Sourkes, T. L., Quik, M., and Falardeau, M., 1974. Effects of iron and copper deficiencies on monoamine metabolism, *Adv. Neurol.* 5:253–258.
Speck, S. H., Neu, C. A., Swanson, M. S., and Margoliash, E., 1983. Role of phospholipid in the low affinity reactions between cytochrome c and cytochrome oxidase, *FEBS Lett.* 164:379–382.
Spiro, T. G., (ed.), 1981. *Copper Proteins*, John Wiley and Sons, New York.
Spoerl, V. R., and Kirchgessner, M., 1975. Veränderungen des Cu-Status und der Coeruloplasmin-aktivitat von mutterlichen und saugenden Ratten bei gestaffelter Cu-Versorgung, *Z. Tierphysiol., Tierernahrg. Futtermittelkde.* 35:113–127.
Starcher, B. C., 1969. Studies on the mechanism of copper absorption in the chick, *J. Nutr.* 97:321–326.
Starcher, B., and Hill, C. H., 1965. Hormonal induction of ceruloplasmin in chicken serum, *Comp. Biochem. Physiol.* 15:429–434.
Starcher, B., Madaras, J. A., Fish, D., Perry, E. F., and Hill, C. H., 1978. Abnormal cellular copper metabolism in the blotchy mouse, *J. Nutr.* 108:1229–1233.
Stassen, F. L. H., 1976. Properties of highly purified lysyl oxidase from embryonic chick cartilage, *Biochim. Biophys. Acta* 438:49–60.
Steffens, G. C. M., and Buse, G., 1988. Integral complexes of cytochrome c oxidase contain three coppers, in *Oxidase and Related Redox Systems*, (T. E. King, H. S. Mason, and M. Morrison, eds.), Alan R. Liss, New York, pp. 687–705.
Steffens, G. C. M., Buse, G., Oppliger, W., and Ludwig, B., 1983. Sequence homology of bacterial and mitochondrial cytochrome c oxidases, *Biochem. Biophys. Res. Commun.* 116:335–340.
Steffens, G. C. M., Buse, G., Oppliger, W., and Ludwig, B., 1984. Cytochrome c oxidases, *Eur. Bioenerg. Conf. Rep.* 3A:195–196.
Steffens, G. C. M., Biewald, R., and Buse, G., 1987. Cytochrome c oxidase is a three-copper, two-heme-A protein, *Eur. J. Biochem.* 164:295–300.
Steinbrucke, P., Steffens, G. C. M., Panskus, G., Buse, G., and Ludwig, B., 1987. Subunit II of cytochrome c oxidase from *Paracoccus denitrificans*, *Eur. J. Biochem.* 167:431–439.
Steinman, H. M., Naik, V. R., Abernathy, J. L., and Hill, R. L., 1974. Bovine erythrocyte superoxide dimutase. Complete amino acid sequence, *J. Biol. Chem.* 249:7326–7338.
Stemmer, K. L., Petering, H. G., Murthy, L., Finelli, V. N., and Menden, E. E., 1985. Copper

deficiency effects on cardiovascular system and lipid metabolism in the rat; the role of dietary proteins and excessive zinc, *Ann. Nutr. Metab.* 29:332–347.

Stern, R. V., and Frieden, E., 1990. Detection of rat ceruloplasmin receptors using fluorescence microscopy and microdensitometry, *Anal. Biochem.* 190:48–56.

Sternlieb, I., Morell, A. G., Tucker, W. D., Greene, M. W., and Scheinberg, I. H., 1961. The incorporation of copper into ceruloplasmin *in vivo*: Studies with copper[64] and copper[67], *J. Clin. Invest.* 40:1834–1840.

Sternlieb, I., van den Hamer, C. J. A., and Alpert, S., 1967. Role of intestinal lymphatics in copper absorption, *Nature* 216:824.

Sternlieb, I., van den Hamer, C. J. A., Morell, A. G., Alpert, S., Gregoriadis, G., and Scheinberg, I. H., 1973. Lysosomal defect of hepatic copper excretion in Wilson's disease (hepatolenticular degeneration), *Gastroenterology* 64:99–105.

Stevens, M. D., DiSilvestro, R. A., and Harris, E. D., 1984. Specific receptor for ceruloplasmin in membrane fragments from aortic and heart tissues, *Biochemistry* 23:261–266.

Stevenson, E. J., 1986. *Cycles of Soil*. Wiley-Interscience, New York.

Stewart, L. C., and Klinman, J. P., 1988. Dopamine beta-hydroxylase of adrenal chromaffin granules: Structure and function, *Annu. Rev. Biochem.* 57:551–592.

Stockert, R. K., Haimes, H. B., and Morell, A. G., 1980. Endocytosis of asialoglycoprotein enzyme conjugates by hepatocytes, *Lab. Invest.* 43:556–563.

Strain, J. J., and Lynch, S. M., 1990. Excess dietary methionine decreases indexes of copper status in the rat, *Ann. Nutr. Metab.* 34:93–97.

Strain, W. H., Macon, W. L., Pories, W. J., Perim, C., Adams, F. D., and Hill, O. A., Jr., 1974. Excretion of trace elements in bile, in *Trace Element Metabolism in Animals—2* (W. G. Hoekstra, J. W. Suttie, H. E. Ganther, and W. Mertz, eds.), University Park Press, Baltimore, pp. 644–646.

Strawinsky, J., Kurek, H., and Walawski, K., 1979. Ceruloplasmin polymorphism in blood serum of lowland black-and-white cattle, *Genet. Pol.* 20:127–133.

Strickland, G. T., Beckner, W. M., Len, M. L., and O'Reilly, S., 1972. Turnover studies of copper in homozygotes and heterozygotes for Wilson's disease and control: Isotope studies with 67-Cu, *Clin. Sci.* 43:605–615.

Strothkamp, K. G., Jolley, R. L., and Mason, H. S., 1976. Quarternary structure of mushroom tyrosinase, *Biochem. Biophys. Res. Comm.* 70:519.

Strydom, D. J., Fett, J. W., Lobb, R. R., Alderman, E. M., Bethune, J. L., Riordan, J. F., and Vallee, B. L., 1985. *Biochemistry* 24:5486–5494.

Studnitz, W., and Berezin, D., 1958. Studies on serum copper during pregnancy, during menstrual cycle and after administration of estrogens, *Acta Endocrinol.* 27:245–253.

Sugawara, N., and Sugawara, C., 1984. Effect of silver on ceruloplasmin synthesis in relation to low-molecular-weight protein, *Toxicol. Lett.* 20:99–104.

Sugiura, Y., Tanaka, H., Mino, Y., Ishida, T., Ota, N., Inoue, M., Nomoto, K., Yoshioka, H., and Takemoto, T., 1981. Structure, properties, and transport mechanism of iron(III) complex of mugineic acid, a possible phytosiderophore, *J. Am. Chem. Soc.* 103: 6979–6982.

Sullivan, P. A., Robinson, W. E., and Morse, M. P., 1988. Subcellular distribution of metals within the kidney of the bivalve *Mercenaria mercenaria* (L.), *Comp. Biochem. Physiol.* 91C:589–595.

Sunderman, F. W., Jr., Nomoto, S., Gillies, C. G., and Goldblatt, P. J., 1971. Effect of estrogen administration upon ceruloplasmin and copper concentrations in rat serum, *Toxicol. Appl. Pharmacol.* 20:588–598.

Sunderman, Jr., F. W., Hohnadel, D. C., Evenson, M. A., Wannamaker, B. B., and Dahl, D. S., 1974. Excretion of copper in sweat of patients with Wilson's disease during sauna bathing, *Ann. Clin. Lab. Sci.* 4:407–412.

Sung, S. M., and Topham, R. W., 1973. Lipid components of human ferroxidase-II. 2, *Biochem. Biophys. Res. Commun.* 53:824–829.

Sutherland, I. V., and Wilkinson, J. F., 1963. Azurin: A copper protein found in Bordetella, *J. Gen. Microbiol.* 30:105–112.

Suttle, N. F., 1974. Recent studies on the copper-molybdenum antagonism, *Proc. Nutr. Soc.* 33:299.

Suttle, N. F., 1975b. Changes in the availability of dietary copper to young lambs with age and weaning, *J. Agric. Sci.* 84:255–261.

Suttle, N. F., 1975a. Role of organic sulphur in copper-molybdenum-S interrelationship in ruminant nutrition, *Brit. J. Nutr.* 34:411–420.

Suttle, N. F., 1980. The role of thiomolybdates in the nutritional interactions of copper, molybdenum, and sulphur: fact or fantasy? *Ann. N.Y. Acad. Sci.* 355:195.

Suttle, N. F., 1987. The nutritional requirements for copper in animals and man, in *Copper in Animals and Man* (J. M. Howell and J. M. Gawthorne, eds.), CRC Press, Boca Raton, Florida, pp. 22–38.

Suzue, G., Vezina, C., and Marcel, Y. L., 1980. Purification of human plasma lecithin: cholesterol acyltransferase and its activation by metal ions, *Can. J. Biochem.* 58: 539–541.

Suzuki, H., Nagai, K., Akutsu, E., Yamaka, H., and Umezawa, H., 1970. On the mechanism of action of bleomycin. Strand scission of DNA caused by bleomycin and its binding to DNA *in vitro*, *J. Antibiot.* 23:473–480.

Suzuki, K. T., 1980. Direct connection of high-speed liquid chromatography (equipped with gel permeation column) to atomic absorption spectrophotometer for metalloprotein analysis: Metallothionein, *Anal. Biochem.* 102:31–34.

Suzuki, K. T., and Karasawa, A., 1990. Calcium- and cysteine-participatory oxidative formation of albumin–copper complex, *Arch. Biochem. Biophys.* 278:120–124.

Suzuki, K. T., Karasawa, A., and Yamanaka, K., 1989. Binding of copper to albumin and participation of cysteine in vivo and *in vitro*, *Arch. Biochem. Biophys.* 273:252–577.

Suzuki, M., Tanemoto, Y., and Takahashi, K., 1972. The effect of copper salts on ovulation, especially on hypothalamic ovulatory hormone releasing factor, *Tohoku J. Exp. Med.* 108:9–18.

Suzuki, Y., and Ogiso, K., 1973. Development of ascorbate oxidase activity and its isoenzyme pattern in the roots of pea seedlings, *Physiol. Plant.* 29:169.

Suzuki, Y., and Yoshikawa, H., 1981. Cadmium, copper, and zinc excretion and their binding to metallothionein in urine of cadmium-exposed rats, *J. Toxicol. Environ. Health* 8:479–487.

Suzuki, Y., Toda, K., Koike, S., and Yoshikawa, H., 1981. Cadmium, copper and zinc in the urine of welders using cadmium-containing silver solder, *Ind. Health* 19:223–230.

Sweeney, S. C., 1967. Alterations in tissue and serum ceruloplasmin concentration associated with inflammation, *J. Dent. Res.* 46:1171–1176.

Swerdel, M. R., and Cousins, R. J., 1986. Induction of ceruloplasmin by retinoic acid and retinol in vitamin A deficient and vitamin A adequate rats, *Fed. Proc.* 35:994 (Abstract no. 3409).

Syed, M. A., Coombs, T. L., Goodman, B. A., and McPhail, D. B., 1982. The nature of the copper(II) components of caeruloplasmin, *Biochem. J.* 207:183–184.

Szerdahelyi, P., and Kasa, P., 1984. Histochemistry of zinc and copper, *Int. Rev. Cytol.* 89:1–33.

Tainer, J. A., Getzoff, E. D., Beam, K. M., Richardson, J. S., and Richardson, D. C., 1982. Determination and analysis of the 2. ANG. structure of copper, zinc superoxide dismutase, *J. Mol. Biol.* 160:181–217.

Tainer, J. A., Getzoff, E. D., Richardson, J. S., and Richardson, D. C., 1983. Structure and mechanism of copper, zinc superoxide dismutase, *Nature* 306:284–287.

Takahashi, K., Yoshioka, O., Matsuda, A., and Umezawa, H., 1977. Intracellular reduction of the cupric ion of bleomycin copper complex and transfer of the cuprous ion to a cellular protein, *J. Antibiot.* 30:861–869.

Takahashi, N., Ortel, T. L., and Dwulet, F. E., Wang, C.-C., and Putnam, F. W., 1983. Internal triplication in the structure of human ceruloplasmin, *Proc. Natl. Acad. Sci. USA* 80:115–119.

Takahashi, N., Ortel, T. L., and Putnam, F. W., 1984. Single-chain structure of human ceruloplasmin: The complete amino acid sequence of the whole molecule, *Proc. Natl. Acad. Sci. USA* 81:390–394.

Takahashi, N., Takahishi, Y., Blumberg, B. S., and Putnam, F. W., 1987. Amino acid substitutions in genetic variants of human serum albumin and in sequences inferred from molecular cloning, *Proc. Natl. Acad. Sci. USA* 84:4413–4417.

Takamiya, K., 1960. Anti-tumor activities of copper chelates, *Nature* 185:190–191.

Takita, T., Muraoka, Y., Nakatani, T., Fujii, A., Itaka, Y., and Umezawa, H., 1978. Chemistry of bleomycin. XXI. Metal complex of bleomycin and its implications to the mechanism of bleomycin action, *J. Antibiot.* 31:1073–1077.

Tanaka, M., Iio, T., and Tabata, T., 1987. Effect of cupric ions on serum and liver cholesterol metabolism, *Lipids* 22:1016–1019.

Tanaka, T., and Knox, W. E., 1959. The nature and mechanism of the tryptophan pyrrolase peroxidase-oxidase reaction of pseudomonas and of rat liver, *J. Biol. Chem.* 234:1162–1170.

Taneja, S. R., and Sachar, R. C., 1977. Effect of auxin on multiple forms of O-diphenolase in germinating wheat embryos, *Phytochemistry* 16:871–873.

Tartakoff, A. M., and Vassalli, P., 1978. Comparative studies of intracellular transport of secretory proteins, *J. Cell Biol.* 79:6934–707.

Tauxe, W. M., Goldstein, N. P., Randall, R. V., and Gross, J. B., 1966. Radioisotope studies in patients with Wilson's disease and their relatives, *Am. J. Med.* 41:375–380.

Tavassoli, M., 1985. Liver endothelium binds, transports, and desialates ceruloplasmin which is then recognized by galactosyl receptors of hepatocytes, *Trans. Assoc. Am. Physicians* XCVIII:370–377.

Tavassoli, M., Kishimoto, T., and Kataoka, M., 1986. Liver endothelium mediates the hepatocyte's uptake of ceruloplasmin, *J. Cell Biol.* 102:1298–1303.

Taylor, G. O., and Williams, A. O., 1974. Lipid and trace metal content in coronary arteries of Nigerian Africans, *Exp. Mol. Pathol.* 21:371–381.

Taylor, G. O., Williams, A. O., Resch, J. A., Barber, J. B., Jackson, M. A., and Paulissen, G. A., 1975. Trace metal content of cerebral vessels in American blacks, Caucasians and Nigerian Africans, *Stroke* 6:684–690.

Tegtmeyer, K., and Rittschof, D., 1988. Synthetic peptide analogs to barnacle settlement pheromone, *Peptides* 9:1403–1406.

Templeton, D. M., and Sarkar, B., 1985. Peptide and carbohydrate complexes of nickel in human kidney, *Biochem. J.* 230:35–42.

Templeton, D. M., and Sarkar, B., 1986. Low molecular weight targets of metals in human kidney, *Acta Pharmacol.* 59 (Suppl. 7):416–423.

Terao, T., and Owen, C. A., Jr., 1973. Nature of copper compounds in liver supernate and bile of rats: Studies with ^{67}Cu, *Am. J. Physiol.* 224:682–686.

Terao, T., and Owen, C. A., Jr., 1974. Copper in supernatant fractions of various rat tissues, *Mayo Clin. Proc.* 49:376–381.

Terao, T., and Owen, C. A., Jr., 1976. Effects of copper deficiency and copper loading on ^{67}Cu in supernatants of rat organs, *Tohoku J. Exp. Med.* 120:209–217.

Terao, T., and Owen, C. A., Jr., 1977. Copper metabolism in pregnant and post-partum rat and pups, *Am. J. Physiol.* 232:E172–E179.

References

Tetaz, T. J., and Luke, R. K. J., 1983. Plasmid-controlled resistance to copper in *Escherichia coli*, *J. Bacteriol.* 154:1263–1268.
Thiele, D. J., Walling, M. J., and Hamer, D. H., 1986. Mammalian metallothionein is functional in yeast, *Science* 231:854–856.
Thiers, R. E., and Vallee, B. L., 1957. Distribution of metals in subcellular fractions of rat liver, *J. Biol. Chem.* 226:911–920.
Tholey, G., Ledig, M., Mandel, P., Sargentini, L., Frivold, A. H., Leroy, M., Grippo, A. A., and Wedler, F. C., 1988. Concentrations of physiologically important metal ions in glial cells cultured from chick cerebral cortex, *Neurochem. Res.* 13:45–50.
Thomas, P., and Summers, J. W., 1978. The biliary excretion of circulating asialoglycoproteins in the rat, *Biochem. Biophys. Res. Commun.* 80:335–339.
Thomas, T., 1989. Ph.D. thesis, University of Melbourne, Australia.
Thomas, T., and Schreiber, G., 1986. Acute phase response of plasma protein synthesis during experimental inflammation in neonatal rats, *Inflammation* 9:1–7.
Thompson, D. A., and Ferguson-Miller, S., 1983. Lipid and subunit III depleted cytochrome *c* oxidase purified by horse cytochrome *c* affinity chromatography in lauryl maltoside, *Biochemistry* 22:3178–3187.
Thornton, B., and Macklon, A. E. S., 1989. Copper uptake by ryegrass seedlings; contribution of cell wall adsorption, *J. Exp. Bot.* 40:1105–1111.
Tibell, L., Hjalmarsson, K., Edlund, T., Skogman, G., Engstroem, A., and Marklund, S. L., 1987. Expression of human extracellular superoxide dismutase in Chinese hamster ovary cells and characterization of the product, *Proc. Natl. Acad. Sci. USA.* 84:6634–6638.
Tilghman, S. M., and Belayew, A., 1982. Transcriptional control of the murine albumin/α-fetoprotein locus during development, *Proc. Natl. Acad. Sci. USA* 79:5254–5257.
Till, G. O., Johnson, J. J., Kunkel, R., and Ward, P. A., 1982. Intravascular activation of complement and acute lung injury: Dependence on neutrophils and toxic oxygen metabolites, *J. Clin. Invest.* 69:1126–1135.
Ting, B. T. G., Kasper, L. J., Young, V. R., and Janghorbani, M., 1984. Copper absorption in healthy young men: Studies with stable isotope ^{65}Cu and neutron activation analysis, *Nutr. Res.* 4:757–769.
Tohyama, C., Shaikh, Z. A., Ellis, K. J., and Cohn, S. H., 1981. Metallothionein excretion in urine upon cadmium exposure: Its relationship with liver and kidney cadmium, *Toxicology* 22:181–191.
Tomita, Y., Hariu, A., Kato, C., and Seiji, M., 1983. Transfer of tyrosinase to melanosomes in Harding-Passey mouse melanoma, *Arch. Biochem. Biophys.* 225:75–85.
Topham, R. W., and Frieden, E., 1970. Identification and purification of a non-ceruloplasmin ferroxidase of human serum, *J. Biol. Chem.* 245:6698–6705.
Topham, R. W., and Johnson, D. A., 1974. Kinetic studies of the Fe(II) oxidation with human serum ferroxidase-II, *Arch. Biochem. Biophys.* 160:647–654.
Topham, R. W., Sung, C. S., Morgan, F. G., Prince, W. D., and Jones, S. H., 1975. Functional significance of the copper and lipid components of human ferroxidase-II, *Arch. Biochem. Biophys.* 167:129–137.
Topham, R. W., Woodruff, J. H., Neatrour, G. P., Calisch, M. P., Russo, R. B., and Jackson, M. R., 1980. The role of ferroxidase II and a ferroxidase inhibitor in iron mobilization from tissue stores, *Biochem. Biophys. Res. Commun.* 96:1532–1539.
Trackman, P. C., and Kagan, H. M., 1979. The enthalpy of proteolysis of liver alcohol dehydrogenase upon binding nicotinamide adenine dinucleotide, *J. Biol. Chem.* 254:7831–7836.
Tredway, D. R., Umezaki, C. U., Mishell, D. R., and Settlage, D. S. F., 1975. Effect of intrauterine devices on sperm transport in the human being: Preliminary report, *Am. J. Obstet. Gynocol.* 123:734–735.

Trip, J. A. J., Que, G. S., Botterweg-Span, Y., and Mandema, E., 1969. The state of copper in human lymph, *Clin. Chim. Acta* 26:371–372.

Trueblood, C. E., and Poyton, R. O., 1988. Identification of RE01, a gene involved in negative regulation of COX5b and ANB1 in aerobically grown *Saccharomyces cerevisiae*, *Genetics* 120:671–680.

Trueblood, C. E., Wright, R. M., and Poyton, R. O., 1988. Differential regulation of the two genes encoding *Saccharomyces cerevisiae* cytochrome *c* oxidase subunit V by heme and the HAP2 and REO1 genes, *Mol. Cell. Biol.* 8:4537–4540.

Tsuchiya, K., and Iwao, S., 1978. Interrelationships among zinc, copper, lead, and cadmium in food, feces, and organs of humans, *Environ. Health Perspec.* 25:119–124.

Tsujimato, M., Yokota, S., Vilcek, J., and Weissmann, G., 1986. Tumor necrosis factor provokes superoxide anion generation from neutrophils., *Biochem. Biophys. Res. Commun.* 137:1094–1100.

Tu, J.-B., Blackwell, R. Q., and Hou, T.-Y., 1963. Tissue copper levels in Chinese patients with Wilson's disease *Neurology* 13:155–159.

Tukendorf, A., Lyszcz, S., and Baszynski, T., 1984. Copper binding proteins in spinach tolerant to excess copper, *J. Plant Physiol.* 115:351.

Turecky, L., Kalina, P., Uhlikova, E., Namerova, S., and Krizko, J., 1984. Serum ceruloplasmin and copper levels in patients with primary brain tumors, *Klin. Wochenschr.* 62:187–189.

Turnlund, J. R., 1991. Copper absorption and gastrointestinal excretion in humans: The impact of adaptation to dietary copper intake on copper deficiency and toxicity, in *Trace Elements in Man and Animals* (TEMA-7), (B. Momcilovic, ed.), in press.

Turnlund, J. R., Michel, M.C., Keyes, W. R., Schutz, Y., and Margen, S., 1982. Copper absorption in elderly men determined by using stable ^{65}Cu, *Am. J. Clin. Nutr.* 36:587–591.

Turnlund, J. R., Keyes, W. R., Anderson, H. L., and Acord, L. L., 1989. Copper absorption and retention in young men at three levels of dietary copper using the stable isotope, ^{65}Cu, *Am. J. Clin. Nutr.*, 49:870–878.

Turnlund, J. R., Keen, C. L., and Smith, R. G., 1990. Copper status and urinary and salivary copper in young men at three levels of dietary copper, *Am. J. Clin. Nutr.* 51:658–664.

Tyrala, E. E., Manser, J. I., Brodsky, N. L., Tran, N., Kotwall, M., and Friehling, L., 1985. Distribution of copper in the serum of the parenterally fed premature infant, *J. Pediatr.* 106:295–296.

Uauy, R., Castillo-Duran, C., Fisberg, M., Fernandez, N., and Valenzuela, A., 1985. Red cell superoxide dismutase activity as an index of human copper nutrition, *J. Nutr.* 115:1650–1655.

Udom, A. O., and Brady, F. O., 1980. Reactivation *in vitro* of zinc-requiring apo-enzymes by rat liver zinc-thionein, *Biochem. J.* 187:329–335.

Umezawa, H., Ishizuka, M., Kimura, K., Iwanaga, J., and Takeuchi, T., 1968. Biological studies on individual bleomycins, *J. Antibiot.* 21:592–602.

Underwood, E. J., 1971. *Trace Elements in Human and Animal Nutrition*, 3rd Ed., Academic Press, New York.

Underwood, E. J., 1977. *Trace Elements in Human and Animal Nutrition*, 4th Ed., Academic Press, New York.

Ursini, M. V., and de Franciscis, V., 1988. TSH regulation of ferritin H chain messenger RNA levels in the rat thyroid, *Biochem. Biophys. Res. Commun.* 150:287–295.

Valberg, L. S., Flanagan, P. R., and Chamberlain, M. J., 1984. Effects of iron, tin, and copper on zinc absorption in humans, *Am. J. Clin. Nutr.* 40:536–541.

Valentine, J. S., and Mota de Freitas, D., 1985. Copper-zinc superoxide dismutase. A unique biological "ligand" for, bioinorganic studies, *J. Chem. Educ.* 62:990–997.

Valentine, J. S., and Mota de Freitas, D., 1986. NMR studies of the anion binding sites of oxidized and reduced bovine copper-zinc superoxide dismutase, in *Superoxide Dismutase in Chemistry, Biology, and Medicine, Proc.* 4th *Int. Conf.*, 1985 (G. Rotilio, ed.), Elsevier, Amsterdam, pp. 149–154.

Valentine, J. S., and Pantoliano, M. W., 1981. Protein–metal ion interactions in cuprozinc protein (Superoxide dismutase), in *Copper Proteins* (T. G. Spiro, ed.), Wiley, New York, pp. 292–358.

Vallee, B. L., Wacker, W. E. C., Bartholomay, A. F., and Hoch, F. L., 1957. Zinc metabolism in hepatic dysfunction. II. Correlation of metabolic patterns with biochemical findings, *N. Engl. J. Med.* 257:1055–1065.

Valsala, P., and Kurup, P. A., 1987. Investigations on the mechanism of hypercholesterolemia observed in copper deficiency in rats, *J. Biosci.* 12:137–142.

Van Barneveld, A. A., and van den Hamer, J. A., 1984. Intestinal passage and absorption of simultaneously administered ^{64}Cu and ^{65}Zn and the effect of feeding in mouse and rat, *Nutr. Rep. Int.* 29:173–182.

van Berge Henegouwen, G. P., Tangedahl, T. N., Hofmann, A. F., Northfield, T. C., LaRusso, N. F., and McCall, J. T., 1977. Biliary secretion of copper in healthy man: Quantitation by an intestinal perfusion technique, *Gastroenterology* 72:1228–1231.

Van Campen, D. R., 1969. Copper interference with the intestinal absorption of zinc-65 by rats, *J. Nutr.* 97:104–108.

Van Campen, D. R., and Gross, E., 1968. Influence of ascorbic acid on the absorption of copper by rats, *J. Nutr.* 95: 617–622.

Van Campen, D. R., and Mitchell, E. A., 1965. Absorption of Cu^{64}, Zn^{65}, Mo^{99}, and Fe^{59} from ligated segments of the rat gastrointestinal tract, *J. Nutr.* 86:120–124.

van den Berg, G. J., and van den Hamer, C. J. A., 1984. Trace metal uptake in liver cells. 1. Influence of albumin in the medium on the uptake of copper by hepatoma cells, *J. Inorg. Biochem.* 22:73–84.

van den Berg, G. J., Le Clerco, E., Kluft, C., Koide, T., Van Der Zee, A., Oldenburg, M., Wijnen, J. T., and Khan, P. M., 1990a. Assignment of the human gene for histidine-rich glycoprotein to chromosome 3, *Genomics* 7:276–279.

van den Berg, G. J., Van Wouwe, J. P., and Beynen, A. C., 1990b. Ascorbic acid supplementation and copper status in rats, *Biol. Trace Elem. Res.* 23:165–172.

van den Hamer, C. J. A., Morell, A. G., and Scheinberg, I. H. 1970. Physical and chemical studies on ceruloplasmin. IX. The role of galactosyl residues in the clearance of ceruloplasmin from the circulation, *J. Biol. Chem.* 245:4397–4402.

van der Meer, R. A., and Duine, J. A., 1986. Covalently bound pyrroloquinoline quinone is the organic prosthetic group in human placental lysyl oxidase, *Biochem. J.* 239:789–791.

van der Meer, R. A., Jongejan, J. A., and Duine, J. A., 1988. Dopamine beta-hydroxylase from bovine adrenal medulla contains covalently-bound pyrroloquinoline quinone, *FEBS Lett* 231:303–307.

Van Hien, P., Kovacs, K., and Matkovics, B., 1974. Properties of enzymes. I. Study of superoxide dismutase activity change in human placenta of different ages, *Enzyme* 18:341–347.

Van Holde, K. E., and Brenowitz, M., 1981. Subunit structure and physical properties of the hemocyanin of the giant isopod *Bathynomus giganteus*, *Biochemistry* 20:5232–5239.

Vänngard, T., 1967. Some properties of ceruloplasmin copper as studied by ESR spectroscopy, in *Magnetic Resonance in Biological Systems* (A. Ehrenberg, B. G. Malmström, and T. Vänngard, eds.), Pergamon, Oxford, pp. 213–219.

Van Wyk, C. P., Linder-Horowitz, M., and Munro, H. N., 1971. Effect of iron loading on non-heme iron compounds in different liver cell populations, *J. Biol. Chem.* 246:1025–1031.

Vardi, P., Hidvegi, J., Lindner-Szotyori, K., Konrad, S., and Somos, P., 1979. Analysis of microelements in amniotic fluid, *Magy. Noorv. Lapja* 42:429–432 (CAS Abstr.).

Vasak, M., and Kagi, J. H. R., 1981. Metal thiolate clusters in cobalt(II)-metallothionein, *Proc. Natl. Acad. Sci. USA* 78:6709–6713.

Vasak, M., Galdes, A., Hill, H. A. O., Kägi, J. H. R., Bremner, I., and Young, B. W., 1980. Investigation of the structure of metallothioneins by proton nuclear magnetic resonance spectroscopy, *Biochemistry* 19:416–425.

Vasak, M., Kägi, J. H. R., Holmquist, B., and Vallee, B. L., 1981. Spectral studies of cobalt(II)- and nickel(II)-metallothionein, *Biochemistry* 20:6659–6664.

Vasak, M., Hawkes, G. E., Nicholson, J. K., and Sadler, P. J., 1985. Cadmium-113 NMR studies of reconstituted seven-cadmium metallothionein: Evidence for structural flexibility, *Biochemistry* 24:740–747.

Vassiletz, I., Derkatchev, E., and Neifakh, S., 1976. The electron transfer chain in liver cell plasma membrane, *Exp. Cell Res.* 46:419–427.

Vassos, A. B., and Newman, R. A., 1988. Isolation and characterization of a membrane associated receptor for human ceruloplasmin, *FASEB J.* 2:Abstract 1896.

Vaughn, K. C., and Duke, S. O., 1984. Function of polyphenol oxidase in higher plants, *Physiol. Plant* 60:106.

Veen-Baigent, M. J., Ten Cate, A. R., Bright-See, E., and Rao, A. V., 1975. Effects of ascorbic acid on health parameters in guinea pigs, *Ann. N.Y. Acad. Sci.* 258:339–355.

Versieck, J., Barbier, F., Speecke, A., and Hoste, J., 1975. Influence of myocardial infarction on serum manganese, copper, and zinc concentrations, *Clin. Chem.* 21:578–581.

Vierling, J. M., Shrager, M. D., Warren, M. A., Rumble, F., Aamodt, B. A. R., Berman, M. D., and Jones, E. A., 1978. Incorporation of radiocopper into ceruloplasmin in normal subjects and in patients with primary biliary cirrhosis and Wilson's disease, *Gastroenterology* 74:652–660.

Vignais, P. M., Henry, M. F., Terech, A., and Chabert, J., (1980). Production of superoxide anion and superoxide dismutases in *paracoccus dentitrificans*, in *Chemical and Biochemical Aspects of Superoxide and Superoxide Dismutase* (J. V. Bannister and H. A. O. Hill, eds.), Elsevier, New York, pp. 154–159.

Vilter, R. W., Ozian, R. C., Hess, E. V., Zellner, D. C., and Petering, H. G., 1974. Manifestations of copper deficiency in a patient with systemic sclerosis on intravenous hyperalimentation, *N. Engl. J. Med.* 291:188–191.

Vir, S. C., Love,, A. H. G., and Thompson, W., 1981. Serum and hair concentrations of copper during pregnancy, *Am. J. Clin. Nutr.* 34:2382–2388.

Vogel, F. S., 1960. Nephrotoxic properties of copper under experimental conditions in mice; with special reference to the pathogenesis of renal alterations in Wilson's disease, *Am. J. Pathol.* 36:699–711.

Volbeda, A., and Hol, W. G. J., 1988. Structure of arthropodan hemocyanin, in *Oxidases and Related Redox Systems* (T. E. King, H. S. Mason, and M. Morrison, eds.), Alan R. Liss, New York, pp. 291–307.

von Ebeling, H., 1975. Enzymatisch und immunologisch bestimmtes Coeruloplasmin: Geschlechts- und Methodenunterschiede unter Ostrogeneinnahme, *Z. Klin. Chem. Klin. Biochem.* 13:445–451.

von Studnitz, W., and Berezin, D., 1958. Studies on serum copper during pregnancy, during the menstrual cycle, and after the administration of oestrogens, *Acta Endocrinol.* 27:245–252.

Vyas, D., and Chandra, R. K., 1983. Thymic factor activity, lymphocyte stimulation response and antibody producing cells in copper deficiency, *Nutr. Res.* 3:343–349.

Waalkes, M. P., Ross, S. M., Craig, C. R., and Thomas, J. A., 1982. Induction of metallothionein in the rat brain by copper implantation but not by cobalt implantation, *Toxicol. Lett.* 12:137–142.

References

Waalkes, M. P., Chernoff, S. B., and Klaassen, C. D., 1984. Cadmium-binding proteins of rat testes, apparent source of the protein of low molecular mass, *Biochem. J.* 220:819–824.

Wahlborg, A., and Frieden, E., 1965. Comparative interaction of thyroxine and analogues with Cu(II), *Biochim. Biophys. Acta* 111:702–712.

Wahle, K. W. J., and Davies, N. T., 1975. Effect of dietary copper deficiency in the rat on fatty acid composition of adipose tissue and desaturase activity of liver microsomes, *Br. J. Nutr.* 34:105–112.

Waisman, J., and Carnes, W. H., 1967. Cardiovascular studies on copper-deficient swine, Part 10, the fine structure of the defective elastic membranes, *Am. J. Pathol.* 51:117.

Waisman, J., Cancilla, P. A., and Coulson, W. F., 1969a. Cardiovascular studies on copper-deficient swine. XIII. The effect of chronic copper deficiency on the cardiovascular systems of miniature pigs, *Lab. Invest.* 21:548–554.

Waisman, J., Carnes, W. H., and Weisman, N., 1969b. Some properties of the microfibrils of vascular elastic membranes in normal and copper-deficient swine, *Am. J. Pathol.* 54:107.

Wakabayashi, T., Asano, M., and Kurono, C., (Nagoya), 1975a. Mechanism of the formation of megamitochondria induced by copper-chelating agents. I. On the formation process of megamitochondria in cuprizone-treated mouse liver, *Acta Pathol. Jpn.* 25:15–37.

Wakabayashi, T., Asano, M., Kurono, C., and Ozawa, T., 1975b. Mechanism of the formation of megamitochondria induced by copper-chelating agents II. Isolation and some properties of megamitochondria from cuprizone-treated mouse liver, *Acta Pathol. Jpn.* 25:39–49.

Wakatsuki, T., Imahara, H., Kitamura, T., and Tanaka, H., 1979. On the absorption of copper into yeast cells, *Agric. Biol. Chem.* 43:1687–1692.

Waldman, T. A., Morell, A. G., Wochner, M. R., Strober, W., and Sternlieb, I., 1967. Measurement of gastrointestinal protein loss using ceruloplasmin labeled with 67 copper, *J. Clin. Invest.* 46:10–20.

Waldrop, G. L., and Ettinger, M., 1990a. Effects of albumin and histidine on the kinetics of copper transport by fibroblasts, *Am. J. Physiol.* 259:G212–G218.

Waldrop, G. L., and Ettinger, M. J., 1990b. The relationship of excess copper accumulation by fibroblasts from the brindled mouse model of Menkes disease to the primary defect, *Biochem. J.* 267:417–422.

Waldrop, G. L., Palida, F., Hadi, M., Lonergan, P., and Ettinger, M., 1988. Differences in Cu-transport by hepatocytes and fibroblasts, in *Trace Elements in Man and Animals* (TEMA-6) (L. S. Hurley, C. L. Keen, B. Lonnerdal, and R. B. Rucker, eds). Plenum Press, New York, pp. 145–146.

Waldrop, G. L., Palida, F. A., Hadi, M., Lonergan, P. A., and Ettinger, J. J., 1990. Effect of albumin on net copper accumulation by fibroblasts and hepatocytes, *Am. J. Physiol.* 259:G219–G225.

Walker, C. D., and Webb, J., 1981. Copper in plants: Forms and behaviour, in *Copper in Soils and Plants* (J. F. Loneragan, A. D. Robson, and R. D. Graham, eds.), Academic Press, Sydney, pp. 189–212.

Walker, C. D., and Welch, R. M., 1987. Low molecular weight complexes of zinc and other trace metals in lettuce leaf, *J. Agric. Food Chem.* 35:721–727.

Walker, F. J., and Fay, P. J., 1990. Characterization of an interaction between protein C and ceruloplasmin, *J. Biol. Chem.* 265:1834–1836.

Walker, W. R., 1982. The results of a copper bracelet clinical trial and subsequent studies, in *Inflammatory Diseases and Copper* (J. R. J. Sorenson, ed.), Human Press, Clifton, New Jersey, pp. 469–478.

Walker, W. R., and Keats, D. M., 1976. An investigation of the therapeutic value of the "copper bracelet"—dermal assimilation of copper in arthritic/rheumatoid conditions, *Agents Actions*, 6:454–459.

Walker, W. R., Beveridge, S. J., and Whitehouse, M. W., 1980. Anti-inflammatory activity of a dermally applied copper salicylate preparation (Alensal), *Agents Actions,* 10:3–47.

Walker, W. R., Reeves, R., and Kay, D. J., 1975. The role of Cu^{2+} and Zn^{2+} in the physiological activity of histamine in mice, *Search* 6:134–135.

Walker-Smith, J., and Blomfield, J., 1973. Wilson's disease or chronic copper posioning? *Arch. Dis. Child.* 48:476–479.

Wallace, E. F., Krantz, M. J., and Lovenberg, W., 1973. Dopamine β-hydroxylase: A tetrameric glycoprotein, *Proc. Natl. Acad. Sci. USA* 70:2253–2255.

Walshe, J. M., 1956. Penicillamine, a new oral therapy for Wilson's disease, *Am. J. Med.* 21:487–495.

Walshe, J. M., 1964. Endogenous copper clearance in Wilson's disease: A study of the mode of action of penicillamine, *Clin. Sci.* 26:461–469.

Walshe, J. M., 1975. Wilson's disease (hepatolenticular degeneration), in *The Treatment of Inherited Metabolic Disease* (N. Raine, ed.), MTP: Lancaster, pp. 171–190.

Walshe, J. M., 1981. Penicillamine and the SLE syndrome, *J. Rheumatol.* (Suppl. 7) 155–160.

Walshe, J. M., 1983. Hudson Memorial Lecture: Wilson's Disease: Genetics and biochemistry—Their relevance to therapy, *J. Inher. Metab. Dis.* (Suppl.) 6:51–58.

Walshe, J. M., 1989. Wilson's disease presenting with features of hepatic dysfunction: A clinical analysis of eighty-seven patients, *Quart. J. Med.* New Series 70:253–263.

Walshe, J. M., and Potter, G., 1977. The pattern of whole body distribution of radioactive copper (^{67}Cu, ^{64}Cu) in Wilson's disease and various control groups, *Quart. J. Med.* 46:445.

Wang, H., Koschinsky, M., and Hamerton, J. L., 1988. Localization of the processed gene for human ceruloplasmin to chromosome region 8q21.13 → q23.1 by *in situ* hybridization, *Cytogenet. Cell Genet.* 47:230–231.

Wang, Z. Y., and Mason, J., 1988. Studies on the uptake and subsequent tissue distribution of [^{35}S] trithiomolybdate in rats: Effects on metallothionein copper in liver, kidney, and intestine, *J. Inorg. Biochem.* 33:19–29.

Wapnir, R. A., and Balkman, C., 1990. Intestinal absorption of copper: Effect of amino acids, *Nutr. Res.* 10:589–595.

Wapnir, R. A., and Stiel, L., 1987. Intestinal absorption of copper: Effect of sodium, *Proc. Soc. Exp. Biol. Med.* 185:277–282.

Ward, P. A., Till, G. O., Kunkel, R., and Beachamp, C., 1983. Evidence for role of hydroxyl radical in complement and neutrophil-dependent tissue injury, *J. Clin. Invest.* 72:789–801.

Wattenberg, L. W., 1985. Chemoprevention of cancer, *Cancer Res.* 45:1–8.

Webb, J., Kirk, J. A., Jackson, D. H., Niedermeier, W., Turner, M. E., Rackley, C. E., and Russell, R. O., 1976. Analysis by pattern recognition techniques of changes in serum levels of 14 trace metals after acute myocardial infarction, *Exp. Mol. Pathol.* 25:322–331.

Weber, D. N., Shaw, C. F., III, and Petering, D. H., 1987. *Euglena gracilis* cadmium-binding protein-II contains sulfide ion, *J. Biol. Chem.* 262:6962–6964.

Webster, M. H., Waitkins, S. A., and Stott, H. A., 1981. Impaired bacteriological responses in babies after maternal iron dextran infusion, *J. Clin. Pathol.* 34:651–654.

Weibel, E. R., Staübli, W., Gnägi, H. R., and Hess, F. A., 1969. Correlated morphometric and biochemical studies on the liver cell. I. Morphometric model, stereologic methods, and normal morphometric data for rat liver, *J. Cell. Biol.* 42:68–91.

Weinberg, E. D., 1974. Iron and susceptibility to infectious disease, *Science* 184:952–956.

Weinberg, E. D., 1978. Iron and infection, *Microbiol. Rev.* 42:45–66.

Weiner, A. L., and Cousins, R. J., 1980. Copper accumulation and metabolism in primary monolayer cultures of rat liver parenchymal cells, *Biochim. Biophys. Acta* 629:113–125.

Weiner, A. L., and Cousins, R. J., 1983. Hormonally produced changes in caeruloplasmin synthesis and secretion in primary cultured rat hepatocytes, *Biochem. J.* 212:297–304.
Weiner, H. L., Weiner, L. H., and Swain, J. L., 1987. Tissue distribution and developmental expression of the messenger RNA encoding angiogenin, *Science* 237:280–282.
Weisenberg, E., Harbreich, A., and Mager, J., 1980. Biochemical lesions in copper-deficient rats caused by secondary iron deficiency. Derangement of protein synthesis and impairment of energy metabolism, *Biochem. J.* 188:633–641.
Weisiger, R. A., 1973. The superoxide dismutases from chicken liver: Organelle specificity: New evidence for the symbiotic origin of mitochondria, PhD. thesis, Duke University.
Weisiger, R. A., 1985. Dissociation from albumin: A potentially rate-limiting step in the clearance of substances by the liver, *Proc. Natl. Acad. Sci. USA* 82:1563–1567.
Weisiger, R. A., and Fridovich, I., 1973. Mitochondrial superoxide dismutase. Site of synthesis and intramitochondrial localization, *J. Biol. Chem.* 248:4793–4796.
Weiss, K. C., 1983. Evidence for the existence of a new copper transport protein in the plasma, Master's thesis, California State University, Fullerton, California.
Weiss, K. C., and Linder, M. C., and the Los Alamos Radiological Medicine Group, 1985. Copper transport in rats involving a new plasma protein, *Am. J. Physiol.* 249:E77–E88.
Weitkamp, L. R., 1983. Evidence for linkage between the loci for transferrin and ceruloplasmin in man, *Ann. Hum. Genet.* 47:293–297.
Welch, J. W., Fogel, S., Cathala, G., and Karin, M., Industrial yeasts display tandem gene iteration at the CUP1 region, *Mol Cell. Biol.* 3:1353–1361.
West, G. B., 1980. Diet and adjuvant-induced arthritis in the rat, *Int. Arch. Allergy Appl. Immunol.* 63:347–350.
Wester, P. O., 1971. Trace elements in the coronary arteries in the presence and absence of atherosclerosis *Atherosclerosis* 13:395–412.
Whanger, P. D., and Ridlington, J. W., 1982. Role of metallothionein in zinc metabolism, in *Biological Roles of Metallothionein* (E. C. Foulkes, ed.), Elsevier, North-Holland, Amsterdam, pp. 263–277.
Whicher, J. T., Bell, A. M., Martin, M. F. R., Marshall, L. A., and Dieppe, P. A., 1984. Prostaglandins cause an increase in serum acute-phase proteins in man, which is diminished in systematic sclerosis, *Clin. Sci.* 66:165–171.
White, G. A., and Krupka, R. M., 1965. Ascorbic acid oxidase and ascorbic acid oxygenase of *Myrothecium verrucaria*, *Arch. Biochem. Biophys.* 110:448–461.
White, I. G., 1956. The interaction between metals and chelating agents in mammalian spermatozoa, *J. Exp. Biol.* 33:422–430.
Whitehouse, M. W., 1976. Ambivalent role of copper in inflammatory disorders, *Agents Actions* 6:201–210.
Whitehouse, M. W., and Walker, W. R., 1978. Copper and inflammation, *Agents Actions* 8:85–90.
Whittaker, M. M., and Whittaker, J. W., 1988. The active site of galactose oxidase, *J. Biol. Chem.* 263:6074–6080.
Widdowson, E. M., 1969. Trace elements in human development, in *Mineral Metabolism in Pediatrics* (E. M. Widdowson, ed.), Blackwell Scientific Publications, Oxford, pp. 85–97.
Widdowson, E. M., Dauncey, J., and Shaw, J. C. L., 1974. Trace elements in foetal and early postnatal development, *Proc. Nutr. Soc.* 33:275–284.
Wikström, M., and Saraste, M., 1984. The mitochondrial respiratory chain, in *Bioenergetics* (L. Ernster, ed.), Elsevier, Amsterdam, pp. 49–94.
Wikström, M., Krab, K., and Saraste, M., 1981. *Cytochrome Oxidase. A Synthesis*, Academic Press, London.

Williams, D. M., Lee, G. R., and Cartwright, G. E., 1974. Ferroxidase activity of rat ceruloplasmin, *Am. J. Physiol.* 227:1094–1097.

Williams, D. M., Loukopoulos, D., Lee, G. R., and Cartwright, G. E., 1976. Role of copper in mitochondrial iron metabolism, *Blood* 48:77–85.

Williams, D. M., Burk, R. F., Jenkinson, S. G., and Lawrence, R. A., 1981. Hepatic cytochrome P-450 and microsomal heme oxygenase in copper-deficient rats, *J. Nutr.* 111:979–983.

Williams, D. M., Kennedy, F. S., and Green, B. G., 1983. Hepatic iron accumulation in copper-deficient rats, *Br. J. Nut.* 50:653–660.

Williams, D. M., Kennedy, F. S., and Green, B. G., 1985. The effect of iron substrate on mitochondrial haem synthesis in copper deficiency, *Br. J. Nutr.* 53:131–136.

Williams, L. M., Cunningham, H., Ghaffar, A., Riddoch, G. I., and Bremner, I., 1989. Metallothionein immunoreactivity in the liver and kidney of copper injected rats, *Toxicol.* 55:307–316.

Williams, M. A., and Kagan, H. M., 1985. Assessment of lysyl oxidase variants by urea gel electrophoresis: Evidence against disulfide isomers as bases of the enzyme heterogeneity, *Anal. Biochem.* 149:430–437.

Williams, R. B., Davies, N. T., and McDonald, I., 1977. The effects of pregnancy and lactation on copper and zinc retention in rat, *Br. J. Nutr.* 38:407–416.

Williamson, P. R., and Kagan, H. M., 1986. Reaction pathway of bovine aortic lysyl oxidase, *J. Biol. Chem* 261:9477–9482.

Williamson, P. R., Kittler, J. M., Thanassi, J. W., and Kagan, H. M., 1986a. Reactivity of a functional carbonyl moiety in bovine aortic lysyl oxidase, *Biochem. J.* 235:597–605.

Williamson, P. R., and Kagan, H. M., 1986b. Reaction pathway of bovine aortic lysyl oxidase, *J. Biol. Chem.* 261:9477–9482.

Willingham, W. M., and Sorenson, J. R. J., 1986. Physiologic role of copper complexes in antineoplasia, *Trace Elem. Med.* 3:139–152.

Wilson, M. L., Iodice, A. A., Schulman, M. P., and Richert, D. A., 1959. Studies on liver δ-aminolevulinic acid dehydrase, *Fed. Proc.* 18:352 (Abstract).

Wilson, S. A. K., 1912. Progressive lenticular degeneration: A familial nervous disease associated with cirrhosis of the liver, *Brain* 34:295–509.

Winge, D. R., and Miklossy, K., 1982. Differences in the polymorphic forms of metallothionein, *Arch. Biochem. Biophys.* 214:80–88.

Winge, D. R., Premakumar, R., Wiley, R. D., and Rajagopalan, K. V., 1975. Copper-chelatin: Purification and properties of a copper-binding protein from rat liver, *Arch. Biochem. Biophys.* 170:253–266.

Winge, D. R., Nielson, K. B., Zeikus, R. D., and Gray, W. R., 1984. Structural characterization of the isoforms of neonatal and adult rat liver metallothionein, *J. Biol. Chem.* 259:11419–11425.

Winge, D. R., Nielson, K. B., Gray, W. R., and Hamer, D. H., 1985. Yeast metallothionein. Sequence and metal-binding properties, *J. Biol. Chem.* 260:14464–14470.

Winge, D. R., Byrd, J., and Thrower, A. R., 1988. Metallothionein: structure and metal cluster stability, in *Essential and Toxic Trace Elements in Human Health and Disease* (A. S. Prasad, ed.), Alan R. Liss, New York, pp. 381–392.

Winyard, P. G., Pall, H., Lunec, J., and Blade, D. R., 1987. Non-caeruloplasin-bound copper ("phenanthroline copper") is not detectable in fresh serum or synovial fluid from patients with rheumatoid arthritis, *Biochem. J.* 247:245–247.

Wirth, P. L., and Linder, M. C., 1985. Distribution of copper among multiple components of human serum, *J. Natl. Cancer Inst.* 75:277–284.

Wissler, J. H., Logeman, E., Meyer, H. E., Kautzfeldt, B., Hockel, M., and Heilmeyer, L. M.

G., Jr., 1986. Structure and function of a monocytic blood vessel morphogen—(angiotropin) for angiogenesis *in vivo* and *in vitro*: A copper-containing metallopolyribonucleo-polypeptide as a novel and unique type of monokine, *Protides Biol. Fluids* 34:525–535.

Wolf, B., 1982. Therapy on inflammatory diseases with superoxide dismutase, in *Inflammatiory Diseases and Copper* (J. R. K. Sorenson, ed.), Humana Press, Clifton, New Jersey, pp. 453–467.

Wollenberg, P., Mahlberg, R., and Rummel, W., 1990. The valency state of absorbed iron appearing in the portal blood and ceruloplasmin substitution, *Bio. Met.* 3:1–7.

Wood, F. L., Wood, D. L., and Roberts, R. J., 1978. Effect of diethyldithiocarbamate on oxygen toxicity and lung enzyme activity in immature and adult rats, *Biochem. Pharmacol.* 27:251–254.

Wood, P. M., 1978. Interchangeable copper and iron proteins in algal photosynthesis. Studies on plastocyanin and cytochrome c-552 in *Chlamydomonas*, *Eur. J. Biochem.* 87:9–19.

Woolhouse, H. W., and Walker, S., 1981. The physiological basis of copper toxicity and copper tolerance in higher plants, in *Copper in Soils and Plants* (J. F. Loneragan, A. D. Robson, and R. D. Graham, eds.), Academic Press, Sydney, Chapter 11.

Worwood, M., Taylor, D. M., and Hunt, A. H., 1968. Copper and manganese concentrations in biliary cirrhosis of liver, *Br. Med. J.* 3:344–346.

Wosilait, W. D., Nason, A., and Terrell, A. J., 1954. Pyridine nucleotide–quinone reductase II. Role in electron transport. *J. Biol. Chem.* 206:271–282.

Wright, C. F., McKenney, K., Hamer, D. H., Byrd, J., and Winge, D. R., 1987. Structural and functional studies of the amino terminus of yeast metallothionein, *J. Biol. Chem.* 262:12912–12919.

Wu, B. N., Medeiros, D. M., Lin, K. N., and Throne, B. M., 1984. Long term effects of dietary copper and sodium upon blood pressure in the Long-Evans rat, *Nutr. Res.* 4:305–314.

Yagle, M. K., and Palmiter, R. D., 1985. Coordinate regulation of mouse metallothionein-I and metallothionein-II genes by heavy metal glucocorticoids, *Mol. Cell. Biol.* 5:291–294.

Yajima, K., and Suzuki, K., 1979. Neuronal degeneration in the brain of the brindled mouse. An ultrastructural study of the celebral cortical neurons, *Acta Neuropathol.* (Berl.) 45:17–25.

Yamanaka, N., and Deamer, D., 1974. Superoxide dismutase activity in WI-38 cell cultures: Effects of age, trypsinization and SV-40 transformation, *Physiol. Chem. Phys.* 6:95–106.

Yamashita, K., Liang, C.-J., Funakoski, S., and Kobata, A., 1981. Structural studies of asparagine-linked sugar chains of human ceruloplasmin, *J. Biol. Chem.* 256:1283–1289.

Yamashita, K., Ohno, H., Kondo, T., Kawamura, K., Mure, K., Yorozu, Y., Ishikawa, M., Shimizu, T., and Taniguchi, N., 1987. Maternal blood distribution of zinc and copper during labor and after delivery, *Gynecol. Obstet. Invest.* 24:161–169.

Yang, F., Luna, V. J., McAnelly, R. D., Naberhaus, K. H., Cupples, R. L., and Bowman, B. H., 1985. Evolutionary and structural relationships among the group-specific component, albumin and α-fetoprotein, *Nucleic Acids Res.* 13:8007–8017.

Yang, F., Friedrichs, W. E., Cupples, R. L., Bonifacio, M. J., Sanford, J. A., Horton, W. A., and Bowman, B. H., 1990. Human ceruloplasmin. Tissue-specific expression of transcripts produced by alternative splicing, *J. Biol. Chem.* 265:10780–10785.

Ya-You, J., Yan-Fang, L., Bo-Yun, W., and De-Yun, Y., 1989. An immunocytochemical study on the distribution of ferritin and other markers in 36 cases of malignant histiocytosis, *Cancer* 64:1281–1289.

Yeo, T.-K., Yeo, K.-T., Parent, J. B., and Olden, K., 1985. Swainsonine treatment accelerates intracellular transport and secretion of glycoproteins in human hepatoma cells, *J. Biol. Chem.* 260:2565–2569.

Yoshida, S., Tanaka, R., and Kahimoto, T., 1982. Copper-binding components in the water-soluble low molecular weight fraction of soybeans, *J. Food Hyg. Soc. Jpn.* 62:367.

Yoshida, T., Lorence, R., Choc, M. G., Tarr, G. E., Findling, K. L., and Fee, J. A., 1984. Respiratory proteins from the extremely thermophilic aerobic bacterium, *Thermus thermophilus*, Purification procedure for cytochromes $c_{552,549}$ and chemical evidence for a single subunit cytochrome aa_3, *J. Biol. Chem.* 259:112–123.

Yoshikawa, S., Tera, T., Takahashi, Y., Tsukihara, T., and Caughey, W. S., 1988. Crystalline cytochrome *c* oxidase of bovine heart mitochondrial membrane: Composition and x-ray diffraction studies, *Proc. Natl. Acad. Sci. USA* 85:1354–1358.

Young, S. N., and Curzon, G., 1974. Neonatel human caeruloplasmin, *Biochim. Biophys. Acta* 336:306–308.

Yunice, A. A., and Lindeman, R. D., 1975. Effect of estrogen-progestogen administration on tissue cation concentrations in rats, *Endocrinology* 97:1263–1269.

Yunice, A. A., Lindeman, R. D., Zcerwinski, A. W., and Clark, M., 1974. Influence of age and sex on serum copper and ceruloplasmin levels, *J. Gerontol.* 29:277–281.

Zagzag, D., and Brem, S., 1986. Control of neoplastic development in the brain: Copper depletion prevents neovascularization and tumor growth, *Surg. Forum* 37:506–509.

Zetter, N. R., 1988. Angiogenesis. State of the art, *Chest* 93:159S–166S.

Zgirski, A., and Frieden, E., 1990. Binding of copper(II) to non-prosthetic sites in ceruloplasmin and bovine serum albumin, *J. Inorg. Biochem.* 39:137–148.

Ziche, M., Jones, J., and Gullino, P. M., 1982. Role of prostaglandin E_1 and copper in angiogenesis, *J. Natl. Cancer Inst.* 69:475–482.

Zidar, B. L., Shadduck, R. K., Zeigler, Z., and Winkelstein, A., 1977. Observations on the anemia and neutropenia of human copper deficiency, *Am. J. Hematol.* 3:177–185.

Zielske, F., Koch, U. J., Badura, R., and Ladeburg, H., 1974. Copper release from copper-T (T-Cu 200) and its influences on sperm migration *in vitro*, *Contraception* 10:651–666.

Zlotkin, S. H., and Casselman, C. W., 1987. Percentile estimates of reference values for total protein and albumin in sera of premature infants (<37 weeks of gestation), *Clin. Chem.* 33:411–413.

Zubay, G., 1983. *Biochemistry*, Addison-Wesley, Reading, Massachusetts.

Index

Absorption: *see* Copper absorption
ACTH
  and ceruloplasmin, 254
  effect on ceruloplasmin, 251
Adrenal tissue, copper-dependent enzymes in, 227
Agouti trait, and tyrosinase, 232
Albocuprein
  in brain, 222
  ceruloplasmin and, 222
Albumin
  abundance, 94
  binding of Cu to the amino-terminal amino acids, 97
  copper binding, 94, 117
    affinity, 94
    ternary complex with histidine, 94
  exchange of copper with transcuprein, 110
  gene expression in development, 308
  molecular weight, 94
  multiple copper sites, 98
  N-terminal amino sequences from various species, 94
  percent serum copper with, 119
  pI, 94
  ternary complex with copper and histidine, 45
Alcoholic liver cirrhosis, copper in, 174
Algae
  metal detoxification of, 404
  phytochelatins in, 404
  plastocyanin in, 388, 390
    induction of, 390
Amicyanin, properties of, 370
α-Amidating enzyme
  in brain, 223

α-Amidating enzyme (*Cont.*)
  cofactors of, 240
  copper in, 240
  functions of, 223, 240
  isolated from bovine, 223
  in pituitary, 223
  substrates of, 240
Amine oxidase
  copper binding, 113, 377
  functions of in blood, 113
  in fungi, 376
    copper binding of, 377
    localization of, 378
    structure and functions of, 377, 378
    substrates of, 377, 378
  6-hydroxydopa and, 113, 376–377
  occurrence, 114
  in plants, 376
    copper binding of, 377
    localization of, 377, 378
    structures and functions of, 376
    substrates of, 377, 378
  properties of fungal and plant enzymes, 376
  pyrroloquinolinequinone and, 113
  sources, 114
  structure, 113, 376–378
Amino acids
  copper binding affinity of, 100
  copper binding to in plasma, 99
Aminolevulinate synthetase: *see* δ-amino levulinate synthetase
Amniotic fluid
  ceruloplasmin in, 123, 307–308
  copper in, 123, 308
  transport of copper entering, 307
Androgens, effect on ceruloplasmin, 251

505

Angiogenesis
  ceruloplasmin and, 298
  copper and, 297–300
  copper–heparin and, 299
  glycylhistidyllysine (GHL) and, 298–299
Angiogenin, 297
Angiotensin, transcription of, and copper, 170
Apoceruloplasmin
  in plasma, 87
  occurrence of, 170
Arthritis
  copper bracelets and, 356
  copper metabolism in, 354
  see also Inflammation
Arthropods, copper binding proteins in, 405
Ascorbate oxidase
  ceruloplasmin and, 83, 88, 393–396
  copper in, 88, 399
  EPR parameters and oxidation/reduction potentials of, 89
  evolutionary relationships of, 393
  function of, 399–400
  in plants, 399
  reviews on, 398
  structure and composition of, 393–396, 399–400
  substrates of, 393
  trinuclear copper cluster in, 399
  UV-visible absorption spectrum of, 394
Ascorbic acid, ceruloplasmin and, 24
Aspirin, as a copper chelator, 357
Azurin
  amino acid sequences of, 374
  in bacteria, 368–375
  ceruloplasmin and, 83
  copper binding of, 371
  evolutionary relationships of, 376
  properties of, 370–371, 374–375
  spectroscopic and redox properties of, 375
  x-ray structure of, 372–373

Bacteria
  amicyanin in/of, 370
  azurin in/of, 368, 369–70
  blue copper proteins in, 369–371
  copper in, 367
  copper binding proteins of, 368–376
  copper in cell walls of, 369
  copper sensitive and resistant mutants of, 367

Bacteria (Cont.)
  copper transport in, 368
  cytochrome oxidase of, 371; see also Cytochrome oxidase
  electron transport in, 371
  infections by
    response of copper metabolism to, 355
    ceruloplasmin response to, 355
  metallothionein-like proteins of, 373
  plastocyanins in/of, 370
  resistance to copper of, 368
  superoxidase dismutase in/of, 373
  tyrosinase in, 231
Betamethasone, effect on ceruloplasmin, 251
Bile
  absorbability of copper from, 141
  ceruloplasmin and, 146, 147, 148
  ceruloplasmin in, 145, 154
  copper absorption from, 140
  copper binding components of, 141, 144, 146
  copper concentrations and, 138
  copper content of in Wilsons' disease, 338–339
  copper dialyzability of, 140
  copper in, 136, 137, 138, 144
  daily copper losses in, 137, 159, 338
  daily excretion of, 136
  glutathione, 151
  importance of bile acids, 144
  inhibition of copper absorption, 140
  liver and, 138
  metallothionein and, 147
  sources of copper for excretion, 149
  sources of copper for liver and, 150
  taurochenodeoxycholic acid and, 146
Bioactive neuropeptides, α-amidating neuropeptides, 223
Bleomycin, 361
Blood–brain barrier, copper transport across, 133
Blood clotting factors Va and VIII
  abundance, 116
  activation, 116
  copper content, 116
  formation, 116
  homologies with ceruloplasmin, 116
Blood plasma
  copper binding components of, 54, 55, 56, 57, 73, 94, 98, 103, 105, 108, 113, 116
  albumin, 94, 119

Blood plasma (*Cont.*)
  copper binding components of (*Cont.*)
    amine oxidase, 11
    amounts of copper in void volume
      fractions separated in gel
      chromatography, 118
    blood clotting factor, 113
    ceruloplasmin, 57, 73, 119
    copper found with, 116, 119
    elution, 103
    ferroxidase II, 108
    in gel permeation chromatography, 103
    glycylhistidyllysine (GHL), 54
    histidine-rich glycoprotein (HRG), 55, 105
    low-molecular-weight copper ligands, 98
    low-molecular-weight fraction, 119
    α_2-macroglobulin, 57, 109
    superoxide dismutase, 113
    transcuprein fraction, 119
  copper in, 122
    albumin, 122
    amino acids and small peptides, 122
    after birth, 313
    ceruloplasmin, 122
    α_2-macroglobulin, 122
    see also Copper
  during copper deficiency, 271
    ferroxidase II, 109
    free copper, 122
    histidine-rich glycoprotein, 122
    transcuprein, 122
  elution of copper in gel permeation
    chromatography, 120
  exchangeable copper pool, 129
  transcuprein, 56
  see also Plasma; *and under individual
    plasma component names*
Blotchy mice, as models for Menkes' disease, 343
Blue copper proteins
  evolutionary relationships among, 376, 392–393
  structural model for blue copper centers in, 395
  substrates of, 393, 397
  see also specific proteins, including: Azurin;
    Ceruloplasmin; Galactose oxidase;
    Laccase
Body fluids, nongastrointestinal, nonblood
  copper and ceruloplasmin in, 123

Bone
  copper in
    during copper deficiency, 271
    during development, 329
    cytochrome oxidase activity in, 211
    SOD activity in, 211
Bovine, superoxide dismutase of, 196
Brain
  α-amidating enzyme, 223
  copper among regions of, 221
  copper and, 55
  copper binding components, 222
  copper in, 168, 175, 229
    during development, 326–329
  copper enzymes in, 272
  copper in cytosol, 222
  cytochrome oxidase activity in, 211, 223
  development, 228
    copper deficiency in, 228
  dietary copper intake and, 175
  distribution of copper in, 220
  dopamine β-hydroxylase in, 220
  involvement of copper proteins/enzymes
    in, 226
  mitochondria in, 168
  neurocuprein in, 222
  SOD activity in, 211, 222
Brindled mice
  copper absorption by, 345
  distribution of radiocopper during suckling
    of, 346–347
  as models for Menkes' disease, 343
  tissue copper levels in, 344

Cadmium
  detoxification by metallothionein, 188
  and metallothionein, 190
  and metallothionein induction, 188
Cancer
  alterations of copper metabolism in, 358
  antitumor action of copper complexes in, 361
  ceruloplasmin and, 61, 248, 358–360, 362–363
  ceruloplasmin synthesis in, 363
  copper absorption and, 30, 363
  copper content of tumors, 359
  copper uptake by tumor cells, 359
  and lysyl oxidase, 220
  serum copper in, 358
  sialyl transferase activity of liver in, 364

Cancer (*Cont.*)
  superoxide dismutase of tumor cells, 360–361
  tissue copper in, 362
  uptake of copper from ceruloplasmin in, 358
  whole body turnover of copper in, 364
Carbohydrate metabolism, in copper deficiency, 275, 291
Cardiac hypertrophy
  copper deficiency and, 285
  dietary factors in, 287
Cardiovascular disease, copper and, 285, 364–366; *see also* Copper deficiency
Catechol, 397
Catecholamine
  copper in formation of, 222
  dopamine β-hydroxylase, 222
  formation in brain, 226
  neurocuprein and, 222
Catechol oxidase
  amino acid composition of, 401
  in plants and fungi, 379, 400–402
  structure of, 401
  substrates of, 379, 401
  see also Tyrosinase
Cell membranes, copper contents, 174
Cell organelles, copper in, 163
Cells, distribution of incoming ^{67}Cu to, 167
Central nervous system, involvement of Cu protein/enzymes in, 226; *see also under specific CNS components*
Cerebrospinal fluid
  albumin and ceruloplasmin in, 124
  ceruloplasmin, activity of, 123
  copper in, 123
Ceruloplasmin
  adrenalectomy and, 255
  ascorbate and, 24, 90
  ascorbate oxidase and, 85
  azurin and, 85
  binding of ^{125}I to cell membranes, 63
  binding to various rat tissue membranes, 67
    effect of Cu-NTA, 67
  in blood plasma during development, 313–315
  bovine, 81, 84
  in cancer, 358, 360, 362
  cancer, and tissue sources of, 61
  characteristics, 73
  of the chicken, 81, 84, 88

Ceruloplasmin (*Cont.*)
  complete amino acid sequence of human, 75
  copper associated with in plasma, 117
  copper atoms in, 87, 88
  in copper-deficient rats, 87, 277
  copper dialyzability, 76
  copper ligands in, 85
  copper release from, 117
  in copper transport, 133
  copper turnover, 152
  Cu(II) loosely bound to, 87
  cytokines and, 243
  desialylation, 66
  discovery of, 76
  effect of inflammation, 243
  effects of contraceptive steroids, 259
  effects of diets on, 252
  effects of estrogens, 260
  endocytosis of, 60
  enzymatic activities of, 89
  enzymatic reactions of, 90
  EPR parameters and oxidation/reduction of, 89
  EPR spectra of, 86
  estradiol and, 255
  estrogen effect on, 256
  evolutionary relationships of, 392–393
  Fenton reaction and, 92
  ferroxidase activity of, 89, 93, 277
  forms of, 77, 79
    effect of ascorbate, 79
  as a free radical scavenger, 91
  function of, 5, 91, 93, 277–280
  galactose receptor and, 66
  gene expression of, 242
  of genetically obese mice, 292
  genetics of, 76, 77, 292
  of the goose, 88
  glycan chains of, 78, 79
  glycosylation of, 242
  heterogeneity in normal and disease states, 248
  high liver copper and, 11
  hydroxide radicals and, 92
  hypophysectomy and, 255
  incorporation of copper into, 335–339
  in inflammation, 243, 249, 252, 254, 351
  intracellular copper binding and, 177
  intracellular copper pathway and, 169
  and iron metabolism, 90, 277–280

Ceruloplasmin (*Cont.*)
  iron oxidation by, 277
  iron release from liver and, 277–278
  lability of, 80
  laccase and, 85
  lectin binding by, 248
  molecular properties of, 81
  molecular weight of, 79, 80
  mRNA concentrations, 243
  nitrilotriacetate (NTA) and, 66
  nondenaturing electrophoresis of, 79
  oopherectomy and, 255
  ovariectomy and, 255, 256
  oxidase activities, 10
    in different species, 10
  oxygen exposure and, 252
  oxygen radicals and, 93
  percent serum copper with, 119
  plasmin and, 80
  plastocyanin and, 85
  of pork, 86
    EPR spectrum, 86
  production and secretion, 242
  production by tumor cells, 363
  rates of transfer from the rough endoplasmic reticulum, 245
  reactions of, 91, 93
  receptors for, 62, 64, 66
    functions of, 64
    K_D, 64
  regulation of, 241, 243, 351
  regulation by glucocorticosteroids and ACTH, 249
  regulation by nutrients, 262
  regulation by steroid hormones, 251
  as a scavenger of free radicals, 354
  sedimentation equilibrium of, 86
  sensitivity to dietary status, 241
  sources of biliary copper and, 154
  as a source of copper for cells, 54, 57, 58, 59, 62, 63
  as a source of copper for lysyl oxidase, 218
  species differences in, 80
  structure of, 58, 76, 77, 80, 396
  subunits of, 80
  superoxide and, 92
  superoxide dismutase activity of, 91
  superoxide dismutation and, 92
  synthesis by liver, 129, 335
  thyroidectomy and, 255

Ceruloplasmin (*Cont.*)
  thyroxine and, 262
  time course of changes in gene expression during inflammation, 246
  tissue expression of, 78, 123
  transcription of, 243
  trinuclear copper cluster in, 85, 88
  in tumor-bearing rats, 362
  tumor implantation effect on, 256
  turnover, 153
    desialylation and, 153
    and entry into bile, 153
  types of copper in, 82, 85, 88
    analogies to other blue copper proteins, 83
  uptake of copper from, 61; *see also* Cancer
  UV-visible spectra of, 84, 394
  in Wilson's disease, 332–337
  X-ray diffraction of, 80
Ceruloplasmin-copper
  apparent half-life, 156
  uptake of by heart, 173
  see also Ceruloplasmin
Ceruloplasmin receptors
  in aorta, 65
  in brain, 65
  in erythrocytes, 65
  in heart, 65
  isolation of, 66
  in leukocytes, 65
  in liver, 65
Cervical mucus, changes in the copper content of during the menstrual cycle, 259
Cholesterol in blood
  in copper deficiency, 275, 286, 289
  in Menkes' disease, 287
Choroid plexus
  ceruloplasmin expression by, 124, 133
  copper in, 220
Chromatin, copper-binding proteins of, 170
Chromosomes, stabilized by copper, 170
Collagen
  cross-linking by lysyl oxidase, 213
  glycylhistidyllysine triplet in, 300
Connective tissue
  copper in, 272
  lysyl oxidase activity of, 212, 216
Copper
  abundance, 1
  in aerosols, 1

Copper (*Cont.*)
abundance (*Cont.*)
in animals, 1
in Arctic snow, 2
in arthropods, 5
in different types of rocks, 2
in the fetus, 305–310
in igneous rock, 2
in insects, 5
in microorganisms, 5
in muck soils, 3
in plants, 1, 5
in sedimentary rock, 2
in shellfish, 5
in soils, 2
in tillable soils and water, 1
see also Appendixes A and B
in algae, reviews of, 383
and catecholamines, 230
angiogenesis and, 297
antiinflammatory actions of, 356–358; *see also* Inflammation; Ceruloplasmin; Superoxide dismutase
in arthropods, 405
atomic number, 1
available from typical American diet, 11
average daily amounts eaten, secreted, absorbed, and excreted, 12
in the average, normal human adult, 12
in bacteria, 367
in the bile, 139
availability for resorption, 139
binding components of intestinal mucosa, 31, 33, 38
in blood plasma, 100, 101; *see also* Blood plasma, copper binding components of
associated with amino acids, 100, 101
in estrous cycle, 258
in menstrual cycle, 258
in brain, 220, 221, 222
in brain regions, 221
and cancer, 358–364
in cell membranes, 174
in choroid plexus, 220
as a co-factor, 1
carbohydrate metabolism and, 291
cardiovascular function and, 284–285
concentration in human body fluids, 8; *see also* Appendix B
content of, 37; *see also* Appendix B

Copper (*Cont.*)
content of (*Cont.*)
in duodenum, 37, 38
in ileum, 37
in intestine, 37
in livers of fetal, newborn, suckling, and adult mammals, 312
in mucosal cell cytosol, 38
in nonmucosal portion of the intestine, 37
in stomach, 37
in tissues of the human body, 7
and coronary artery disease, 364–366; *see also* Cardiovascular disease
in cytochrome oxidase, 204
in cytosol, 167, 222
detoxification by metallothionein, 188
dialysis of
in bile, 142
in gastric juice, 142
in saliva, 142
in diamine oxidase, 239
distribution in living organisms, 4
of DNA, 170
effect of processing foods, 6
as an element essential for life, 1
essential need for, 3
fertility and, 296
in fungi, 376
heme biosynthesis and, 277–280, 282
in hypothalamus, 221
immunity and, 293, 353
induction of metallothionein by, 188
in infant serum, 322
and inflammation, 350
role of copper in, 352
tissue content of in, 352
in inhibition of tumor growth, 360
insulin-like effects on adipocytes, 292
and iron metabolism, 94, 273
lipid metabolism and, 289
in liver, 163
rats, 163
humans, 163
in lysosomes, 174
in milk during lactation, 313–318
and metallothionein, 190
regulation of by copper in yeasts, 380
in microsomes, 174
in milk, 312
in mitochondria, 171

Index 511

Copper (*Cont.*)
  in molluscs, 405
  in nonvertebrate organisms, 367
  oligosaccharide complexes with, 178
  in organelles, 167
  ovulation and, 297
  in phenylalanine hydroxylase, 237
  in plants, 385
    reviews of, 383
  prostaglandin synthesis modulation by, 352
  in quahog, 169
  in roots, 383
  removal in milling of grains, 5
  resistance to toxicity of: *see* Copper resistance
  rheumatoid arthritis and, 354
  roles in brain, 226
  stability constants for, 100
    binding to amino acids, 100
    histidine binding, 100
  stabilization of chromosomes by, 170
  storage of, 168
  subcellular distribution of, 163
  in superoxide dismutase, 199
  in tissues of adult humans and animals, 16; *see also* Appendix B
  tissue concentrations, variation, 10
  in transcription, 170
  in trout, 169
  in tryptophan hydroxylase, 237
  uptake, 19; *see also* Copper absorption
    by brush border membrane vesicles, 19
    by different portions of the gastrointestinal tract, 16
  uterine contractions and, 259
  uterine estrogen receptors and, 260
  in vertebrate cells, 163
  whole body turnover, 154, 155, 158, 364
    in cancer, 364
    ceruloplasmin and, 156
    in different physiological states, 154–158
  in yeast, 379
  see also specific copper-binding components
Copper absorption
  from bile, 140, 142
  by brindled mice, 345
  brindled, mottled, and blotchy traits and, 25, 345
  by the duodenum, 15
  cancer and, 35

Copper absorption (*Cont.*)
  control mechanisms in, 19, 29
  and copper balance, 16
  copper deficiency and, 35
  from Cu-histidine, 142
  effects of anoxia, 27
  effect of dietary zinc supplementation, 28, 33
  effects of dinitrophenol, 27
  effect of dose in rats, 20
  effects of iodoacetate, 27
  energy requirement, 26
  factors affecting, 21
    amino acids, 21
    ascorbate, 25
    ascorbic acid, 23, 24
    bile, 23
    cadmium, 23
    carbonate, 22
    citrate, 21
    cobalt, 34
    competing metals, 21
    EDTA, 21
    exogenous, 21
    fiber, 21
    fructose, 23, 25
    gluconate, 22
    gluten, 23
    high dietary protein, 22
    high zinc, 23, 27
    histidine, 21
    homocysteine, 22
    iron, 23, 24, 34
    methionine, 22
    nitrilotriacetate, 22
    oxalate, 21
    pancreatic juice protein, 23
    phosphate, 21
    phytate, 21
    protein, 21
    saliva, 23
    sodium, 22, 26
    sucrose, 25
    thiomolybdate, 22
    tryptophan, 22
    zinc, 23, 27, 31, 33, 34
  estradiol and, 35, 36
  in estrogen-treated rats, 30
  from gastric juice, 12
  inhibition of by bile, 140
  in the jejunum and ileum, 15

Copper absorption (*Cont.*)
  mechanisms of, 19, 26
  in Menkes' disease, 29, 35, 343, 345
  metallothionein and, 30, 36
  in mottled mutant mice models for Menkes' disease, 345–347
  multiple ion carriers and, 42
  in the newborn, 35, 318–320
  oral contraceptives and, 35
  by plants, 383
  pregnancy and, 35
  by rat jejunum, 18
  reciprocal relationship between intestinal retention and transfer to the blood, 29
  regulation of, 19, 30, 34
  relationship between amount administered and absorbed, 17
  reviewed, 17
  from saliva, 142
  in the stomach, 15, 16
  summary, 40, 41
  time course and apparent kinetics of, 18
  in tumor-bearing rats, 30
  tumor implantation and, 36
  values reported in man, 16
  in Wilson's disease, 341
  zinc and, 28, 33
Copper aspirinate, antiinflammatory effects of, 356
Copper binding components
  in bacteria, 369
  of blood plasma, 73; *see also* Blood plasma
  in cells, 179, 181, 212
    apparent molecular weights of, 182
    in brain, 222
    cytochrome oxidase, 204
    diamine oxidases, 238
    dopamine β-hydroxylase, 212, 224
    glutathione, 181
    lysyl oxidase, 212
    metallothionein, 181
    superoxide dismutase, 181, 194
    see also under specific components
  of cytosol, 175, 178, 180
    after intravenous administration as $^{67}Cu(II)Cl_2$, 180
    apparent molecular weight, 175
    distribution of ^{67}Cu among, 180
    separated by gel permeation chromatography, 175

Copper binding components (*Cont.*)
  of cytosol (*Cont.*)
    superoxide dismutase (SOD) and, 175
  of extracellular fluid, 73
    ceruloplasmin, 73
  fractionation of on Sephadex G-150, 179
  of nonvertebrates, 367
  in vertebrate cells, 182
Copper-binding proteins: *see individual proteins or other binding components of cells*
Copper bracelets, 356
Copper-chelatin, in Cu-deficient rats, 37
Copper chelators: *see specific compounds, including:* Amino acids; Albumin; Aspirin; GHL; Histidine; Penicillamine
Copper complexes
  in treatment of inflammatory disease, 356–357
  reviews of actions, 357
  superoxide dismutase activity of, 360
  see also Penicillamine; *and specific chelating agents*
Copper components in cells: *see* Cytosol; Nuclei; *and under individual names*
Copper deficiency
  amine oxidase activity in blood plasma during, 272
  anemia of, 273, 280
  assessment of copper status in, 271
  blood pressure in, 286
  and brain, 227
  brain copper in, 229
  brain enzymes in, 228, 269, 272
  carbohydrate metabolism in, 275, 291
  cardiac hypertrophy and, 275, 285
  cardiovascular changes in, 285
  catecholamines in, 229
  ceruloplasmin in, 253, 266, 280
  cholesterol synthesis in, 290, 365
  copper chelatin and, 37
  copper contents of liver, kidney, and femurs of rats during, 270
  copper in tissues
    during, 266
    time course of changes in, 267
  cytochrome oxidase and, 171, 266, 379
  and development, 228
  enzymes in, 228, 229
  fertility and, 276, 296

Copper deficiency (Cont.)
  fragility of erythrocytes in, 284
  galactose oxidase and, in fungi, 379
  glucose tolerance and, 291
  gonadotropin-releasing factor secretion in, 296
  granulopoiesis in, 273, 276
  heart weight in, 275
  hematocrit in, 274–277, 280
  heme synthesis in, 283
  hematopoiesis and, 275–277, 280
  human studies, 289
  hypercholesterolemia and, 286, 289, 365
  immunity and, 276, 293–295
  insulin production and, 292
  and iron metabolism, 274–276
  kidney and, 168
  kidney iron in, 277
  lipid metabolism in, 275, 290
  liver and, 168
  liver copper in, 266, 271
  liver iron in, 277
  liver mitochondria in, 284
  lysyl oxidase activity in, 271–272
  and Menkes' disease, 343
  mitochondria and, 171
  neutropenia in, 276
  organelles and, 168
  ovulation in, 296
  plasma copper in, 271, 277
  plastocyanins and, 388
  reticulocytes in, 283
  serum iron in, 277
  sorbitol synthesis in, 293
  sperm motility and, 276
  superoxide dismutase in, 211, 269
  tyrosinase activity in, 272
  wool pigmentation of sheep and, 272
Copper-deficient rat, copper in, 164; *see also* Copper deficiency
Copper-dependent enzymes, structure and activity of, 239; *see also under specific enzymes*
Copper deprivation: *see* Copper deficiency
Copper excretion
  bile and, 137, 138, 150, 151
  ceruloplasmin and, 151
  ceruloplasmin turnover and, 152
  concentrations of copper in, 135
  daily losses from GI tract, 135

Copper excretion (Cont.)
  daily losses from skin, sweat, hair, and nails, 135
  daily losses via the urine, 137
  daily secretions in feces, 135
  daily secretions in GI fluids, 135
  gastric fluid and, 137
  via gastrointestinal routes other than bile, 139
  glutathione and, 159
  liver and, 151, 159
  menstrual losses, 137
  net fecal losses, 139
  pancreatic fluid and resorption, 138
  regulation of, 158
  routes of, 135
  saliva and, 137
  summary, 161
  urine and, 150
  in Wilson's disease, 152, 338
  see also Bile; Copper secretion; Ceruloplasmin
Copper intake
  by adult humans, 12
  in infancy, 318–320
  liver copper and, 166
  see also Copper absorption
Copper metabolism
  in cancer, 359, 362–363
  digestive tract capacity for, 17
  during inflammation, 350, 357
  overview in mammals of, 11–12
  regulation of by hormones, 241, 263
  reviews on, 13
Copper resistance
  in bacteria, 368, 373
  glutathione and, 181
  in plants, 384–402
  in yeast, 368, 382
Copper salicylate, 357
Copper status, changes in human dietary intake and, 168
Copper sulfide, in bacterial resistance, 368
Copper-thionein: *see* Metallothionein
Copper transport, summary model for in the human, 131; *see also* Copper uptake; *and individual components of blood plasma*
Copper uptake
  albumin in, 130

Copper uptake (*Cont.*)
  albumin in (*Cont.*)
    from ceruloplasmin, 60, 61, 62, 69, 70
      effect of ascorbate, 62, 69
      mechanisms, 69
      potential models, 70
      summary, 69
    erythrocytes and, 53, 54
      amino acid stimulation, 54
    exchangeable copper pool and, 130
    by fibroblasts, 49, 51
      from normal and brindled mice, 49
    glutathione and, 181
    by heart, 172
    by hepatocytes, 43, 44, 45, 46, 47, 49, 50, 51, 52
      albumin and, 47, 48
      amino acids and, 45, 47
      apparent rates, 50, 51
      azide and, 44, 45
      in Cu-deficient and brindled mice, 50
      2,4-dinitrophenol and, 43
      DNP and, 44, 45
      effects of amino acids, 43
      histidine and, 43, 44, 47, 48, 49
      kinetics of, in rats, 46, 50
      K_m of, 52
      at 37°C, 51
      mechanism, 44
    histidine in, 130
    by hypothalamic slices, 52, 53
      apparent K_m of, 53
      cysteine and, 53
      histidine, threonine, and, 53
      kinetics, 53
    ionic copper and, 43
      albumin and, 43
    α_2-macroglobulin in, 130
    nonceruloplasmin sources in, 61
      oligomycin effect on, 45
      other divalent metal ions and, 48
      ouabain effect on, 45
      oxygen and, 43
      time course of, 50
      valinomycin and, 45
      V_{max} of transport sytems for, 52
      zinc and, 51
  by nongastrointestinal vertebrate cells, 43
  by nonhepatic cells, 52
    albumin and, 52

Copper uptake (*Cont.*)
  by nonhepatic cells (*Cont.*)
    by fibroblasts, 52
    K_m of transport systems, 52
  by placenta, 307
  by plants, 383
  by yeasts
    carriers for, 382
    kinetics of, 382
    mechanisms of, 382
Cord blood, copper-binding components of, 321
Coronary artery disease: *see* Cardiovascular disease; Copper deficiency
Corticosteroids: *see* Glucocorticosteroids
Cow, copper in milk of, 313, 317
Crab, metallothioneins of, 183
$CuCl_2$, intracellular distribution of copper from, 169; *see also* Copper uptake
Cushing's disease, urinary excretion of copper, 266
Cytochrome oxidase
  activity of, 204
  basic functional unit of, 206
  cofactors of, 240
  configuration of, 208
  copper in, 204, 207, 209, 420
  and copper deficiency, 171, 211
  domains of, 207
  functions of, 240
  of fungi, 379
  gene expression of, 380
  genetics of, 210
  heme in, 209
  heme regulation of, 380
  heme biosynthesis and, 283
  in the hippocampus, 223
  metal binding of, 206
  mitochondrial membrane location of, 208
  in plants, 386
  of *paracoccus denitrificans*, 206
  reaction of, 209
  restoration of activity in tissues of copper-depleted rats, 173
  reviews of, 205
  spectra of, 207
  structure of, 206–210
  structural conservation in evolution, 204

Cytochrome oxidase (*Cont.*)
  substrates of, 240
  subunits of, 204, 205, 206, 209
    amino acid sequences, 206
    apparent molecular weight of, 205
    coordinate synthesis of, 210
    in fungi, 204
    origins of, 380
    in plants, 204
    in procaryotic bacteria, 204
    in protozoa, 204
    sites of synthesis, 205
    in vertebrates, 204
    in yeasts, 204, 380
  in synaptic terminal regions, 223
  synthesis of, 209, 210
  in *Thermus thermophilus*, 206, 207
  in yeast, 380
Cytosol
  copper-binding components of, 175, 177, 178
    carbohydrates and, 178
    fractionated on a larger pore gel, 178
    radioactive copper and, 178
  copper in, 165, 168, 175
  dietary copper intake and, 175
  fractionation of on Sephadex G150, 179
  of liver, 178
  of skeletal muscle, 178
  of small intestine, 178
  see also Superoxide dismutase;
    Metallothionein

δ-Amino levulinate synthetase, copper and, 282; *see also* Copper deficiency, hematopoiesis
Desmosine, and lysyl oxidase, 215
Development
  changes in copper metabolism during, 325
  copper in bone and skin during, 329–330
  copper in brain during, 326–329
  see also Pregnancy
Dexamethasone
  effect on ceruloplasmin, 251
Diabetes, accumulation of copper by liver and kidney in, 292
Diamine oxidase
  cofactors of, 240
  copper in, 239, 240
  functions of, 238, 240
  structure of, 239

Diamine oxidase (*Cont.*)
  substrates of, 240
  tissue distribution of activity, 239
  see also Amine oxidase
Dietary copper
  brain and, 168
  copper content of milk and, 316
  distribution to tissues, 124, 126
    albumin and, 126
    function of, 126
    transcuprein and, 126
  effects of digestive secretions on the availability of, 139
  human intake, 12
  initial uptake by liver and kidney, 126
  liver and, 129, 168
  portal blood and, 130, 132
  time course of distribution to tissues and plasma components, 127
  see also Intravenous copper
Diets
  copper in that of the average human adult, 12
  high-copper, 167
    tissue copper in, 167
    copper retention by liver in, 165
  low-copper, 167
    tissue copper in, 167
Dog
  albumin, 97, 98
  high liver copper, 11
  liver copper concentrations of, 97
DOPA, as substrate of tyrosinase, 235
Dopamine β-hydroxylase
  activity in copper deficiency, 229, 230
  of the adrenal medulla, 224
  in brain metabolism, 224
  in chromaffin granules, 224
  cofactors of, 240
  copper content of, 225, 240
  functions of, 240
  6-OH-dopa and, 225
  overview of, 226
  pyrroloquinoline quinone and, 225
  reaction of, 224
  structure of, 225
  substrates of, 240
Drosophila, metallothioneins of, 182
Duodenum
  copper in, 136

Duodenum *(Cont.)*
  daily excretion of copper by, 136

Efflux of copper, from cells, 49
Elastin
  and copper deficiency, 217
  cross-linking by lysyl oxidase, 213
Electron transport
  azurins in, in bacteria, 371
  plastocyanins in, 371
  see also Cytochrome oxidase
Endoplasmic reticulum, copper contents of, 174
Enkephalin
  in chromaffin granules, 224
  copper binding to, 224
  sensitivity to copper intake, 224
Epinephrine, effect on ceruloplasmin, 252; *see also* Catecholamines
Erythrocuprein: *see* Superoxide dismutase
Erythrocytes
  pink copper protein in, 182
  superoxide dismutase of, 197, 270
  see also Copper deficiency
Erythropoietin, in copper deficiency, 274
Essentiality of copper to living organisms, discovery of, 4
Estradiol
  and copper excretion, 159
  and lysyl oxidase, 220
  see also Estrogen
Estrogen
  ceruloplasmin and, 250, 303
  copper absorption and, 30, 363
  effect on collagen, 264
  effect on serum Cu, 250
  effects on copper contents of rat tissues, 261
  and lysyl oxidase, 220, 263
  serum copper and ceruloplasmin during treatment with, 257
Estrous cycle, and serum copper, 258
Exchangeable serum copper pool: *see* Amino acids; Albumin; Transcuprein
Excretion: *see* Copper excretion
Extracellular superoxide dismutase (EC-SOD)
  activity in plasma, 115
  binding to glycosaminoglycans, 115
  in blood, 203
  copper content of, 115
  in copper deficiency, 270

Extracellular superoxide dismutase (EC-SOD) *(Cont.)*
  distribution of, 114
  occurrence of, 114
  structure, 114
  tissue/cell origin of, 115
  tissue distribution of, 203
  zinc content of, 115

Feces
  copper in, 136
  daily excretion of copper via, 136
Ferritin, in iron metabolism, 280
Ferroxidase I: *see* Ceruloplasmin
Ferroxidase II
  apparent molecular weight of, 108
  copper in, 109
  copper deficiency and, 109
  discovery of, 108
  iron metabolism and, 278
  occurrence of, 108
  subunits of, 108
Fetal circulation
  α-fetoprotein in, 306–308
  ceruloplasmin in, 306–307
  copper in, 306
  low-molecular-weight ligands in, 308
  transport of copper entering, 307
Fetal liver
  copper in, 164, 309
  metallothionein and, 309
Fetal lung, ceruloplasmin production by, 307
α-Fetoprotein
  copper affinity of, 109
  copper binding of, 98
  copper bound to, 118
  copper exchange with albumin, 109
  discovery of, 109
  N-terminal amino acids of, 309
  synthesis of, 98
Fever, biliary copper excretion and, 160
Fungi
  amine oxidases of, 376–378
  ascorbate oxidase of, 379
  catechol oxidase of, 400–402
  copper uptake by, 379
  cytochrome oxidase of, 379
  galactose oxidase in/of, 378–379
  laccase of/in, 379, 393–398
  metallothioneins of, 379

Index

Fungi (*Cont.*)
  tyrosinase in, 379, 398, 400–402

Galactose oxidase
  absorption spectrum of, 378
  copper binding of, 378
  copper deficiency effect on, 379
  in fungi, 378
  structure of, 378
  synthesis and secretion of, 379
Gastric juice
  copper in, 136, 137
  daily secretion of copper into, 136, 137
Gestation, copper metabolism in, 301; *see also* Pregnancy
Glial cells, copper localized in, 221
Glucagon, and metallothionein, 190
Glucocorticosteroids
  biliary copper secretion and, 159
  and ceruloplasmin, 251, 254
  ceruloplasmin regulation by, 249, 251
  and metallothioneins, 190
  and urinary copper excretion, 266
γ-Glutamyl cysteine dipeptidyl transpeptidase, in phytochelatin formation, 402
Glutathione
  copper binding by, 181
  copper resistance and, 181
  copper uptake and, 181
Glycylhistidyllysine (GHL), 54
  angiogenesis and, 300
  collagen has sequence of, 300
  complexes of copper with, 104
    stability constants of, 104
  copper and, 54
  copper binding and, 103, 122
  copper uptake and, 104
  effects of, 54, 104
  growth modulating properties of, 104
  plasma copper in, 212
  possible structure of tripeptide-Cu(II) complex of, 105
  X-ray crystallography of, 104

Hamster, metallothioneins of, 183
Heart
  copper in, 167, 172, 175
  cytochrome oxidase activity in, 211
  dietary copper intake and, 175
  mitochondria of, 172

Heart (*Cont.*)
  SOD activity in, 211
  subcellular copper distribution in, 168
  uptake of copper by, 172
  weight in copper deficiency, 274
Hematocrit, in copper deficiency, 171, 274
Heme oxygenase, in copper deficiency, 274
Heme synthesis
  copper and, 282
  and copper deficiency, 171, 274
  by liver mitochondria, 171
  by reticulocytes, 171
Hemocyanin
  in arthropods and molluscs, 405–407
  binuclear copper center in, 407
  copper binding of, 406
  evolutionary relationships of, 400, 405
  function of, 405, 407
  X-ray structure of, 406
Hepatocytes: *see* Liver cells
Hepatolenticular degeneration: *see* Wilson's disease
Histamine, inactivation by amine oxidase, 113
Histidine
  complex with copper and albumin, 45
  copper binding affinity of, 100, 101, 103
  copper uptake and, 49
  in plasma, 99
    copper binding to, 99
Histidine-rich glycoprotein (HRG)
  amino acid sequence of, 105
  blood concentrations of, 106
    at birth, 106
    in disease, 106
    normal, 106
  carbohydrate content, 106
  copper binding of, 56, 106
  functions of, 55
  heme binding of, 105
  histidine in, 55, 106
  isolation of, 105
  K_D for Cu(II) of, 56
  metal binding of, 105
  metal binding sites of, 56
  molecular weight of, 106
  repeats in, 107
  plasminogen and, 105
  primary structure of, 107
  sedimentation coefficient of, 106
  as a source for copper, 56

Histidine-rich glycoprotein (HRG) (*Cont.*)
  structure of, 55, 106
Hormones, regulation of copper metabolism by, 241
Horse
  metallothioneins of, 183
  superoxide dismutase and, 196
Human
  copper in milk of, 313, 317
  metallothioneins of, 183
Humus, laccase in formation of, 398
Hypercholesterolemia: *see* Cholesterol
Hypothalamus, copper in, 221

IL-1
  ceruloplasmin and, 152, 243, 248, 351
  overview of the regulation and effects of, 247
IL-2, ceruloplasmin synthesis and, 248
IL-6, ceruloplasmin and, 243, 252
Immunity
  copper and, 293–296, 353
  energy and, 295
  see also Infection; Inflammation
Indian childhood cirrhosis
  albumin-copper in, 333
  ceruloplasmin in, 333
  transcuprein-copper in, 333
Infection
  and metallothionein, 191
  hypoferremia of, 354
  see also Inflammation
Inflammation
  ceruloplasmin in, 248, 252, 351, 355
  ceruloplasmin synthesis in, 351
  copper in liver and kidney during, 352
  copper metabolism in, 350, 357
  copper therapy in, 356
  damage and swelling in, 295
  iron metabolism in, 280
  and metallothionein, 190, 191
  roles of copper in, 352–355
Interferon, and metallothioneins, 190
Intestinal juices, nonbiliary, nonpancreatic-derived, copper in, 138
Intestinal mucosa, copper binding components of, 39
  electrophoretic migration of radioactive components, 39

Intestinal mucosa, copper binding components of (*Cont.*)
  see also Copper binding components; Metallothionein
Intestine
  changes in copper content after birth, 324, 326
  copper in, 167
  in metallothionein of after birth, 324
Intravenous copper
  uptake of by tissues, 125
  bone, 125
  brain, 125
  heart, 125
  kidney, 125
  liver, 125
Iron
  bacterial infection and, 354
  of blood hemoglobin in pigs, 25
    as a function of dietary copper and iron, 25
  copper metabolism and, 25
  copper status and, 24
  form released from storage sites, 278
  in inflammation, 254, 280
  metabolism of, 280; *see also* Iron metabolism
Iron metabolism, 280
  ceruloplasmin and, 278
  interactions of copper with, 284
Iron uptake, by reticulocytes, 171

Kidney
  copper in, 167, 169, 175
  copper storage in, 168
  cytochrome oxidase activity in, 211
  dietary copper and, 168, 175
  in initial uptake of dietary copper, 125
  of *Mercenaria mercenaria*, 169
  of molluscs, metal-binding granules in, 407
  SOD activity in, 211

Laccase
  ceruloplasmin and, 83, 393–396
  copper in, 88
  EPR parameters and oxidation/reduction of, 89
  evolutionary relationships of, 393, 397
  functions of, 397, 398
  localization of, in plant tissues, 398

Index

Laccase (Cont.)
  model for blue copper sites in, 395
  in plants, 392–396, 398
  regulation of, 398
  similarities to tyrosinase, 397
  structure and composition of, 393–396
  substrates of, 393, 397
  UV-visible spectrum of, 394
Lactation, copper metabolism in, 315–316
Lignin, 397
Lipid metabolism, in copper deficiency, 275
Liver
  ceruloplasmin synthesis by, 242, 335, 339
  of copper-loaded adult rats, 165
  copper in, 10, 166, 167, 169, 175
    cytosol, 166
    microsomes, 166
    mitochondria, 166
  copper storage in, 165, 168, 310
  copper, iron, and zinc concentrations during development, 310
  copper and metallothionein in, during development of rats, 311
  cytochrome oxidase activity in, 211
  deposition of excess copper in, 165
  dietary copper and, 168
  and dietary copper intake, 175
  during development
    ceruloplasmin secretion by, 318
    copper in, 320
  in initial uptake of dietary copper, 125
  iron release from, 278
  metallothionein in, 265
  as origin of ceruloplasmin and albumin: see Ceruloplasmin; Albumin
  SOD activity in, 211
  in Wilson's disease, 336
Liver cells
  copper in, 163, 164
  distribution of copper in, 164
  distribution of radioactive copper in, 164
  in gestation, 163
  as a repository for copper, 163
  see also Liver
Liver cirrhosis, copper in, 166
Liver copper
  adrenalectomy and, 255
  after birth, 309, 320
  during gestation, 305, 309
  estradiol and, 255

Liver copper (Cont.)
  hypophysectomy and, 255
  oopherectomy and, 255
  ovariectomy and, 255
  thyroidectomy and, 255
  see also Liver
Liver iron, in copper deficiency, 274
Low-molecular-weight fractions binding copper in human serum, 121
Lung
  copper in, 167
  lysyl oxidase activity in during copper deficiency, 271
Luteinizing hormone-releasing factor, and copper, 226
Lysosomes
  excess copper and, 174
  metallothionein in, 173, 174
  in Wilsons' disease, 332
Lysyl oxidase
  activity of, 212
  amine oxidase and, 212
  amino acid composition of, 215, 216
  apparent molecular weight of, 215
  binding to collagen of, 214
  in bone during development, 330
  cardiovascular disease and, 366
  ceruloplasmin as a source of copper for, 220
  cofactors of, 240
  and collagen, 213
  in connective tissue, 219
  and connective tissue formation, 217
  copper as a co-factor for, 217
  copper deficiency and, 217
  copper in, 240
  in cross-linking of collagen and elastin, 213
  and desmosine formation, 215
  and elastin, 213
  functions of, 240, 284
  in hen oviduct, 219
  heparin and, 220
  isoforms of, 215, 217
  6-OH-dopa and, 218
  purification of, 215, 216
  pyridoxal phosphate and, 218
  pyrroloquinoline quinone and, 218
  reactions of, 214
  regulation of, 220
  regulation by estrogen of, 263
  reviews of, 212

Lysyl oxidase (*Cont.*)
  substrates of, 215, 240
  synthesis of, 218
  and testosterone, 264
  tissue distribution of, 216
  in tissues during development, 219

α_2-Macroglobulin
  copper binding of, 58
  potential copper affinity of, 112
Man, copper balance in, 16
Melanin
  formation of, 234
  regulation of production of, 233
  and tyrosinase, 231
Melanocytes, and copper, 233
Membranes of cells, copper found with, 174
Menkes' disease, 342–365
  cholesterol metabolism in, 365
  copper absorption and, 30, 343, 345
  copper content of cells in, 347
  copper metabolism in, 344
  copper uptake of cells in, 348
  defects in, 349
  as a disorder of copper metabolism, 342
  genetics of, 343
  metallothionein in, 347
  metallothionein regulation in, 348
  mouse mutant models for, 343, 345
  tissue copper levels in, 344
Menstrual bleeding, losses of copper through, 137
Mercenaria mercenaria characteristic granules of kidney in, 169
Metallothionein
  as an acute phase reactant, 264
  affinity for copper of, 170
  amino acid composition of, 183
  amino acid sequence of, 183
  amounts in intestines of rats, 29
  in blood plasma, 191, 192
  changes in concentrations after birth, 193
  crystal structure of, 186
  cysteine content of, 183
  detoxification of metals by, 188
  discovery, 182
  distribution in living organisms, 183
  domains of, 185
  excess intracellular copper and, 165
  in fetal liver, 309

Metallothionein (*Cont.*)
  forms of, 187
  functions of, 188, 190, 194
  gene expression of, 190
  genetics of, 187, 191
  glucocorticoid regulation of, 189
  induction of, 188
    by cadmium, 188
    by copper, 188
    by zinc, 188
  inflammation and, 264, 265
  interferon regulation of, 189
  in the intestinal mucosa, 29, 39, 40
    during development, 324
  intracellular copper binding and, 177
  in intracellular transport, 191
  in kidney, 192
  in laparotomy, 265
  in liver, 192
    of the neonate, 273
  in lysosomes, 173
  in Menkes' disease, 347–349
  in Menkes' lymphoblasts, 349
  metal binding of, 185
  as metal donor to enzymes, 192
  metal ions interacting with, 188
  metal regulation of, 189
  metal-regulatory elements in genes of, 189, 191
  metal storage and, 190
  mRNA of, in rats on zinc and copper diets, 29
  in nuclei, 165, 173
  oxidation of, 193
  in physical stress, 265
  of plasma, 193
  polymerization of, 193
  regulation of, 188, 190, 264
    by cadmium, 190
    by copper, 190
    by glucagon (fasting), 190
    by glucocorticosteroids, 190
    by inflammation, 190
    by interferon, 190
    by metal ions, 188
    by zinc, 190
  regulatory sequences in mammalian genes for, 189, 191
  related proteins and, 183, 187
  as a repository for copper, 165, 188, 170

Metallothionein (*Cont.*)
  silver binding to, 263
  similarities to other plant proteins of, 402
  stability constants of, 188
  in starvation, 265
  structure of, 185
  turnover of, 193, 194
  uptake of ^{67}Cu by, 181
  in urine, 193
  variants of, 187
    induced by glucocorticoids, 188
    induced by interferon, 188
  X-ray structure of, 185
  in yeast, regulation of, 380
  see also Copper resistance
Microsomes, copper in, 174
Milk
  copper in, 163, 312–313, 315–318
    in different mammals, 313
  copper absorption from, 318
Mitochondria
  copper binding by, 173
  in copper deficiency, 171
  copper distribution to, 167
  cytochrome c oxidase in, 206
  in heart, 172
  low molecular weight ligands in, 317
  see also Cytochrome oxidase
Mitochondrial fraction: *see* Mitochondria
Molluscs
  copper-binding proteins in, 405
  metal binding granules in kidneys of, 407
  metallothionein in, 407
  superoxide dismutase in, 407
Monkey, metallothioneins of, 183
Mouse, metallothioneins of, 183
Mugineic acid, 381
Myelin formation
  in copper deficiency, 275
  in development, 327–328

Neonatal liver, copper in, 164
Neurospora, metallothioneins of, 183
Newborn, serum copper and ceruloplasmin in, 314
Nickel, oligosaccharide complexes with, 178
Nicotianamine, 381
Nuclei
  copper in, 165, 170
  copper components in, 169

Nuclei (*Cont.*)
  metallothionein in, 165, 169
Nucleus: *see* Nuclei
Nutrients, modulation of plasma ceruloplasmin concentrations by, 262

Oral contraceptives
  and copper in cervical mucous, 259
  effects on ceruloplasmin, 259, 261
Oxidative stress: *see* Superoxide dismutase; Ceruloplasmin

Pancreatic fluid
  copper concentration of, 138
  copper in, 136
  daily secretion of, 136
Penicillamine
  anti-inflammatory activity of copper complex of, 356
  in treatment of Wilson's disease, 342
Phenolase: *see* Tyrosinase; Catechol oxidase
Phenylalanine hydroxylase
  activity of, 235
  in *Chromobacterium violaceum*, 235
  copper in, 237
  iron in, 237
  in liver, 235
  pterin cofactor of, 235
Photosynthetic systems I and II of plants
  components of, 389
  copper ligands in, 387
Phytochelatins
  in copper resistance, 384, 402
  copper storage by, 381–382
  formation from glutathione, 382, 402
  function of in plants, 384, 402
  induction of, 402
  occurrence in plants and yeasts of, 381, 384, 402
  structure of, 381
Phytocyanins
  properties of, 370–371, 375
  X-ray structure of, 391
Pig, copper in milk of, 313, 317
Pink Cu(II) protein, of erythrocytes, 182
Pituitary, copper-dependent enzymes in, 223
Placenta
  ceruloplasmin mRNA in, 306
  copper in, 306
  copper uptake by, 307

Placenta (*Cont.*)
  superoxide dismutase in, 303–304
Plants
  amine oxidases of, 376
  ascorbate oxidase of, 393–396, 398–400
  blue copper proteins in, 389; *see also specific proteins*
  catechol oxidase in, 400–402
  copper binding glycoproteins of, amino acid sequence of, 374
  copper absorption by, 383
  copper ligands in, 384
  copper in photosynthesis by, 385
  copper binding components of, 403
  copper binding proteins of, 374, 384
  copper in respiration of, 385
  copper storage in, 381, 384
  copper tolerance of, 384, 402
  copper transport in, 384
  cytochrome oxidase in/of, 386
  functions of copper in, 384
  laccase in, 393–396
  low-molecular-weight ligands in, 402
  metallothionein in, 384
  metallothionein-like proteins in, 402
  photosynthetic systems of, 387
  phytochelatins in, 381, 384
  phytochelatin induction of, in, 404
  properties of small blue copper proteins of, 370
  small blue copper proteins in, 384–392; *see also specific proteins*
  sizes of, 403
  tissue copper levels in, 385
  tyrosinase in, 231, 400–402
Plasma copper
  adrenalectomy and, 255
  estradiol and, 255–256
  oopherectomy and, 255
  ovariectomy and, 255–256
  thyroidectomy and, 255
  tumor implantation effect on, 256
  see also Blood plasma
Plastocyanins
  amino acid sequences of, 374
  copper binding of, 371, 387–388
  copper deficiency effect on, 388
  evolutionary relationships of, 376, 391
  functions and properties of, 370–376, 386–387

Plastocyanins (*Cont.*)
  induction of, by copper, in algae, 390
  location of in plant tissues, 386
  spectroscopic and redox properties of, 375
  structure of, 388, 391
  X-ray structure of, 373
  see also Plastocyanins
Polyamine biosynthesis, role of amine oxidase in, 378
Polyamine conjugates, copper binding and, 103
Polyphenol oxidase: *see* Laccase
Porphyrin synthesis: *see* Heme synthesis
Portal blood, transcuprein in, 132
Pregnancy
  ceruloplasmin in, 301–303
  copper content of the maternal body in, 303
  effects on ceruloplasmin, 253, 259
  maternal copper stores in, 305
  retention of dietary copper in, 303
  superoxide dismutase in placenta during, 303
  total plasma copper and, 261, 301
  see also Gestation
Progestogen
  effect on ceruloplasmin, 250
  effect on serum copper, 250
  effects on copper contents of rat tissues, 261
Pseudoazurin, properties of, 370
Pyrroloquinoline quinone
  amino oxidase and, 113
  blood plasma, 113
  dopamine β-hydroxylase and, 225
  and lysyl oxidase, 218
  structure of, 219

Quahog, copper in, 169

Rabbit, metallothioneins of, 183
Rat tissues, copper in, 167; *see also* Copper; Liver; Kidney; *individual tissues*; Appendix B
Resorcinol, 397
Rheumatoid arthritis
  ceruloplasmin in, 354
  copper in, 354
  superoxide dismutase activity in, 354
  see also Inflammation
Roots
  copper in, 383

Index   523

Roots (*Cont.*)
  function of in uptake of copper by plants, 383
Rusticyanin, properties of, 370

Saccharomyces, metallothioneins of, 184
Saliva
  copper binding components of, 143
  copper in, 136, 137
  daily copper secretions into, 137
  daily secretion of, 136
Sea urchin, metallothioneins of, 184
Seminal fluid, copper in, 123
Serum: *see* Blood plasma
Serum iron, in copper deficiency, 274
Sheep
  and copper excretion, 159
  copper in milk of, 313, 317
  metallothioneins of, 183
Small intestine
  cytochrome oxidase activity in, 211
  SOD activity in, 211
SOD: *see* Superoxide dismutase
Spermine oxidase: *see* Amine oxidase
Spleen
  cytochrome oxidase activity in, 211
  SOD activity in, 211
Stellacyanin, 390
Stomach, copper in, 167
Subcellular compartments, copper in, 169; *see also* Cytosol; Lysosomes; Mitochondria; Nuclei
Superoxide dismutase
  active site of, 199, 200
  activity of, 194
  as an activity of copper complexes, 360
  amino acid sequence of, 196
  cofactors of, 240
  copper deficiency and, 195, 293
  copper in, 240
  crystal structure of, 198
  in cytosol, 195
  of different species, 197
  discovery of, 194
  EPR spectrum of, 197
  extracellular enzyme, 203; *see also* Extracellular superoxide dismutase
  forms of, 195
  functions of, 194, 201, 240
  inflammation and, 264
  interaction with other antioxidants of, 202

Superoxide dismutase (*Cont.*)
  intracellular copper binding and, 175
  in lysosomes, 195
  measurement of activity of, 91
  metal binding of, 199
  metal ion content (native protein) of, 197
  metal storage and, 203
  molecular weight of, 197
  in molluscs, 407
  oxidative stress and, 201
  in placenta, 303–305
  properties of, 197
  reaction mechanism of, 203
  reaction of, 200
  redox potential of, 197
  regulation of, 195
  structure of, 196
  substrates of, 200, 240
  subunits of, 196
  synthesis of, 195
  tissue distribution of, 203
  in tumors, 360
  turnover of, 195
Sweat
  copper in, 136
  daily excretion of, 136

Testis, copper in, 167; *see also* Appendix B
Testosterone, effects on lysyl oxidase, 264
Thermus thermophilus, cytochrome oxidase in, 207
Thiomolybdates, copper availability in diet and, 272
Threonine in plasma, copper binding to, 99
Thymus
  cytochrome oxidase activity in, 211
  SOD activity in, 211
Thyroid function, copper and, 262
TNF: *see* Tumor necrosis factor
Tumor bearing rats: *see* Cancer
Tumors: *see* Cancer
Tumor necrosis factor
  ceruloplasmin synthesis and, 243, 248
  effect on ceruloplasmin, 252
Transcuprein
  characteristics of, 56, 57, 111, 112
  copper affinity of, 109, 111
  copper associated with, 117
  copper exchange with albumin and, 56, 109, 110

Transcuprein (*Cont.*)
  discovery of, 56, 109
  in Indian childhood cirrhosis, 333
  in infant blood plasma, 321
  properties of, 112
  purification of, 112
  as a source of copper for cells, 56, 57
  in Wilson's disease, 333
Transferrin
  in inflammation, 254
  rates of transfer from the rough endoplasmic reticulum after synthesis, 245
Transport: *see* Copper uptake; Ceruloplasmin; Albumin; Transcuprein; Blood plasma
Triglyceride, in blood in copper deficiency, 275, 289
Trinuclear copper cluster
  in ascorbate oxidase, 85
  in ceruloplasmin, 85
  in laccase, 85
Trout, copper in, 169
Tryptophan dioxygenase: *see* Tryptophan oxygenase
Tryptophan oxygenase
  activity of, 237
  copper in, 237
  of *Pseudomonas*, 238
Tyrosinase
  activity of, 234
  cloning of, 231
  cofactors of, 240
  copper in, 133, 223, 240, 400
  functions of, 234, 240, 401–402
  enzyme family of, 231
  evolutionary relationships of, 393, 400, 405
  genetics of, 231–232
  in copper deficiency, 272
  isoforms of, 232
  localization of in plant tissues, 400
  molecular properties of, 232
  molecular weight of, 232
  in molluscs, 407
  pigmentation and, 231
  reaction mechanism proposed for, 236
  reactions of, 234
  relation to hemocyanin, 405
  similarities with laccase of, 397
  substrates of, 234, 240, 397, 400–401
  subunits of, 231
  synthesis of, 232

Umecyanin, 390
Urine
  copper-binding components of, 142, 143
  copper concentrations in, 143
    cadmium toxicity and copper in, 143
  copper in, 136, 142
  copper losses in, 137
  daily excretion of, 136
  sources of copper for excretion via, 149
Urushiol, 397
Uterus, mRNA for ceruloplasmin is expressed by, 260

Wilson's disease
  absorption of copper in, 340
  albumin-copper in, 333
  apoceruloplasmin in, 336
  biliary copper excretion in, 338
  ceruloplasmin in, 77, 152, 331–334
  ceruloplasmin-bound copper in, 331–333
  ceruloplasmin mRNA in, 334
  ceruloplasmin synthesis in, 334
  ceruloplasmin gene transcription in, 334
  copper absorption in, 341
  copper excretion in, 142
  defects in, 332, 337–339
  gene for, 331, 341
  metabolism of radiocopper in, 336, 339
  metallothionein in, 340
  mitochondria of copper loaded liver in, 340
  penicillamine and, 342
  reviews on, 332
  serum copper in, 331, 333
  symptoms of, 331
  tissue copper levels in, 332
  transcuprein in, 333
  treatment of, 342
  urinary copper in, 332

Yeast
  copper metabolism in, 380
  copper storage in, 381
  copper toxicity to, 382
  cytochrome oxidase of, 380
    regulation and assembly of, 380
    structure of, 380
  metallothioneins of, 183
  phytochelatins in, 381
  resistance to copper of, 382
  superoxide dismutase and, 196

Index

Yolk sac
   albumin and, 124
   ceruloplasmin expression in, 260
   copper in, 306
   α_2-macroglobulin, 124

Zinc
   binding components in cell cytosol for, 179
   copper absorption and, 21, 23, 28, 31, 33, 34

Zinc (*Cont.*)
   detoxification by metallothionein of, 188
   factors affecting absorption of, 34
   hepatocyte uptake of copper and, 51
   induction of intestinal metallothionein by, 29
   and metallothionein, 190
   and metallothionein induction, 29, 188
   in superoxide dismutase, 199